Measuring and Monitoring Biological Diversity

Standard Methods for Mammals

Biological Diversity Handbook Series

Series Editor: Mercedes S. Foster

This series has been established by the National Biological Service (U.S. Department of the Interior) and the National Museum of Natural History (Smithsonian Institution) for the publication of manuals detailing standard field methods for qualitative and quantitative sampling of biological diversity. Volumes focus on different groups of organisms, both plants and animals. The goal of the series is to identify or, where necessary, develop these methods and promote their adoption world-wide, so that biodiversity information will be comparable across study sites, geographic areas, and organisms, and at the same site, through time.

Measuring and Monitoring Biological Diversity

Standard Methods for Mammals

Edited by
Don E. Wilson,
F. Russell Cole,
James D. Nichols,
Rasanayagam Rudran,
and Mercedes S. Foster

SMITHSONIAN INSTITUTION PRESS
Washington and London

Copyeditor and typesetter:
  Peter Strupp/Princeton Editorial Associates
Designer: Janice Wheeler

Library of Congress Cataloging-in-Publication Data

Measuring and monitoring biological diversity.
    Standard methods for mammals / edited by
    Don E. Wilson. . . [et al.]
      p.   cm.
    Includes bibliographical references and index.
    ISBN 1-56098-636-0 (hardback). —
    ISBN 1-56098-637-9 (pbk.)
      1. Mammals—Speciation—Research.   2.
    Biological diversity—Measurement.   I. Wilson,
    Don E.   II. Series.
    QL708.5.M435   1996
    599.052′4828Mé0723—dc20   96-15546

British Library Cataloging-in-Publication Data is
available.

Cover illustration of Humboldt's Monkey
(=*Lagothrix lagothricha*) from C. F. Holder and J. B.
Holder, *Elements of Zoology* (1885), D. Appleton &
Co., New York.

Manufactured in the United States of America
02 01 00 99 98 97 96   5 4 3 2 1

For permission to reproduce illustrations appearing
in this book, please correspond directly with the
owners of the works, as listed in the individual
captions. The Smithsonian Institution Press does not
retain reproduction rights for these illustrations
individually or maintain a file of addresses for photo
sources.

♾ The paper used in this publication meets the
minimum requirements of the American National
Standard for Permanence of Paper for Printed
Library Materials Z39.48-1984.

Contents

Chapter 6. Observational Techniques for Nonvolant Mammals   81

Rasanayagam Rudran, Thomas H. Kunz, Colin Southwell, Peter Jarman, and Andrew P. Smith

Chapter 7. Observational Techniques for Bats   105

Thomas H. Kunz, Donald W. Thomas, Gregory C. Richards, Christopher R. Tidemann, Elizabeth D. Pierson, and Paul A. Racey

Chapter 8. Capturing Mammals   115

Clyde Jones, William J. McShea, Michael J. Conroy, and Thomas H. Kunz

Figures

Tables

Authors and Contributors

Peter August, *University of Rhode Island*

Carol Baker, *University of Rhode Island*

Kyle R. Barbehenn, *Consultant, Bethesda, Maryland*

John W. Bishir, *North Carolina State University*

Stuart C. Cairns, *University of New England*

Timothy F. Clancy, *Queensland Department of Environment and Heritage*

F. Russell Cole, *Colby College*

Michael J. Conroy, *National Biological Service*

Joseph A. Cook, *University of Alaska Museum*

Ronald I. Crombie, *Smithsonian Institution*

Martha L. Crump, *Northern Arizona University*

Martin Denny, *Consultant, Oberon, Australia*

Chris R. Dickman, *University of Sidney*

Michael E. Dorcas, *Idaho State University*

Louise H. Emmons, *Smithsonian Institution*

Mercedes S. Foster, *National Biological Service*

Thomas H. Fritts, *National Biological Service*

William L. Gannon, *University of New Mexico*

Scott Lyell Gardner, *University of Nebraska*

Sarah B. George, *University of Utah*

Thomas Grant, *Cronulla, New South Wales, Australia*

Gregory Gurri-Glass, *Johns Hopkins University*

Lee-Ann C. Hayek, *Smithsonian Institution*

Virginia Hayssen, *University of Nottingham*

Robert F. Inger, *Field Museum of Natural History*

Peter Jarman, *University of New England*

Clyde Jones, *Texas Tech University*

J. Edward Kautz, *New York State Department of Environmental Conservation*

Gordon W. Kirkland, Jr., *Shippensburg University*

Thomas H. Kunz, *Boston University*

Charles LaBash, *University of Rhode Island*

Richard A. Lancia, *North Carolina State University*

Geoffrey Lundie-Jenkins, *Conservation Commission of the Northern Territory*

Roy W. McDiarmid, *National Biological Service*

William J. McShea, *Smithsonian Institution*

Helene Marsh, *James Cook University of North Queensland*

Dale G. Miquelle, *Smithsonian Institution*

James D. Nichols, *National Biological Service*

Thomas J. O'Shea, *National Biological Service*

Elizabeth D. Pierson, *University of California*

Paul A. Racey, *University of Aberdeen*

William E. Rainey, *University of California*

Galen B. Rathbun, *National Biological Service*

Robert P. Reynolds, *National Biological Service*

Gregory C. Richards, *Division of Wildlife and Ecology, CSIRO*

Rasanayagam Rudran, *Smithsonian Institution*

Andrew P. Smith, *University of New England*

Christopher Smith, *University of Rhode Island*

Colin Southwell, *Australian National Parks and Wildlife Service*

Donald W. Thomas, *Université de Sherbrooke*

Christopher R. Tidemann, *Australian National University*

Christen Wemmer, *Smithsonian Institution*

Don E. Wilson, *Smithsonian Institution*

Terry L. Yates, *University of New Mexico*

Foreword

This is a second contribution to what it is hoped will be an ongoing series of manuals prescribing standard methods for measuring and monitoring biological diversity. The various species that together constitute the sum of coexisting organisms in a biological community are often referred to as the *species richness* of the community. Measuring the relative and/or absolute numbers of individuals belonging to each species permits calculation of the *diversity* of species in the community. It is, however, essential that standard methods for determining occurrence (presence or absence) and abundance be employed if any studies are undertaken to compare biological diversity in more than one study area or over successive time intervals. This need has long been recognized by biologists and environmental managers, and this series is the response to that need.

The editors of the first volume (*Measuring and Monitoring Biological Diversity: Standard Methods for Amphibians,* W. R. Heyer et al., 1994, Smithsonian Institution Press, Washington, D.C.) were energized by the recognition at the end of the last decade that many well-known species of amphibians seemed to be abruptly declining in numbers and geographic occurrence. Amphibians began to be regarded as "canaries in the mine." But because many amphibians are cryptic, often nocturnal, and very small, methodological comparability is essential for quantitative evaluation of their numbers and distribution.

Like amphibians, many species of mammals are also in serious decline (some are already extinct, either in the wild or totally). Unlike amphibians, a considerable number of mammal species are well known to the general public—they are larger, they are often diurnal and/or conspicuous, and population levels of many have been monitored for long periods. The vast majority of mammals, however, are, like amphibians, small, nocturnal, and cryptic, and they also require comparability of methodology.

The aim of this book is to develop standard methodology that can be used anywhere in the world on any mammalian faunal group (for example, rodents or large open-country grazers such as gazelles or kangaroos) and yield comparable results in terms of occurrence and of absolute (or more often relative) population sizes. Chapter 2 of the present volume provides an overview of global mammalian diversity, a necessary context for all but the most broadly trained mammalogists.

Subsequent chapters provide extensive and detailed treatment of survey design; data acquisition, for both populations and their environments; and data analysis. Emphasis throughout is on proven techniques, all keyed to the life-history parameters of the targeted species. Knowledge of life histories is critical to the success of a monitoring effort—it is self-evident that a technique designed to enumerate bats will fail to reveal the presence of seals.

Some techniques will, however, serve to capture different kinds of organisms. Pitfall traps, for example, are very effective not only for obtaining some kinds of amphibians but also for capturing some small mammals (shrews, voles) and ground-dwelling arthropods. Similarly, mist nets set for bird sampling during the day can remain deployed for bats at night. The "all-taxa inventory," a goal that is beginning to be considered achievable, will be aided by taking advantage of techniques that sample more than one taxonomic group. Obviously, this "all-taxa" approach will require a diverse team of samplers and standard methods for many more groups.

Another requirement for the effective sampling of biodiversity is a means of identifying all sampled organisms to the species level, or as near to it as possible. This requirement demands identification keys based upon presumptive occurrence at the locality sampled in order to make the job of sample identification manageable. The recent development of illustrated computerized keys points the way in this area. A corollary requirement is for a taxonomic "dictionary," so that a standard nomenclature may be employed. These standards, such as *Amphibian Species of the World* (D. R. Frost, 1985, Allen Press and Association of Systematics Collections, Lawrence, KS) and *Mammal Species of the World* (D. E. Wilson and D. M. Reeder, 1993, Smithsonian Institution Press, Washington, D.C.) must be updated frequently to remain useful; here computer databases should soon replace printed books.

As inventories are conducted in poorly known areas, undescribed species will be encountered. Even fairly large species have recently been described as new—examples include the giant muntjac deer, *Megamuntiacus vuquangensis,* and the forest bovid *Pseudoryx nghetinhensis,* both discovered in north-central Vietnam (Ha tinh and Nghe provinces; 18°20′N 105°54′E and 18°50′N 105°20′E, respectively). New species of small mammals continue to be described. The total number of mammals recognized in 1992 was 4,629, but by now it is probably closer to 4,700, somewhat more than the amphibian total.

In addition to enumerating the structural composition of biotic communities—the focus of this volume and of the series thus far—biologists should be planning ahead for studies of ecosystem function. The integration of structure and function in communities is accomplished through knowledge of the flow of matter and energy in the ecosystems of which communities are components. In amphibians and mammals, as in all heterotrophic organisms, nutrient relationships, or more simply food habits, are key. The literature is venerable and vast, but additional data are needed on many species. As mammalian population data are obtained with the methods described herein, there will be opportunities to obtain quantitative information (on, e.g., cheek pouch and gut contents, feces, feeding observations, food caches) for future studies.

Robert S. Hoffmann
National Museum of Natural History
Smithsonian Institution

Preface

Origin of publication

This is the second book in a series that presents standardized methods for measuring biological diversity. The first (Heyer et al. 1994) focused on Amphibians. The series is under the overall editorship of Mercedes Foster, who developed the program beginning in 1990. Financial and administrative support for this volume was provided by the National Biological Service and the Office of Biodiversity Programs in the National Museum of Natural History, Smithsonian Institution.

Much of our knowledge of mammalian biodiversity is scattered through an extensive literature and, in general, is not available to serve as an adequate baseline against which to measure changes through time. Standardized methods for documenting population changes either do not exist or are not generally known.

Foster asked Don Wilson to set up a core committee to plan the book and oversee the production of a manuscript on mammals. The committee included James D. Nichols and Rasanayagam (Rudi) Rudran. This committee met early in 1991 and subsequently contacted nearly 40 individuals throughout the world who had experience with the inventory and monitoring of mammals, to request their input regarding the value of the proposed book and the techniques and issues that should be covered. In mid-1991, the core committee asked those who had responded positively to draft manuscripts describing the techniques to be included in the volume. The manuscripts would be discussed and a plan for completing the book developed at a subsequent workshop. The workshop was held 9–11 June 1992 in Arlington, Virginia. Participants were Michael Conroy, Mercedes Foster, Thomas

Fritts, Lee-Ann Hayek, Clyde Jones, Thomas Kunz, William McShea, James Nichols, Rudi Rudran, Colin Southwell, Chris Wemmer, Don Wilson (workshop chair), and Terry Yates.

The workshop participants completed three tasks: (1) They identified and recommended a set of standard procedures for measuring mammalian biological diversity by monitoring populations and inventorying sites. As a result of that process, they also developed the essentials of an outline of the book. (2) They wrote all or substantial parts of several chapters. (3) They outlined and assumed responsibility for the tasks needed to complete the book.

Russell Cole, James Nichols, Rudi Rudran, Don Wilson, and Mercedes Foster took on the responsibilities of compiling and editing the publication. Contributions of a few additional experts were sought for the final manuscript. A draft of the manuscript was completed in May 1993. This manuscript was reviewed and revised and the final draft was completed in July 1995.

Authorship

After some discussion, participants at the workshop decided that it would be appropriate and desirable to assign authorship for individual sections of the book. The editors asked the participants to identify as *authors* those who participated significantly in the writing of the manuscript and to identify as *contributors* those who provided information. Considerable parts of Chapter 4 were modified only slightly from similar discussions in the amphibian volume (Heyer et al. 1994). These were changed as needed and the original authorship retained, although in three instances where considerable modification was necessary an additional author was added to indicate responsibility for the changes. Chapter 12 reflects the sentiments of the editors.

Acknowledgments

As with any multiauthored book, the authors and editors were assisted by many people during the manuscript preparation phase. The editors particularly want to thank Fiona Wilkinson, who performed heroically in verifying all bibliographic citations, and the librarians Linda Garret, Alvin Hutchinson, and Polly Lasker, who provided many of the references. Kate Spencer, Kinard Boone, and Angela Greco prepared standardized versions of the figures; John Steiner took many of the photographs of others. Claudia Angle of the National Biological Service drew the cheetah that leaps across the page at the beginning of each chapter. Dale Crawford of Remtech Services, Inc./National Biological Service provided the entertaining artwork for page 326. Adele Conover arranged for most of the permissions for use of figures. Charles R. Mann prepared the table of random numbers. Jeremy Jacobs, Amy Levin, Martha Osborne, Cassandra Phillips, and Fiona Wilkinson incorporated numerous manuscript changes into computer files or provided other assistance. Fiona Wilkinson of the National Biological Service and Marsha Sitnik and Judy Sansburry of the Biodiversity Programs Office at the National Museum of Natural History provided continuous assistance at every stage of the project.

Most importantly, the editors would like to recognize the authors, contributors, and workshop participants for their considerable effort. This book would not have been produced without their collaboration and counsel. James L. Patton and an anonymous reviewer read the entire manuscript and provided valuable, detailed comments. Robert Giles, Michael Griffiths, David Hardy, Eric Hoberg, Paul Joslin, Charles McDougal, Eric Rexstad, P. M. Wijeyaratne, Horacio Zeballos Patron, and an anonymous reviewer also read and commented on specific chapters or sections. To each of these individuals, we are indeed grateful.

Institutional and financial support were provided by the National Biological Service, Department of the Interior, and the Smithsonian Institution. This is contribution number 100 from the Biological Diversity in Latin America (BIOLAT) program.

The Volume Editors

Chapter 1

Introduction

Don E. Wilson, James D. Nichols, Rasanayagam Rudran, and Colin Southwell

Background

Biological diversity, or *biodiversity,* is the term used to describe the variety of life forms, the ecological roles they perform, and the genetic diversity they contain (Wilcox 1984). In recent years, there has been growing recognition that the earth is facing a loss of biological diversity of crisis proportions (Wilson 1992). The ever-increasing pressure of human population growth has led to worldwide habitat degradation that has driven many known, and countless unknown, species of plants and animals to extinction and put numerous others at risk (Wilson and Peter 1988; Cole et al. 1994). This dilemma has focused attention on the need to survey biological resources as a first step in developing manage-ment strategies. Such strategies are essential for establishing legislation and other guidelines to save biological resources for the future.

The first step in biological resource surveys, an assessment of biodiversity, is estimation of diversity (e.g., with respect to species richness) at one time and location. This step frequently leads to a second stage, monitoring of biodiversity, which refers to estimation of diversity at the same location at more than one time, for the purpose of drawing inference about change. Investigating biological diversity continues to be a central theme of ecological, systematic, and evolutionary biology; it is also absolutely critical to the emerging fields of conservation biology and resource management.

Biological diversity encompasses the follow-ing hierarchically related levels of biological or-

1

ganization: genetic diversity within populations, species diversity within communities, and ecosystem diversity within different landscapes and regions of the world. Above these levels of biological organization one can also distinguish landscape and regional diversity on a global scale. Regional or landscape diversity is concerned with spatial organization, the scale of which may vary from a reserve to a biogeographic province. Regions and landscapes include features such as size, shape, topography, and connectivity of habitat patches, which have important influences on the number and composition of species and on their abundance, viability, and genetic makeup. The ecosystem includes abiotic components of the environment, such as soils and climate, with which the biotic community is interdependent. Communities are the biotic components of ecosystems and comprise populations of species coexisting at a site. The species is the level at which most field biologists work, and it forms an important basis for developing management strategies and conservation legislation. A population comprises individuals of a single species. Populations may be isolated geographically and reproductively, or they may exist as metapopulations and consist of many subpopulations connected to varying degrees by dispersing organisms. Genes refer to the hereditary material distributed within and among the individuals in a population.

Biodiversity at each of these levels has recognizable attributes that can be assessed and monitored. At the regional or landscape levels, attributes of interest include the identity, distribution, and proportions of each type of habitat, and the distribution of species within and among those habitats. At the ecosystem level, the identity, richness, evenness, and diversity of species, guilds, and communities are important. At the species level, attributes such as abundance, density, and biomass of each population are of greatest interest. At the genetic level, allelic diversity of individual organisms within a population is important. As the basis for all diversity, genetic diversity is of interest at all levels and provides the impetus for studies of phylogeny and population variation, and for interspecific genetic comparisons. These attributes can be assessed and monitored using tools and techniques relevant to each level of organization.

Noss (1990) provided a hierarchical conceptual framework for identifying specific, measurable attributes of biodiversity to assess and determine status and change. He suggested that biodiversity be assessed and monitored at multiple levels of organization and at multiple spatial scales. Such multifaceted investigations help detect large as well as small changes in biodiversity through time.

Biodiversity at regional and landscape levels

A region is determined by a complex of climatic, physiographic, biological, economic, social, and cultural characteristics. Generally, regions are defined on a scale of hundreds to thousands of square kilometers. They almost always contain more than one landscape. For example, the New England region of the United States contains high mountain landscapes, forested hilly landscapes, agricultural landscapes, suburban landscapes, and urban landscapes. A landscape is defined as a heterogeneous land area composed of interacting ecosystems (Forman and Godron 1986). In each component ecosystem, the biotic community and its environment interact as an ecological unit.

In addition to containing one or more ecosystems, all landscapes are composed of habitat patches, habitat corridors, and a background matrix. Patch characteristics of importance include size, shape, topography, nature of the edge, and location relative to other patches. Important characteristics of corridors include width, connectivity, curvilinearity, narrows, breaks, and composition. The matrix in which patches and corridors occur is the most extensive and connected landscape element, and it plays the predominant role in landscape dynamics. That matrix has been increasingly modified by humans in recent centuries. All three elements (ecosystems, patches, and matrices) may exist in a natural state or may be modified anthropogenically to varying degrees.

Landscapes may be considered in terms of their composition (the elements present), their structure (the spatial relationships of the elements), their function (the interactions among the elements), and their dynamics (the alteration or change of structure or function over time). It is frequently useful to distinguish between two categories of structural elements of landscapes: those that are part of the natural environment and those that are determined mainly by human influence.

Landscape composition and structure can be assessed through existing topographic and vegetation maps, aerial photography, and satellite imagery, and the data can be organized and displayed with a geographic information system (GIS). Plants and animals are usually heterogeneously distributed among the landscape elements, and mobile organisms such as mammals move among these elements. Information on the spatial distribution of organisms, on their movements, and on changes in these patterns enables one to assess landscape function and dynamics. Broad-scale assessments at the landscape level can be used to identify centers of species richness and endemism. Such areas may then be monitored more intensively or considered for protection.

Assessment of mammal diversity at the regional or landscape level focuses on the broad distributional patterns of mammals in relation to both natural and human landscape elements. By obtaining information on mammal distributions, natural landscape elements, and human influences at any given site, many questions relevant to the understanding and conservation of mammal diversity can be addressed. Examples of such questions include the following:

1. Is mammal diversity heterogeneous across the landscape?
2. If so, where are the centers of high diversity?
3. Do such centers correspond with particular patterns in the natural or human elements?
4. Is habitat-patch size or shape important to mammal diversity?
5. Do vegetated corridors of natural habitat effectively join areas of high diversity?
6. Do mammals use vegetation corridors?
7. Do existing protected areas and reserves contain areas of high diversity?
8. Which unprotected areas should be considered for future protection so that maximum biodiversity can be conserved?
9. How might changes in land-use practices affect mammal diversity in protected versus unprotected areas?

Biodiversity at ecosystem, community, and population levels

Ecosystem biodiversity can be considered in terms of composition, structure, and function. Composition is the identity and variety of biotic communities in an ecosystem. Structure is the physical and biological organization or pattern of an ecosystem. Function involves ecological or evolutionary processes such as gene flow, disturbance, nutrient cycling, and energy flow.

Ideally an investigator should consider the composition, structure, and function of ecosystems, but from a practical view we often favor composition, because it is both more tangible and easier to measure than the other traits. A knowledge of the composition of biological communities within ecosystems is essential to the development of conservation measures such as land management plans. The need for such conservation measures is becoming acute in tropical rain forest regions, where deforestation is rapidly changing ecosystem composition.

Quantitative information on mammalian species comes only from detailed surveys at the community and population levels. Such information is essential for the development of conservation and management strategies for species and communities. Techniques for collecting quantitative information vary tremendously depending on the taxa of concern. The techniques are used to determine the species present at a selected site and to estimate species richness and absolute or relative abundance of species at that site. Most of this handbook focuses on techniques appropriate for collecting quantitative information and for assessing mammal diversity at community and population levels.

Concepts of biodiversity at the community level

More has been written about diversity at the community level than about diversity at any other level of organization (see, e.g., the bibliography of Dennis et al. 1979), and measures of species diversity are commonly reported in studies of biological communities. The basis for all ideas about species diversity, and the raw material for all diversity measures, is the *species abundance distribution*. This distribution specifies the number of individuals belonging to each of the species represented in the community. Many diversity measures can be computed from a knowledge of the *relative* abundances of the different species. In other words, for many diversity measures it is sufficient to know that the community contains twice as many individuals of species A as of species B, without knowing the exact numbers of animals in each species. In practice, however, individuals from different species will not appear in our samples with equal probability. In order to estimate relative abun-

dance properly, therefore, we must estimate species-specific sampling probabilities and, thus, absolute abundances.

The concept of species diversity contains two conceptually distinct components. The first component is *species richness,* defined simply as the total number of species in a community. The second component has been described by the terms *equitability* and *species evenness* and refers to the degree to which relative abundances of individuals among the different species are similar. Consider a two-species community with 100 individuals. If 50 individuals belong to each species, then evenness or equitability is high (actually, it is maximum for this scenario). If 99 individuals belong to one species and only one animal to the other species, however, then evenness or equitability is low (the minimum for this particular community).

Diversity measures, or "indices," represent attempts to extract relevant information from the species abundance distribution and to encode this information in a single statistic (sometimes two or three statistics are used). People have proposed different uses of diversity indices and, thus, have different ideas about what information is relevant. Such differences have led to the development of a variety of diversity indices (e.g., Pielou 1977; Grassle et al. 1979). Values of most diversity indices increase with increased species richness and with increased evenness or equitability. Beyond this general idea of what constitutes more and less diverse, however, diversity indices differ substantially.

Much of the ecological literature on species diversity does not report or deal with the identities of the species in the community. Naturally, we must identify individual animals in order to assign them to species categories correctly. However, the contribution of a particular species to most diversity indices is determined solely by the number (or relative number) of individuals in the species. It is possible to "weight" the relative contributions of different species by factors other than abundance (e.g., notions of "importance" to community function, endangered status). Such alternative weighting schemes require knowledge of species identity, in addition to species richness and abundance.

In our view, there is no single, "best" diversity index for all purposes. Therefore, we consider only the estimation of the two components of the species abundance distribution, species richness and abundance of individuals within species. The investigator who estimates these two components has obtained all the information relevant to most concepts of species diversity and can then select any index consistent with his or her goals. The process of computing a single diversity statistic from a species abundance distribution necessarily entails a cost in terms of loss of information. If the statistic reflects exactly those characteristics of the distribution that are essential to the question(s) posed by the investigator, then use of the statistic is justified. In many other cases, however, it will be more reasonable to base comparisons of communities at different points in space or time on the entire species abundance distribution.

Previous work

Investigators have employed a variety of methods to estimate abundance and to compute community-level measures, such as diversity, for mammal populations and communities. Some of these methods represent state-of-the-art approaches to estimation. Most are simply ad hoc approaches that yield estimates that are subject to large potential biases (see critiques of such methods in Jolly and Dickson 1983; Nichols and Pollock 1983; Pollock et al. 1990) and that cannot be used reasonably for comparative or other purposes (Nichols 1986; Skalski and Robson 1992). Compilations of estimation methods (e.g., Davis 1982) typically contain a mixture of (1) ad hoc methods that are not very useful;

(2) estimators that were appropriate historically, but that do not reflect recent improvements; and (3) useful estimators that represent the current state of the art. Such compilations provide little guidance regarding the relative utility of the different estimation approaches presented.

Exceptions to the typical compilations of methods include the excellent book by Seber (1982; also see updates, Seber 1986 and 1992) and the review by Lancia et al. (1994). These compilations of abundance-estimation techniques present the most useful methods available at the time of their writing, although they do not focus specifically on mammals. In addition, they contain little discussion of methods for estimating species richness.

Although we urge the adoption of the most productive techniques possible for measuring and monitoring biodiversity, an accompanying caveat is absolutely necessary. Modern solutions to any scientific problem tend to be complex, detailed, high-tech, and expensive. Initial approaches to biodiversity surveys in developing countries where resources are limited may well need to be simple, straightforward, low-tech, and inexpensive. Such techniques provided the basis for our knowledge of biodiversity everywhere on the planet and will continue to play an important role. The important factor to stress is a quantitative approach that will ensure comparability across space and time, regardless of level of complexity, technology, or expense.

Purpose of this volume

In this book we present methods for developing species lists and for estimating abundance and species richness. We focus specifically on mammals, and bring together in one place the disparate methods for quantitative and qualitative sampling now in practice for various mammalian taxa. Although we discuss indices of abundance and species richness, we emphasize formal estimation approaches. These estimation methods provide abundance and richness measures that are comparable across habitats and over time. Such comparability is a goal shared by the International Union of Biological Sciences (di Castri et al. 1992).

As indicated earlier, a biodiversity survey can be a two-stage process, beginning with an assessment of the biota at a particular site and followed by periodic monitoring to detect changes at that site, including natural changes such as the impacts of seasonality. Assessment and monitoring can be carried out at any level of biological or spatial organization. The techniques used for assessment or monitoring will differ, however, depending on the level of organization that is investigated. In this manual we outline field techniques that facilitate effective assessment and monitoring of mammalian diversity at all levels of organization.

Detecting changes in mammalian diversity across space or time has been difficult in the past, because data sets or estimates have often been obtained using disparate techniques that preclude comparisons. A set of standardized methods is needed to provide data that can be compared and thus lead to a greater understanding of mammalian biodiversity. With this in mind, the authors of different chapters have assessed the utility of various methods and made recommendations for standardization where appropriate. The standardized techniques presented in this volume will facilitate comparisons and the detection of real change in future investigations of mammalian biodiversity.

Intended audience

Like its companion volume for amphibians (Heyer et al. 1994), this book on mammals was written to help biologists and conservationists worldwide in the design and implementation of biodiversity projects, in both developed and developing countries. It is targeted at individuals with a primary university degree in biology or a higher level of training. We expect that such people will be able to design a study and estimate mammal diversity in a given area, using this volume as the primary methodological reference. The volume may also serve faculty members and students at universities as an aid to teaching or biological field research. Others who will benefit from this book include protected area managers, planners, and wildlife consultants.

Organization

In Chapter 2 we provide an overview of mammalian diversity and brief comments on the natural history of each of the 26 orders of living mammals. Inventory planning should be based on as much knowledge as possible of the organisms one expects to encounter and the habitats they occupy. Unfortunately, those areas of the planet most in need of inventory are also those for which the least information is available.

We provide guidelines for designing a mammalian diversity study in Chapter 3. Proper study design will improve the ability to draw reasonable inferences from resulting data, making the study both repeatable by future investigators and comparable with other such studies.

Chapter 4 includes a wide-ranging review of the many kinds of ancillary data that could prove useful to basic inventory projects. Because inventory work requires a significant investment of resources and time, careful planning is essential in order to make the results as broadly useful as possible. The planning and other preliminary aspects of conducting a mammalian biodiversity survey are discussed in Chapter 5.

Chapters 6 through 9 cover the bulk of the actual field techniques for mammalian inventories. These techniques are treated under the broad categories of observation, capture, and sign. In general, we recommend capture techniques mainly for small mammals such as rodents and bats. We recommend observational techniques primarily for mammals of medium and large size. Indirect techniques, such as those based on mammalian sign, are appropriate for only a few species.

In Chapter 10, we describe statistical techniques for estimating population size and species richness from data collected in the field. We provide a unified conceptual framework for these different estimation methods and emphasize the intuition underlying the methods.

In Chapter 11, we summarize geographic information systems that can be used for storage and analysis of biodiversity data. Methods for translating the field data into the proper electronic formats and methods for analysis of those data are outlined. Chapter 12 presents our conclusions and a series of recommendations regarding the design and implementation of mammalian biodiversity projects.

Finally, in a series of appendices, we provide information on a variety of additional issues bearing on biodiversity studies of mammals. These include ethics in research; human health concerns; preservation of voucher specimens; tissues, cell suspensions, and chromosomes; assessment of sex, age, and reproductive condition in mammals; field parasitology techniques for use during mammal surveys; methods for marking mammals; methods for recording mammal calls; vendors; and a table of random numbers.

The publications listed in the Literature Cited section of this volume provide detailed treatments of the described techniques and also provide an introduction to available knowledge of the biology of mammals.

Chapter 2

Mammalian Diversity and Natural History

F. Russell Cole and Don E. Wilson

Introduction

The 26 orders of living mammals occur in a wide variety of terrestrial and aquatic habitats throughout the world (Nowak 1991). Mammals are found on all continents, occurring from above the Arctic Circle in the northern hemisphere to the southern tips of the continents and to the large islands in the southern hemisphere (Vaughan 1978). All 26 orders include species that generally occupy terrestrial or freshwater habitats (Cole et al. 1994). Two orders, Cetacea and Sirenia, include species that are highly adapted for life in aquatic (primarily marine) environments. The order Carnivora also includes three families of marine mammals: Odobenidae,

Otariidae, and Phocidae. Three additional orders (Didelphimorphia, Insectivora, and Rodentia) include species that are semiaquatic or aquatic (Nowak 1991).

What do we know about patterns of mammalian diversity? Latitudinal gradients in diversity, changes in the general structure of mammalian communities at different latitudes, changes with elevation in mountainous areas, and differences associated with the size of the land area that support them are all important to the design of inventory and monitoring programs. A complete treatment of these important topics is beyond the scope of this book, but an excellent introduction to the rich literature available on these topics can be found in Anderson and Patterson (1994).

Knowledge of the geographic distribution and natural history of the mammalian species of interest and cognizance of the environment in which it occurs are basic to planning an inventory or monitoring program. Such information aids in the selection of appropriate sampling techniques and increases the effectiveness of sampling and the precision with which mammalian species diversity and abundance can be estimated. In this chapter, we summarize the geographic distributions of mammalian taxa, including their zoogeographic affinities. We also describe the natural history of living mammals and outline the primary habitats occupied by various taxa. In addition, we supply a tabular summary of the mammalian taxa that might be found in each zoogeographic region or in the world's oceans and seas, including information on body size (small, medium, and large) and mode of life (aquatic, fossorial, terrestrial, or arboreal) for each family of mammals. Once the objectives of a mammalian diversity study have been delimited, information contained in these tables may assist in the choice of inventory methods.

Ordinarily a survey will be better if the investigator reads about the ecology and natural history of mammals that may occur in the area to be surveyed before conducting the survey. Good summaries of relevant material are available in widely available general works (e.g., Macdonald 1984; Nowak 1991; Wilson and Reeder 1993) or in regional publications (e.g., *Nearctic:* Banfield 1974; Hall 1981; Chapman and Feldhammer 1982; Ramírez-Pulido et al. 1982; González and Leal 1984; *Neotropical:* Goodwin 1942; Handley 1966; Hershkovitz 1972; Wilson 1983; Eisenberg 1989; Emmons and Feer 1990; Redford and Eisenberg 1992; *Palearctic:* Zhenhuang 1962; Corbet 1978, 1984; Corbet and Ovenden 1980; Gromov and Baranova 1981; Niethammer and Krapp 1990; *Ethiopian:* Kingdon 1971–1982; Haltenorth and Diller 1980; Smithers 1983; *Oriental:* Lekagul and McNeely 1977;

Roberts 1977; Phillips 1980–1984; Medway 1983; Payne et al. 1985; Heaney et al. 1987; Musser 1987; Corbet and Hill 1992; *Australia:* Ziegler 1982; Strahan 1983; Flannery 1990).

Geographic considerations

For the purposes of this review we have divided the mammal-inhabited, large land masses of the world into six regions: Nearctic and Neotropical in the New World and Ethiopian, Palearctic, Oriental, and Australian in the Old World. The regions correspond to the zoogeographic realms (Darlington 1957), with a few changes noted below.

We include the land from the Canadian Arctic south to the Mexican border with Guatemala in the Nearctic region, which is roughly equivalent to the Nearctic realm. The Neotropical region encompasses the mainland of Middle and South America and the islands of the West Indies. The Ethiopian region includes sub-Saharan Africa and the island of Madagascar. The Palearctic encompasses the British Isles and the area bounded by the Atlantic Ocean in the west; the Arctic Ocean in the north; Mediterranean Africa, the Arabian Peninsula, the Middle East, and the Himalayan Front in the south; and the Pacific Ocean in the east. The Indo-Malayan region and the Philippines compose the Oriental region. Roughly separated from the Oriental region by Wallace's line, the Australian region includes New Guinea, Australia, Tasmania, and adjacent islands. Large islands often possess a more diverse fauna than small islands, and we have included them in the region with which they share the greatest part of their mammal fauna. In most instances they are included within the adjacent continental region.

Seals, sea lions, walruses, freshwater dolphins, manatees, dugongs, marine otters, and the polar bear frequent aquatic areas adjacent to the regions in which they are recorded. Our analysis

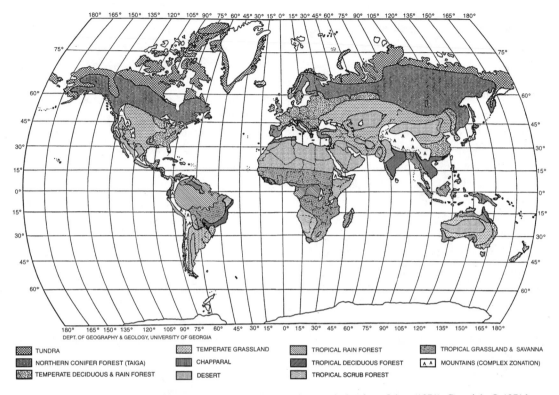

180° 165° 150° 135° 120° 105° 90° 75° 60° 45° 30° 15° 0° 15° 30° 45° 60° 75° 90° 105° 120° 135° 150° 165° 180°

DEPT. OF GEOGRAPHY & GEOLOGY, UNIVERSITY OF GEORGIA

	TUNDRA		TEMPERATE GRASSLAND		TROPICAL RAIN FOREST		TROPICAL GRASSLAND & SAVANNA
	NORTHERN CONIFER FOREST (TAIGA)		CHAPPARAL		TROPICAL DECIDUOUS FOREST		MOUNTAINS (COMPLEX ZONATION)
	TEMPERATE DECIDUOUS & RAIN FOREST		DESERT		TROPICAL SCRUB FOREST		

Figure 1. Distribution of the major biomes of the world. Adapted with permission from Odum (1971). Copyright © 1971 by Saunders College Publishing, reproduced by permission of the publisher.

includes coastal areas for each New World and Old World zoogeographic region. We group the mainland, islands, and ice floes of the poles into the Arctic and Antarctic regions. Whales, porpoises, and oceanic dolphins are also categorized by the ocean (Atlantic, Pacific, and Indian) or polar region (Arctic and Antarctic) in which they occur.

We have placed the world's oceans into a seventh category, the Oceanic region. The Oceanic region includes the world's small islands that are isolated from major land masses. Because the mammal faunas of such islands are typically depauperate and usually include only bats, introduced murids, and other mammals associated with human activity (Vaughan 1978), we do not include them in our analyses.

A rich array of climatic conditions and vegetation types occurs in the New World. The Nearctic includes environments ranging from Arctic tundra in the north to tropical forest in the south (Fig. 1). Northern coniferous or boreal forest dominates the northern part of the continent, extending into tundra in the far north. Moving east to west across the central part of the continent, temperate forest, temperate grassland, and desert environments predominate. A complex of vegetation types occurs in mountainous areas, particularly the western mountains, which extend north to south through the Nearctic region. Boreal forest, chaparral, and desert habitats are encountered as one moves south along the west coast of the United States. Desert and grasslands extend into the southern Nearctic, where tropical scrub and rain forest habitat also occur.

Tropical rain forest environments dominate much of Middle America and northern South America (Fig. 1). Amazonia is also bounded to the south by tropical savanna and scrub forests.

Tropical deciduous forests occur along parts of the north and east coasts. The southern part of the continent, where the climate is more temperate and rainfall more moderate, is dominated by temperate grassland and desert habitat. The Andes provide a range of climatic conditions and a complex of vegetation types from warm, wet montane forests at lower elevations to cold, dry alpine tundra at higher levels. Chaparral changing to desert can be found as one moves north along the west coast.

The Old World possesses vegetation types roughly similar to those found in the New World, although each type often covers larger areas (Fig. 1). This generalization of similar vegetation types with larger coverage per type in the Old World is particularly true for the comparison between Palearctic and Nearctic environments. Boreal forest extending into tundra dominates the northern Palearctic. Temperate forests are typical of central Europe and extend into north Asia; they also occur at similar latitudes along the west coast of the Palearctic. Central Palearctic environments include desert surrounded by temperate grasslands. Mountainous areas covered by complex vegetation types separate the Palearctic from the Oriental region. Chaparral and desert environments dominate the southeastern Palearctic boundary.

The Ethiopian region is dominated by desert and arid grassland in the north extending southward to tropical savanna or scrub forest, and ultimately to tropical rain forest habitat in equatorial Africa (Fig. 1). Tropical savanna or scrub forest habitat dominates much of southern Africa. The most extensive tropical savanna ecosystems in the world are found in this region. Tropical scrub forest in the southwest, chaparral in the south, and grassland in the southeast dominate the coastline.

The Oriental region includes desert, tropical scrub, and tropical deciduous forest moving west to east across India (Fig. 1). Southeast Asia is dominated by tropical rain forests in the west

and by tropical deciduous forests in the east. Tropical rain forest habitat is prevalent throughout the Philippines and adjacent archipelagos. The mountainous areas separating Palearctic and Oriental regions possess a complex of vegetation types.

The northern part of the Australian region, which includes New Guinea, adjacent islands, and parts of the east coast of Australia, is covered by tropical rain forests (Fig. 1). Temperate forest occurs along the remainder of the east coast of Australia. The northern coast of Australia is characterized by tropical scrub forest. Central Australia is dominated by grasslands surrounding desert habitat. Chaparral is found along parts of the southern coast.

Human activity has affected naturally occurring terrestrial and marine environments throughout the world. Large tracts of native forest habitat in tropical and subtropical areas of the Neotropical, Ethiopian, and Oriental regions have been deforested and converted to agriculture, cattle ranching, or human settlements. In addition, extensive areas of native habitat in many temperate regions of the world, especially the Nearctic and Palearctic, have been converted to agriculture, grazing, or human use. The growth of major urban areas and accompanying pollution have destroyed or degraded native habitats in both temperate and tropical areas. In general, boreal and Arctic habitats have been less affected by human activity than habitats in temperate and tropical regions. In addition, mountainous areas appear to have been altered less than areas with moderate to little topographic relief. Accessible habitats in almost all regions have been affected by human activity, particularly habitats along the coasts of continents and large islands, where population growth, resource exploitation, and pollution are often most severe.

The anthropogenic activity has contributed to current threats to mammalian diversity. Deforestation and other habitat destruction, over-hunting, introduction of exotic organisms, animal damage control projects, and removal of predators have all helped to reduce mammalian diversity in every area where one species of mammal, *Homo sapiens,* has become established.

Distributions and zoogeographic relationships of mammals

The second edition of *Mammal Species of the World* (Wilson and Reeder 1993) provides a recent comprehensive review of the taxonomy and distribution of mammalian species and is an important resource for planning mammalian ecological or conservation biology studies. The taxonomy used in this handbook follows that volume. *Mammal Species of the World* identifies 4,629 species of living and recently extinct mammals. These living and extinct species are distributed among 26 orders, 136 families, and 1,135 genera.

The great majority of mammal species inhabit terrestrial environments; only 2.5% of all species occupy marine habitats. Rodentia is the largest mammalian order with 2,015 species, more than twice the 925 species described for the next largest order, Chiroptera. The diversity of these orders is illustrated by the fact that Rodentia is represented by 28 families and 443 genera and that Chiroptera is divided into 17 families and 177 genera. Both orders are found naturally on all major land masses except Antarctica. These groups are typically represented in the mammalian faunas of isolated islands as well. Members of five other orders are also found in terrestrial environments of most regions: Insectivora (6 families, 420 species), Carnivora 11 families, 271 species), Primates (13 families, 233 species), Artiodactyla (10 families, 220 species), and Lagomorpha (2 families, 80 species). Only two of these orders (Primates and Carnivora) are represented in Australia, by one and two species, respectively; both are non-native. The Perissodactyla (3 families, 18 species), which are native to both the New World and the Old World, are also widely distributed.

The mammalian fauna of the Nearctic includes 10 orders, 37 families, and 643 species (14% of all species; Cole et al. 1994). This region shares approximately half of its families (46%) with the Palearctic (Table 1). Many genera in the Soricidae, Vespertilionidae, Canidae, Felidae, Mustelidae, Ursidae, Cervidae, Sciuridae, and Muridae are found in both regions. This affinity probably reflects the dispersal avenues (e.g., Bering land bridge) that connected the Palearctic and the Nearctic in the geologic past. Thirty Nearctic families (81%) extend into the Neotropical region. The fauna of the Nearctic has been influenced only to a limited extent by species of Neotropical origin, primarily some bats, a porcupine, and the armadillo. Only two families are endemic to the Nearctic: Antilocapridae (pronghorn) and Aplodontidae (mountain beaver).

The greatest diversity of mammals in the New World is found in the Neotropical region, where 12 orders and 50 families (19 endemic) are represented. Approximately 60% of the 309 genera and 80% of the 1,096 species are endemic to the region (Table 1; Cole et al. 1994). Although representatives of the Xenarthra and the Didelphimorphia are found in both New World regions, species richness within these orders is greatest in the Neotropical region. The Neotropical mammal fauna also includes two endemic orders (Paucituberculata and Microbiotheria) and a diverse group of Hystricognath rodents (10 endemic families). In addition, three species of the genus *Lama* live in the Neotropical region; they are the only New World representatives of the Camelidae.

The Palearctic possesses 13 orders, 42 families, and 843 species (18% of all mammal species; Cole et al. 1994). This region shares 67% of its families with the Ethiopian region and 90% of its families with the Oriental region (Table 1). Approximately 30% of the 262 genera and 60%

Table 1. Distributions of Mammal Families and Genera among Six World Regions[a]

Region	Number of families (genera)	Percent families (genera) endemic	Number of families (genera) in common					
			Nearctic	Neotropical	Palearctic	Ethiopian	Oriental	Australian
Nearctic	37 (184)	5.4 (26.6)	—	30 (113)	17 (32)	13 (13)	15 (20)	5 (6)
Neotropical	50 (309)	38.0 (63.1)	30 (113)	—	12 (10)	11 (8)	12 (11)	5 (7)
Palearctic	42 (262)	0 (26.7)	17 (32)	12 (10)	—	34 (63)	36 (158)	9 (17)
Ethiopian	52 (298)	32.7 (77.5)	13 (13)	11 (8)	34 (63)	—	33 (61)	9 (17)
Oriental	43 (254)	7.0 (31.1)	15 (20)	12 (11)	36 (158)	33 (61)	—	10 (38)
Australian	28 (135)	42.9 (63.0)	5 (6)	5 (7)	9 (17)	9 (17)	10 (38)	—

[a]Data from Cole et al. (1994); taxonomy and distributions follow Wilson and Reeder (1993).

of the 843 species are endemic to the region. The greatest diversity of genera and species within this region occurs in East Asia (Cole et al. 1994).

The mammal fauna of the Ethiopian region is diverse and includes 13 orders, 52 families, and 1,045 species (almost 25% of all species; Cole et al. 1994). One order is unique to the Ethiopian fauna: the Tubulidentata, or aardvarks. Three orders (each with one family) are shared with the Oriental region: Pholidota (7 species), Hyracoidea (6 species), and Proboscidea (2 species). Almost 35% of the families, almost 80% of the genera, and over 90% of the species are endemic (Cole et al. 1994). The mammalian fauna of this region is noteworthy for its impressive array of endemic and savanna-inhabiting ungulates; its diverse groups of cercopithecid primates, viverrid carnivores, and Hystricognath rodents; and the lemuroid primates of Madagascar (Vaughan 1978).

The fauna of the Oriental region is also diverse and includes 43 families and almost 20% of all described mammal species. More than 60% of the Ethiopian families extend to the Ori-

ental region (Table 1; Cole et al. 1994). Rhinoceroses, elephants, great apes, lorises, and viverrid carnivores are present in both areas, but the lemuroid primates, the distinctive African Hystricognath rodents, and the rich assemblage of antelope are absent from the Oriental region. Families endemic to this region include the hognosed bat (Craseonycteridae—possibly the world's smallest mammal), tarsiers (Tarsiidae), gibbons (Hylobatidae), and musk deer (Moschidae). The mammal fauna of the Oriental region also includes two endemic orders: the flying lemurs (Dermoptera; 2 species) and the tree shrews (Scandentia; 19 species).

The native mammal fauna of the Australian region includes 28 families, 135 genera, and 472 species (10% of the world's mammal species). Over 60% of the genera and almost 90% of the species are endemic (Cole et al. 1994). Sixteen families of marsupials, occupying a wide array of ecological niches, two families of monotremes, and 10 families of placental mammals are represented in nine orders. The mammal fauna of the region has its greatest affinities with the fauna of the Oriental region, with which it shares almost 30% of its families (Table 1). One marsupial order (Diprotodontia) extends to this region. Although this order is found in both zoogeographic regions, the species richness is greatest in the Australian region. Monotremata and three marsupial orders (Dasyuromorphia, Peramelemorphia, Notoryctemorphia) are endemic to the Australian region.

Marine mammals are included in three orders: Cetacea, Sirenia, and Carnivora. Most marine carnivores (34 species) are included in the families Odobenidae, Otariidae, and Phocidae. Three additional species of Carnivora that inhabit marine environments include two otters (Mustelidae) and the polar bear (Ursidae). Species of Cetacea (10 families, 78 species) and marine Carnivora inhabit all oceans and adjoining seas, although some members of Platanistidae (Cetacea) occur in freshwater lakes and rivers. Representatives of the Sirenia (2 families, 4 species) are present in coastal and estuarine waters of most continents, and some species are found in freshwater lakes and rivers as well.

Natural history of mammals

In this section we present an overview of the natural history of each of the 26 orders of mammals, including general geographic distribution, preferred habitat types, typical activity pattern, mode of locomotion, food habits, and notes on social behavior. We also include tabular summaries of the distribution of the 26 mammalian orders and their families in each zoogeographic region or in the world's oceans and seas and information on the size range and mode of life for each family of mammals (Tables 2–4). Information for this section was obtained from a number of sources, particularly Nowak (1991) and Vaughan (1978).

Order Monotremata

The two families, three genera, and three species of monotremes are arguably the most primitive and distinctive of mammals (Groves 1993a). They differ from all other mammals in laying shell-covered eggs that hatch outside the female's body, much like the eggs of reptiles. Male monotremes have horny spurs on the ankles, and at least in the duck-billed platypus, these spurs are grooved and carry a poisonous secretion. Monotremes are restricted to the Australian region.

Echidnas or spiny anteaters (Tachyglossidae) are generally solitary and terrestrial, and they are powerful diggers. These animals occur in a variety of habitats, including forests, grass meadows and plains, and hilly or rocky areas. They are typically crepuscular or nocturnal, sheltering in hollow logs, in cavities under rocks or roots, or in burrows during the day. When disturbed, echnidas may use their spines to anchor themselves inside their shelters. Echidnas eat ants, termites, other insects, and worms.

The duck-billed platypus (Ornithorhynchidae) occupies freshwater habitats. This animal is a proficient swimmer and diver but also digs burrows along the shore. This species feeds on aquatic insect larvae, crustaceans, worms, snails, frogs, and small fishes; it often probes the substrate with its bill, searching for food. The platypus is crepuscular and generally solitary.

Monotremes are hard to census because they are either aquatic or mostly nocturnal and seldom seen. They have been studied extensively, however, because of their interesting differences from other mammals and their primitive position in mammalian evolution (Griffiths 1978; Groves 1993a).

Order Didelphimorphia

This taxon traditionally was grouped with six other marsupial taxa into the single order Marsupialia. Current consensus among taxono-

Table 2. Distributions of New World Mammal Families among Nearctic and Neotropical Regions by Animal Size Range and Habit[a]

Taxon	Nearctic				Neotropical			
	Aq	Fo	Te	Ar	Aq	Fo	Te	Ar
DIDELPHIMORPHIA								
Didelphidae	SM		SM	S	SM		SM	S
PAUCITUBERCULATA								
Caenolestidae							S	
MICROBIOTHERIA								
Microbiotheriidae							S	S
XENARTHRA								
Bradypodidae								ML
Megalonychidae								ML
Dasypodidae			SM				SML	
Myrmecophagidae			ML	SML			ML	SML
INSECTIVORA								
Solenodontidae							SM	
Soricidae	S		S				S	
Talpidae	S	S	S					
CHIROPTERA								
Emballonuridae				S				S
Noctilionidae				S				S
Mormoopidae				S				S
Phyllostomidae				S				S
Natalidae				S				S
Furipteridae								S
Thyropteridae				S				S
Vespertilionidae				S				S
Molossidae				S				S
PRIMATES								
Callitrichidae								SM
Cebidae				ML				SML
Hominidae			L				L	
CARNIVORA								
Canidae			ML				ML	
Felidae			ML	ML			ML	ML
Mustelidae	ML		SML	M	SML		SML	
Procyonidae			ML	ML			ML	ML
Ursidae			L				L	

Table 2. (*Continued*)

Taxon	Nearctic				Neotropical			
	Aq	Fo	Te	Ar	Aq	Fo	Te	Ar
PERISSODACTYLA								
Tapiridae			L				L	
ARTIODACTYLA								
Tayassuidae			L				L	
Camelidae							L	
Cervidae			L				L	
Antilocapridae			L					
Bovidae			L					
RODENTIA								
Aplodontidae			M					
Sciuridae			SM	SM			SM	SM
Castoridae	L							
Geomyidae		SM				SM		
Heteromyidae			S				S	
Dipodidae			S					
Muridae	SM	S	SM	S	S	S	S	S
Erethizontidae			ML	ML				M
Chinchillidae						SML		
Dinomyidae						L		
Caviidae						SML		
Hydrochaeridae					L	L		
Dasyproctidae			ML			ML		
Agoutidae			L			L		
Ctenomyidae						S		
Octodontidae						S	S	
Abrocomidae						S		
Echimyidae						S	SM	SM
Capromyidae							M	SM
Myocastoridae					ML			
LAGOMORPHA								
Ochotonidae			S					
Leporidae			SML				SM	

[a]*Habit:* Aq = aquatic; Fo = fossorial or subterranean; Te = terrestrial; Ar = arboreal; based on Nowak (1991). *Animal size:* S = head-body length (HBL) < 300 mm; M = HBL 300–600 mm; L = HBL > 600 mm. Taxonomy follows Wilson and Reeder (1993).

Table 3. Distributions of Old World Mammal Families among Palearctic, Ethiopian, Oriental, and Australian Regions by Animal Size Range and Habit[a]

Taxon	Palearctic				Ethiopian				Oriental				Australian			
	Aq	Fo	Te	Ar	Aq	Fo	Te	Ar	Aq	Fo	Te	Ar	Aq	Fo	Te	Ar
MONOTREMATA																
Tachyglossidae															ML	
Ornithorhynchidae													M			
DASYUROMORPHIA																
Thylacinidae[b]															L	
Myrmecobiidae															S	
Dasyuridae															SML	S
PERAMELEMORPHIA																
Peramelidae															SM	
Peroryctidae														S	SM	
NOTORYCTEMORPHIA																
Notoryctidae														S		
DIPROTODONTIA																
Phascolarctidae																L
Vombatidae															L	
Phalangeridae												M				M
Potoroidae															SM	
Macropodidae															ML	ML
Burramyidae																S
Pseudocheiridae																SM
Petauridae																S
Tarsipedidae																S
Acrobatidae																S
INSECTIVORA																
Tenrecidae					SM	S	SM	S								
Chrysochloridae						S										

Taxon						
Erinaceidae	S			S	SM	
Soricidae	S	S		S	S	
Talpidae	S	S		S	S	
SCANDENTIA						
Tupaiidae	S			S	S	
DERMOPTERA						
Cynocephalidae					M	
CHIROPTERA						
Pteropodidae		SM	SM	SM	SM	SM
Rhinopomatidae		S	S		S	
Craseonycteridae		S	S	S	S	S
Emballonuridae		S	S	S	S	
Nycteridae		S	S	S	S	
Megadermatidae		S	S	S	S	S
Rhinolophidae		S	S	S	S	S
Myzopodidae						
Vespertilionidae		S	S	S	S	S
Molossidae		S	S	S	S	S
Mystacinidae						S
PRIMATES						
Cheirogaleidae			S			
Lemuridae			SM			
Megaladapidae			M			
Indridae			ML			
Daubentoniidae			M			
Loridae		SM	SM		SM	
Galagonidae			SM		S	
Tarsiidae					S	
Cercopithecidae	M	ML	ML		SM	ML

(Continued)

Table 3. (*Continued*)

Taxon	Palearctic				Ethiopian				Oriental				Australian			
	Aq	Fo	Te	Ar	Aq	Fo	Te	Ar	Aq	Fo	Te	Ar	Aq	Fo	Te	Ar
Hylobatidae												ML				
Hominidae			L	ML			L	L			L	L			L	
CARNIVORA																
Canidae			ML				ML				L					
Felidae			ML				ML				ML					
Hespestidae			SM				SM				ML					
Hyaenidae			L				L				SM					
Mustelidae	ML		SML	ML	ML		SML	ML			SML	ML				
Ursidae			L				L				L					
Viverridae			ML	ML	M		ML	M			ML	ML				
PROBOSCIDEA																
Elephantidae			L				L				L					
PERISSODACTYLA																
Equidae			L				L				L					
Tapiridae											L					
Rhinocerotidae							L				L					
HYRACOIDEA																
Procaviidae			M				M	M								
TUBULIDENTATA																
Orycteropodidae							L									
ARTIODACTYLA																
Suidae			ML				ML				ML					
Hippopotamidae							L									
Camelidae			L													
Tragulidae			ML				L				ML					

	Aq	Fo	Te	Ar
Giraffidae			L	
Moschidae			L	
Cervidae			L	
Bovidae			L	
PHOLIDOTA				
Manidae			ML	ML
RODENTIA				
Sciuridae		SM	SM	SM
Castoridae	L			
Dipodidae		S	S	
Muridae	S	SM	SM	SM
Anomaluridae				SM
Pedetidae		M	M	
Ctenodactylidae			S	
Myoxidae			S	S
Bathyergidae		S		
Hystricidae			ML	ML
Petromuridae			S	
Thryonomyidae			M	
LAGOMORPHA				
Ochotonidae			S	
Leporidae			ML	
MACROSCELIDEA				
Macroscelididae			SM	

[a]*Habit*: Aq = aquatic; Fo = fossorial or subterranean; Te = terrestrial; Ar = arboreal; based on Nowak (1991). *Animal size*: S = head-body length (HBL) < 300 mm; M = HBL 300–600 mm; L = HBL > 600 mm. Taxonomy follows Wilson and Reeder (1993).

[b]Probably extinct.

Table 4. Distributions of Mammal Families among Coastal Areas and Oceans of the World by Animal Size Range[a]

Taxon	New World coastal waters		Old World coastal waters				Arctic region	Atlantic Ocean	Pacific Ocean	Indian Ocean	Antarctic region
	Nearctic	Neotropical	Palearctic	Ethiopian	Oriental	Australian					
CARNIVORA											
Odobenidae	SM		SM				SM	SM	SM		
Otariidae	SM	S	S	S		S			S	S	S
Phocidae	SM	SM	S	SM		SM	S	SM	SM	SM	SM
Mustelidae	S	S									
Ursidae							S				
CETACEA											
Balaenidae	L	L	L	L		L	L	L	L	L	L
Balaenopteridae	L	L	L	L		L	L	L	L	L	L
Eschrichtiidae	L		L				L		L		
Neobalaenidae		M		M		M		M	M	M	
Delphinidae	SML	SML	SML	SML	SML	SML		SML	SML	SML	SML
Monodontidae	ML		ML				ML	ML	ML		
Phocoenidae	S	S	S	S	S	S	S	S	S	S	S
Physeteridae	SML	SML	SML	SML	SML	SML		SML	SML	SML	SML
Platanistidae		SM			SM						
Ziphiidae	ML	ML	ML	ML	ML	ML		ML	ML	ML	L
SIRENIA											
Dugongidae			SM	SM	SM	SM			SM[b]	SM[b]	
Trichechidae	SM	SM		SM				SM[b]			

[a]Families of marine mammals occurring in coastal environments are listed under the appropriate continent; those occurring in deep-water environments are listed under the appropriate ocean or region. Animal size: S = head-body length (HBL) < 300 mm; M = HBL 300–600 mm; L = HBL > 600 mm; based on data from Nowak (1991). Taxonomy follows Wilson and Reeder (1993).

[b]Species found in the coastal waters of some islands.

mists, however, favors recognition of seven separate orders (Wilson and Reeder 1993). The order Didelphimorphia includes a single family, Didelphidae, which includes 15 genera and 63 species and is split into two subfamilies, Caluromyinae and Didelphinae (Gardner 1993a). Both subfamilies occur from southern Canada to northern Argentina and reach their greatest species diversities in tropical regions (Eisenberg 1989). Body sizes range from 75 mm head-body length (HBL) in tiny species of *Monodelphis* (Anderson 1982) to more than 500 mm in the largest *Didelphis* (McManus 1974). There are both terrestrial and arboreal species, and even an aquatic form, the yapok or water opossum, *Chironectes minimus* (Marshall 1978b).

Opossums occur in a variety of habitats, including forests and grasslands, often near lake shores or stream banks; the yapok inhabits freshwater streams and lakes in tropical and subtropical areas. Species in the Didelphimorphia are mostly solitary and nocturnal or crepuscular. Although some are primarily insectivorous or carnivorous, most are omnivorous. Opossums have short gestation periods and long postnatal developmental times. Some species have an external pouch that the young occupy during early postnatal development.

Because the natural histories, behaviors, and habitat preferences of opossums are diverse, a variety of techniques must be used to census these animals and to estimate their population sizes. Except for some cryptic tropical species, opossums are reasonably easy to sample.

Order Paucituberculata

Shrew opossums are small (HBL 90–135 mm), crepuscular or nocturnal carnivores. The single family Caenolestidae includes three genera and five species (Gardner 1993b), all known from habitats in the Andes of South America. Relatively few specimens are available for study, and the species' habits are poorly known (Patterson and Gallardo 1987). Shrew opossums appear to be mainly terrestrial but climb well. These animals move over the ground in runways and typically inhabit montane forests and meadows. Shrew opossums eat a variety of invertebrates and small vertebrates and may occasionally take plant material. Because of their small size and mainly terrestrial habits, it should be possible to sample them with standard trapping regimes.

Order Microbiotheria

The single species, *Dromiciops gliroides* (Microbiotheriidae), is restricted to parts of Chile and Argentina (Gardner 1993c). These animals live in densely vegetated, moist forests and build nests under fallen trunks or rocks, in tree hollows, or on branches (Marshall 1978a). This species, commonly known as "monito del monte," is small (HBL 83–130 mm), nocturnal, and scansorial (Redford and Eisenberg 1992). These animals often are found in pairs during the breeding season and hibernate during periods of adverse weather or food shortage. They eat mostly insects and invertebrates but may also take plant matter. Presumably they are susceptible to most standard trapping methods.

Order Dasyuromorphia

This order contains three families (Groves 1993b): Thylacinidae (thylacine or Tasmanian wolf), Dasyuridae (marsupial "mice," "cats," and the Tasmanian devil), and Myrmecobiidae (numbat or banded anteater). They are restricted to the Australian region. There are 17 genera and 63 species, 10 of which have been described in the past 10 years; voucher specimens of this poorly known group are needed. These taxa should be amenable to standard trapping techniques.

The thylacine or Tasmanian wolf (*Thylacinus cynocephalus*) is probably extinct, although unconfirmed sightings are still reported occasionally (Ride 1970). This large (HBL 1,000–1,300 mm), nocturnal carnivore appears to have inhabited grasslands and open to dense forests and sheltered under rocky outcrops and in hollow logs (Nowak 1991). The thylacine, primarily a solitary species, may have hunted in pairs or small groups. Its diet consisted mostly of mammals and birds.

The Dasyuridae are considered to be closely related to the Didelphidae of the New World (Nowak 1991). These animals are mainly small and medium-size (5-g marsupial mice to the 10-kg Tasmanian devil), and typically crepuscular or nocturnal. Most species are terrestrial, although some marsupial mice are arboreal. Quolls typically inhabit dense lowland and montane forests of the Australian region, but some species may be found in more arid, brush-covered areas or grasslands. Dasyurids nest in burrows, in hollow logs, or on the ground under brush; marsupial mice may also build arboreal nests. They are agile runners and climbers and eat a wide range of invertebrates, small vertebrates, and even plant material at times. The Tasmanian devil (*Sarcophilus laniarius*), now restricted to Tasmania, occupies an array of habitats with good cover but may favor coastal heath and sclerophyll forest (Guiler 1970). Its activity pattern and food habits are similar to those of other dasyurids, although it may also eat carrion.

The numbat (*Myrmecobius fasciatus*) inhabits open scrub woodland, especially where eucalyptus trees are common, as well as desert habitats (Nowak 1991). Unlike most members of the order, this small (HBL 175–275 mm), terrestrial, and usually solitary species is active during the day. It rests at night in nests made of leaves or grass that are placed in shallow burrows that it digs, in burrows dug by other animals, or in hollow logs. The numbat feeds primarily on termites but may also take ants and other invertebrates.

Order Peramelemorphia

This order of bandicoots and bilbies includes two families (Groves 1993c): Peramelidae (4 genera, 10 species) and Peroryctidae (4 genera, 11 species). These animals range in size from 150 mm to 500 mm HBL and are restricted to the Australian region. Standard methods should be used for the capture of bandicoots and bilbies.

All but one species of peramelid (*Isoodon macrourus*) occur in Australia, Tasmania, and adjacent islands. Peramelids typically occupy dry habitats such as woodland, savanna, shrub-covered land, grassland, and sparsely vegetated desert where soils are suitable for burrowing (Nowak 1991). On the other hand, the Peroryctid bandicoots generally inhabit lowland and montane rain forests of New Guinea and adjacent islands; one species (*Echymipera rufescens*) is also found in northeastern Australia (Nowak 1991). Peramelid bandicoots and bilbies are powerful diggers; the Peroryctid bandicoots are not. Peramelids often live in burrow networks of their own construction; they may plug the burrow entrance after entering or leaving. Species living in grasslands use runways. Typically, representatives of both families rest during the day in burrows, ground nests made of vegetation, rock crevices, or hollow logs. These species range from mostly insectivorous or carnivorous to omnivorous. They appear to be solitary.

Order Notoryctemorphia

The single genus of two species of marsupial "moles" (family Notoryctidae) is restricted to Australia (Groves 1993d). These moles, approximately the size of placental moles (HBL 90–180 mm), are well adapted for fossorial life and provide an example of evolutionary convergence between marsupial and placental mammals. *Notoryctes* species frequent dry, sandy habitats, especially along river flats. They are less subterranean than true moles (Talpidae) and do not usually leave a long-lasting tunnel, although females may construct deep, permanent burrows in which to bear young. Individuals appear to be active both day and night and are, apparently, solitary. They eat larval and adult arthropods and some seeds (Nowak 1991).

Order Diprotodontia

Diprotodontia is the largest order of marsupials, encompassing 10 families, 38 genera, and 113 species of koalas, possums, cuscuses, wombats, "rat" kangaroos, wallabies, and kangaroos (Groves 1993e). Nine families are endemic to the Australian region; the remaining family is found, in addition, on islands within the Oriental region. Animals in this group range from the small honey possums (7–12 g, HBL 70–85 mm) and pygmy possums (12–71 g, HBL 60–120 mm) to the largest living marsupials, the wallaroos and kangaroos (20–90 kg, HBL 850–1,600 mm for the red kangaroo; Nowak 1991). Koalas, possums, and cuscuses are primarily arboreal, whereas wombats, rat kangaroos, wallabies, and kangaroos are generally terrestrial.

The koala, *Phascolarctos cinereus* (Phascolarctidae), is restricted to the eucalyptus forests of Australia. It is largely nocturnal and completely arboreal, although individuals may occasionally venture to the ground to change trees or to lick soil or gravel as a digestive aid (Nowak 1991). This species is usually solitary and has a small home range, which sometimes includes only a few food trees. The koala's diet is comprised almost exclusively of the leaves and young bark of *Eucalyptus* species.

Cuscuses (Phalangeridae) are mainly inhabitants of tropical forests and thick scrub (Nowak 1991). These nocturnal and arboreal animals also descend to the ground occasionally. Cuscuses are slow moving, sluggish, and generally solitary. They eat mainly fruits and foliage but may also take insects and small vertebrates.

The possums are a diverse group of small to medium-size species distributed among six families. Typically, dormouse possums and mountain pygmy possums (Burramyidae) dwell in forests or shrublands. The dormouse possums

are arboreal, but the mountain pygmy possum is mainly terrestrial, living in holes among and under rocks. Both of these possums are active climbers and have prehensile tails. They are omnivorous, generally rest during the day, and may hibernate during cold weather.

The honey possum, *Tarsipes rostratus* (Tarsipedidae), is common in tree and shrub heaths. Its tongue and mouth are well adapted for probing into flowers and lapping nectar and pollen. This species often hangs upside down from a flower or an adjacent branch while feeding. Honey possums appear to be gregarious, and several animals may feed in close proximity.

The Acrobatidae includes two species of possums: the feather-tailed possum (*Distoechurus pennatus*) and the pygmy gliding possum (*Acrobates pygmaeus*). The feather-tailed possum inhabits rain forests throughout New Guinea, where it feeds on blossoms, fruits, insects, and other invertebrates (Nowak 1991). The pygmy gliding possum inhabits the sclerophyll forests and woodlands of eastern Australia. This species is an acrobatic glider. Insects dominate its diet, although it may take nectar and other plant matter. Both acrobatid species make small, spherical nests of leaves in trunk and branch hollows.

The three families of the somewhat larger possum species include the Phalangeridae (brush-tailed possums), Pseudocheiridae (ring-tailed and greater gliding possums), and Petauridae (striped and lesser gliding possums). Species in these families are found in rain forest, sclerophyll forest, woodland, and shrubland (Smith 1973). These mostly arboreal possums are typically nocturnal and active. Many members of this group are adept climbers and have a prehensile tail; others are proficient gliders, with well-developed gliding membranes and flattened tails. These possums may be solitary or paired or may occur in groups. They eat a variety of plant materials, and some species may also take insects, other invertebrates, and small vertebrates.

Wombats (Vombatidae) are terrestrial and rapid, powerful diggers. These species inhabit upland forest (especially near rocky areas), savanna woodland, grassland, and steppe with low shrubs. Wombats live in burrows that may branch into a complex network of tunnels. A network of trails may connect individual burrow systems and feeding areas. Wombats frequently venture from their burrows to feed on grasses, roots, bark, and fungi. These animals are shy and generally nocturnal and, therefore, difficult to observe in the wild. Nevertheless, individuals have been observed sunning themselves near the entrances to their burrows.

The rat kangaroos and potoroos (Potoroidae) and wallabies and kangaroos (Macropodidae) are generally nocturnal, although some individuals may sunbathe or be active during the day. Rat kangaroos and potoroos occur in rain forest, sclerophyll forest, woodland, savanna, thick scrub, and grassland (Nowak 1991). Most representatives of these two families are terrestrial and saltatorial, although some species, such as tree kangaroos, are agile climbers. Some of the species are solitary; others are gregarious. Wallabies and kangaroos are primarily grazers and browsers, feeding on a variety of plant species (Poole 1982). The presence of ruminant-like, bacterial digestion enables these species to eat plant materials of relatively low nutritional content and, therefore, to occupy environments where such plants dominate the flora. The rat kangaroos and potoroos eat tubers, seeds, and foliage but may take insects and worms as well.

Order Xenarthra

Representatives of this order are found exclusively in the New World from the southern United States to South America (Gardner 1993d). They comprise four families, 13 genera, and 29 species and fall into three distinct and highly specialized subgroups: sloths, armadillos, and anteaters (Montgomery 1985). Sloths (Bradypo-

didae and Megalonychidae) are exclusively arboreal and inhabit subtropical and tropical forests. They venture to the ground only occasionally to urinate or defecate. These herbivorous animals spend much of their life hanging from branches and move only slowly along branches using a hand-over-hand motion. They are active primarily at night. Armadillos (Dasypodidae) and the giant anteater (Myrmecophagidae) are ground dwellers and powerful diggers (McBee and Baker 1982). Armadillos and the giant anteater are active day or night and inhabit savannas, grasslands, and forests. Lesser and silky anteaters are found in tropical forests and savannas, where they are usually nocturnal and arboreal. Anteaters and some armadillos are insectivorous; other armadillos are more omnivorous (Nowak 1991). Xenarthrans are generally solitary animals, but individuals of some species may form loose associations.

Order Insectivora

This relatively large order encompasses six families, 65 genera, and 420 species of extant mammals (Hutterer 1993). Most species are found in the Palearctic, Ethiopian, and Oriental regions, but a few also occur in the New World. These mouse-like animals are generally small (HBL <180 mm) and inhabit terrestrial, often moist, habitats (Nowak 1991). Over 70% (312 species) of the Insectivora are shrews (Soricidae). Larger terrestrial members of this order include the hedgehogs of Eurasia and Africa (Erinaceidae), the solenodons (Solenodontidae) of the West Indies, and the tenrecs (Tenrecidae) of Madagascar and Africa.

Insectivora inhabit an array of habitats. Terrestrial species are found in grasslands, scrub, and forests and on cultivated lands. They often shelter in and under logs, branches, and leaf litter; among rocks and roots of trees; in burrows; and under dense vegetation (George 1989). Some species, such as the desmans (Talpidae) and the water or otter shrews (Tenrecidae), are semiaquatic or aquatic, seeking protection in the water or in burrows located in banks above the high water level of streams, ponds, and lakes. Other species, such as the moles (Talpidae) of the Palearctic and Nearctic (Hartman and Yates 1985) and the golden moles (Chrysochloridae) of the Ethiopian region, are fossorial and spend most of their life underground. These species frequent areas where the soil is loose and often sandy.

Many Insectivora are nocturnal, but some shrews and aquatic forms may be active day or night. Insectivores are generally solitary, but individuals of some species occur in small groups. Members of this order generally feed on immature or adult insects, other invertebrates, and small vertebrates that they encounter while moving over the ground or through their burrows, or while digging or swimming (Nowak 1991).

Order Scandentia

The one family (Tupaiidae), five genera, and 19 species of tree shrews (Wilson 1993a) in this order are primarily restricted to the forested areas of the Oriental region. Most tree shrews are swift runners, adept climbers, and diurnal. Typically, they forage on the ground or in the tree canopy, eating mostly insects and fruits, but also taking animal and plant foods (Nowak 1991). Most representatives of this order spend the majority of their time on the ground or in low shrubs. These squirrel-like species hide and nest in holes among rocks, in hollow logs, and in burrows. One species, the pen-tailed tree shrew (*Ptilocercus lowii*) of Malaysia and Indonesia, is, however, nocturnal and almost exclusively arboreal, feeding and sheltering in nests within the tree canopy (Whittow and Gould 1976).

Order Dermoptera

This order is restricted to the forests of the Oriental region and encompasses one family (Cyn-

ocephalidae), one genus, and two species of colugos or flying lemurs (Wilson 1993b). These animals are adept, but slow, climbers and seldom descend to the ground. They have a large gliding membrane attached to the neck and body on either side and are renowned for their ability to glide long distances between trees. These animals are generally nocturnal, resting during the day in holes or hollows of trees. At dusk, colugos glide off to feed, often in the same tree night after night. Although relatively little is known about colugo social behavior, these animals appear to share shelters and feeding trees. They eat plant material, including fruits, buds, flowers, and leaves (Lekagul and McNeely 1977).

Order Chiroptera

Bats comprise the second largest order of mammals, with 177 genera and 925 species (Koopman 1993). This diverse group of mammals is divided into two suborders: Megachiroptera (one family of Old World fruit bats, Pteropodidae, containing 166 species) and Microchiroptera (16 families and 756 species). Chiroptera occur throughout the temperate and tropical regions of both hemispheres, and on all but the most isolated oceanic islands (Nowak 1991). Bats are the only true flying mammals, and they occupy a wide array of ecological niches. Bats shelter in tree cavities, crevices, caves, and buildings, and some rest exposed on trees. Bats in temperate regions may hibernate or migrate to warmer regions during the cold seasons (Fenton 1983). Extensive summaries of methods for studying bats can be found in Kunz (1988) and Handley et al. (1991).

The Megachiroptera inhabit forests, woodlands, and savannas in the subtropical and tropical areas of the Old World (Mickleburgh et al. 1992; Wilson and Graham 1992). This group, which includes small nectar- and pollen-feeding species (subfamily Macroglossinae) as well as large, fruit-eating bats, is restricted to areas that possess a reliable supply of ripe fruit, pollen, or nectar. Some species make long flights from roosting sites to feeding areas and may return to the same feeding areas repeatedly.

Microchiroptera are renowned for their ability to orient and hunt using echolocation. The majority of these bats are insectivorous, catching insects while flying using echolocation. Other species feed on mammal blood, fishes, or other small vertebrates (Wilson 1973). Some Microchiroptera, however, feed on plant materials, including pollen, nectar, and particularly fruits. Fruit- or blood-eating bats appear to use vision and olfaction in addition to echolocation to find prey (Eisenberg and Wilson 1978).

Three families of microchiropteran bats—Vespertilionidae, Molossidae, and Emballonuridae—occur in both the Old World and the New World (Wilson 1989). The insectivorous family Vespertilionidae (little brown bats) occurs throughout temperate and tropical areas and is the largest and most widely distributed family of bats (318 species). Many species living in temperate regions hibernate during the winter, often migrating long distances to their hibernation caves (Nowak 1991). The free-tailed bats (Molossidae) are generally limited to subtropical and tropical environments. Sac-winged or sheath-tailed bats (Emballonuridae) are also widespread in tropical and subtropical areas.

Seven families of microchiropteran bats—Rhinopomatidae, Craseonycteridae, Nycteridae, Megadermatidae, Rhinolophidae, Myzopodidae, and Mystacinidae—are restricted to the Old World; the Rhinolophidae (horseshoe bats) make up the most diverse family in this group with 130 species. Species living in temperate areas hibernate during the winter. The remaining families are usually found in tropical and subtropical areas. The monotypic Craseonycteridae are endemic to the Oriental region, where individuals have been observed near bamboo stands and teak trees. Slit-faced bats (Nycteridae) and false vampire and ghost bats (Megadermatidae)

are found in forested and open areas and roost in caves, rock crevices, buildings, hollow trees, and vegetation (Hudson and Wilson 1986). The single species of sucker-footed bat (Myzopodidae) is restricted to Madagascar, and the two species (one probably extinct) of short-tailed bats (Mystacinidae) are restricted to New Zealand forests. These three species appear to be omnivorous, taking both arthropods and plant material. Unlike many of the Old World bat species, the mouse-tailed bats of the southern Palearctic, Ethiopian, and Oriental regions (Rhinopomatidae) usually inhabit treeless, arid regions.

Six families of Microchiropteran bats are restricted to the New World. The New World leaf-nosed bats (Phyllostomidae) occur in subtropical and tropical areas from the southern United States to Argentina. These bats number 141 species and dominate the bat fauna in the Neotropical region. They are the most diverse family of Microchiroptera and have exploited the widest variety of food types. They range from small insectivorous species, through blood-feeding vampires, to large carnivorous species.

The remaining bat families endemic to the New World are primarily tropical in distribution. Leaf-chinned bats (Mormoopidae) are among the most abundant. These bats are exclusively insectivorous and gregarious cave dwellers (Nowak 1991). Fishing bats (Noctilionidae) are proficient at catching fish but may also eat insects and crustaceans. The five species of funnel-eared bats (Natalidae), two species of smoky bats (Furipteridae), and two species of disk-winged bats (Thyropteridae) are usually found in forested areas of the tropics.

Order Primates

The Order Primates comprises 13 families, 60 genera, and 233 species (Groves 1993f). One species in this order, *Homo sapiens,* is nearly worldwide in distribution. Other representatives of the order occur in the New World, from Mexico to Argentina, most of Africa, Madagascar, part of the Arabian Peninsula, south and southeastern Asia, and the East Indies (Nowak 1991). Many members of this group are arboreal, although some species are partly or mostly terrestrial. Primates range in size from the small mouse lemurs (38–98 g, HBL 125–150 mm) to the large *Gorilla gorilla* (70–275 kg, 1.25–1.75 m in height when standing on two feet). Most primates are opportunistic omnivores, taking fruits, nuts, flowers, leaves, insects, and small vertebrates. In addition, some terrestrial species feed on agricultural crops. Although some primates are solitary, many species live in family groups or large troops.

The New World monkeys (Cebidae) inhabit primary and secondary tropical forests as well as cultivated areas (Nowak 1991). These species are mainly diurnal; they are agile jumpers and runners and have prehensile tails. Although these animals are arboreal, they may descend to the ground to forage or to move long distances. Marmosets or tamarins (Callitrichidae) are also restricted to the New World. These small primates possess keen sight and are adept at running, jumping, and leaping through the tree canopy. Marmosets are diurnal and live in small groups; they rest in tree cavities (Kleiman 1977).

The Old World monkeys (Cercopithecidae) are mostly diurnal and generally medium-size to large animals (Table 3). Baboons and patas monkeys are mainly terrestrial. Macaques are adept at moving in the tree canopy but also descend to the ground to feed and travel long distances. Other species, such as langurs and colobus monkeys, are primarily arboreal (Nowak 1991). Gibbons (Hylobatidae) are highly arboreal and brachiate with agility through the canopies of eastern Palearctic and Oriental forests. Other Old World primates include the small to medium-size lorises and pottos (Loridae), and the bush babies and galagos (Galagonidae) of the Ethiopian region. These nocturnal animals are arboreal, moving along branches using a deliberate, hand-over-hand motion. Tarsiers (Tarsiidae), small primates similar in habit to the Loridae and Galagonidae, and some galagos are able to leap long distances with precision. These animals feed largely on insects (MacKinnon and MacKinnon 1980).

The largest of the primates are the great apes (Hominidae): orangutans, gorillas, chimpanzees, and humans. The great apes (excluding humans) are generally herbivorous but may take animal foods (e.g., chimpanzee). The gorilla and chimpanzee are mostly terrestrial and quadrupedal but can climb; orangutans are primarily arboreal (Groves 1971). The great apes are largely diurnal.

Madagascar has a rich primate fauna with five endemic families of lemurs. Mouse and dwarf lemurs (Cheirogaleidae) are the smallest. These forest dwellers are adept at moving through the canopy, usually quadrupedally. They are nocturnal and rest during the day in leaf nests. Nocturnal sportive lemurs (Megaladapidae) are arboreal, hopping from one perch to the next. Lemurs (Lemuridae), on the other hand, are primarily diurnal, and most species appear to be territorial. Like the mouse and dwarf lemurs, these animals walk or run along horizontal branches with ease, or leap from perch to perch. The arboreal aye-aye (Daubentoniidae) is found in dense forests and bamboo stands. Leaping lemurs (Indridae) are fairly large, diurnal (indri and sifakas) or nocturnal (avahi) herbivores that are deliberate climbers and may occur solitarily or in association with other individuals (Nowak 1991).

Order Carnivora

This order is native to all continents except Australia and Antarctica, and to the world's oceans (Wozencraft 1993). The Order Carnivora encompasses 11 families, 129 genera, and 271 species. The otters frequent freshwater habitats; others, such as sea otters, seals, sea lions, and walruses, are marine organisms. Most species have well-developed senses of smell, sight, and hearing. Typically, these animals eat vertebrate prey, which they stalk or pounce on after lying in wait.

Canidae, Felidae, and Mustelidae are found naturally on most large land masses from the Arctic to the tropics. They feed primarily on freshly killed vertebrate prey (Nowak 1991). Canids are the most cursorial of the carnivores and frequently travel in packs (Bueler 1973). Canids may be active anytime during the day and have excellent sight, smell, and hearing. Cats are also excellent hunters and may stalk prey until close enough for a brief dash and the kill. Many species are nocturnal, but some are diurnal. Most mustelids search aggressively for prey in burrows or dense cover. Others, such as the mink and otter, are good swimmers and hunt in aquatic environments.

Bears (Ursidae) occur worldwide from the Arctic to the tropics. In colder regions, bears enter a deep sleep during winter. Bears may shelter in holes they dig, in caves or crevices, in tree hollows, or in dense vegetation. Most bears are omnivorous, but the polar bear takes little other than fish and seals, and the panda feeds primarily on bamboo. Procyonids (Procyonidae) are also omnivorous. They typically are active at night, and all species appear to be good climbers (Grzimek 1975).

Civets, genets, and mongooses (Hespestidae and Viverridae) typically inhabit the forests of the Ethiopian, Oriental, and eastern Palearctic regions, but they may also occur in grasslands and shrublands (Schreiber et al. 1989). Many representatives of these families climb, and some restrict most of their activities to tree canopies. Species are diurnal or nocturnal. Civets, genets, and mongooses eat small invertebrates or vertebrates, and occasionally plant material or carrion.

Hyenas (Hyaenidae) are also restricted to the Old World, where they typically inhabit open grassland, brushland, and forest (Mills 1982a); they live in groups or as solitary individuals.

Their diet is dominated by mammalian carrion, but some species also take insects, vertebrates, and plant material. The aardwolf takes mostly insects, especially termites and larvae.

Carnivora includes approximately 37 species of marine mammals in five families: Phocidae (earless seals), Otariidae (eared seals and sea lions), Odobenidae (walrus), Mustelidae (two species of marine otter), and Ursidae (the polar bear). They are well adapted for life in aquatic environments and are excellent swimmers and divers. Most species are solitary while at sea but form large colonies during the breeding season, usually on isolated islands. Marine Carnivora consume invertebrate and vertebrate prey (Riedman 1990).

Order Cetacea

Cetaceans are highly adapted for aquatic life and occur in all oceans of the world (Table 4); some species (e.g., dolphins—Platanistidae) are restricted to freshwater lakes and rivers (Mead and Brownell 1993). There are 10 families, 41 genera, and 78 species of cetaceans, which are grouped into two suborders: Odonticeti, or toothed cetaceans (Delphinidae, Monodontidae, Phocoenidae, Physeteridae, Platanistidae, and Ziphiidae), and Mysticeti, or baleen whales (Balaenidae, Balaenopteridae, Eschrichtiidae, and Neobalaenidae). Most species of Odonticeti use echolocation for orientation and for locating food. They generally feed on fish and invertebrates, especially cephalopods and crustacea. The killer whale (*Orcinus orca*) regularly takes vertebrate prey such as penguins, seals, and other cetaceans (Heyning and Dahlheim 1988). The Mysticeti filter zooplankton from the water column using baleen plates.

Dolphins are perhaps the fastest and most agile of the cetaceans; they often are observed following ships. Whales are also strong swimmers and may travel long distances during migrations between feeding and breeding areas.

Generally, the smaller species of the Odonticeti appear to be shallow divers; some large Odonticeti and the Mysticeti can perform deep dives. Many members of this order can be found in coastal waters, at least during part of the year; some species may enter bays and estuaries. Other species are exclusively deep-water inhabitants, often occurring near cold-water upwellings where food may be abundant. Cetaceans communicate by producing a variety of underwater sounds, and most species are to some degree gregarious (Gaskin 1982).

Order Sirenia

This order contains two families of aquatic mammals: Dugongidae (dugongs) and Trichechidae (manatees). The dugong inhabits shallow coastal waters, moving into freshwater zones of estuaries and rivers in tropical areas of the Indian and western Pacific oceans (Wilson 1993c). Manatees live along the coast and in coastal rivers of western Africa (one species) and of the New World, from the southeastern United States to northern South America (two species).

Sirenians are large, relatively slow-moving mammals that graze on aquatic plants (Domning 1982). Dugongs appear to take sea grasses and algae typical of the marine habitats in which they occur, and manatees eat aquatic plants common to their generally freshwater habitats. Sirenians propel themselves through the water with their fluke-like tails and steer with their flippers. Some species are solitary; others live in small groups. These aquatic mammals, especially dugongs, may migrate from resting areas to forage in shallow waters. Movement patterns of these species can be influenced by weather conditions, changes in water level or current, and food availability.

Order Proboscidea

The order includes one family, the Elephantidae. The Asian elephant (*Elephas maximus*) is found in the Palearctic (East Asia) and the Oriental regions, and the African elephant (*Loxodonta africana*), the largest living terrestrial mammal, is endemic to the Ethiopian region (Wilson 1993d). Elephants occupy riverine, savanna, thornbush, and forest habitats and may venture into agricultural areas. They are gregarious and form herds of varying sizes. Both species may migrate seasonally over considerable distances in search of water, food, and shade; they also may move aseasonally between water and feeding areas. Elephants follow regular paths during these trips. Agricultural development has restricted the movements of elephants, and ivory hunting has reduced elephant numbers throughout their original ranges, resulting in patchy distributions. Elephants are herbivores that feed on a variety of grasses, roots, small stems, and leaves, as well as on cultivated crops.

Order Perissodactyla

The odd-toed ungulates include the horses, zebras, and asses (9 species) as well as tapirs (4 species) and rhinoceroses (5 species). Perissodactyls occur in the Ethiopian, Palearctic, and Oriental regions and in subtropical and tropical areas of the New World (Grubb 1993a). The members of this order are of medium to large size and generally adapted for cursorial locomotion.

Horses, zebras, and asses (Equidae) inhabit grasslands, steppes, and arid shrublands. These species are the most highly cursorial representatives of the Perissodactyla. Asses are more sure-footed and possibly slower runners than other Equids, which reflects their occurrence, typically, in rocky habitats. Some members of this family migrate long distances to find food or water, and most form herds. This gregarious behavior appears less pronounced in colder regions. Equidae are primarily grazers and generally diurnal. Wild equids are now restricted to Africa and parts of Asia, but feral equids occur throughout the world (Churcher and Richardson 1978).

Tapirs (Tapiridae) are tropical or subtropical in distribution and inhabit moist forests or open areas near water (Eisenberg et al. 1987). These animals typically rest in forests or shrub thickets during the day and browse at night. Tapirs, which are adept at moving in terrestrial or aquatic habitats, often connect feeding areas, watering holes, and resting sites with trails. Tapirs are usually solitary animals and communicate with each other by whistling and by scent marking with urine.

Rhinoceroses (Rhinocerotidae) inhabit grasslands, savannas, brushlands, and dense forests in tropical and subtropical regions. African species generally live in more open habitats than do the rhinos of the Oriental region. Like tapirs, rhinos are usually found near water. These crepuscular or nocturnal herbivores usually rest in heavy cover or in wallows during the day. Some species are browsers, but others graze. Rhinos often make trails between feeding and watering sites and may scent mark their paths with dung (Laurie et al. 1983).

Order Hyracoidea

The Procaviidae or hyraxes (3 genera, 6 species) occur in the southwestern Palearctic and throughout most of the Ethiopian region (Schlitter 1993a). The terrestrial genera (*Procavia* and *Heterohyrax*) inhabit rocky, scrub-covered areas and grasslands. Terrestrial hyraxes are agile and adept at running, jumping, and climbing over rough terrain. They are diurnal and live in colonies among rocks, where they may be observed basking. Tree hyraxes (*Dendrohyrax*), on the other hand, are mostly arboreal (Nowak 1991). They are nocturnal and shelter in tree hollows and dense canopy. Tree hyraxes feed mostly in the tree canopy but may descend to the ground and move about. Hyraxes are primarily grazers and browsers. Tree hyraxes are less gregarious and have smaller home ranges than terrestrial hyraxes, and their activities may center around a single tree.

Order Tubulidentata

This order includes one family (*Orycteropidae*) and a single species, the aardvark (*Orycteropus afer*). Aardvark are endemic to the Ethiopian region and occur in grassland, savanna, scrub, and woodland habitats (Schlitter 1993b). Their distribution among habitats appears to reflect the availability of ants and termites, their preferred food items. They forage by digging, exposing nests, and searching on the ground. Although primarily nocturnal, these animals may walk about during the day and even sunbathe near the entrance to their burrows. Aardvarks are powerful diggers and construct an elaborate network of tunnels with multiple entrances and chambers. They may follow a regular route from nest to nest or systematically search a patch of ground before moving on to the next patch (Nowak 1991).

Order Artiodactyla

Even-toed ungulates are found worldwide on all major land masses except Australia and Antarctica (Grubb 1993b). The pigs and warthogs (Suidae), peccaries (Tayassuidae), and hippopotamuses (Hippopotamidae) are separated from the rest of the order by common morphological characteristics, including the absence of a ruminant stomach.

Pigs (5 genera, 16 species) inhabit primarily shrub-covered or forested areas. The warthog (*Phacochoerus aethiopicus*), in contrast, is diurnal. Some suids use the snout or tusks to dig for food. Peccaries (3 genera and 3 species) are restricted to the New World; they are found in arid scrub, woodland, and rain forest, usually near water holes or streams (Sowls 1984). They are agile runners, are gregarious, and may use group defense against predators.

Hippopotamuses (2 genera, 4 species) are found only in Africa and seem to prefer areas with permanent water and adjacent marsh or grasslands (Kingdon 1979). These animals are good swimmers and divers, and during the day, they usually rest partially or mostly submerged in water. At night, they venture out to forage, taking leaves, stems, and fruits.

Most artiodactylids, including camels (Camelidae), chevrotains (Tragulidae), giraffes (Giraffidae), musk deer (Moschidae), deer (Cervidae), pronghorns (Antilocapridae), and cows, sheep, goats, and antelope (Bovidae), are ruminants (Vaughan 1978). Representatives of this order, especially some deer and antelope species, are noted for their long seasonal migrations in search of water and food.

Camels are native to the Palearctic region and occupy arid and semiarid habitats. They are renowned for their ability to withstand extremes in temperature and to go without water for extended periods. Consequently, they have been widely domesticated. Their South American relatives, guanacos and vicuñas, live on grassy plains or in mountainous areas and are adept at moving over rough terrain. These animals live alone or in small groups. They also have been domesticated, as have llamas and alpacas.

The smallest Artiodactyla are the chevrotains of the Ethiopian, Palearctic (East Asia), and Oriental regions (3 genera and 4 species). These shy, nocturnal animals typically live near water in the rain forest undergrowth and often move along tunnel-like paths. Musk deer (Moschidae), with one genus and four species, are found only in the Palearctic and Oriental regions. These animals lack antlers, but their upper canine teeth are developed as tusks. Musk deer usually hide in dense vegetation during the day and forage at night.

Deer (16 genera, 43 species) are distributed almost worldwide and occur in a variety of habitats from forests to deserts and tundra (Wemmer 1987). Most species graze or browse. The pronghorn (*Antilocapra americana*), which is closely related to deer, is the only species in the family Antilocapridae and is endemic to the grasslands and deserts of the Nearctic.

The Bovidae (45 genera, 137 species) is the largest and most diverse family of the Artiodactyla. Most bovid genera are native to the Old World; the fauna of East Africa is particularly diverse. Nonbranching but elaborately curved or spiraled horns may be present in both sexes. The Giraffidae (2 genera, 2 species) also occupy savannas and open woodlands and are endemic to the Ethiopian region. The smaller and diurnal okapi lives in dense humid forests and usually travels along obvious paths.

Order Pholidota

The Manidae or pangolins (1 genus, 7 species) inhabit forests, savannas, and grasslands in the Ethiopian, Palearctic (East Asia), and Oriental regions (Schlitter 1993c). Most species of pangolins are nocturnal, but a few are active during the day. Some species are terrestrial and live in burrows that they dig or that have been abandoned by other animals. Typically, pangolins walk slowly; they often use only the hind legs, with the tail as a support. If disturbed, pangolins may roll up into a tight ball so that the armored limbs and tail protect the soft underbody. Some species are arboreal, have prehensile tails, and rest in tree hollows. The pangolin diet is dominated by ants and termites, but these animals may also take other adult and larval arthropods.

Order Rodentia

Rodentia is the largest order of mammals, comprising 28 living families and more than 2,000 species (Wilson and Reeder 1993). This worldwide group is ecologically diverse, with aquatic or semiaquatic, fossorial, terrestrial, and arboreal representatives (Tables 2 and 3).

Muridae is the largest family of mammals (281 genera, 1,326 species) and includes the mice, rats, hamsters, voles, lemmings, and gerbils (Musser and Carleton 1993). It is the only family of Rodentia native to Australia. These rodents are most diverse in the tropical and sub-tropical areas of the New World and Old World. They exhibit a wide range of food habits and may eat plant material, invertebrates, and small vertebrates.

The Sciuridae (50 genera, 273 species) includes the squirrels, chipmunks, marmots, and prairie dogs (Hoffmann et al. 1993). Sciurids tolerate a wide range of environmental conditions and are found from the Arctic to the tropics and in arid as well as humid areas. The Sciuridae includes terrestrial (e.g., chipmunks), arboreal (e.g., tree and flying squirrels), and semifossorial species (e.g., ground squirrels, marmots, and prairie dogs). They are usually herbivorous, but occasionally take insects and small vertebrates.

The jerboas (Dipodidae) are also widespread, with most species occurring in the Old World (Holden 1993a). The 51 species inhabit deserts, semideserts, and steppes (Nowak 1991). The jerboas are most numerous in the Palearctic region (45 species). They are also proficient burrowers and may attempt to regulate environmental conditions within the burrow by periodically plugging the entrance hole. Their diet is dominated by seeds, other plant parts, and insects (Nowak 1991).

The semiaquatic beavers (Castoridae), two of the largest species of rodents, occur throughout the northern hemisphere (Wilson 1993e). They feed on the leaves, twigs, roots, and bark of trees, often storing logs under water for a winter food supply (Jenkins and Busher 1979).

NEW WORLD RODENTIA

Although the Nearctic rodent fauna is dominated by murids and sciurids, Geomyidae (pocket gophers, 35 species) and Heteromyidae (pocket mice, kangaroo mice, and kangaroo rats, 59 species) are also important families (Patton 1993a, 1993b). Pocket gophers (Geomyidae) are fossorial, solitary animals that occur in localized or isolated areas with soil suitable for digging (Nowak 1991). Their tunnel systems may be extensive and are marked by closed mounds of dirt. Pocket mice, kangaroo mice, and kangaroo rats inhabit a range of environments from arid deserts to humid tropical forests (Nowak 1991). They generally rest in their burrows during the day, often plugging the entrance to create a more favorable burrow environment. The heteromyid diet is dominated by seeds and vegetation, but may also include insects and other invertebrates (Nowak 1991).

The mammal fauna of the Nearctic also includes one medium-size and two large rodents: the beaver (Castoridae, described previously), the North American porcupine, *Erethizon dorsatum* (Erethizontidae), and the mountain beaver, *Aplodontia rufa* (Aplodontidae), respectively. Unlike its Neotropical relatives, the North American porcupine is mostly terrestrial, although it frequently climbs trees to seek food or shelter (Woods 1973). Porcupines are active throughout the year and eat leaves, twigs, seeds, nuts, berries, roots, and bark. The mountain beaver is endemic to North America (Wilson 1993f). This rodent occurs from southwestern British Columbia to northern California and inhabits forests and other densely vegetated areas. The species burrows and constructs tunnel systems with multiple openings, often connecting nesting and feeding areas. These animals eat plant matter but take twigs and bark when green forage is not available.

The Neotropical region has more rodent families than any other region, including four families in the suborder Sciurognathi and 13 in the suborder Hystricognathi, 10 of which are endemic (Woods 1993). The Neotropical Sciurognath rodents include the Sciuridae (23 species), Geomyidae (9 species), Heteromyidae (9 species), and Muridae (321 species), families also represented in the Nearctic region.

Three families of medium- to large-size Hystricognath rodents occupying tropical and subtropical environments of the New World are the Agoutidae (agoutis), Dasyproctidae (pacas), and

Erethizontidae (porcupines). Agoutis inhabit grass, brush, and forest. They eat plant material, particularly fruits and foliage. Pacas frequent forested areas near streams (Nowak 1991). They follow well-worn paths when traveling between their burrows and feeding areas or water sources. They eat leaves, stems, seeds, fruits, and roots. Porcupines, except the terrestrial North American porcupine, are nocturnal and arboreal. These rodents shelter in tree hollows or dense foliage and eat a variety of plant materials, including the bark of some tree species.

The most common Hystricognath family is the Echimyidae (spiny rats). These animals are most diverse in South America (71 species), but the distributions of three species extend to Middle America. Spiny rats are herbivorous and may be terrestrial or arboreal. Another common Hystricognath group found only in the Neotropical region is the Ctenomyidae (tuco tucos). Tuco tucos occupy sandy, often dry soils of coastal plains, grasslands, and forests from the subtropics to the sub-Antarctic (Nowak 1991). The Caviidae (cavies) are restricted to South America. Cavies are terrestrial and crepuscular and often live in groups.

The largest living rodent, the capybara (*Hydrochaeris hydrochaeris,* Hydrochaeridae), is restricted to the Neotropical region (Mones and Ojasti 1986). These animals are active on land but are good swimmers and may seek shelter in aquatic habitats. The South American pacarana, *Dinomys branickii* (Dinomyidae), is another large Hystricognath species (White and Alberico 1992). It is typically terrestrial but may climb; it is less cursorial than members of Caviidae or Hydrochaeridae.

The Capromyidae (hutias) are Hystricognath rodents that are restricted to the West Indies. Hutias are mostly herbivorous but may take some animal matter (Morgan 1989). The nutria, *Myocastor coypus* (Myocastoridae), is another medium-size to large Hystricognath rodent endemic to South America (Woods et al. 1992). It

appears to be strictly herbivorous and has been introduced broadly in many areas.

Three additional endemic families of Hystricognath rodents inhabit rocky and mountainous areas of the Neotropical region. The octodonts (Octodontidae) are herbivorous, primarily terrestrial mammals that shelter in crevices in rock piles, in hedgerows, and in burrows that they dig or that have been abandoned by other animals (Woods and Boraker 1975). The viscachas and chinchillas (Chinchillidae) inhabit pampas as well as mountainous areas. They may be diurnal, nocturnal, or crepuscular, and they live in burrows or rock crevices. Chinchilla rats (Abrocomidae) are mostly terrestrial. These animals appear to be colonial and shelter among rock crevices or in burrows with entrances located under shrubs or rocks (Nowak 1991).

OLD WORLD RODENTIA

The Old World rodent fauna, like that of the New World, is dominated by the Muridae and Sciuridae. Murid rodents are most numerous in the Ethiopian, Palearctic, and Oriental regions (Musser and Carleton 1993). As noted earlier, murids are the largest single family of mammals, with more than a quarter of the currently recognized species (1,326 of 4,629). As such, they are the dominant members of most rodent faunas worldwide and can be expected to form an important component of any biodiversity inventory. Their dominance makes generalizations about their natural history superfluous. Although less numerous, the sciurid rodents show the same distribution pattern. The jerboas (Dipodidae) are also important components of the Palearctic rodent fauna. The rodent faunas of the Palearctic and Oriental regions encompass several of the same families (Sciuridae, Muridae, Myoxidae, and Hystricidae). Nevertheless, almost six times as many species are found in the Oriental region as in the Palearctic region.

Two families of rodents, the dormice (Myoxidae) and the Old World porcupines (Hystricidae), are distributed in the Ethiopian, Palearctic, and Oriental regions (Holden 1993b; Woods 1993). Dormice are nocturnal, scansorial, and arboreal; inhabit shrub-covered and wooded areas; and are adept at moving through the tree canopy (Nowak 1991). These species hibernate during the winter in the colder parts of their ranges. The Old World porcupines inhabit deserts, savannas, and forests (Nowak 1991). They typically eat vegetation but also may take carrion. Gundis (Ctenodactylidae) occur in the Ethiopian and Palearctic (West Asia) regions (Dieterlen 1993a). These small, nonburrowing mammals are diurnal and may be seen foraging or sun basking near their living sites.

Five families of Rodentia are endemic to Africa. The Sciurognath group (Dieterlen 1993b, 1993c) includes the scaly-tailed squirrels (Anomaluridae) and the springhares (Pedetidae). Scaly-tailed squirrels are found in the tropical and subtropical forests of western and central Africa (Nowak 1991). These arboreal animals run along branches and glide from tree to tree. They eat flowers, fruits, bark, nuts, and insects. The springhares of southern Africa inhabit dry sandy soils in arid or semiarid country (Nowak 1991). They generally use a saltatorial locomotion and shelter in burrows located near trees or shrubs.

The other endemic families are Bathyergidae (mole rats), Petromuridae (dassie rats), and Thryonomyidae (cane rats) in the suborder Hystricognathi (Woods 1993). The five genera and 12 species of African mole rats are typically found in loose, sandy soil. Mole rats feed on bulbs, roots, and tubers that they encounter as they tunnel. Dassie rats (*Petromus typicus*) inhabit rocky areas on hills or mountains (Nowak 1991) and shelter under rocks; they feed on seeds, berries, and other plant matter. There are two species of cane rats. One species is semiaquatic and lives in marshes; the other species is terrestrial and inhabits moist savannas (Nowak 1991). Both species appear to live in groups of mixed age and sex for at least part of the year. Despite their size, they are relatively fast and agile on land. These animals are mostly grazers but include a variety of plant matter in their diet.

Order Lagomorpha

The Lagomorphs are terrestrial animals native to most major land masses except Australia (to which they have been introduced). They are absent from southern South America, the West Indies, Madagascar, and many islands of the Ethiopian region (Hoffmann 1993). They are primarily grazers but eat a wide variety of food items. When grasses and other forage are scarce, individuals may browse, taking shrub stems and tree bark.

The 54 species of hares and rabbits (Leporidae), which are native to or have been introduced into most regions of the world, inhabit grasslands, shrublands, and forests as well as tundra and alpine habitats (Chapman and Flux 1990). They appear to be most active at dusk or at night. Rabbits may rest in burrows or in surface nests, which are connected by obvious trails. Hares, in contrast, prefer rock crevices and caves; they run to escape predators. Rabbits are small, are solitary to gregarious, and give birth to altricial young. Hares are generally larger, are often solitary, and produce precocial young.

The pikas (Ochotonidae) of the Palearctic and western Nearctic regions (1 genus, 25 species) inhabit open plains, deserts, steppes, and forests. Many species are found in rock outcrops, where they shelter in burrows under the rocks and forage nearby. Pikas are active throughout the day and may be seen sunning themselves on rocks or harvesting hay for winter food supplies (Nowak 1991). These animals do not hibernate during the winter but feed on stored hay as food. Many of the Palearctic and Oriental species of pika are gregarious and live in colonies, but the Nearctic species are highly territorial (Smith and Weston 1990).

Order Macroscelidea

This order contains the single family Macroscelididae with four genera and 15 species; 14 species are endemic to the Ethiopian region and one is found in a small part of the Palearctic region (Schlitter 1993d). Elephant shrews are small to medium-size organisms (HBL 95–315 mm) that are mainly diurnal (Nowak 1991). They occur in forest undergrowth, scrub thickets, grasslands, and rocky outcrops. These animals seek shelter in ground depressions, under logs, in rock crevices, in burrows of other animals, or in burrows they construct. They move along runways through grass and under brush. Elephant shrews are primarily insectivorous, appearing to favor ants, termites, and beetles, but they may also eat other animals and plants.

Information Needs

The foregoing brief summary of our knowledge of the biodiversity of mammals begs the question of what we do not know about mammal biodiversity. In spite of their relative conspicuousness and obvious importance, our knowledge of detailed distribution and natural history data for many species of mammals is still rudimentary.

Conservation status

Mammal species may be at risk from a variety of factors, but perhaps the most serious threat to their diversity is the rapidly growing human population. Population growth and the accompanying increase in resource consumption result in increased pressures to convert wildlife habitat to other uses as well as increased levels of pollution. These problems probably are most serious in subtropical and tropical areas, where the richness of mammal species is high and the rates of human population growth are rapid. Tropical species, which frequently occur in low numbers and in restricted areas, are often at greater risk of extinction than temperate species, which are often less habitat specific and more common.

In this section we assess the conservation status of mammalian species. We have considered all species listed in the Convention on International Trade in Endangered Species of Wild Fauna and Flora, the 1990 Red Data Book of the International Union for the Conservation of Nature and Natural Resources (IUCN 1990), and the U.S. Endangered Species Act to be at risk. Additional information provided by specialists on selected taxa was gleaned from the second edition of *Mammal Species of the World* (Wilson and Reeder 1993). We have also included species designated by these individuals as rare, vulnerable, threatened, or endangered in the conservation need category. Our assessment is biased by the fact that more is known about some taxa than about others. The statuses of large, charismatic, diurnal species, for example, are better known than those of small, nocturnal mammals. Generally, more is known about temperate species than tropical species. More study is required for an accurate determination of the statuses of species that inhabit remote terrestrial and aquatic environments. Nevertheless, the information presented in this analysis is probably conservative; that is, the level of threat to members of the Class Mammalia is probably greater than we indicate.

All representatives of three terrestrial orders of mammals (Primates, Proboscidea, and Pholidota) and two marine orders (Cetacea and Sirenia) warrant conservation status (Cole et al. 1994). Within the remaining 23 orders, 43 families (32%) are at risk, and protective measures should be considered. Approximately 30% of Old World families and 15% of New World families are included in the conservation need category. An additional 20% or so of all mammalian genera and 20% of all species require conservation efforts.

A wide array of mammalian species should be considered for conservation programs (Cole et al. 1994). For example, many large ungulates with extensive home ranges or seasonal migratory routes are susceptible to population declines. Our survey indicates that approximately 50% of the Artiodactyla and 90% of the Perissodactyla deserve conservation status. Although the Lagomorpha are smaller and less mobile, almost 25% of the species are listed in this category. Populations of large predators are also likely to be affected by environmental perturbations. Our assessment indicates that approximately 50% of the species in the order Carnivora are threatened by declining population size. On the other hand, less than 10% of the species in the two large orders of small mammals, Rodentia and Chiroptera, warrant conservation status. Finally, the Australian mammal fauna, noteworthy for its richness of marsupials, is threatened by habitat destruction and the introduction of alien species. Conservation needs exist for 50% of the species in the Diprotodontia, Notoryctemorphia, and Peramelemorphia, and 25% of the species of Dasyuromorphia are at risk. The New World marsupials are not threatened as a group.

Habitat destruction, fragmentation, and degradation are extremely severe threats to the survival of many mammal species. Such activities detrimentally affect migratory routes (e.g., woodland caribou and many African ungulates), food sources, and breeding areas (e.g., seals). Worldwide deforestation, particularly of subtropical and tropical areas, may be the greatest cause of habitat degradation and fragmentation. Conversion of deforested land to agricultural uses or cattle grazing may also result in significant loss of wildlife habitat. The expansion of human settlements also threatens habitats throughout the world, but especially in tropical and subtropical areas.

Hunting may also threaten the survival of many mammal species. Although subsistence hunting has declined in most developed parts of the world, it still exists in most developing countries. Sport hunting is closely regulated in many countries, but illegal hunting still occurs. Species of large predators in families such as Felidae and Ursidae, and large ungulates, which often possess elaborate horns or antlers, are often sought for trophies. Legal and illegal commercial hunting also threatens many species of mammals. Commercial hunting of cetaceans has led to reduced populations of many species of whales and some seal species (Miller 1992). Large furbearers may also be susceptible. Some mammals such as large cats (e.g., cheetah, jaguar, tiger, and snow leopard) are hunted for their pelts, and rhinoceroses (horns), black bears (gallbladder), Asian pangolins (scales), some deer (antlers), and seals (tail) are killed illegally for body parts with alleged medicinal properties. Elephants and walruses may be taken for their ivory. Legal and illegal commercial hunting for the pet trade, entertainment, or medical research as well as habitat destruction have had a negative impact on primates and other mammals.

Predator and pest control programs threaten many mammal species. Potential livestock predators and large mammals perceived as threats to human populations—including species such as wolves, grizzly bears, polar bears, mountain lions, and tigers—are targets for extermination. Large mammals that invade croplands, and other agricultural pests, are also subject to systematic shooting, trapping, or poisoning. African elephants, for example, have been shot to prevent crop destruction. Tuco tucos (South America) and prairie dogs (North America) are threatened by poisoning because their burrows in grazing lands may harm cattle. Pest control programs may also affect nontarget species such as African pangolins (insecticide poisoning) and black-footed ferrets (poisoning and loss of prairie dog prey), which are now endangered.

The introduction of alien species is a potential threat to native animals that has gained attention in recent years. Without natural predators and competitors, populations of alien species can expand rapidly and ultimately threaten the survival of native mammal species. Among the alien mammals that can cause serious ecosystem destruction and affect the survival of native mammal species are European rabbits, rats and house mice, and pigs, goats, and other ungulates.

Certain biological traits may make some mammal species more susceptible to extinction than other species (Miller 1992). Species that possess several such characteristics are perhaps at greatest risk. Examples of these traits include low reproductive rates due to delayed age at first reproduction, small litter size, and long gestation and lactation periods; specialized feeding habits or location at the top of the food chain; and narrow habitat requirements or limited availability of appropriate habitat, including along migratory routes.

Traditional and successful approaches to the conservation of biotic diversity—such as the establishment of zoos, botanical gardens, refuges, and reserves; habitat restoration; and captive breeding—should be continued and enhanced. The UNESCO program of establishing biosphere reserves in representative habitats throughout the world appears to be an effective conservation strategy. These reserves address conservation priorities as well as the need to establish sustainable forms of development.

Scientists must educate and train the general public and policy makers to broaden the support for conservation efforts. Novel programs like the pioneering effort of the Instituto Nacional de Biodiversidad in Costa Rica to employ and train local people as "parataxonomists" to inventory and monitor the fauna and flora of the country should be established in other areas. Research programs should also be initiated or expanded, particularly in the tropics, where the threat to biological diversity appears most severe. The ecological requirements of different mammal species must be studied and understood before effective conservation strategies can be designed. Wildlife managers must also investigate the relationship between the population sizes of different mammal species and habitat degradation, physiological stress from pollution, and habitat change due to succession and fragmentation.

The conservation of mammal species will require an ongoing and intensive effort. Although programs to conserve mammal species should be increased worldwide, threatened areas of high species richness, or "hot spots," should receive priority. Governmental policies that strengthen conservation efforts and reduce exploitation of biological resources must be instituted. Economic policies that lead directly or indirectly to the destruction of biodiversity must be changed. New integrative approaches to conservation, such as the bioregional management philosophy that involves local people and incorporates ecological, cultural, economic, and administrative considerations on a regional scale, should be pursued. Innovative solutions to the problem of funding for conservation efforts throughout the world must also be found. The task is daunting, but the integration of scientific and cultural knowledge, governmental policies, and economic resources is needed to confront the current and critical problem of loss of mammal diversity.

Chapter 3

Designing a Study to Assess Mammalian Diversity

Michael J. Conroy and James D. Nichols

Introduction

In this chapter we provide users with general
guidelines for properly designing a study to assess mammalian diversity. Proper study design is
important to ensure that the results of a particular
biodiversity survey will be repeatable by future
investigators and to ensure that the results from
one survey will be comparable to those from
surveys by other investigators in other geographic regions or at other times.

Much emphasis in the ecological literature
has been placed on diversity indices and other
measures of community structure. All of these
measures are ultimately derived from two basic
parameters: abundance (numbers or densities of

animals) by species, habitat, and so forth, and
total number of species. Given knowledge of the
number of species in a community and of the
abundance values for each species, derivation of
any of a variety of community indices, such as
diversity or evenness, is straightforward (Pielou
1977). Ecologists, however, are usually faced
with imperfect knowledge of communities and
typically must rely on estimates based on samples rather than on exact parameter values. Most
of the material in this handbook deals with estimation of either abundance or species richness
from sample data.

Estimates of abundance are used to draw inferences about the abundance of a particular species over time (e.g., seasonally or year-to-year)

or space (e.g., among habitats). Species richness estimates are used to draw inferences about the number of mammalian species present at different times or locations. Estimates of abundance for all species in the community can be used to compute species evenness or equitability, the distribution of abundance among species. From a sampling standpoint, the goal of estimating richness is very different from that of estimating abundance; the same method that provides for a good estimate of the number of species in an area may not provide a good estimate of either absolute or relative abundance. For individuals interested in monitoring and conserving biological diversity, both goals are important: an investigator would like to know the number of species present in an area and the status and population viability of each species.

Statistical concepts

It is beyond the scope of this manual to offer a detailed treatment of statistical concepts. Rather, we present a few major points of particular relevance to the design and implementation of surveys. We refer interested readers to more detailed coverage of these topics in Cochran (1977) and White et al. (1982:14–44).

Inferential statistics has two distinct, but closely related, goals. The first is estimation, and explanation of this concept requires some definitions. An *estimator* is a statistic (a function of sample data) thought to convey information about some unknown quantity of interest; it is also the mathematical formula used to compute an estimate from sample data. Estimators are usually based on an underlying statistical model (a mathematical statement of assumptions). An *estimate* is the value of an estimator computed from a particular set of sample data. If N denotes the quantity of interest, then it is standard notation to denote both the estimator of N and a specific estimate of N as $\hat{N}$. The goal of estima-

tion is to use sample data in conjunction with an estimator to obtain an estimate for some quantity of interest.

The other principal goal of statistics is to use sample data and estimates to *test hypotheses* or to compare different populations represented by the samples. This is accomplished by computing test statistics that permit formal statements about the likelihood of incorrectly rejecting a null hypothesis (i.e., incorrectly concluding that the populations differ with respect to the quantity of interest).

Designing a sample survey

Target and Sampled Populations

In sampling ecological communities, it is important to be clear about the extent to which one can extrapolate from the sample to the "real world." The *target population* is the real world entity or collection that the investigator wishes to sample and about which he or she wishes to draw inferences. The target population might be a population of white-footed mice on a particular island or the entire mammalian fauna of a region or country. The *sampled population* is the portion of the target population that is available or exposed to sampling efforts (Fig. 2).

An investigator, for example, may actually be interested in all the mammal species in a 100-km^2 park, but only 20 km^2 of the park may be acces-

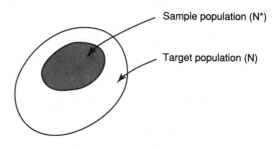

Figure 2. Conceptual representation (Venn diagram) of sample and target populations.

sible for sampling. Samples are taken from the sampled population (i.e., from the 20-km^2 area) to compute an estimate, for example, of species richness, $\hat{S}$. The estimate of S strictly applies only to the sampled population, that is, to the 20-km^2 area (assuming the sampling scheme and statistical model are adequate). If the investigator wishes to extend his or her inferences about species richness to the target population, he or she does so without a basis in statistical inference, because much of that population never had the potential to be part of the sample. The investigator may have legitimate reasons, for example, to assume that species richness for the target population will be larger than that for the sampled population by some specified amount (e.g., 50% larger), but those reasons are outside the realm of objective statistical inference. One can define sample and target populations in a number of ways. In ecological studies they will ordinarily be delineated by reference to some geographic space (e.g., vegetative community or geopolitical boundary) and time (e.g., the breeding season in a particular year).

Corresponding examples in abundance estimation occur when certain individual animals in a population are untrappable (i.e., have no chance of ever being captured), because of size, age, previous trap experience, and so forth. The target population is N, the true number of animals in the population, but the sampled population is N^*, the trappable portion. No matter how good the sampling scheme and capture-recapture model (statistical models used to estimate population size and other quantities from trapping data; see "Capture-Recapture Methods," Chapter 10), if $N - N^*$ animals are untrappable, then one can only obtain unbiased estimators of N^*, not of N. If the investigator wishes to estimate N, he or she will need some additional data, independent of the sample, in order to extrapolate.

Obviously the ideal situation is for the target and sampled populations to correspond com-

pletely, so that the sample estimates permit inference about the entity of interest, rather than some subset of it. The questions one should always keep in mind when designing a sampling scheme for a biodiversity survey are "What inferences do I wish to make based on this sample?" and "Will these inferences be justified?"

Selection of Sampling Units

An investigator, having delineated a sampled population, must then decide how to sample it. From a statistical viewpoint, the best approach frequently is to select the sampling elements (e.g., animals, quadrats) at *random* from the sampled population, ordinarily with equal probability of selecting any particular element. This type of sampling is known as *simple random sampling* and is the basis for the most straightforward use of inferential statistics. Many other types of sampling are possible, however, and indeed are logical choices in some cases. *Systematic sampling* (for example, selecting every fourth of 100 quadrats in the sampled population) will provide satisfactory results in cases in which the members of the population are distributed at random.

In trapping studies, entry of an animal into a trap is by no means a random event, and the more realistic capture-recapture models explicitly incorporate such nonrandom features as previous capture experience and individual animal characteristics into the estimation process. The question to ask of sampling is not necessarily "Is it random?" but rather "Will the results enable me to make general statements about the sampled population from the sample in hand?"

Determining Sample Size

Determination of sample sizes needed to achieve some goal generally requires several kinds of information. First, the goal of the sampling effort must be established; often this is in the form of a statement about the amount of imprecision one is willing to tolerate in the resulting estimate. Sometimes this is stated in a probabilistic manner, for example, that one is 95% "confident" that the true value of population size will be within 10% of the estimate. This stated goal further implies that one has identified a parameter of interest and a sampling procedure and estimation model appropriate to estimating that parameter. A different type of goal would be the ability to test a hypothesis about a population, for example, to detect a 10% difference in population density over time or between habitats with high (e.g., >0.90) probability. Both goals would ordinarily require some preliminary estimate of sampling variability (reflecting variation among the different samples in the quantity of interest) before the sample size needed to achieve the goal could be approximated. Often this estimate can be an educated guess or a figure based on a previous study. Sometimes it is necessary to conduct a small pilot study to estimate sampling variation. The pilot estimates can then be used as a basis for deciding how many additional samples may be needed.

Cochran (1977:76–77) provided a formula for approximating sample sizes needed for sampling from a finite population. This formula would be appropriate, for example, for deciding how many quadrats to sample to estimate mean density of animals per quadrat to a desired precision. The formula is written as follows:

$$n_o = \left(\frac{CV^2}{CV_o} \right) \quad (1)$$

and

$$n = \frac{n_o}{1 + (n_o/N)} \quad (2)$$

where CV is the observed (pilot sample) coefficient of variation and CV_o is the desired coefficient of variation; and N is the total number of possible quadrats from which a sample of n quadrats is selected.

Sample size recommendations are also available for specific estimation methods. These include line-transect methods (Buckland et al. 1993) and capture-recapture methods for both closed (Otis et al. 1978; Seber 1982) and open (Pollock et al. 1990) populations.

Stratification

In many situations, the sampled population will comprise several identifiable components or *strata*. Stratification involves dividing the sampled population into such components and then sampling and obtaining estimates for each stratum. For example, a 10-km^2 island might include five types of habitat (five strata): beach (1 km^2), dune (2 km^2), shrub (2 km^2), maritime forest (3 km^2), and salt marsh (2 km^2). An investigator might wish to estimate the abundance of small mammals on this island, but small mammal densities are likely to differ substantially among the habitats. The investigator could take a random sample from the entire island and estimate an average mammal density, but this would be relatively inefficient (i.e., the variance of the resulting density estimate would be relatively large).

Stratification has at least three advantages. First, the investigator is assured of representative coverage of the island if he or she breaks the island into relatively homogeneous strata and samples from each stratum. Second, stratified samples will, in general, result in estimates (in this case of island-wide abundance) having greater precision (smaller variance) than estimates from simple random samples. Finally, adequate sampling of the individual strata may provide reliable stratum-specific estimates, which may be of greater interest than a single estimate. For example, estimates of small mammal densities for each of the five habitat types may be more useful than a single estimate for the island.

Additional discussion of stratified sampling is provided by Cochran (1977). Various schemes can be used to allocate a given total sample size among strata. In *proportional allocation* the total sample (*n*) is divided among the strata in direct proportion to the fraction of the sampled population's total sampling units or area located in each stratum. For the island example, if it were determined that 100 total samples were needed, then with proportional allocation 10, 20, 20, 30, and 20 samples would be taken from the five habitat types, respectively. Proportional allocation is the method of choice when the cost per sampling unit and the variances of the stratum means are the same or when there are no data to the contrary. *Optimal allocation* (that producing estimates with minimum variance) formulas (Cochran 1977:98) should be used when stratum variances differ or when the cost of sampling varies among strata.

Interpreting animal count statistics

Observability

Information on either mammalian abundance or species richness generally involves some sort of count statistic (e.g., number of animals caught, seen, or heard). Presumably, count statistics are related to the quantity of interest, although the exact nature of the relationship is never known. Individual animals present in an area being sampled have certain probabilities of being observed (*observation probabilities;* these are not known but are nearly always less than 1) and, thus, of appearing in a count statistic. Some knowledge of observability is needed in order to use count statistics to draw reasonable inferences about abundance or species richness. We discuss calculation of observation probabilities in Chapter 10 (see "Observability" under "Conceptual Framework" in the section on "Estimation of Mammal Abundance").

Sampling Variability

Investigators are seldom able to count every animal in a population or every species in a community and must, instead, *sample* the population or community and estimate population size or species richness. When this is the case, the resulting estimate is subject to *sampling variability*. That is, different samples from the same population or community would result in estimates that differ from one another. *Precision* is a relative term reflecting the magnitude of variation. Estimators producing estimates that show little variation among samples are said to be *precise,* whereas estimators yielding more variable estimates are considered imprecise. Typically, only one sample is available. It is then important for the investigator to obtain an estimate of this sampling variability, called the *sample variance*. The estimate of sample variance, in conjunction with assumptions about the statistical distribution of the estimator, can be used to obtain estimates of *confidence intervals*. Confidence intervals define a range of values that encompasses the true abundance, richness, or other parameter with a specified degree of certainty. All procedures for estimation that we consider enable researchers to obtain this essential information.

Use of Count Statistics in Estimation

Estimators can be evaluated based on two criteria: bias and precision. *Bias* is the difference between the expected value of an estimator [$E(\hat{\Theta})$, where E denotes expectation and $\hat{\Theta}$ is the estimator] and the true quantity being estimated (Θ):

$$\text{Bias} = E(\hat{\Theta}) - \Theta \qquad (3)$$

For example, consider an estimator $\hat{N}$ used to estimate N, or population size. One can take a very large number of samples (e.g., 10,000) and

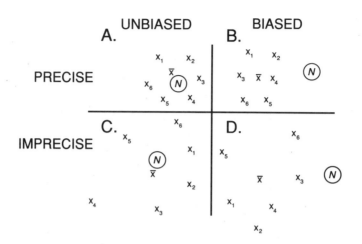

Figure 3. Conceptual representation of bias and precision. N is the true value for the population; x_i are sample estimates; $\bar{x}$ is the mean of the sample estimates and serves as an estimator $\hat{N}$ of N. A. $\hat{N}$ = unbiased and precise. B. $\hat{N}$ = biased and precise. C. $\hat{N}$ = unbiased and imprecise. D. $\hat{N}$ = biased and imprecise.

compute $\hat{N}$ for each sample. If the average of the 10,000 $\hat{N}$s is very close to N, then $\hat{N}$ is an *unbiased estimator* of N or, at least, its bias is very small (Figs. 3A and 3C). If an investigator uses a count statistic as an estimator, then the bias of this estimator is related to the observation probability. When the observation probability is 1.0 (i.e., when all of the animals or species are counted exactly), then the count statistic is an unbiased estimator. Almost all animal count statistics are biased when used as estimators.

If an investigator can develop a means of estimating the observability associated with a count statistic, then he or she can use this estimate, together with the count statistic, to compute an unbiased estimate of N (see "Observability" under "Conceptual Framework" in the section on "Estimation of Mammal Abundance," Chapter 10). Use of such an estimator can be thought of as "correcting" the count statistic for observability. This type of estimator can be used to estimate animal abundance and species richness.

As noted in the previous section on sampling variability, it is also important to consider the variation among the $\hat{N}$s. If the variation is small, so that the various samples each give values of

$\hat{N}$ that are close to one another, then the estimator gives highly repeatable results and is precise. If the sample values of $\hat{N}$ are scattered over a wide range, then the estimator has a higher variance and is less precise (Figs. 3C and 3D). Note that an estimator can be biased but precise (Fig. 3B), unbiased and imprecise (Fig. 3C), both biased and imprecise (Fig. 3D), or (ideally) both unbiased and precise (Fig. 3A). If an investigator estimates N using a count statistic corrected with an estimated observability fraction, then the precision of $\hat{N}$ will be strongly influenced by the estimate of the observation probability. $\hat{N}$ is most precise when the observation probability is estimated precisely and is close to 1.0.

Use of Count Statistics in Comparative Studies

Count statistics relating to species richness and abundance are often used for comparative purposes. An investigator would generally like to compare the quantity of interest (e.g., abundance or species richness) at two or more locations or times. Variation in observability naturally causes problems in any such comparison. For example, assume that an investigator wishes to compare the abundance of a particular species in a particular area in a specific month in two successive years. Further assume that the investigator traps 25 animals in the first year and 50 animals in the second year. If the observability (capture probability in this situation) is exactly the same in the two years, then the investigator can infer that the abundance in year 2 was twice that in year 1. Suppose, however, that the actual abundance in both years was 125 animals, but that the capture probability in year 1 was 0.2 (i.e., about 20% of the animals in the area were caught) and the capture probability in year 2 was 0.4. If the investigator inferred a change in abundance on the basis of the raw count statistics, then he or she would be misled. The investigator needs to estimate the capture probabilities in order to interpret the trapping data reasonably. Although this example concerns animal abundance, variation in observability can also produce misleading inferences about species richness.

Investigators interested in comparing species richness or abundance are, thus, forced to deal with variation in observability. The ability to deal adequately with variation in observability depends upon a knowledge of underlying sources of variation and on the ability to control such variation. If an investigator can both identify and control the important sources of variation in observability, then he or she can use standardization to obtain count statistics that can be used for comparing abundance and species richness. *Standardization* refers to attempts to obtain count statistics using similar methods under similar conditions. For example, if an investigator used trapping to compare abundance of small mammals in two different areas, he or she might use the same number and type of traps arrayed in the same spatial configuration and set for the same period of time with the same baits in both areas. The investigator would also like to trap under very similar environmental conditions (e.g., temperature, moisture, moon phase) in both areas. By standardizing methods, the investigator attempts to produce equal capture probabilities in the two areas. If the investigator has correctly identified the important sources of variation in observability and is able to exert sufficient control over them, then the count statistics can be regarded as measures of "relative abundance" (i.e., the counts made at different areas and/or times will reflect about the same fraction of the population) and can be used for comparing abundance.

Unfortunately, one cannot always control the important sources of variation in observability that have been identified. For example, logistical difficulties may force the investigator to sample at certain times without regard to weather conditions. Habitat differences, microclimatic differences, and differences in predator communities could also produce differences in observability (e.g., different capture probabilities) in two areas. In situations in which observability varies as a function of variables that can be identified but not controlled, the investigator can record information on the influencing variables and use such data as covariates in analyses of the count statistics. Such uses of covariates may be reasonable in some cases but require the rigid assumption that the covariates influence observability in exactly the same way among the different samples being compared.

We suspect that in many cases the investigator will not be able to identify all factors that influence observability (e.g., animals may exhibit different behavioral responses to traps in different places or at different times). Neither standardization nor use of covariates can produce useful comparative inferences from count statistics in this case. The only way to deal with unknown sources of variation in observability and, thus, to ensure reasonable inferences in a comparative study based on count statistics is to estimate the observation probabilities for the samples being compared. The formal methods for estimating mammal abundance and species richness that are presented later in this volume (see Chapter 10) can be thought of as methods for estimating observability. Once observability is estimated, it can be used to translate a count statistic into an estimate of abundance or species richness. Resulting estimates can then be used to compare abundance or richness over time and space.

Standardization

As discussed previously, standardization can be an effective means of dealing with sources of variation in observability that one can identify. However, use of count statistics obtained using standardized methods can yield misleading inferences in comparative studies because of uncontrolled or unidentified sources of variation in observability. If possible, therefore, it is best to use a sampling method that permits estimation of the observation probability and that thus provides an unbiased estimate of the quantity of interest. Such estimates can be used directly in comparative studies, and resulting comparisons do not depend on standardization.

Standardization is not necessary when formal estimation methods are used. Nevertheless, we believe that standardization can be useful in two classes of situations. The first class involves cases in which formal estimation methods cannot be employed for some reason. In such circumstances, it may be necessary to standardize over all variables suspected to influence sampling probability and simply compare resulting count statistics. The disadvantage of this approach is that the investigator never knows how similar the underlying observabilities are and thus how accurate or useful the resulting comparison may be.

The second situation in which standardization is useful involves use of formal estimation methods. As stated previously and discussed extensively in Chapter 10, such methods are designed

to permit estimation of observation probabilities. The data required for such estimation can be used in direct tests for equality of observation probabilities among areas or times for which comparisons are desired. If such tests indicate different observation probabilities, then subsequent comparisons should be based on abundance estimates (count statistics "corrected" for observability). If tests indicate similar observation probabilities, then comparisons of abundance can often be based on the count statistics themselves. Use of count statistics in such situations is recommended because estimation carries a cost in terms of increased variance and decreased test power. Raw count statistics, unadjusted for observability, have smaller variances and result in comparative tests with greater "power" (greater probabilities of detecting differences that really exist) than estimates of abundance or richness (see Skalski and Robson 1992). Standardization should increase the likelihood of obtaining equal observation probabilities, so we recommend standardizing, even when using a formal abundance estimation method.

Standardization can also be useful with formal estimation methods as a means of reducing the sources of variation in observability requiring consideration in a formal model. For example, some capture-recapture models consider variation in capture probability from one sampling period to the next. If such variation can be eliminated (by expending similar effort on each occasion and by trapping only under a certain range of weather conditions), then variances of resulting abundance estimates will be smaller, and tests will be more powerful. Both conditions are desirable in comparative studies.

Developing a Species List

We have focused, thus far, on using count statistics to estimate and to compare abundance or species richness. In addition to estimating the number of species in an area, however, an investigator may wish to identify the species. A shift of interest from the number of species occupying an area to a list of those species requires a change in both perspective and preferred methods.

Our previous discussion on animal count statistics was general by design, and we have said little about how these count statistics are actually obtained. As we will discuss in subsequent chapters (especially Chapter 10), specific sampling protocols are required by many of the methods used to estimate observation probabilities. However, if the primary objective is to obtain as complete a species list for an area as possible, then a different approach may be needed. In such cases, the investigator should use a wide variety of methods for detecting animals (e.g., observations of animals and associated sign, captures in traps and nets, perhaps interviews with people living in the area). In addition, the investigator should use intuition and past experience to direct search efforts to specific locations (e.g., trap sites, observation posts) where the largest numbers of species are likely to be recorded. Directed searches with a variety of methods will not always yield count statistics that are useful in estimating species richness. Nevertheless, directed efforts are appropriate when the primary objective of the investigation is to develop a species list. Such efforts will frequently "miss" species, and the number missed will be un-

known, but the resulting list is still a useful goal of some investigations.

Conclusions

Ecologists often view statistics as something that is applied after data are collected, to help sort out interesting facts and address interesting questions. A more efficient and ultimately more productive use of statistics is possible when (1) study goals are objectively stated before any data are collected and (2) a sampling design is constructed that will allow those goals to be achieved. In biodiversity studies, two closely related goals are the determination of abundance and species richness, and the examination of differences in abundance and richness over space and time. In practice, data used to achieve these goals are hardly ever complete censuses but instead are usually count statistics based on samples. Proper consideration of the principles of statistical inference, including sampling design, estimation, and estimate reliability, is essential to the proper interpretation of ecological patterns based on samples. These principles are the same, regardless of whether one is counting animals, capturing them, or observing indirect evidence of their occurrence. Following these principles will not prevent all errors of interpretation, but it will go a long way toward assuring results that at least will be repeatable among investigators.

Chapter 4

Keys to a Successful Project:
Associated Data and Planning

Introduction

Although obviously the direct collection of inventory or monitoring data will be the major focus of a biodiversity study, a proper study plan will include the collection of ancillary data according to some preplanned, logical scheme. Because of the general applicability of similar information in the first volume of this series, on Amphibia, appropriate sections from that volume have been modified to refer to mammals and reprinted here with the permission of the authors.

Field biologists traditionally keep notes on weather conditions during fieldwork, but in many cases it may be possible to make accurate measurements of climatic variables with a minimal investment in equipment. Long-term monitoring studies may warrant investment in more

expensive automated data loggers to gather associated information. Just as standardized methods of gathering basic biodiversity information are valuable, data standards for habitat characterizations and sampling schemes will increase the comparability of studies. Mammal collectors sometimes record microhabitat information on specimen labels and in field notes, but standard descriptions allowing comparability are rare. Identification of mammals in the field, particularly poorly known groups such as rodents and bats, is difficult at best, requiring the preparation of voucher specimens to allow future verification. Frequently, the collection of such specimens will require permits from the local authorities.

Some attention to these details in the planning stages of any study will greatly increase the value of the data collected. Most biodiversity studies are done in remote areas, increasing the necessity for all equipment and supplies to be gathered beforehand, and for careful logistical planning from the outset.

Climate and environment

MARTHA L. CRUMP

Weather data are especially critical for interpretation of results in mammalian studies because seasonal cycles are so dependent on climate. Although different species have different ranges of tolerance, all mammals are affected by local weather conditions. Temperature, precipitation, and other climatic factors influence the geographic and ecological distributions of mammals and the timing and intensity of feeding, reproduction, and migration. Climatic conditions also affect population densities and assemblage-wide interactions.

Another rationale for collecting weather data concerns human-induced changes around the world, even in remote and protected areas that may otherwise appear to be natural or undisturbed. These changes include air and water pollution, acidified precipitation, habitat destruction or modification, introduction of alien predators or competitors, and changes in global climatic conditions, such as increased temperature and decreased rainfall. It is useful to document environmental conditions with the hope that the factors responsible for mammalian distributions can be identified. We know that the geographic limits of some species have shifted with changing climates in historic as well as geologic time. It may be important to differentiate between major, long-term distributional shifts and recent extirpations or invasions caused by anthropogenic perturbations.

Basic Weather Data

Maximum and minimum temperatures and precipitation should be recorded for every inventory or monitoring project. Information on general weather patterns and long-term climatic characteristics can be obtained from a standard weather station if one is located near the study site (often available at airports or universities). Often, however, the only way an investigator can obtain the data is to gather them onsite. Whenever possible, weather data should be collected for several weeks prior to as well as during a survey, because these data often provide insights for the interpretation of the inventory or monitoring results.

TEMPERATURE

Temperature data are useful because temperature significantly influences mammalian activity patterns and often controls reproductive cycles and behavior (particularly for temperate zone species). Temperature changes can affect predation, parasitism, and susceptibility to disease. Cooling or warming trends can initiate migrations and thus influence distribution and activity patterns. Changes in temperature can affect food supplies, thus influencing growth, development, and survivorship. Depending on the goals of a study, any or all of the following temperatures may be relevant: air, water, soil, or substrate.

For a general inventory, one should record the temperature continuously or maximum and minimum temperatures at a specified time each day. Often, recording temperature at the beginning and end of a sampling period will provide information useful in evaluating mammalian activity. More frequent measurements of temperatures may yield additional information, but if the time needed for these recordings decreases the habitat area that can be sampled, then the data may not be worth the effort. Before any temperature data are recorded, the investigator must consider the exact questions to be answered, the statistical analyses to be done, and the cost-benefit ratio of recording various types of weather data.

Instruments for measuring temperature range from standard mercury thermometers to elaborate recording devices. Hand-held thermocouples are often preferable to standard thermometers for measuring air temperature because thermocouples are more durable. Maximum-minimum thermometers provide the high and low temperature for any time interval (usually 24 hr is used). Thermometers can be placed at any height above the ground; 2 m above ground is a standard reference height for meteorological stations. Placement of thermometers in shrub or tree canopies may yield relevant data for arboreal mammals. Thermometers also can be inserted into burrows or tunnels to monitor changes in the thermal environment of fossorial mammals. Temperatures that affect aquatic mammals may be measured at most depths within aquatic environments by using thermometers attached to stakes or floats. Accurate temperatures for various microhabitats can be obtained with resistance thermometers or with thermistors and microprobes. Continuous recordings of temperature can be made with recording thermographs or with sensors interfaced to data loggers (see "Measuring Weather Variables" below).

PRECIPITATION

Precipitation, like temperature, strongly influences mammalian activity, distribution, dispersion patterns, and reproductive cycles. Because the seasonal distribution of rainfall is more relevant than average annual precipitation, daily precipitation should be recorded.

The simplest way to measure rainfall is with a rain gauge. If the data desired are measures of the actual amounts of precipitation, the gauge should be set in an open area. On the other hand, if one wishes to know how much water falls through the canopy onto the forest floor, the rain gauge should be installed so that through-fall precipitation is measured. Gauges range from simple plastic devices that must be manually emptied, to automatic electronic devices that measure rainfall, forward the information to a remote recorder, and then empty themselves. Automatic gauges can accumulate the total amount of rainfall over any specified period. The obvious advantage of this type of gauge is that, except for maintenance, it never needs to be checked or emptied.

Additional Environmental Data

Depending on the goals of the study and available resources and personnel, specific microclimatic data for each animal encountered may be desirable. Following are some relevant factors that are known to influence distribution and activity of mammals.

WIND SPEED AND DIRECTION

Because most mammals are visually oriented, wind currents may affect their ability to detect prey or predators. Wind speed can be determined easily with hand-held anemometers. If data are recorded at the site where each mammal is observed, correlations between wind speed, individual presence, and activity can be determined. If general trends are desired, daily mean wind velocity and direction can be obtained from a standard weather station, if one is located nearby.

RELATIVE HUMIDITY

The combination of temperature and humidity determines the rate of water loss from a mammal's lungs. For this reason, the amount of moisture in the air may affect distribution and activity patterns of mammals, especially in arid environments. The simplest method of obtaining humidity measurements in the field is with a sling psychrometer or battery-operated, hand-held thermohygrometer (the latter is convenient because one gets a digital readout of both temperature and humidity). Air temperatures should always be recorded in conjunction with measurements of relative humidity. A hygrothermograph continuously records both temperature and humidity.

BAROMETRIC PRESSURE

The environmental factors that trigger migratory behavior and that stimulate changes in hormone levels preparatory for breeding activity are not clearly understood in all mammals. Whenever field conditions permit, barometric pressures should be recorded and analyzed in conjunction with patterns of mammalian activity. Barometric pressure can be measured with hand-held barometers or with automatic recording devices (see "Measuring Weather Variables" below).

SUBSTRATE MOISTURE

Moisture levels of substrates such as soil and leaf litter can affect mammal distribution and activity patterns. Soil moisture can be measured with a tensiometer, and sensors are available that give both temperature and wetness readings for litter surfaces. For continuous readings, moisture sensors can be interfaced to data loggers.

WATER LEVEL

Physical characteristics of aquatic environments influence the distributions and activity patterns of aquatic mammals. Water depth of a sampled area should always be measured or estimated. In an inventory conducted in freshwater environments, perhaps the only points of interest are maximum and minimum depths. On the other hand, in a monitoring study, changing profiles of water depth in the lake, pond, river, or stream may be useful. The number of points at which water depth should be measured depends upon the size of the habitat. In a monitoring study, water depth should be measured at the same

points or read from previously located depth markers (marked poles) each time the habitat is sampled. Water depths are easy to obtain with a collapsible meter stick. For continuous readings, mechanical recorders are available, or sensors can be connected to data loggers.

For marine mammals, it is difficult to measure depth directly, although captured animals can be fitted with depth sensors. Depths may be obtained from navigational charts in many coastal or deep-water areas. Changes in salinity profiles or in the velocity and location of tidal and ocean currents may be more important than depth changes in most marine environments.

pH

Because excessive acidification of water may have detrimental effects on mammals inhabiting freshwater environments, we encourage investigators doing inventories and long-term monitoring studies to document pH conditions at their study sites. Most experts agree that pH indicator paper gives unreliable and misleading results that often are worse than no data at all. Many types of portable, battery-operated pH meters appropriate for field use are available.

Measuring Weather Variables

Whenever possible, weather and microclimatic data should be collected automatically using recording instruments because these instruments increase the accuracy of the data collected and reduce the field time required to collect them. A record of the overall variation in an environmental factor is preferable to individual measurements taken at predetermined times. Continuously recording machines provide a picture of that variation. If recording equipment cannot be used, manual instruments can be employed successfully, given sufficient time and personnel.

Data should always be recorded in the field using actual numbers rather than codes. The reasons for this are many and include confusion when multiple persons are involved in data collection, difficulty of remembering the codes used, and ease of making mistakes under adverse field conditions.

Digital recorders are generally more accurate and reliable than are mechanical, battery-powered, or electrical recording devices. A drawback to use of data-acquisition systems in the field is that typically they must interface directly with a computer. In contrast, field data loggers can operate in a stand-alone mode, because they typically have internal memories; data loggers can collect the data as integrated, averaged, or point values over logging periods ranging from 1 minute to 24 hours. Data stored in the memory can then be transferred to a compatible computer or printer. In recent years, rapid advances have been made in the development of portable data acquisition systems and data loggers suitable for use in the field, with new models continually being introduced (Pearcy 1989). Investigators should consult with manufacturers (see Appendix 9) prior to purchase regarding the suitability of a particular digital recorder for use in connection with environmental monitoring systems.

Automated weather recording equipment will not be an option for all field studies because of cost (recording equipment, especially automated devices, is expensive), security considerations (in many instances the risk of theft precludes the use of expensive instruments), and risk of equipment failure (a serious consideration if the study site is a long way from the nearest repair shop). Backup, manually operated instruments should always be available in case recording devices fail or are stolen during a study.

The following are merely examples of the sorts of instruments available from scientific suppliers. Anyone seriously contemplating the purchase of equipment is advised to search through catalogues for those instruments with prices and specifications best suited to the study (see Appendix 9).

The most efficient field method of obtaining weather data is to set up a portable weather station at each survey site. Machines that measure maximum and minimum temperatures, precipitation, relative humidity, barometric pressure, wind speed, and wind direction can be purchased for U.S. $1,000–$1,500. These units run on size-D batteries and, thus, are convenient for field use. More restrictive recording units include spring-wound, 7-day recording thermometers (U.S. $250–$275), automatic electronic rain gauges (U.S. $75–$100), spring-wound and battery-run hygrothermographs for 1-, 7-, 31-, or 62-day continuous recording (U.S. $600–$1,500), electric-powered anemometers for continuous 30-day recording of wind velocity (U.S. $700), and 7-day electric-powered barometers for continuous recording of data (U.S. $325–$350). Data acquisition systems cost about U.S. $500 or more, and data loggers about U.S. $1,500 or more.

. Numerous nonrecording instruments are available: maximum-minimum thermometers (U.S. $30–$50); digital thermocouple thermometers (U.S. $175–$200); standard rain gauges (U.S. $10–$25); sling psychrometers (U.S. $40–$75); battery-operated, hand-held thermohygrometers (U.S. $125–$400); anemometers (U.S. $15 for hand-held portable wind meters), tensiometers (U.S. $75–$250); soil moisture and leaf wetness sensors (U.S. $50–$100); barometers (U.S. $30–$300); and battery-powered pH meters (U.S. $250–$400).

Weather Conditions and Study Design

Weather conditions can influence sampling results, as illustrated in the following example. Imagine that the goal of a study is to compare mammalian species richness between two seasons (2 weeks in the warmer, wetter season and 2 weeks in the cooler, drier season) within two forest types in a region. Ten persons spend one week in forest A and a second week in forest B during the wet season. Heavy thunderstorms occur every day during the first sampling week, but no rain falls during the second week. During the dry-season surveys, week 1 is approximately 10° C warmer than week 2. In the wet-season inventories, 18 species are recorded in forest A and 25 species in forest B. During the dry season, 13 species are found in forest A and 22 in forest B. Based on the number of species found, one concludes that the mammalian assemblage in forest B is considerably larger than that in forest A during both seasons. Actually, the assemblages may be equal, or the assemblage in A may be larger. The data may not reflect the true species richness because of uncontrolled weather variables: many mammals, particularly bats, are more difficult to sample in wet than in dry periods, and temperature differences may affect activity patterns. If the weather data are not recorded, it may be difficult to evaluate the mammal data obtained.

The effect of weather can be minimized in several ways, depending on time and personnel constraints. In the previous example, a better design would have been to have five persons work in forest A at the same time that five persons surveyed forest B. If personnel had been limited (e.g., a field crew of three persons), another option would have been to carry out half-day inventories using all personnel, thus surveying both sites each day (alternating sampling times for each site) for two weeks. If the sites were too far apart to reach within one day and still have time to survey both areas, inventories in two sites could have been done on alternate days. (This design does not solve the problem entirely, but alternating days is preferable to surveying for 7 days at one site followed by 7 days at the other.) If time were not a constraint (i.e., if the survey could have been done over several months each season), it might have been better to carry out many replicate inventories in the two sites; increased samples should minimize effects caused by differences in

weather. Potential confounding effects of weather would have been minimized if the investigator had used a method for estimating species richness (see Chapter 10), rather than simply obtaining counts with no means of estimating detection probabilities.

Acknowledgments. I thank Maureen Donnelly, Frank Hensley, Ron Heyer, and Roy McDiarmid for helpful comments on the manuscript and Steven Oberbauer for information concerning weather instruments.

Data standards

ROY W. McDIARMID AND DON E. WILSON

The many individual mammals encountered during the course of an inventory or monitoring project will have to be identified to species. Depending on the goals and sampling method(s) used, some individuals will be identified from a distance by sight or by their calls; others will be captured. At the same time, some will be marked for reobservation or recapture, and others will be sampled as vouchers. For each individual, certain minimum data should be recorded. In this section we consider data pertaining to locality and sampling methodology. Information on microhabitats and specimen vouchers is covered in sections that follow. The data outlined here should be the *minimum* for any project. Investigators with specific goals may require additional types of data as well.

Standardized, printed sheets containing the required data categories provide a convenient, inexpensive, and effective way to ensure that all the desired information is recorded in a consistent format. Data sheets should be well organized, be printed on good-quality paper (75%–100% cotton content), and include extra space (e.g., backside of sheet) for notes that do not fit pre-established categories.

Data should be recorded in the field with permanent (waterproof) ink as simply and directly as possible. The use of numerical codes in the field should be avoided; it is easy to forget codes or to enter the wrong code, and subsequent users may be unfamiliar with the codes used. Original data sheets can be photocopied for security, but they should not be copied by hand. If data are to be coded for computer analysis, the original or photocopied sheets should be used for data entry to minimize transcription errors. Some workers prefer recording information on small tape recorders; this also works well if a list of the standard data categories is checked during taping to ensure that all required information is recorded. Information recorded on tapes should be transcribed to data sheets or into a computer within 24 hours of taking the sample.

Geographic Characterization

Specific information about a locality should include a geopolitical characterization of the study site and a description of the habitats sampled. The geographic and political descriptions of the locality minimally should include the following information:

1. *Country or island group.* The country name is normally equivalent to the political unit, but substituting island names for country may be of value in some instances.
2. *State or province.* A secondary political unit should be part of every locality record.
3. *County, district, or other tertiary division.* For specimens collected in the United States and certain other countries, a tertiary political unit should be included. In countries in which tertiary divisions exist but are infrequently used or rarely mapped, this category may not be useful.
4. *Mountain range and other geographic data.* Some reference to the closest mountain range is important, especially in remote

areas for which detailed maps are not readily available. Inclusion of other geographic information may also be extremely helpful (e.g., drainage system, savanna, zoogeographic region).
5. *Specific locality.* The locality should be as detailed and specific as possible. Distances and compass directions from easily located places (e.g., towns, mouths of rivers, mountain peaks) are essential. Whether the distances are by road or straight-line on a map should be specified. Inclusion of a map or gazetteer reference is helpful.
6. *Latitude and longitude.* This geographic attribute is independent of political units. It is the only generally recognized locator that allows for universal retrieval of data from any geographic area and for electronic mapping. Workers should include coordinates for each locality as specifically as possible. However, approximate coordinates, clearly identified as such, are also of value if specific coordinates cannot be obtained. Latitude and longitude are reported with the standard notation of degree, minute, and second, rather than with a decimal. Portable global positioning devices that provide accurate measures of latitude and longitude are available for field use (about U.S. $1,000; see Appendix 9). Such devices record seconds as a decimal; such measures can be converted to the standard notation.
7. *Elevation.* Elevation should be noted. Approximate elevation, clearly indicated as such, is better than none. Elevations and distances should be given in standard metric units.

Habitat

Mammals occupy both terrestrial and aquatic habitats. Habitat descriptions should include the following information.

TERRESTRIAL HABITATS

1. Moderately detailed description of the kind(s) of vegetation (e.g., lowland tropical evergreen forest, temperate deciduous forest, thorn scrub, savanna-woodland) at each site. For forests, some mention of percent canopy cover and stratification, as well as abundance of ground and shrub cover. Height of canopy, abundance of vines and epiphytes, and average size of leaf can also be highly informative about habitat type. For savanna-woodland habitats, designation as natural, agricultural, or fire-maintained; indication of extent and regularity of seasonal flooding. For other terrestrial sites, plant type and cover. If plant species are known, a list of some of the dominant forms is useful. Published references to vegetation at the site should be noted.

   Descriptive lists of vegetation types exist for most regions of the world (e.g., Walter 1973) and can be used as a foundation for specific site descriptions. Representative vegetation types for tropical and subtropical forests in Southeast Asia might include the following: primary rain forest, hilly; primary rain forest, flat; evergreen oak/chestnut montane forest; mossy montane forest; coniferous forest; deciduous forest; gallery forest; selectively logged forest; rubber plantation; secondary growth; large clearing; camp.

2. Description of the climate at each site, including details of weather, with distribution and abundance of rainfall and annual and diel variations in temperature.

3. Some indication of the degree of disturbance. For forests, designation as primary, secondary, or plantation may be adequate. For grasslands, some mention of the influence of grazing, agricultural use, or frequency of fire or flooding may be important. Sampling done near or through a forest edge should be indicated.

4. Brief mention of other habitat factors (e.g., substrate type, soil type, soil compaction, type and abundance of litter layer, presence of rock piles and outcroppings, general topography, elevation, and other features) potentially important to mammals is helpful.

FRESHWATER HABITATS

Details of surrounding vegetation (see item 1 under "Terrestrial Habitats," above), climate and weather (item 2 above), water temperature, water clarity, and information for the type of water body sampled.

LENTIC—PONDS, LAKES, AND WETLANDS

1. Habitat type (e.g., lake, pond, swamp), size (surface areas in hectares or length × width), and depth (minimum, maximum, and average); percentages of the water surface that are open or occupied by emergent or surface vegetation; notation of whether the site is open above or covered by forest canopy.

2. Some indication of the relative duration of the habitat (e.g., is permanent, has water most years, results from flooding).

3. Nature of any shoreline or emergent aquatic vegetation; species or types of vegetation (e.g., reeds, water lilies), if known.

4. Bottom type (e.g., silt, sand, leaf pack).

5. Evidence of habitat disturbance from natural causes or human activities.

LOTIC—STREAMS AND RIVERS

1. Habitat type (e.g., river, stream, spring, creek), width, and depth (e.g., pools and shallows, riffles); some indication of the flow rate (e.g., cascades and falls, white water–high gradient, moderate current, slow and meandering, meters per second).

2. Some indication of the relative duration (life) of the habitat (e.g., flows all year, only in the wet season).

3. Nature of any bordering vegetation (e.g., trees, shrubs); plant types and species, if available.

4. Substrate type (e.g., rocks, boulders, gravel, sand, mud, leaf pack).
5. Evidence of habitat disturbance from natural causes or human activities.

TRANSITION HABITATS

ESTUARIES

1. Habitat type (e.g., coastal plain, bar-built, fjord), size (surface area), and depth (minimum, maximum, and average).
2. Some indication of the physical characteristics of the habitat (e.g., salinity patterns, wave action, tidal currents, turbidity, temperature, and oxygen profiles).
3. Nature of any shoreline vegetation or subtidal algal communities; types of vegetation or species, if known.
4. Bottom type (e.g., mud, sand, gravel, organic material).
5. Evidence of habitat disturbance from natural causes or human activities.

SALT OR TIDAL MARSHES

1. Habitat type (e.g., associated with an estuary or along sheltered, open coast), size (surface area and percentage of the surface that is covered by tidal creeks and streams or shallow pools).
2. Some indication of the relative degree and duration of flooding at high tide, topographic patterns within the marsh, and runoff from adjacent terrestrial habitats.
3. Nature and zonation of marsh vegetation, presence of algal mats, type of shoreline vegetation; species or types of vegetation, if known.
4. Substrate type (e.g., mud, sand, litter pack).
5. Evidence of habitat disturbance from natural causes or human activities.

MANGROVE FORESTS

1. Type (e.g., dominant tree species), size of forest (area: length × width).
2. Some indication of the relative duration and stability of the habitat (e.g., degree of tidal flooding, evidence of terrestrial runoff).

3. Nature of zonation, including species, if known; canopy stratification and cover.
4. Bottom type (e.g., mud, litter pack).
5. Evidence of habitat disturbance from natural causes or human activities.

MARINE HABITATS

NEAR-SHORE SUBTIDAL

1. Habitat type (e.g., seagrass, kelp forest, oyster reef) and size (surface area).
2. Some indication of physical characteristics (e.g., tidal range, depth, currents, salinity patterns, turbidity, distance from shore, evidence of terrestrial runoff).
3. Nature of the community (e.g., zonation and stratification of community components, density).
4. Bottom type (e.g., sand, gravel, rock).
5. Evidence of habitat disturbance from natural causes or human activities.

COASTAL AND CONTINENTAL SHELF

1. Location, slope, and depth (minimum, maximum, and average) of the habitat.
2. Some indication of important physical characteristics (e.g., depth, turbulence, temperature, oxygen, salinity profiles, light penetration, distance from shore, evidence of freshwater runoff).
3. Nature of phytoplankton, zooplankton, nekton, and benthic communities, if known.
4. Bottom type (e.g., mud, sand, gravel, ledge).
5. Evidence of habitat disturbance from natural causes or human activities.

DEEP WATER

1. Location relative to continents.
2. Some indication of relevant physical characteristics (e.g., salinity, temperature, depth, currents, light penetration).
3. Nature of phytoplankton, zooplankton, and nekton communities, if known.
4. Bottom type (e.g., topographic relief, presence of ridges and mountains, if known).

5. Evidence of habitat disturbance from natural causes or human activities.

Sampling Methodology

Information pertinent to sampling procedures should be recorded, with reference to the specific method(s) used. In addition, the following information should be taken for each specimen encountered during an inventory or monitoring project (see also "Microhabitat Description" and "Voucher Specimens," below):

1. Date and time of encounter.
2. Identification of specimen (e.g., *Cryptotis mexicana, Neotomys* sp., brown rat of type A).
3. Size of specimen. Standard measurements for mammals include total length, tail length, hindfoot length, ear length, and weight. Adult, subadult, and juvenile may be convenient size categories for use in monitoring studies of well-known species, but the use of these terms can present problems (e.g., adult-size mammals are not necessarily mature nor are juvenile-size mammals necessarily immature, as the names imply).
4. Sex. Recorded only if the determination is confirmed. If in doubt, a voucher should be collected.
5. Position in environment. The horizontal and vertical position of each individual, in as much detail as possible.

Microhabitat description

ROBERT F. INGER AND DON E. WILSON

Mammals typically are irregularly, often patchily, distributed in a habitat, particularly in complex habitats. Individual species occur in microhabitats, which are limited subsets of habitats at each site. Microhabitats, as used here, are the precise places where individual mammals occur within the general environment. Although simple species richness at a site can be determined without knowing the microhabitats used by the mammals living there, recording microhabitat data for each individual mammal observed will result in data that are scientifically richer. For example, differential microhabitat use by the same species at different sites can be determined, as can seasonal differences of microhabitat use at a given site. Knowing that certain mammal species are restricted to given microhabitats might have profound conservation implications.

Recording microhabitat data requires advance planning, especially in the design of an appropriate checklist for registering microhabitat features. Taking such data can be time-consuming and may result in a decrease in the number of specimens captured and preserved. However, the general utility of specimen records that include microhabitat data is so superior to the utility of those without them that the trade-off in reduced numbers of specimens preserved overwhelmingly favors collection of the data. Microhabitat information is useful for determining ecological distributions in a manner that is repeatable from site to site. By combining all data from a microhabitat classification scheme, it should be possible to describe the ecological distribution of each species at a site and to compare distributions across sites.

Each major biome type has its unique environmental features and will, therefore, require a distinct descriptive checklist, with two important caveats. First, no paper scheme can duplicate the actual complexity of the real world; consequently, one must expect to amplify certain records with supplementary notes. Second, the use of a microhabitat checklist does not obviate the need to record gross aspects of the environment, such as vegetation type, elevation, general topography, and weather. Nevertheless, it should be possible to create a microhabitat classification scheme for every major environment in which mammals occur. A microhabitat checklist will have both unique and general characteristics and will vary in complexity depending on the habitat sampled. For example, tropical wet forest sites presumably will require a more complex microhabitat classification scheme than temperate grassland sites.

Whatever checklist is assembled must balance detail and generality. The goal is to achieve generality without undue loss of information. Another important characteristic of a good microhabitat checklist is expandability; it should be possible to add elements as local situations demand. For example, an investigator should be able to add vegetation or habitat types as mammals are encountered in them.

Characteristics of a Microhabitat Checklist

Analysis of the information recorded with each observation leads to an understanding of the ecological distribution and habitat use of mammal species. It is important, therefore, that the data with each specimen be complete and recorded in a standard way. Generally, six major elements of the microhabitat of each individual observed are described. For each element, there is a checklist of environmental features about which information should be noted, as well as a series of standard descriptions for each feature. The notion is that for every mammal encountered a single notation for each feature of each element will describe that microhabitat. Use of the checklist of features and the standard descriptors facilitates complete and standard notation of data.

The six elements to be recorded for each observation are as follows:

1. Date and time of observation (24-hr clock).

2. General location, vegetation type, and elevation (refer to descriptions and standards in the section "Data Standards," above).
3. Horizontal position, with reference to bodies of water, shade-casting vegetation, fallen logs, rock faces, and, in the case of some aquatic environments, the shore. Each position needs to be qualified in detail (see checklist below).
4. Vertical position. In terrestrial environments, defined as height. In aquatic environments, defined as depth.
5. Substrate, usually mineral soil, dead leaves, log, rock, or vegetation. Each substrate often requires finer subdivision (see checklist below).
6. Special information that does not fit easily into the preceding categories—for example, branch projecting over water, under exfoliating rock, in termite mound.

Basic Descriptors for a Microhabitat Checklist

Investigators will need to develop descriptors for microhabitat checklists to be used in biomes such as temperate forest, grassland, desert, and so forth. The following descriptive categories were devised for use in tropical and subtropical forests. We include them here, to illustrate the method.

DESCRIPTORS FOR MAMMALS IN TROPICAL AND SUBTROPICAL FORESTS

DATE

HOUR (24-hr clock)

VEGETATION (Use separate descriptors for each major vegetation and habitat type at the site. See the section "Habitat" under "Data Standards.")

HORIZONTAL POSITION
  Terrestrial habitats
    Rock outcrops or piles
    Elevation
    Canopy gap (presence or absence, size)
    Proximity to ecotone (including canopy
      gaps)
    Proximity to water source
    Density of canopy cover
    Density of ground and shrub cover
  Aquatic habitats
    Midstream on bar or snag
    On bank; distance (m) to water
    On exposed dry bed; distance (m) to water
    On overhanging vegetation; distance (m)
      above water

VERTICAL POSITION
  Under surface of soil; depth (cm)
  In or under dead leaves
  In log; diameter (cm) of log
  On surface of leaf litter
  On rock; maximum dimensions (cm) of rock
  On log; diameter (cm) of log
  In grass
  On shrub or sapling (1–7 m); height (m)
    above ground or water
  On tree or large vine (>7 m); height (m)
    above ground or water; diameter (cm)
    at breast height (DBH) for woody plant
  On stump; height (m) above ground

SUBSTRATE
  Stem of shrub or tree
  In or under log, stump, or tree
  Bank of mud, sand or small gravel, rock

SPECIAL ATTRIBUTES OF MICROHABITAT
  Tree hole
  Burrow
  Bank: flat (<20°), moderately sloping (20–
    45°), or steep (>45°)
  Between tree buttresses
  In or on building

Other (describe on back of field sheet or elsewhere)

Depending on the nature of the study, the following information also may be appropriate:

PLOT OR LOCAL GRID NUMBER

TYPE OF ACTIVITY
  Quiescent or resting
  Disturbed by investigator
  Active and alert
  Uncovered by investigator
  In nest

DETECTION METHOD
  Observed
  Heard
  Pitfall trap
  Other trap

Field Methods

Recording microhabitat information in the field can be simplified greatly with temporary data sheets. Such sheets are ruled into columns corresponding to the major categories of information required by the microhabitat descriptor checklist being used. As animals are observed, appropriate information is entered. Upon return to camp, data are transferred into permanent field catalogues or notebooks. Data should be transferred within a few hours of collection. A computer should be used in the field only if hard copy can be produced at the site, because total reliance on disk storage in the field can be risky. In either case, original data sheets along with field notes and other records (e.g., maps, photos) should become part of the permanent archival record at the institution where the voucher specimens are deposited and should be maintained indefinitely.

If vouchers are collected, each should be numbered individually, and that number should be included as part of the temporary field record.

Mammals should be processed as soon as possible to avoid mixing of data and loss of specimens.

Voucher specimens

*ROBERT P. REYNOLDS, RONALD I. CROMBIE,
ROY W. McDIARMID, AND TERRY L. YATES*

Specimens that permanently document data in an archival report are called *voucher specimens.* Such specimens and corresponding data assembled during field studies of mammals, particularly the small and medium-size species that are difficult to identify and often poorly known, are critical for accurate identification of the animals studied and for verification of the data gathered and reported as resulting from the investigation. In addition, voucher specimens are critical for a wide array of future studies. Voucher specimens with extensive associated materials, such as tissue samples, chromosomes, and parasites, are particularly valuable. Such complete vouchers allow many different research projects to be linked in a network through the primary voucher. Primary specimen identification is important not only for research documentation, but also for assessment of change caused by natural or human perturbation. Voucher specimens will, therefore, play an increasingly important role in research on biological diversity. We highly recommend that vouchers with high-quality ancillary data be preserved.

Vouchers physically and permanently document data by (1) providing for confirmation of the identities of mammals accumulated and used in a study and (2) assuring that the study can be repeated, reviewed, and reassessed accurately (Yates 1985). Vouchers are the only reliable means of corroborating provenance of data accumulated during a study and documented in any reports of that study. The accurate identification of mammals is essential for providing credibility to the studies of these animals and to the publications that result from such investigations.

It is convenient to view mammalian voucher specimens in three groups: (1) type specimens, upon which names of taxonomic units are based; (2) taxonomic support specimens, which document identifications in taxon-based studies other than nomenclatural studies; and (3) biological documentation specimens, which document identifications of individuals obtained for genetic studies or environmental impact projects. Through time, a voucher specimen collected in a biodiversity study may serve all three functions. Photographs may suffice to document observations of especially large mammals or of protected species. Handling of these valuable materials should follow procedures similar to those described for actual specimens.

Standardized requirements exist for the data that should be affixed to each voucher specimen (see Appendix 3 and Yates et al. 1987). In addition, a considerable array of ancillary materials and data may be obtained and recorded for each voucher specimen (see Appendices 4–6). Recording and handling of data associated with voucher specimens of mammals must be taken as seriously as the handling of the specimen itself, because voucher specimens without accurate data are either suspect or worthless.

Field Identification

Accurate specific identification of small mammals in the field is rarely possible except in areas for which the fauna has been studied in detail. Even there, diagnostic characters are often subtle and difficult to see without magnification or, sometimes, dissection. Even mammalogists with considerable experience in an area commonly provide only generic or tentative specific identifications of specimens in the field. These names serve for bookkeeping purposes rather than for identification, and they facilitate tracking of numbers of species and specimens sampled.

Accurate species identifications are such an integral part of all aspects of comparative biology that studies without voucher specimens violate a basic premise of scientific methodology, that is, the ability of subsequent workers to repeat the study. Only voucher specimens provide a basis for verification of identifications and thereby duplication of a study. The literature is replete with examples of comparative studies in physiology, ecology, behavior, morphology, and systematics for which research results are questionable or even useless because of species misidentifications or failure to recognize that more than one species was involved. Most decisions relating to the management and conservation of species also depend on accurate species identifications. Voucher specimens are the only means to verify or, if necessary, correct specimen identifications and, therefore, are essential to scientific investigation in the above-mentioned disciplines.

All field identifications should be verified by a person with experience with the group, through the use of reliable and authoritative keys, or by comparison with specimens in museum collections. Vouchers should be deposited in appropriate repositories, usually a natural history museum. With erroneous field identifications, specimens of poorly known species may be overlooked, and important data may not be collected because the investigator assumes the species involved is well known. For purposes of sampling in little-studied regions, we recommend that all field identifications be treated as tentative and that all species be considered equally important.

Except for well-studied areas, such as North America and Europe, few useful field guides or identification manuals for mammals exist, and for many countries even lists of the recorded species are not available. Many of the older monographs on mammal faunas (e.g., Cabrera 1957–1961; Ellerman and Morrison-Scott 1966; Kingdon 1971–1982; Prater 1980) were based almost entirely on (often poorly) prepared museum specimens and are of limited utility for field identifications or even as sources of general information on geographic and habitat distributions. We suggest, therefore, that investigators become familiar with available primary literature before commencing an inventory and, whenever possible, that they examine museum specimens of species from the area of interest prior to beginning the fieldwork. Notes on the mammal fauna of the region with a list of the species and their diagnostic features should allow the worker to identify the more common species, focus on those of specific interest, and recognize any taxa that may be protected (see the section "Permits," below).

Because vouchers serve as the sole means of verifying data collected during investigations of biodiversity and provide critical information for future studies, the importance of voucher materials should be generally recognized and their preparation considered essential to good science. We acknowledge, however, that the removal and preservation of specimens for scientific purposes can be an emotional issue. Therefore, it is essential that field investigators carefully plan their studies in advance, clearly identify their objectives, and evaluate the need to collect voucher specimens. It is also essential that governmental and nongovernmental agencies requiring and supporting biodiversity assessments recognize the critical need for vouchers and provide support in both field and museum budgets for their preservation and maintenance.

Sample Size

What constitutes an adequate or optimal sample for the purposes of identification is not easily determined. For some species identification is possible from a single specimen (although this is rare); for other species, 20 individuals would not adequately sample the variation in the population, and a larger sample would be necessary.

Some species are polymorphic; some have striking sexual, ontogenetic, geographic, and/or individual variation; and others are relatively uniform even across broad geographic areas. Modern systematics takes into account this potential for variation and the significance of ancillary biological data in attempting to determine species limits. Gone are the days of running a single specimen through a key and magically achieving a reliable specific identification. This "cookbook" approach and the idea that a single specimen could be "typical" of a deme or a population, much less an entire species, are scientifically unsound. Keys, if properly constructed, can be useful tools in providing identifications, but these preliminary identifications must be tested by comparisons with descriptions in the literature and with museum specimens. The quality of keys, however, varies widely on a global scale; in some areas of the United States, keys are adequate, and in some poorly studied areas, they are nonexistent.

We agree with Frith (1973:3) that the number of animals removed from a population "really has no [biological] significance unless it is related to the total number of animals in the population and their rate of replacement." Many mammals are prolific, with reproductive potentials sufficient to accommodate increased levels of predation, although some species have low rates of reproduction. As predators on small mammals, scientists usually are singularly inefficient compared to snakes, birds, and other organisms. Furthermore, preparing specimens and documenting species (Appendix 3) are time-consuming tasks, and when done correctly, discourage human collectors from random "oversampling" (see also Foster 1982:6–7).

It would be convenient if we could provide an absolute value for, or formula to calculate, the number of vouchers of a given species that should be collected, but science is rarely convenient. Providing a meaningful formula for all 4,629 species of mammals is beyond our capability. For areas where the mammal fauna is well known, a single representative adult specimen of each population at each site will suffice minimally as a voucher for an inventory or monitoring study, unless the objective is to measure genetic diversity. Normally, the first adult of every species encountered during a project is suitable. For monitoring studies, we recommend that a voucher be prepared at the initiation of the study. If additional vouchers are required, they can be taken at the end of the study or from an area adjacent to the study site. As an operational figure, we recommend that 10 to 20 specimens would better represent the species at each site in well-studied areas.

Because we are in the early discovery phase and do not understand the taxonomic relationships of many tropical forms, and because many tropical areas are poorly known and numerous species are undescribed or inadequately represented in systematic collections, we usually recommend collecting many more than one voucher specimen when working in tropical areas. Generally speaking (and with an awareness of the frailties of any generalization), we recommend a sample of 20 individuals (ideally 10 adult males and 10 adult females) for identification purposes. We strongly encourage additional sampling of polymorphic species and those known to be inadequately understood taxonomically or suspected to include several taxa; for such species, samples of up to 25 males and 25 females may be adequate. A researcher who is interested in assessing genetic diversity within and among sites should prepare tissue samples for biochemical analysis (Appendix 4) and preserve voucher specimens of a minimum of 10 to 20 males and 10 to 20 females from each site.

Factors other than sample size can also affect the potential for accurate identification of specimens. Improperly or carelessly prepared specimens are often difficult or impossible to identify because diagnostic features are obscured or modified. Anyone collecting material for scientific purposes should be intimately familiar with proper techniques for specimen preparation and documentation. Ecological information also often aids identification. Generally speaking, a small number of carefully prepared specimens with detailed data is preferable to a large, carelessly prepared sample with inadequate biological data. Instructions for preparing and preserving mammal specimens as vouchers are provided in Appendix 3.

Specimen Data

To fulfill their function as vouchers of monitoring or inventory studies, all specimens must be thoroughly documented with locality and relevant associated data. Data associated with voucher specimens enhance the value of the vouchers and potentially make identifications easier, but those data must be accurate.

In addition to full locality data in a standard format and information on sampling procedures and habitat (see the section "Data Standards," above), the minimum information required for each voucher specimen includes the following:

1. *Unique sample designation.* This unique field number is assigned by the collector to a specimen obtained at one place and time during the inventory. The number is noted on a field tag that is tied to the specimen.
2. *Date and time of collection.* The date and time (24-hr clock) that the specimen was collected and the date it was prepared (if different) are essential. The month should be written out (i.e., numeric designations or abbreviations are not used).
3. *Name of collector.* The collector is the person (or persons) making the collection. The collector's name is never abbreviated, and the middle initial is included when available.
4. *Taxonomic identification.* Ideally each specimen should be identified to genus and species. This level of identification often is

impossible in the field; a family or other taxon name (murid, mouse, *Mus*) can be substituted for the scientific name until the animal is identified.
5. *Standard measurements.* The sex of the specimen should be entered both on the specimen label and in the field notes. In addition, the traditional measurements of total length (head-body length), tail length, length of hindfoot, ear length, and weight should be included in both places (see Appendix 3).
6. *Other information.* The existence of an associated special preparation (e.g., tissue sample) or other specimen data (e.g., behavioral observation, photograph) should be entered in the field notes and associated with the unique field number of the voucher specimen. Maps of the study area and trip itineraries are always useful for identification, cataloguing, and historical or archival purposes.

Most institutions require that the original or clear photocopies of a collector's field notes and catalogue accompany any incoming collection. The importance of good field notes to all subsequent use of the collection cannot be overemphasized. Poorly recorded field data can seriously mislead the specialist and reduce the usefulness of specimens. If the data accompanying the collection are a secondary compilation from the original field notes, they should be clearly labeled as such.

Selection of a Specimen Repository

Voucher specimens of mammals, including the data associated with the specimens, must be placed in an appropriate, recognized repository. The repository must adhere to at least the minimal standards for collection care and maintenance recommended by the American Society of Mammalogists (Yates et al. 1987). The specimens must be managed according to standardized collection management procedures, made available for use by researchers, and protected for use by future generations. Voucher specimens and their associated data should be transferred to a permanent repository as soon as possible after collection in order to avoid their deterioration in the field or in inadequate temporary storage facilities. All publications involving the specimens should provide the name and location of the repository that houses them. This is true for both accessory material, such as frozen tissues, and the actual voucher specimen.

Voucher specimens from faunal surveys that are accompanied by detailed field notes and associated documentation have almost incalculable scientific value. Given the inevitable widespread habitat destruction that may preclude collection of additional material from many areas, and the rapid technological advances that allow for previously unsuspected uses of specimens, we can only guess at the possible significance of such specimens in the future. Consequently, this often irreplaceable "time capsule" of information should be permanently stored in a secure institutional collection with a documented long-term commitment to conserving specimens and making them available for study by qualified researchers.

The amount of time, space, and money required to maintain a museum collection is enormous, and relatively few institutions are able to provide the long-term security necessary for large research collections. Therefore, selection of an appropriate institution for the deposition of field vouchers is of critical importance. Establishing a private collection unavailable for study by qualified researchers does a disservice to the scientific community and often imperils the long-term survival of the study specimens. Many important collections are lost or destroyed when the collector dies or retires and his or her home institution loses interest in them or realizes it no longer can provide the space or funds required for their maintenance.

When a researcher from one country carries out a study in another country that involves the collection of specimens, it is highly appropriate (and often a requirement of the collecting permit) for representative material to be returned, after identification, to designated institutions in the country of origin for the purpose of establishing functional reference collections. All such studies should involve appropriate in-country collaborators, a practice that will facilitate specimen deposition. The primary concern of all responsible biologists should be the long-term maintenance of specimens and associated data and their availability to qualified scientists for study.

Several variables influence the choice of a deposition site for collections; they are discussed by Lee et al. (1982). If identifications are required, an institution that has a history of research in the geographic area, an appropriate specialist on the staff, and access to extensive library facilities is optimal. Prospective donors should, however, obtain a statement of the museum's policies regarding acquisition, preservation, maintenance, and deaccessioning of collections to determine if these policies meet their needs. Most institutions will honor reasonable requests from the donor, but policy is determined by many factors.

The identification, distribution, and cataloguing of voucher collections is a service provided by museums to the scientific community. Many museums are currently suffering from budget cuts and staff shortages. The identification of a large collection often occupies many hours of staff time. It may require a curator to borrow specimens or visit other institutions so that pertinent materials may be compared directly, to lend specimens to specialists for identification, and to search the literature. Altruism, if it exists, has its limits. The donor must keep in mind that few museums can afford to invest the time and energy required to identify a major collection without the complete cooperation of the donor. If assistance with identifications is requested of an institution but the collection is to be deposited elsewhere, the requester should offer at least to deposit representative material in the institution that provides the service.

Donors often expect institutions to maintain a voucher collection as a discrete unit, separate from the main collection. This desire is understandable, but most institutions cannot accommodate such requests, because of limited space and curatorial support. Whether a voucher collection should be maintained in a single institution or distributed among several is also debated. Each option has merit. The first obviously simplifies future study of the collection; the latter provides for greater access by researchers in many areas. Donors concerned about this issue should ask about an institution's exchange policy before depositing specimens there.

Permits

ROY W. McDIARMID, ROBERT P. REYNOLDS, AND RONALD I. CROMBIE

During the past few decades, the number of laws regulating the collection, acquisition, study, transport, and disposition of wildlife and wildlife products has increased significantly. These laws have been proposed and promulgated in an effort to control activities that are deemed harmful to animals and plants. Although habitat loss generally is acknowledged to be the primary factor affecting species' distributions, abundances, recruitment, and extinctions, commercial exploitation also has had a detrimental effect on certain species of wildlife.

Some species considered to be endangered, threatened, or otherwise in need of protection have been protected by international treaty (e.g., Convention on International Trade in Endangered Species of Wild Fauna and Flora [CITES]), or

various federal (Federal Register 1973, 1995, and published amendments), state, and local laws. Additional regulations are stipulated by the International Union for the Conservation of Nature and Natural Resources. The laws and regulations contained in the U.S. Endangered Species Act and in CITES are those of primary concern, but many other foreign, federal, state, and local regulations may also apply to users of this manual. Many states, for example, require permits for the use of traps; permission to use such devices to sample mammals should be clarified with the local authority. Other regulations with which travelers should be familiar restrict the transport of liquid nitrogen, alcohol, and formalin, or the possession and transport of syringes and certain killing agents, drugs, or chemicals used in specimen preparation.

Laws regulating scientific collecting vary widely among states and countries and change constantly. Furthermore, the government agencies responsible for issuing collecting permits sometimes change or are restructured. Current information on most international and federal regulations and responsible agencies can be obtained by writing to or calling the U.S. Fish and Wildlife Service, Office of Management Authority (4401 N. Fairfax Drive, Arlington, VA 22203 USA; telephone: 703-358-1708). Interpretations of laws and regulations designed to protect animals in the United States are provided in the Code of Federal Regulations (1973, 1979) and a report from the National Research Council (1985). Information on state and local regulations can be obtained from the appropriate conservation or management agency in the jurisdiction of interest. The variation in requirements often makes obtaining collecting and export permits a trying process. Nevertheless, it is the responsibility of the individual collector to learn about and comply with the relevant regulations as they apply to mammals. Although certain provisions of a collecting permit may appear to have little bearing on the conservation of species or protection of habitats and in some instances may even restrict the conduct of scientific research, all of us are obliged to abide by the regulations.

Because obtaining the necessary permits is a crucial step in ensuring the success of a field study and often is the most difficult part of the preliminary work, it is essential that the investigator present a carefully planned proposal with clearly defined objectives to the permit-granting agency. We recommend that investigators be prepared for delays, which often are inevitable, by allowing a long lead time between the request for permits and the initiation of the field study.

Most institutions cannot or will not accept voucher material unless it is accompanied by documents verifying that the specimens were legally collected and, where appropriate, exported and imported. In many countries, permits for specimen collection and export are issued by different government agencies. In addition, some countries require an animal health permit, issued by a third agency, before specimens can be legally exported. In other countries collection and export, at least for noncommercial purposes, are unregulated. In these cases, a letter on official stationery from the most appropriate government agency stating that such permits are not required may suffice for purposes of importation.

Endangered and protected species require special permits beyond the normal collecting and export permits. In addition, in CITES-member countries, export permits for any species covered by CITES must be issued by the designated CITES official. The U.S. Fish and Wildlife Service (see address given previously) maintains an international directory of CITES Management Authorities, that is, of offices authorized to issue permits or equivalent documentation in accordance with CITES regulations. It is the responsibility of the researcher to ensure that he or she has complied with all laws governing the collection and export of scientific specimens and to secure the appropriate permits.

For import into the United States, a completed Fish and Wildlife Service form 3-177 (available from a Fish and Wildlife Service agent at a designated port of entry or from the U.S. Fish and Wildlife Service, Division of Law Enforcement, P.O. Box 3247, Arlington, VA 22203-3247 USA) accompanied by the above documents (copies are sufficient) from the country of origin must be presented at the port of entry. It is prudent to notify the agent at the port of entry of your anticipated date and time of arrival. If it is not possible to meet with a Fish and Wildlife agent at the time of arrival, the completed 3-177 form should be left with the customs inspector and a copy sent to the address specified on the form within the specified time. For purposes of declaration, scientific specimens, by definition, have "no commercial value." Importation of specimens into countries other than the United States and shipments through other countries will require other permits. In these instances local agencies should be consulted for information regarding regulations and appropriate procedures.

Chapter 5

Conducting a Survey to Assess Mammalian Diversity

Rasanayagam Rudran and Mercedes S. Foster

Introduction

Before launching a mammal biodiversity survey, an investigator must have a clearly defined objective. The objective is used to guide the survey through all stages of planning and execution. Once the objective is established, planning for the survey can begin. The importance of planning should never be underestimated. It enhances efficiency of data collection, improves the quality of the information gathered, and allows for effective allocation of resources.

An estimate of funds required is an important element for planning a survey. Funds are usually requested through research proposals submitted to an institution interested in underwriting bio-

logical investigations. Advice on preparing proposals is beyond the scope of this book, but any request for funds to conduct a biodiversity survey should specify the objective of the study, its location, and its duration and should describe other aspects of the investigation that may help ensure financial support. A survey should not be launched until necessary funds have been obtained.

There are three distinct stages in planning a mammalian biodiversity survey. In the first stage, the investigator defines the scope of the investigation in terms of the species selected for study. This selection depends on the study objectives, the time (duration of the study) and money available for the survey, and the characteristics of the survey area, particularly its size. Although biodiversity surveys often involve the study of many little-known species, it is important to review whatever information is available and use it as a basis for the survey plan.

In the second stage, the investigator selects the most appropriate techniques for estimating the richness or abundance of the study species from the wide range of methods discussed in this handbook. The choice of techniques depends on the above-mentioned factors and other variables. The final stage of planning involves the integration of the theory and practice, that is, the tailoring of the selected techniques to the specifics of the field situation.

Recruitment of personnel and purchase of equipment and supplies should commence as their need is identified, during the planning stages. Personnel can conduct preliminary surveys, which will also be useful for identifying suitable sites for base camps within the study area. Once base camps are established and equipment and supplies are deployed, the investigator can begin the survey.

Objectives

The primary goals of a mammalian biodiversity survey are to estimate species richness (the number of different species) and species abundance (the number of individuals within a species) within a particular area. A particular survey may have important secondary objectives as well. Often, a survey may be undertaken to obtain information for a specific purpose, such as the comparison of biodiversity among areas, establishment of a protected area, or conservation or management of species populations. Surveys that address such issues should consider them from the early planning stages.

Plans should also be made for the collection and preservation of voucher specimens (see "Voucher Specimens," Chapter 4, and Appendix 3). Species may be tentatively identified in the field, but final confirmation of species presence must be based on detailed examination of specimens collected from a survey area.

Defining the scope of a survey

Species Lists

The first step in preparing for a survey is to review the scientific literature for mammalian investigations conducted within the study area or at nearby sites. The information obtained is used to develop a preliminary list of species that may be encountered at the study site. Such lists are important for defining the scope of a survey but should not be considered complete. The investigator should anticipate the occurrence of "new" species within the survey area, especially in highly diverse areas such as tropical forests. Alternatively, an investigator can conduct preliminary surveys at the study site to develop a species list. Indeed, such surveys are strongly recommended regardless of the amount of information gleaned from the scientific literature. Preliminary surveys do not have to be extensive, but they should be conducted in a way that will confirm the presence of as many mammal species as possible in the shortest time. Some of the techniques described in the following chapters may be used in preliminary surveys or rapid assessment programs.

Selecting Target Species

With a preliminary species list in hand, an investigator can decide which species to include in a proposed survey. If time and money and the characteristics of the survey area are not limiting, an investigator may decide to include all species that are likely to occur within the study area in the investigation of species richness and abundance. Often, however, time and money impose limitations, even if the study area characteristics do not, and the investigator is compelled to select target species.

A number of criteria can be used to select such species. One is expected frequency of occurrence, although this criterion can be applied only if the investigator has some knowledge of the abundance of different species. If such information is available, target species can be selected in order of their absolute abundance, or in order of their abundance within different mammalian orders or different life-style categories. Thus, although carnivores in the survey area may not be abundant, the most numerous species might be selected as a target species along with the most common herbivore, arboreal species, nocturnal form, and so forth.

Species may also be selected for a survey on the basis of their size, vocalizations, signs left in the habitat, or other characteristics that make them fairly easy to detect. Special interest in a species can also be used as a criterion for selection. Such interest may relate to the collection of specimens for detailed taxonomic studies. It can relate to the conservation of a species that is declining in other localities or is found in small numbers within the survey area. There also may be interest in reducing populations of pest species or in harvesting a population on a sustainable basis.

Choosing field techniques

Numerous field techniques are available for investigating species richness and the abundance of mammals. These techniques may be broadly classified as observational techniques, capture techniques, and techniques based on animal signs. The techniques are discussed in detail in Chapters 6 through 9. A number of factors that

influence choice of field techniques are considered in this section.

Appropriateness

Several techniques may be employed in a multi-species mammal survey. The techniques are selected on a species-specific basis, considering their applicability and appropriateness for each target species. Both sign and observational techniques may be equally suitable for measuring the abundance of a burrow dweller, for example, but if the latter technique provides more reliable information, then it becomes the technique of choice. Field and estimation techniques must also produce information appropriate for meeting the objectives of the survey.

Physical Characteristics and Behavior of Species

An investigator must have a good knowledge of the behavior and physical characteristics of the target species in order to choose the right techniques. An important aspect of species behavior that affects the choice of techniques is the daily activity pattern of a species. Observational techniques are most appropriate for diurnal species but can sometimes be applied to nocturnal animals when used in conjunction with night observation devices. More often, nocturnal species are surveyed with capture techniques or by detection of their signs. Similarly, observational techniques may be more applicable to social species or those that form temporary aggregations. In aquatic habitats, observational techniques may not be appropriate for species, such as whales and hippos, that remain submerged for long periods. Small body size and cryptic coloration can also make a species difficult to detect and rule out observational techniques. Seasonal activities, such as migration, can also affect the choice of techniques.

Size of the Survey Area

The size of the area about which inferences are to be drawn is an extremely important determinant of the estimation techniques that are used. If the target population inhabits a relatively small area, then it may be possible to cover the entire area using one or more estimation techniques. If the target population inhabits a very large area (e.g., ranging up to an entire region or country), then *spatial sampling* will be required. In spatial sampling, estimation techniques are applied to sampling units selected from the total area of interest, and an overall estimate is based on the estimates from these units. The size of the survey area can also influence selection of the sampling method. For example, aerial surveys are especially useful when large areas are to be covered.

Habitat and Climate

Features of an animal's natural environment can influence the choice of field techniques. For example, density of the vegetation and degree of habitat heterogeneity can affect direct observations of animals. In highly heterogeneous and dense habitats aerial counts are often inapplicable. Cloud cover, fog, rain, wind, and heat haze can similarly affect aerial surveys as well as other observational techniques. Rainfall, snow, soil conditions, and the presence of animals that feed on carrion and feces (e.g., dung beetles) can affect the choice of techniques based on animal signs. The steepness of the terrain can make certain areas inaccessible and preclude the use of techniques based on animal capture or animal signs. In such areas only aerial surveys may be possible.

Personnel and Time

Some foot surveys can be conducted by a single investigator; many other techniques require the involvement of more than one person. The array of field techniques available for a survey increases as the number of people involved in the investigation increases. The expertise of these individuals can also enhance the array of techniques used. Thus, aerial surveys are possible if a pilot, a navigator, and at least one person experienced in counting animals from the air are included in a team. If a team includes people with experience in detecting numerous mammals by sight, sound, or smell, the quality of foot surveys increases greatly. People who depend on a survey area for their livelihood often have such experience and can make significant contributions to biodiversity surveys, even though they may not have a formal education.

Several field techniques have to be repeated in order to provide the best possible estimates of species richness and abundance. Considerable time is needed for such techniques as well as for surveying areas, such as a tropical rain forest, that are large or rich in species. If time is a limiting factor, investigators may have to rely on techniques that only provide indices of abundance, or to restrict the number of species surveyed. Both situations can limit the value of a survey.

Budget and Equipment

Funds available for a biodiversity survey can sometimes limit the choice of techniques. For example, the potentially high cost of some trapping equipment may preclude its use in capturing animals. For similar reasons, techniques that require costly electronic equipment may not be feasible. Even inexpensive equipment (e.g., compass, binoculars) can limit the choice of field techniques if it is unavailable. Every effort must be made to acquire the funds and equipment necessary to ensure that the most appropriate field techniques can be used. In addition, legal considerations regarding use of certain types of equipment (e.g., guns) may limit the choice of techniques in some countries.

Integrating theory and practice

After completing the first two planning stages, the investigator can decide how to implement the selected survey and estimation techniques, given the realities of the field situation. General guidelines for such implementation are provided in Chapters 3 and 10. Here we describe some practical techniques that may facilitate such implementation.

Use of Maps

Maps of the survey area are the key to identifying and measuring the sampling units (e.g., points, lines, quadrats) where the selected field techniques will be implemented. Before identifying and measuring the sampling units on a map, an investigator should know the practical aspects of statistical design and the proper techniques for using maps in conjunction with a compass.

An important feature of any map is its scale. The scale provides the relationship between the size of any entity on a map and its dimensions at the site where it is actually located. For example, 1 unit on a map with a scale of 1:100,000 represents 100,000 units in the area mapped. Thus, a line of 1 cm between two points on a map of a survey area represents 100,000 cm or 1 km between the same points in that area. A distance of 2.5 km within the survey area is represented by 2.5 cm on the map. Likewise, 1 cm on a map with a scale of 1:250,000 represents a distance of 2.5 km.

Another feature of a map is its declination, which is the angle between true north (the direction of the north pole) and magnetic north (the direction toward which the compass needle sets) in the area represented by the map. This angle is usually shown on a map by the intersection of two lines. One line is vertical and points toward true north. The other line, representing magnetic north, lies to the left (west) or to the right (east) of the vertical line and indicates a westerly or easterly declination. Knowledge of the type and angle of declination is important for transferring measurements from the map to the survey area and vice versa. If there is a westerly declination, the angle of declination must be added to the bearing (angular measurement) between two points on the map when transferring this information to the survey area. When transferring information from the survey area to a map, however, the angle of declination is subtracted. Exactly the reverse is done if there is an easterly declination at the survey site: The angle of declination is subtracted when transferring angles from map to survey area and added when going from survey area to map.

Measuring the Size of a Sample Unit

A map is usually the only source of information that provides area measurements of a study site or localities within it. Therefore, the areas of the localities surveyed (sample units) usually are calculated from a map. This is done by using a planimeter, an instrument that translates the linear measurements of the perimeter of a sample unit into area dimensions. To obtain the actual size of a sample unit, an area of known size should be drawn to scale on a map of the survey area. For example, on a map of 1:100,000 a square 2 cm on a side represents an area of 4 km^2. Several measurements of the perimeter of the square are made with a planimeter, in order to calibrate the instrument. The average of these measurements, A, represents the planimeter reading for 4 km^2 within the study site. Next, the perimeter of the sample unit on the map is repeatedly measured with the planimeter, and an average of these measurements, B, is calculated. Because 4 km^2 of the study site is represented by A on the planimeter, the area, S, of the sample unit is calculated as follows:

$$S = 4 \cdot \frac{B}{A}$$

If a planimeter is not available, a "dot grid" may be used to estimate the size of the sample unit within the survey area. A dot grid is a transparent sheet with a regular array of black dots. This sheet is first placed over an area of known dimensions on the map (e.g., a square), and the number of dots falling within the known area is counted repeatedly to obtain an average number of dots. The procedure is then repeated by placing the dot grid over the sample unit and calculating the average number falling within it. This information along with the map's scale can then be used as before to obtain the size of the sample unit.

Choosing a Random Sample

Under some sampling designs (see "Selection of Sampling Units," Chapter 3), sampling units are randomly selected from the entire study area or from a stratum. A random numbers table can be used for such selections and for a variety of other purposes during a biodiversity survey. A table of random numbers consists of several rows and columns (see Appendix 10). It is essentially a list of single digits from 0 to 9, in which each digit has the same probability of occurring anywhere in the table's columns or rows. Thus, one important feature of a random numbers table is that each digit occurs with the same overall frequency in the table. Another feature of the table is that the occurrence of a digit anywhere in the table is independent of the occurrence of its neighbors. Consequently, the digits that occur beside each other may be combined to provide random numbers consisting of several digits. There is no restriction on the pattern of combining adjacent digits in the table. One could combine digits from the right side of the table moving left or vice versa along any row, and from the top of the table to the bottom or vice versa along any column. The only stipulation in using the table is that the pattern of combining adjacent digits must be decided before looking at the table.

Suppose an investigator wishes to select three quadrats randomly from the total of 50 covering the entire survey area. First, the 50 quadrats are assigned numbers from 1 to 50 on a map. The random numbers table is then read in pairs using a predetermined pattern. Reading the digits in pairs confers all 50 quadrats with the same probability of being selected in the random sample of three. However, it also permits double-digit numbers greater than 50 (i.e., 51–99) to be selected from the table. When this occurs, the numbers greater than 50 are rejected, and the investigator continues to read the table until the required sample of three quadrats is randomly selected.

When reading the table, one may come upon a suitable random number (i.e., 1–50 in the above example) more than once before completing the random selection of quadrats. In this case, the repeated number is usually rejected, and the investigator continues to read the table until the selection is complete. This type of sampling is referred to as *simple random sampling without replacement* and is the most common technique of random selection used in biodiversity surveys. If a random number is included in a sample each time it is repeated in the table, the process of selection is referred to as *simple random sampling with replacement*.

Instead of quadrats, an investigator may wish to select five random directions in which to establish census transects. In that case, the digits from the random numbers table should be read in triplets, in whatever pattern is predetermined by the investigator. Reading the random numbers in triplets permits all angles from 0° to 360° to have an equal probability of occurring in the sample of five random directions. Again, numbers greater than 360 are ignored when selecting the five random directions.

Choosing Random Points

The random selection of angles and areas is often preceded by the random selection of points in space. In selecting random points one should remember that in mathematical terms a point has no dimensions. In practical terms, however, a pencil point on a map has some dimensions and actually represents a linear measurement in the survey area, which is dependent on the map's scale. For example, a pencil point is about 0.5 mm wide, and on a map of 1:100,000 scale it represents 50 m within the survey area. Thus, it is not possible to locate a random point along a line on a map of this scale with less than 50 m accuracy. To choose random points along a line, the line is first divided into convenient intervals with equally spaced points. For example, a line of 5 cm on a map of 1:100,000 scale (representing 5 km within the survey area) could be divided with 26 equally spaced points, which are 2 mm (equivalent to 100 m within the survey area) apart. Next, the points are assigned numbers starting at one end of the line from 00; a random numbers table is used to select the points needed to make up the sample size. If points 05 and 11 were randomly selected, they would be located 500 m and 1100 m from the starting point of the line, respectively, when it is established within the survey.

One may also have to select random points from a line already established within the survey area. Thus, to select points from a 5-km line, the investigator may read the random numbers in groups of four to enable all distances from 0000 m to 5000 m to occur in the randomly selected sample. A practical problem with this type of selection is the accuracy of locating two points 1 m apart that may be randomly selected. This problem can be avoided by identifying points on the line that are equally spaced at greater distances (e.g., 25 m, 50 m, or 100 m apart) and then making a random selection.

The first step in selecting a random point in space is to locate the space (e.g., a forest patch) on a map and establish a pair of axes (x and y) at right angles to each other, so that they include the entire space. The axes are then divided into convenient units (as shown in Fig. 4), and the random numbers table is used to select random points along the axes. The table is read in groups of four digits. The first pair gives a location of the point along the x axis, and the second pair gives a location along the y axis. These coordinates indicate a random point in the space. The process is repeated until the required number of points is identified within the space. Points that fall outside the space are discarded. Forty-five groups of four-digit numbers were read in order to identify the four random points within the space illustrated in Figure 4.

After the random points are identified on a map, they must be located within the space (e.g., the forest patch) where they actually occur. To do this, one normally uses compass bearings and distances measured from landmarks indicated on the map. A single landmark can be used, but use of measurements from two or more landmarks increases accuracy. Sophisticated global positioning system (GPS) devices, which provide readings of longitude and latitude, are becoming increasingly popular for locating points within survey areas and are likely to become standard equipment in future biodiversity surveys. Some GPS devices are accurate to within 50 m.

Minimizing Error

Care should be taken to avoid errors that may lead to inaccurate or imprecise estimates. For example, all equipment used during a survey, especially that used for measurement, must perform accurately. Investigators should purchase well-designed, well-made equipment, which should be properly maintained and calibrated. Newcomers to mammalian surveys must learn the proper use of measuring equipment, such as

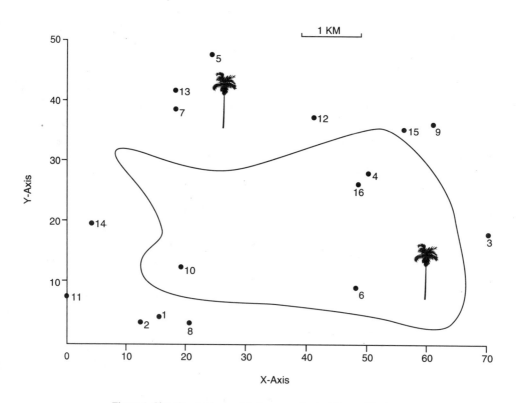

Figure 4. Choosing random points in space. See text for explanation.

compasses, planimeters, range finders, and vernier calipers. The accuracy of a compass can be affected by even small metallic objects carried or worn by an investigator.

Investigators using observational techniques should learn the cues and methods for detecting different species, preferably with the help of an experienced observer. The proper search image for a species can be developed by repeatedly observing it in its natural habitat. This exercise can greatly enhance accuracy in counting. Experience in counting large aggregations in a systematic manner also helps to minimize error. Whenever possible photographic records should be made when observational techniques are used to survey gregarious or social species. Such records are useful in establishing the relationship between an investigator's visual counts and the actual number of animals in the survey area.

Surveys using observational techniques should last no more than 3 to 4 hours, even under the best field conditions. During a period of this length, the activity of a species can change considerably, and investigator fatigue can affect animal detection and counting accuracy.

Correct interpretation of scats, tracks, nests, burrows, and food caches depends on features such as size, shape, composition, and age. Detection of animal signs can be greatly enhanced by developing the proper search image, especially for animal structures such as nests and burrows. Experience in distinguishing between animal calls is also an asset in detecting different species and minimizing error.

Captured animals should be treated (e.g., marked and released or removed) as stipulated for the method used. Efforts should be made to minimize animal escapes and handling deaths.

In addition, marks placed on animals should not disappear between capture sessions or increase animal mortality.

Conclusions

Planning and organizing a mammalian biodiversity survey requires considerable knowledge and experience. An investigator should be thoroughly familiar with statistical design for sampling animal populations and have considerable experience in using different field techniques, in addition to having organizational skills. Experienced field biologists with a good knowledge of the survey area, the target species, and statistical and research design should be involved in the planning and organization of a mammalian biodiversity survey. Their input is invaluable.

Chapter 6

Observational Techniques for
Nonvolant Mammals

Rasanayagam Rudran, Thomas H. Kunz, Colin Southwell,
Peter Jarman, and Andrew P. Smith

Introduction

Several observational techniques can be used to confirm the presence of mammalian species within a survey area and to estimate their population sizes. All of these techniques involve detection of animals by sight, sound, or smell, followed by visual identification of the species and counting of individuals. Mammals can also be detected, identified, and counted indirectly with the help of equipment such as cameras, tape recorders, and ultrasonic detectors. In this chapter, we describe techniques that rely on direct sightings of diurnal and nocturnal species and also some methods that use accessory equipment.

Visual identification of mammals in their natural environment is influenced by several factors (see "Choosing Field Techniques," Chapter 5). These factors affect the various observational techniques differently and should be considered carefully when choosing among them. Certain factors may preclude the use of a particular technique in some situations.

Observational techniques can be used to survey all but the smallest mammals (e.g., most rodents and insectivores) or those that live in inaccessible areas, difficult terrain, or dense vegetation, or under extreme climatic conditions. Such techniques have been used to survey small arboreal or terrestrial mammals, including squirrels (Healy and Welsh 1992) and agoutis (Cant 1977); medium-size species such as white-tailed deer (Floyd et al. 1982; Teer 1982); and large mammals such as rhinoceroses (Kiwia 1989; Dinerstein and Price 1991), bison (Wolfe and Kimball 1989), and elephants (Buss and Savidge 1966; Dublin and Douglas-Hamilton 1987; Ottichilo et al. 1987). Observational techniques have also been used to survey bats (Barclay and Bell 1988; A. P. Smith et al. 1989) and marine mammals such as cetaceans, sirenians, and pinnipeds (Irvine and Campbell 1978; Braham et al. 1980; Braham 1982; Drummer et al. 1990; Lefebvre and Kochman 1991). They are most appropriate for surveys of medium-size to very large terrestrial mammals that live in relatively open habitats on fairly flat terrain. These include many species of ungulates, primates, carnivores, and marsupials.

Density and Relative Abundance

Population estimates can be used in conjunction with information on the size of a survey area and the distribution of species to calculate two measures of density. One measure is *crude density,* which is the number of individuals of a species per unit area of the entire study site. The second measure is *ecological density,* which is the number of individuals of a species per unit area of the habitat that is utilized by the species. The latter measure is a more meaningful estimate of the density of a species, especially in heterogeneous survey areas where the species may occupy only certain habitat types.

Often mammalian surveys are concerned with relative abundance (or relative density) of species, which can be expressed as the ratio of the population size (or density) of a species to the total population size (or total density) of sympatric species in the group of interest. Relative abundance (or relative density) estimates reflect the comparative sizes of different populations with respect to the entire mammalian fauna (or some other defined group of taxa) of a study site. When surveys are conducted at the same site at different times, relative abundance estimates reflect temporal changes in a population with respect to the populations of sympatric species. When surveys are conducted at different sites at the same time, these measurements can provide comparisons of spatial differences in the relative sizes of populations that are common to the different sites. Spatial and temporal comparisons of the relative abundance of a species using count statistics are valid only when the observability is the same for the different species, times, or locations being compared (see "Use of Count Statistics in Comparative Studies," Chapter 3).

Total Counts and Sample Counts

Observational techniques provide total counts of individuals inhabiting a survey area if the entire area is searched, if all animals are located during searches, and if located animals are counted accurately (Norton-Griffiths 1978). It is not always possible to meet these requirements. For instance, some animals may go undetected despite a thorough search, particularly when survey areas are relatively large. Also, within large survey areas, it may only be possible to count animals in a portion of the total area because of time or other constraints. Therefore, methods involving direct sightings frequently provide incomplete counts of individuals occupying a survey area.

Drives and Silent Detection Methods

Techniques for directly observing and counting animals within an area are of two types. In one type, animals are driven, or made to run, and counted as they flee. Drive techniques are most appropriate for surveys of diurnal, medium-size, terrestrial mammals with a conspicuous flight reaction that live on flat terrain with relatively good visibility. These techniques are inappropriate for surveys of species that hide rather than flee and for large predators, fossorial species, and arboreal species. Drives can be conducted most effectively when survey areas are relatively small (i.e., a few square kilometers).

The other type of observational technique is based on silent detection. The observer approaches animals as silently as possible and counts them while they remain undisturbed. Silent detection methods can be used to survey species over a wide range of size classes, from relatively small to very large, including species that can be surveyed with drives. Such methods are also used to survey marine mammals, arboreal species, fossorial species, diurnal species, and nocturnal species. Detection of nocturnal mammals requires the use of a spotlight or other light source. Silent detection methods are advantageous over drives because they are not stressful to animals.

Approaching animals silently on foot and counting them is a relatively slow and time-consuming process. Therefore, foot surveys are only suitable for counting animals in areas of less than 10 km². For animal counts over larger areas, investigators frequently use mobile platforms such as horses or mules, land vehicles, airplanes, and ships or boats (Schaller 1967; Goddard 1969; Norton-Griffiths 1975; Wilson and Wilson 1975; Sinnary and Hebrard 1991). Although transport vehicles may cause disturbance, they allow the investigator to survey relatively large areas rapidly and to count animals in their natural state. In this respect techniques that use mobile platforms differ from drives and resemble foot surveys and other silent detection methods.

Drives

Drives for Total Counts

In drives for total counts, animals occupying an area are completely surrounded and counted as they are forced to leave the area. To obtain accurate total counts, a relatively large number of people are needed to survey even small areas, and the target species must be clearly visible within the survey area.

RESEARCH DESIGN

The technique involves a group of stationary observers and a group of mobile "drivers," who are initially positioned at the periphery of the survey area and surround it completely. All observers and drivers only count animals to one side during the drive (i.e., to their right or to their left). Spacing between team members is critical to ensure that all animals within the survey area are flushed, sighted, and counted accurately. Spacing distance is obtained by comparing *twice* the minimum flight distance of the target species and the maximum distance at which it is visible in its natural habitat. The lesser of the two distances is used. Drivers and observers should maintain a clear view of their neighbors on the counting side when conducting a drive.

DATA COLLECTION

From the periphery of the survey area, the drivers move toward the stationary observers, driving animals ahead and always maintaining visual contact with neighbors on the counting side. The observers, who must remain hidden from the fleeing animals, record animals moving ahead and out of the survey area. The drivers count animals that break through the line of advance and escape between them. Both drivers and observers also record animals that enter the survey area on the counting side after the drive commences. The drive terminates when all animals are driven out of the survey area and the drivers reach the line of stationary observers.

DATA ANALYSIS

A total count of the animals that occupied the survey area prior to the drive (i.e., population size) is obtained by summing the individual counts of drivers and observers, as follows:

Total count = (No. of animals moving ahead)
+ (No. of animals escaping between drivers)
− (No. of animals entering survey area)

The crude density of the species surveyed can be calculated by dividing the total count by the area covered by the drive. Ecological density can be obtained by subtracting the area of habitat not utilized by the target species (e.g., lakes, rocky outcrops) from the total area of the drive before calculating density.

VARIATIONS

Animals can be driven toward an impenetrable barrier, such as a fence or a body of water (rather than toward a group of observers) and counted as they break through the line of advancing drivers and escape (Short and Hone 1988). Most often animals break through the line of advance when they are near the barrier. If animals group together and escape simultaneously, counts may be inaccurate. To minimize this possibility, a barrier with a narrow opening or escape route can be used. Observers are stationed close to the opening to count animals moving out of the survey area; drivers count animals that escape through the line of advance.

PERSONNEL NEEDS

The number of people required to obtain accurate total counts during a drive increases with the size of the survey area and often limits the size of the area that can be surveyed. In general, it is difficult to carry out total count drives in areas larger than 10 km^2, even with adequate personnel. This is mainly due to the difficulties of maintaining proper spacing and constant visual contact with neighbors, especially in relatively dense vegetation or undulating terrain. Nevertheless, drives have been used to obtain total counts of several solitary African duikers (Koster and Hart 1988), wild pig (Barrett 1982), and kangaroos (Short and Hone 1988).

SPECIAL CONSIDERATIONS

The possibility of stressing or injuring animals is an important concern with any drive technique, particularly when barriers are used. For example, female eastern gray kangaroo, common wallaroo, and swamp wallaby have been known to eject young from their pouches during drives. These problems are minimized if drives are conducted at times least disruptive to animals (e.g., outside the birth season) and if the speed of drives and the behavior of drivers is regulated. We also recommend that the time interval between drives be long enough to permit the animals to recover from the temporary disturbance. Frequent drives can cause stress and injury and can induce animals to leave a survey area for extended periods, leading to underestimates of total numbers.

Drives for Sample Counts

Drives for sample counts are used to obtain animal counts from relatively small plots or sampling units (e.g., square quadrats, rectangular transects) selected from a large survey area. They require fewer people than drives for total counts. All drive team members are mobile and record all animals leaving and entering the sample plots.

RESEARCH DESIGN

Before conducting drives for sample counts, plots of equal size are selected according to a specified sampling design (e.g., random, stratified random, systematic) from the survey area. The size of a plot, particularly its width, will depend on the number of people available to conduct the drives. If the target species occupies only one habitat type within a heterogeneous survey area, then the most efficient estimates will be obtained by selecting plots only from that particular habitat.

Drives for sample counts can be conducted with a minimum of two people. In this case, the width of a plot (and the distance between the team members) should be no more than twice the minimum flight distance or the maximum visibility distance, whichever is less. The distance between team members should be set so that individuals can walk along the boundaries demarcating the length of the plot. Ideally, these boundaries should be marked before the drive commences, so that animals moving across them while leaving or entering the plot can be accurately determined. If the boundaries are unmarked, team members must use compass bearings to record animals entering or leaving the plot.

DATA COLLECTION

Team members move along the parallel boundaries demarcating the length of the plot and maintain constant visual contact while driving animals. Each member records the number of individuals that leave or enter the plot across the boundary. Team members also record the animals that pass between them, in one of two ways. Because observers maintain constant visual contact, both could record animals moving in either direction across the entire length of the imaginary line between them. Team members would then compare records to confirm the number of animals leaving and entering the plot between them. Alternatively, team members could record animals moving in both directions up to the midpoint of the imaginary line between them and later sum the animals entering or leaving the plot. This procedure can lead to double counting of animals if team members cannot accurately gauge the midpoint of the imaginary line. Thus, the first option is preferable. The second procedure may be feasible if visibility is less than ideal for sighting animals and maintaining constant visual contact between team members. Team members repeat the counting procedure in other plots selected randomly or systematically within the survey area.

DATA ANALYSIS

The procedure produces a count of animals that occupied the plot before the drive commenced. This count is divided by the size of the sample plot to compute a density estimate for the species being surveyed. The precision of the resulting estimate for the entire area of interest is increased by increasing the number of plots on which drives are conducted (see "Determining Sample Size" and "Use of Count Statistics in Estimation," Chapter 3). A mean density estimate is calculated and multiplied by the area of the habitat type occupied by the target species to estimate population size.

VARIATIONS

If more than two individuals are available for drives, they can be used to increase the width of the plot. The additional drivers are positioned between those at the boundaries demarcating the plot's length. The maximum distance between observers is determined in the same manner as for drives with two people. The spacing between drivers is then used to divide the plot into several imaginary parallel strips of equal width.

Double counting by team members becomes a potentially serious problem when more than two individuals are involved in drives. To avoid double counting animals that break through the line of advance, team members should walk along the imaginary boundaries of the strips using a compass and should count animals only on one side. The person at one end of the drive line does not need to count animals to one side, but this person and the one at the other end of the drive line must monitor the boundary ahead for animals leaving or entering the plot. To avoid double counting of animals that move ahead in the strips, each team member uses a map of the area to record the exact time and the location and direction of movement of each animal when it is first sighted in the strip on his or her counting side. These records are compared later to determine whether an animal has been recorded by more than one team member. Corrected records can be used to calculate the total number of animals moving ahead and entering or leaving the plot from the side. Double counting of Chinese water deer (*Hydropotes inermis*) was effectively avoided with this system of data collection during sample count drives involving about a dozen people (R. Rudran, unpubl. data).

In India, chital, barasingha, sambar, blackbuck, and gaur were counted during drives involving several people (Schaller 1967). To overcome the problems of poor visibility and wet conditions when surveying for barasingha in tall-grass swamps, drivers were mounted on elephant back to count animals flushed from the survey area.

SPECIAL CONSIDERATIONS

One of the potential problems with drives for sample counts is that animals that move ahead of the drivers may escape unnoticed. To determine whether this problem could affect a particular survey, the average flight distance of the species should be calculated (van Lavieren 1982). If this distance exceeds the maximum distance at which the species is visible, it is likely that a sample count will be substantially less than the actual number of animals occupying a plot. Despite this potential bias, van Lavieren (1982) obtained fairly accurate population estimates of several large mammals in moderately dense woodland savanna in northern Cameroon, and of sambhur and wild pigs in a lowland forest of Indonesia using drives for sample counts. This technique has also been used to estimate densities of hares and rabbits (Chapman and Wilner 1986).

Silent detection for total counts

Individual Identification

With the technique of individual identification, physical characteristics of animals, such as body markings, size and shape of antlers or horns, or scars and other natural deformities are used to identify specific individuals that are then used as "marker" animals to obtain total counts from the population. The technique has been used to provide total counts of arboreal and semiterrestrial primates that live in relatively stable social groups (Struhsaker 1967, 1975; Sugiyama 1967; Altmann and Altmann 1970; Rudran 1973, 1978, 1979; Dittus 1975, 1977; Robinson 1988; Rumiz 1990). It has also been used to count social species such as lions (Schaller 1972), wild dogs (Bertram 1979), feral horses (Garrot and Taylor 1990), and kangaroos and wallabies (Jarman et al. 1989), and solitary species such as leopards and cheetahs (Bertram 1979), one-horned rhinos (Dinerstein and Price 1991), and red deer (Clutton-Brock et al. 1982). Individual leopards and cheetahs were identified mainly on the basis of spotting patterns.

RESEARCH DESIGN

Social groups living within the study site must first be habituated through prolonged and repeated contact. Habituation permits close scrutiny for the identification of individuals on the basis of their physical features. Identifiable individuals are then used to distinguish between social groups.

DATA COLLECTION

After habituation has been accomplished, all social groups within the survey area are contacted and censused regularly. With highly dispersed groupings, censuses are most easily accomplished when animals come together to rest or socialize. When censusing some species, it may be necessary to add extragroup individuals to the group counts to obtain total counts of animals living within the survey area (Rudran 1979; Crockett and Eisenberg 1986; Rumiz 1990). When censusing a group, the locations of its members are plotted on a map of the study site. The period of census and location plotting continues until areas used by groups no longer increase and all parts of their home ranges have been established.

DATA ANALYSIS

The boundary of a group's home range is obtained by connecting the most peripheral locations where its members have been seen. The size of the home range is measured using a planimeter or the "dot grid" method (see "Measuring the Size of a Sample Unit," Chapter 5). The same methods can also be used to determine the size of any areas of range overlap between neighboring social groups.

If the home range sizes of all social groups inhabiting a survey area are known, density may be obtained by dividing the total count during a census (i.e., population size) by the sum of all the home range sizes less the sum of any areas of range overlap. Group sizes can sometimes change during the prolonged period required to determine home range sizes, leading to fluctuations in measures of population size. To compensate for such fluctuations, a mean population size is calculated from the total counts of each census and used in the calculation of density.

If home ranges overlap and the home range and group size of only one group (or a few groups) within the survey area are known, the calculation of density becomes slightly more complicated. In such cases the home range size of the known group(s) is reduced by an amount that reflects the number of groups whose home ranges overlap with that of the known group and the degrees of overlap (National Research Council 1981). For example, if the known group has a home range of 100 ha, of which 30 ha overlap with one group and 18 ha overlap with two groups, the adjusted home range of the known group is given by the sum of the non-overlapping area (52 ha), half the overlap area of two home ranges (15 ha) and one-third the overlap area of three home ranges (6 ha). The resulting adjusted home range size of 73 ha and the group

size (or mean group size) are then used to calculate the density of the species surveyed. Population size is estimated by multiplying the density estimate by the total area occupied by the target species. Density estimates based on home range sizes are ecological densities rather than crude densities.

SPECIAL CONSIDERATIONS

The time required to habituate and identify individuals and determine their home ranges is an important consideration when using this technique. These tasks can be accomplished relatively quickly only in localities (such as protected areas) where animals have positive interactions with humans. Thus, we do not recommended this technique for preliminary surveys or rapid assessments of animal populations. This technique is useful because it provides accurate density estimates as well as demographic and ecological information, which are essential for correctly assessing the status of populations. This assessment is important if the objectives of the survey include long-term monitoring and management of populations that are endangered or can be harvested.

Observation of Emergence from Dens and Burrows

Observation of emergence is used to obtain total counts of species in which all individuals emerge from their dens or burrows within a restricted time period. Observers simply count all individuals as they emerge. The technique requires initial location of all tree hollows or burrow sites occupied by the target animals within the survey area. It has been used to survey nocturnal, tree hollow–dependent arboreal mammals (e.g., A. P. Smith et al. 1989), and also the diurnal banded mongoose in Africa, which shelters in termite mounds and aardvark holes (Rood 1975). The technique has been evaluated for surveys of many Australian possums and gliders and found to be most useful for small to medium-size species (A. P. Smith et al. 1989).

Two assumptions must be met to obtain unbiased density estimates of the target species using this technique. The first is that all individuals in the population den in tree hollows or burrows. The second is that all individuals are visible and detectable in natural light when they emerge.

RESEARCH DESIGN

The first step is to divide the survey area into square or rectangular plots of known dimensions. Each plot is then thoroughly searched, and locations of tree hollows or burrows potentially used by the target species are mapped.

DATA COLLECTION

Observers continuously watch the shelter entrances at all mapped sites over the period when animals are known to emerge and count animals that emerge. If, for example, a target mammal is nocturnal and arboreal, observers should be stationed at the base of each tree with a shelter in a position that best silhouettes the shelter entrance against the evening sky. Observers must ensure that they can see the emerging animals and that their presence does not hinder emergence. In some circumstances it may be possible to replace human observers with automated recording devices.

DATA ANALYSIS

At the end of the observation period the counts of all observers are tallied to obtain the population size of the target species. Density is obtained by dividing the population size by the size of the survey area.

SPECIAL CONSIDERATIONS

The major constraint on the use of this technique is the extensive knowledge an investigator must have about the species being counted. For example, to be confident of an unbiased count, an investigator must know the time of emergence of

all individuals in a population and must be able to find all dens and burrows. The method can be labor intensive. For example, one observer per 0.25 ha was needed to count possums emerging from tree hollows (A. P. Smith et al. 1989)

Total Counts from Mobile Platforms

As indicated earlier, mobile platforms can be used to facilitate surveys of relatively large study sites. The speed of most mobile platforms is such that it "freezes" animals in their habitat in relation to an investigator's rate of travel, so that double counting is avoided. However, the speed of the survey imposes certain limitations on the sizes of animals and types of habitats that can be surveyed realistically. Mobile platforms are suitable for use when making total counts of medium-size to large mammals living in relatively open habitats. They are generally inappropriate for total counts of species that are small or occupy dense habitats or mountainous terrain. Light aircraft may be used for making total counts of large mammals such as elephants (Buss and Savidge 1966; Dublin and Douglas-Hamilton 1987; Jachmann 1991), hippos (Tembo 1987), and buffalo (Sinclair 1973) and of medium-size mammals such as bighorn sheep (Hudson 1982). Slower-moving land vehicles have been used for total counts of medium-size species such as howler monkeys (Rumiz 1990), mule deer and white-tailed deer (Martinka 1968), elk (Martinka 1969), and many species of African ungulates (Foster and Kearney 1967; Field and Laws 1970).

RESEARCH DESIGN

The survey area is divided into irregularly shaped blocks on the basis of easily distinguishable natural boundaries or landmarks such as mountain ranges, roads, and rivers, or it is divided into equal-size quadrats (e.g., square subplots). In the latter case, the boundaries of the quadrats must be identified within the survey area. For aerial surveys the boundaries can be identified with the help of physical features that are visible from the air. In land surveys the subplot boundaries are usually marked with stakes, paint, or plastic tape. To obtain accurate counts, a large-scale map (1:50,000) of the subplots must be used to record the locations of animals seen.

DATA COLLECTION

Each subplot is searched thoroughly until all target animals are located and counted. A mobile platform usually traverses a block across its width, rather than along its length, because the travel path can be monitored more accurately across the shorter distance. The travel path must be mapped exactly in order to reveal areas that have been overlooked; those areas must be searched before the survey is terminated. When animals are counted, their locations should also be mapped to reduce the likelihood of double counting. Double counting is also minimized if all subplots are surveyed simultaneously and the time that animals were observed is recorded. Large aggregations counted from an aircraft should be photographed so that the accuracy of visual counts can be verified (see "Aerial Surveys," below). The key to success when using mobile platforms is to count the observed animals accurately and to map animal locations and areas searched exactly.

DATA ANALYSIS

The size of the survey area can be measured with a planimeter or by the "dot grid" method (see "Measuring the Size of a Sample Unit," Chapter 5). The population size of the target species within the survey area can be obtained by summing the total counts from each subarea. Crude density is obtained by dividing population size by the size of the survey area.

SPECIAL CONSIDERATIONS

To obtain accurate total counts, mobile platforms must be able to traverse all parts of a survey area. Aircraft, for example, can be flown systematically along predetermined flight paths over an entire survey area in relatively flat terrain (see "Aerial Surveys" below). It may be difficult, however, to access all parts of a survey area with land-based mobile platforms unless the area is relatively open (e.g., grassland) and suitable for cross-country travel or crossed with a network of trails. Investigators should open trails, if necessary, before commencing a survey, so that all areas will be accessible.

Silent detection for sample counts

Often an area is too large (or resources are too limited) to be surveyed in its entirety. In such instances, animals are surveyed in a selection of subplots (blocks, quadrats, or strips) within the larger study area. Usually, the plots surveyed constitute a relatively small proportion of the survey area. Sample count techniques based on silent detection of animals are numerous and also are the most frequently used in mammalian surveys.

Line Transects

In the line-transect technique, an observer travels along a straight line recording either (1) the perpendicular distance from each sighted animal to the transect line or (2) the radial distance from the observer to the animal and the sighting angle between the line of sight to the animal and the transect line at the moment of detection. The length of the transect line must be known. Four conditions must be met for accurate density estimation with the line transect technique. Most critically, all animals that are directly on the transect line must be seen. Second, distances and angles must be measured accurately. Third, animals must not move large distances before being detected, nor can they be counted twice. Fourth, the sightings of animals must represent independent events.

RESEARCH DESIGN

The line-transect technique was developed and is most appropriate for surveys conducted on foot or from land vehicles. Land-based transects are discussed first, and aerial and shipboard transects are presented later as variations. Foot transects can be conducted under a variety of budgetary, personnel, topographic, and vegetation conditions, and they can be used to survey small as well as medium-size and large mammals. However, sampling intensity (i.e, number of transects sampled) per person per hour of effort is limited with foot surveys. Vehicle surveys allow for greater sampling intensity, but they can be conducted only in areas that are relatively open and permit cross-country driving.

The most important condition for line-transect estimation, that all animals on the line be counted, is most likely to be violated when the observer and target animals are on different horizontal planes. This can occur in land (foot and vehicle) surveys when sampling over undulating terrain or when sampling arboreal or fossorial species. Violation of this condition can be avoided if the rate of travel along the line is controlled and if personnel are experienced in the technique and have a well-developed search image of the target species. Distances and angles can be measured accurately if personnel are experienced and if equipment and measuring techniques are appropriate. The assumption that animals have not moved before being detected is a special problem with highly mobile species such as deer, antelope, and large kangaroos. If the observer moves along the line as silently as possible, animals are less likely to flee from one area to another and be counted twice. In addition, silent detection reduces the chance of animals becoming alarmed and disturbing other animals as they flee.

The length of a transect line (also the timing and duration of the survey) depends on several factors that influence the observer's ability to detect and count animals accurately. Surveys should be conducted when target species are most active, and they must not last more than 2 to 3 hours. These conditions could translate into transects of 3 to 5 km for foot surveys and of 12 to 20 km for land vehicle surveys, if ideal field conditions prevail. Under less than ideal conditions, especially when a transect cannot be maintained as a straight line over long distances, the length of the transect should be substantially less.

Under most sampling designs, the placement of transects within an area should be independent of the distribution of the animals to be surveyed. Transect lines must not be so close that animals can move from one line to another line that is being surveyed or that will be surveyed soon afterward. The necessary separation between transects will depend on the displacement distances of the target species when mildly disturbed. In most instances a separation of about 2 km between transect lines will be adequate. A sufficient number of transect lines must be surveyed using standardized procedures to calculate precise density estimates.

DATA COLLECTION

Ideally a transect line should be straight and well defined. In land surveys some deviation from the line may be inevitable, but these deviations must be kept to a minimum. Large deviations from the line to avoid difficult physical features of the habitat can lead to biased sampling of microhabitats and inaccurate measurement of distances and angles. Marking of the line ensures minimal deviations when surveys are repeated along the same transect. Even though marking is time consuming, we recommend it, especially in long-term monitoring studies. In short-term surveys, one could use a global positioning system (GPS) or follow a compass bearing. However, observers conducting foot surveys in densely vegetated, rugged country may be distracted from the task of searching for animals by efforts to follow the compass line. This problem can be minimized if two people participate in each survey, one observing and the other navigating and recording data. They must move and communicate quietly, however, to avoid violating assumptions of the technique.

When an animal is sighted, the sighting distance and sighting angle must be measured accurately. A properly calibrated range finder should be used for distance measurements, and either a GPS, a good-quality compass, or an angle board should be used for measuring sighting angles. When a compass is used, the sighting angle is obtained from the difference between the bearings along the sighting line (the imaginary line from observer to animal) and the direction of travel along the transect line (Table 5). Specialized distance and angle measuring equipment has been developed for shipboard surveys of aquatic mammals in featureless environments (Joyce et al. 1985; Thompson and Hiby 1985). In aerial surveys a sighting frame, comprising parallel rods attached to the wing strut or fuselage, or marks on the perspex window or bubble can be used to group perpendicular distances accurately.

Table 5. Examples of Sighting-Angle Measurements in Relation to Transect Direction and Animal Bearing

Transect direction (°)	Animal bearing (°)	Sighting angle (°)
0	30	30
0	335	25
0	75	75
60	90	30
60	35	25
60	135	75
240	270	30
240	215	25
240	315	75

Ideally, distance and angle measurements are made specifically for each animal or cluster sighted (such data are said to be *ungrouped*). This may not be possible, however, in aerial surveys, in which the observer moves quickly past the animals; in shipboard surveys, in which animals are visible on the surface of the water for short periods and no obvious alternative features to which one can measure are available nearby; and in acoustic surveys, in which animals are detected only by sound (e.g., during foot surveys in fairly dense vegetation). In such situations animals detected can be classified according to distance intervals (e.g., 0–10 m, 11–20 m, 21–30 m, and so on), and density can be estimated from the frequency data.

For mammals that live in social groups or form clusters, the sighting distance and sighting angle should be measured to the geometric center of the cluster (or group), and individuals in each cluster should be counted. Obtaining such data can be difficult. Investigators should practice counting the individuals in clusters of the target species rapidly and accurately before conducting line-transect surveys. Such experience enables them to assess the time required to count clusters of different sizes accurately under various field conditions. To facilitate identifying the geometric center of a cluster, investigators should develop mental pictures of the center points of different geometric shapes.

Investigators should record weather conditions and start and end times of each survey and use these data to standardize the application of the technique between surveys. The time of each animal sighting and related vegetational features can be useful for data interpretation or for investigation of habitat preferences. The age and sex of animals sighted during surveys provide useful information on population structure. We recommend that investigators design a standard data sheet covering all information to be collected in a particular survey. This sheet can be duplicated, preferably on waterproof paper, for use on re-

Line-Transect Data Sheet

Date:_____ Survey Site:_____ Transect #:_____

Weather:_____ Start Time:_____ End Time:_____

Investigator(s):_____ Target Species:_____ Transect Direction:_____

Observation	Number of Individuals	Age/Sex	Sighting Distance	Sighting Angle	Comments

Figure 5. Suggested format for a standardized line-transect data sheet.

peated surveys. A sample data sheet is presented in Figure 5.

DATA ANALYSIS

Analysis methods for line-transect data follow the distance sampling estimation methods of Buckland et al. (1993). These methods and the associated data-analytic software are described in Chapter 10 (see "Line-Transect Sampling").

VARIATIONS

Line-transect surveys can be conducted from an aircraft or a ship, as well as from land. Transect lines can be followed fairly easily from an aircraft with reference to a map of the survey area and known features on the ground, or with satellite navigation aids. Such surveys allow coverage of large areas. Determining the geometric center of a cluster is not a problem in aerial surveys, and photographs can provide accurate counts of the number of individuals in each cluster. Aerial line-transect surveys do, however, have some limitations. They are useful only for medium-size to large mammals living in flat, relatively open areas. Aerial surveys are also relatively expensive. Furthermore, given the speed of the aircraft, it is difficult to measure sighting distances and angles accurately. Measurement inaccuracy can also be a problem in shipboard surveys, which are usually conducted in a featureless environment, unless specialized equipment for measuring angles and distances is used. The problems associated with surveying animals on a different horizontal plane affects all aerial line transects of terrestrial and marine mammals and shipboard surveys of mammals that are frequently submerged. The assumption of seeing all animals on the transect line is likely to be violated often. Given these disadvantages, aerial and shipboard line-transect surveys are relatively uncommon, although increasing use of GPSs may alleviate some of the problems.

PERSONNEL AND MATERIALS

Foot surveys on marked transects can be conducted by one person. On unmarked transects, however, foot surveys are best undertaken by two people, one acting as observer and the other as navigator cum data recorder. Surveys from mobile platforms require at least one person to maintain transect direction in addition to one or two observers. The number of teams used within a study site will depend on the size of the area, the availability of time and funds, and the intended sampling intensity. For ground surveys, materials required in the field include a vehicle (to traverse and/or to travel between transects), binoculars (to facilitate species identification), range finder, compass, maps, data sheets, and any safety or survival gear considered necessary if the area is remote or inaccessible. For aerial and shipboard surveys, the same equipment is used along with either a two- or a four-seat aircraft, a helicopter, a small boat, or a seagoing vessel for transport, and standard navigation gear.

SPECIAL CONSIDERATIONS

Investigators using line-transect surveys usually collect data on only one or a few target species. Counting several species during a single survey can be confusing and is likely to lead to inaccurate counts. Nevertheless, all species sighted along a transect can be recorded, although a novice should not attempt such counts. Even experienced personnel should avoid counting all species seen along a transect, unless the number of species and the abundance of each are relatively small.

Strip Transects

The strip-transect technique differs from the line-transect method in assuming that all animals within the sample strip are seen (i.e., observability is 1). A strip is surveyed from a line that runs through the middle of it. All animals within a predetermined distance w (half the strip width) on either side of the midline of the transect are counted. Animals beyond this distance are ignored. The data obtained are the numbers of animals seen within the strip.

Density estimation using the strip-transect technique is based on several assumptions. Animals must be accurately designated as in or out of the strip, and all animals within the strip must be counted. Sightings must be independent events, and animals must not move long distances before being detected and must not be counted twice. Determining whether an animal is in or out of a strip and counting all animals within a strip accurately are easier with aerial surveys than with land (foot or vehicle) surveys. Therefore, the strip-transect technique is used more often in aerial surveys than in land surveys. In this section we discuss use of the strip-transect technique during land surveys. Its application in aerial surveys is presented under "Specialized Techniques," below.

RESEARCH DESIGN

In general, strip transects are used to survey narrow, rectangular areas. The half-width of a strip is determined with respect to the visibility of the target species within the survey area and should ensure that all individuals within the strip will be located by the investigator. The strip half-width is generally fixed along the entire transect. When vegetation density varies along a transect, however, the half-width can be fixed at different distances (i.e., variable fixed width) based on assured visibility of the target species in each segment of the transect. Strip width can also be determined by measuring the visibility profile along the transect (fixed visibility profile) or by calculating the mean perpendicular distance to the animals sighted (variable visibility profile) after a survey. We recommend only the fixed-width methods of determining strip width, because the two visibility profile methods are open to many sources of bias (Norton-Griffiths 1978).

The length of a strip is determined by the same guidelines used in the line-transect technique. The midline and boundaries of the strip should be marked at regular intervals, particularly when the strip will be used repeatedly in long-term monitoring studies. The statistical aspects of research design for this technique are the same as those for the line-transect technique (see "Research Design" under "Line Transects," above).

DATA COLLECTION

An observer should record the number of animals seen within a strip from points along the midline. Unless a strip is marked, however, it is difficult to determine whether an animal seen close to a "boundary" is in the strip. Locations can be determined by measuring sighting distances and sighting angles to animals and calculating corresponding perpendicular distances from the midline to the animal after the survey. Animals sighted less than 0.5 strip-width from the midline are included in the strip counts. Alternatively, an observer can move along the midline until the sighting line is perpendicular to the midline and then measure the distance to the animal. If this distance is half the strip width, the animal is in the strip. In either case, a compass and a range finder must be used to determine the positions of animals seen near the strip boundary accurately. The length of each transect, along with the number of animals seen within the strip, is recorded.

DATA ANALYSIS

Density of the target species is obtained by dividing the number of animals seen by the area of the strip. Data from several strips can be used to calculate a mean density and improve the precision of the density estimate.

SPECIAL CONSIDERATIONS

Accurately determining whether an animal is in a strip is a major problem with strip transects conducted on foot or from land vehicles. To overcome this difficulty considerable time must be spent marking the strip boundaries before the

surveys or measuring distances and angles during the surveys.

Nonlinear Frequency-Density Plot

The proportion of plots or other sample units that contain at least one animal (or animal sign) is a count statistic that is sometimes called a frequency index. Under certain assumptions (see below) about the spatial distribution of animals over the entire sampled area, frequency indices can be used to estimate animal density.

Multiple plots within an area are selected randomly and then surveyed to determine presence or absence of a target species. Sometimes, it is possible to record the number of individuals present on each plot. The proportion of plots with animals (f) is estimated directly. When animals are randomly distributed, the number of plots containing 0, 1, 2, . . . , n animals fits a Poisson distribution with parameter x, the mean number of animals per plot (Seber 1982:55–58). The proportion of plots containing no animals is

$$1 - f = e^{-x}$$

where e is the base of natural logarithms. In the above equation f can be calculated from the data, and x can be estimated as $-\ln(1 - f)$, where ln denotes the natural logarithm.

The validity of the frequency-density estimator depends on the assumption that individuals are randomly distributed and do not group. Gerrard and Chiang (1970) discussed consequences of nonrandom distribution of animals for the frequency-density estimator and presented some corresponding modifications of this method using the number of plots containing 0, 1, 2, . . . , n animals.

RESEARCH DESIGN

Pilot investigations are carried out before sample plots are selected. The average distance at which the target species is visible within the survey area is determined by measuring the distances at which individuals or groups of the target species are seen. Movements of individuals or social groups of the target species are also monitored to provide estimates of home range size and group size, unless such data are available from previous studies of the species in the same or similar areas. Based on this information, sample strips (plots) are established, typically on a random basis. The width of each strip is twice the visibility distance, and its length is set so that the total area of the strip is half the area of an average home range (National Research Council 1981).

DATA COLLECTION

Each random strip is surveyed to determine the presence or absence of the target species. The number of individuals seen on the plot should also be noted if the Gerrard and Chiang (1970) analysis is to be used. An investigator walking along the midline of the strip should use a GPS or a range finder and compass to determine accurately whether animals are within the strip (see "Strip Transects," above).

DATA ANALYSIS

When all strips have been surveyed, the proportion of plots with animals and the mean number of animals per plot are determined. For species that live in social groups or clusters, mean density refers to density of groups or clusters and not to individual density. Individual density is obtained by multiplying group (or cluster) density by mean group size.

The frequency-density estimator was used to estimate the density of black howler monkeys in Panama (National Research Council 1981) and red howler monkeys in Venezuela (Eisenberg 1979). In both cases long-term studies provided data on mean home range and group sizes needed to calculate individual densities.

Road Counts

With the road-count technique an investigator counts all animals within a specified distance on either side of a road. The technique resembles a strip-transect count, except that roads are usually built along contours and are, therefore, nonrandom in their distribution within a habitat. In addition, animals tend to occur along contours, and the construction of roads usually opens areas along road edges that are the preferred habitats of certain species (Norton-Griffiths 1978; Sinnary and Hebrard 1991). As a result, road counts are likely to produce highly biased estimates of density and population size (Norton-Griffiths 1978; Southwell and Fletcher 1990).

In general, road counts should be avoided. If such counts are the only feasible approach, as, for example, when the habitat and terrain prevent cross-country driving, then the following precautions must be taken. The investigator must classify the vegetation along the road transect by type and then calculate density estimates for the target species in each type (see "Strip Transects," above). These estimates can then be multiplied by the total area of each vegetation type in the survey area and summed to produce the population estimate for the entire area.

Quadrat Sampling

Quadrat (typically square sample plots) sampling is simply a minor variant on strip sampling, and the two approaches require the same assumptions. All animals within the quadrat must be counted, they must not move over large distances (i.e., leave the quadrat) before being sighted, and they must not be counted twice.

RESEARCH DESIGN

Quadrats may be surveyed on foot, or from a land vehicle or an aircraft. The design for surveying them is the same as that described earlier for total counts from mobile platforms. Quadrat boundaries must be clearly marked, and all animals within a quadrat should be counted accurately. Quadrats selected for sampling typically represent only a portion of the entire survey area. In contrast, the entire area is surveyed when total counts are made from mobile platforms. Because animal counts within quadrats must be accurate, the use of land vehicles and aircraft is limited to surveys in relatively open habitats or to seasons of good visibility. Only foot travel is appropriate in rather dense habitats and relatively small quadrats.

DATA COLLECTION

Quadrats are searched thoroughly in the same manner used to obtain total counts from mobile platforms. Foot surveys on relatively small quadrats can benefit from maps (made by the investigator if necessary) that include trails, topographic features, vegetation characteristics, animal structures, and other landmarks. These features help to ensure accurate plotting of animal locations and of areas searched by the investigator. At the end of the search the number of animals seen within the quadrat is tallied to obtain the sample count. The entire procedure is repeated in other quadrats selected for sampling.

DATA ANALYSIS

Density is estimated by dividing the number of animals seen within a quadrat by its area. Density estimates from several quadrats are then used to calculate a mean density. The precision of the mean density estimate can be maximized by increasing the number of quadrats surveyed. The mean density is multiplied by the size of the survey area to obtain the population size of the target species.

VARIATIONS

A variation of the technique just described is used in relatively dense habitats. It incorporates some of the principles of strip sampling and requires several people. The first step is to calculate the average visibility distance (W) of the target species within the survey area. A quadrat selected on a random or systematic basis is then divided into several strips of width $2W$. A team member is positioned at the midpoint of each strip along one side of the quadrat. The observer moves along the midline of her or his strip following a compass bearing and maps the location of animals seen on either side within the strip. Observers should record the time of detection and direction of movement of each animal. At the end of the survey, data from all team members are pooled so that the number of animals seen within the quadrat can be assessed and species density can be calculated. With knowledge of the normal speed of travel of the target species, one can assess the likelihood that an animal has moved from one strip to the next and, thus, minimize double counting. If the target species is social, it is also necessary to know the size of groups and the dispersion of individuals within groups, so that the number of groups seen is not overestimated.

This technique differs from the similar drive technique (see "Drives for Sample Counts," above) in at least three ways. First, with drives investigators are spaced twice the minimum flight distance apart or at maximum visibility distance. Here they are placed twice the average visibility distance apart. Second, this technique depends on silent detection, and animals are less likely to flee unnoticed than in a drive. Finally, this technique has been used to survey semi-arboreal and arboreal primates (Southwick et al. 1961; National Research Council 1981), whereas the drive technique is unsuitable for tree-dwelling species.

SPECIAL CONSIDERATIONS

One of the problems with quadrat sampling arises from the tendency for animals to be clumped rather than randomly distributed. As a result, all the animals in an area can be present in one quadrat, while adjacent quadrats have none. In contrast, animals are often more equitably distributed among adjacent strips. Therefore, animal counts from a random sample of quadrats tend to be more variable than counts from the same number of random strips of the same size. In practice, then, one needs to sample more quadrats than strips of equal size to obtain the same level of precision in a density estimate.

Another problem with quadrat sampling is that an investigator does not have a predetermined line (as in strip sampling) from which to assess whether animals are within the quadrat. It is essential, therefore, that quadrat boundaries be clearly marked, to ensure that only animals within sample plots are counted. Overall, quadrat sampling involves greater effort than strip sampling, all else being equal.

Special techniques

Aerial Surveys

In aerial surveys a fixed-wing aircraft or helicopter is used to obtain total or sample counts of mammals that are fairly easily detectable. Such surveys permit rapid counting of animals and are most useful for surveying very large areas. They have been used to survey marine mammals such as cetaceans, sirenians, and pinnipeds (Irvine and Campbell 1978; Leatherwood et al. 1978; Smith 1981; Odell 1982; Drummer and Mcdonald 1987; Drummer et al. 1990) and a large number of terrestrial species living in open or patchy habitats (Bayless 1969; Norton-Griffiths 1975; Hudson 1982; Dublin and Douglas-Hamilton 1987; Ottichilo et al. 1987; Tembo 1987; Johnson et al. 1991). Aerial surveys can also be used with groups whose habits, size, and morphology allow location, identification, and enumeration from the air (e.g., carnivores).

RESEARCH DESIGN

The first step in launching an aerial survey is to define the area to be surveyed. This area can be defined by the boundaries of a study site such as a protected area, or by the distribution of a target species selected for the aerial survey. In either case, the total area covered by an aerial survey should not be more than 10,000 km^2 (Norton-Griffiths 1978). Larger areas should be divided into smaller ones, each of which is surveyed separately.

The next step is to decide between total counts or sample counts. Total counts are appropriate when a target species is terrestrial and fairly widely dispersed within a relatively small (a few hundred square kilometers) survey area. Under such conditions, total counts can be obtained by dividing the land area into blocks or quadrats and counting all animals within each plot (see "Total Counts from Mobile Platforms," above). In most other situations, including surveys of featureless environments (e.g., the sea) and of very large land areas, aerial sample counts are usually the only alternative.

Aerial sample counts can be obtained using a line-transect, strip-transect, block, or quadrat technique. For surveys at sea, block and quadrat sampling are inappropriate, unless a GPS is used, because there are few reference points with which to establish plot boundaries. The line-transect technique has certain shortcomings when used with aerial surveys; in particular it is difficult to meet the assumption of all animals on the transect line being detected and counted. The best option for sample counts at sea is the strip-transect technique. For similar counts over land, an investigator can choose among strip, block, or quadrat sampling.

The quadrat sampling technique has been used in aerial surveys to estimate densities of moose (Evans et al. 1966), caribou (Siniff and Skoog 1964), and mule deer (Kufeld et al. 1980). However, the quadrat technique is not often used for aerial sample counts on land. The block sampling technique is appropriate when the target species occurs in large aggregations, when the vegetation is dense, or when the topography of

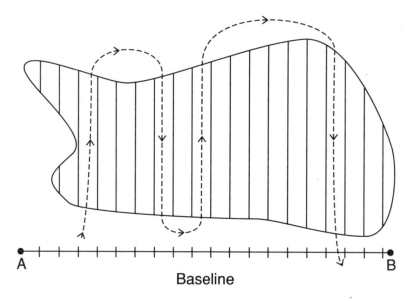

Figure 6. A survey area divided into strips of predetermined width, using a baseline. The zigzag flight path of an aircraft from one end of the survey area to the other, over randomly selected strips, is indicated by the dashed line.

the survey area is broken and mountainous. Under all other conditions strip sampling is preferred (Firchow et al. 1990), because there is no need to search for boundaries and because navigation is easier (see below). Strip counts also tend to be less variable than block or quadrat counts.

Variability of animal counts in heterogeneous survey areas can be reduced for all three techniques with stratification. Sampling effort and sampling intensity within the resulting homogeneous areas either should be proportional to their representation in the survey area (proportional allocation) or, in the case of unequal stratum variances or unequal stratum sampling costs, should follow optimal allocation principles (see "Stratification," Chapter 3, and Cochran 1977: 98). The number of plots (strips, quadrats, or blocks) sampled should be selected to ensure reasonable precision of the density estimates (see "Determining Sample Size," Chapter 3). Sizes of quadrats and blocks should allow for thorough searching for animals in 3 hours or less to avoid observer fatigue. A 3-hour survey limit is appropriate for strip sampling as well.

Before strips are selected, strip width must be fixed based on the size of the target species, the vegetation density, and also the altitude and speed of the aircraft. The entire survey area is then divided into parallel strips of equal width, which run from one end of the survey area to the other (Fig. 6). A survey area can be divided across its width rather than along its length to provide shorter transects, which facilitate navigation and reduce variability in animal counts. Variability can also be reduced by placing the transects across gradients of animal density rather than along them. Factors that influence an aircraft's flight path, such as crosswinds and direction of the sun, must also be considered before determining the direction of transect placement. Strips often are of unequal length in survey areas with irregular boundaries (Fig. 6). Lengths can be equalized to facilitate data analysis, but this is not essential. A random numbers table is used to select strips for the survey.

Unlike blocks or quadrats, strips traversed in aerial surveys do not have identifiable physical boundaries, although a GPS can be used to define them. Boundaries can be indirectly defined

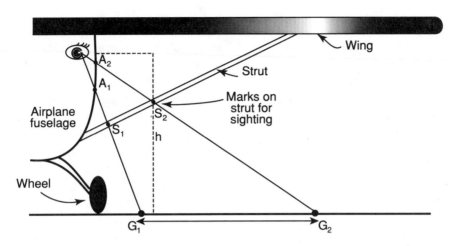

Figure 7. Placement of markers on the window (A_1 and A_2) and wing struts (S_1 and S_2) of an aircraft to define a strip of pre-determined width for aerial survey. See text for explanation.

on the ground by marking the observation window and wing struts of the aircraft before conducting the survey. The aircraft is jacked up into a flying position while it is on the ground. A person sits in the observer's seat and assumes a comfortable position looking out the window. The height (h) of the observer's eye above the ground is measured (Fig. 7). A streamer (S_1) is attached to the wing strut and is adjusted, so that the observer's line of sight through the streamer is clear of the aircraft's wheel but still as close to the body of the aircraft as possible. Markers are placed on the window (A_1) and on the ground (G_1) along this line of sight and aligned with the streamer on the wing strut. A second marker is then placed on the ground at a point G_2, at a distance (w) from G_1 and away from the body of the aircraft. The distance (w) is obtained from the following formula:

$$w = \frac{W}{H} \cdot h$$

where W is the predetermined strip width and H is the predetermined flying height. The value of W can be adjusted to half-strip width if the aircraft can accommodate two observers to look for animals on either side of the aircraft. When w is

determined, G_2 is marked on the ground and a second streamer (S_2) is placed on the wing strut in the observer's line of sight to G_2, when his or her eye is still aligned from A_1 to G_1 through the first streamer (S_1). While in this position, the observer makes another mark (A_2) on the window. Thus, the observer's eye is aligned from A_1 to G_1 through S_1 and from A_2 to G_2 through S_2. The marks on the aircraft window can be made of colored adhesive tape and should be relatively permanent. Streamers are usually made of heavy rope and are kept horizontal during flight by plastic funnels attached to their ends. Each streamer is of a distinct size or color so that it can be distinguished in photographs taken from the air.

The survey crew consists of a pilot and one or more observers. The pilot's primary responsibility is to navigate the aircraft while the observers count animals rapidly and accurately, photograph them when necessary, and record information using data sheets or tape recorders. The accuracy of counting is influenced by aircraft altitude and speed. Altitudes between 100 m and 200 m and speeds of 100 km/hr to 150 km/hr permit observers to count land animals accurately and also conform to standards of aircraft safety. The altitude and speed used in marine mammal surveys tend to be somewhat higher, to ensure the safety of the aircraft.

Large-scale (1:50,000) topographic maps indicating the locations of sample plots and the aircraft's flight path are used for navigation during aerial surveys. Flight paths are arranged so that the time required to fly between sample plots ("dead time") is minimal. They are also arranged so that the sun is not directly ahead and so that crosswinds do not cause the aircraft to drift away from a transect or force its wings to tip up or down from the horizontal (banking). Banking changes the width of the strip seen between the streamers and causes the observer to overestimate or underestimate the number of animals within a strip.

DATA COLLECTION

Aerial surveys should be conducted at times of day when the target species is most active, so that accurate counts of all animals can be obtained. The pilot must maintain the aircraft at the predetermined altitude and prevent the aircraft from banking. To measure altitude accurately a radar altimeter must be used, and an observer should periodically record the altitude to provide for corrections of strip width during analysis. The pilot must also maintain a constant direction and speed and inform the observers when each strip begins and ends. This indicates to the observers when to start and stop counting animals.

Observers count only the animals within a strip. They must remember to align the window markers with the wing strut markers in order to define the strip before counting animals. All animals occurring within the strip must be counted even when the aircraft drifts or banks; corrections for such navigational errors can be made during data analysis. If the streamers pass through a cluster of individuals, only those within the streamers are counted. When the target species occurs in clusters of more than 5 to 10 individuals, counting accuracy can be greatly enhanced with aerial photography. A 35 mm camera with a zoom lens and motor drive for automatic film advance is ideal for aerial photography. The focus of the camera should be fixed at the infinity position. Each film cassette and leader must be numbered, and the number on the leader should be scratched into the photographic emulsion so that the information is not lost during film processing. When large clusters are encountered, several overlapping photographs must be taken so that all individuals are included in the photographic series. The photographs should include the upper and lower strip markers so that animals occurring within the strip can be distinguished and counted accurately. Each frame must be unambiguously identified on the data sheet or on the tape recorder with reference to film cassette number, exposure number, species photographed, and the visual count of animals in the frame. Because cameras, film, and film processing are sometimes faulty, photographic records must never be substituted for visual counts.

The start and end times of surveys along different strips and the data collected from each strip must be recorded separately. Any temporary interruption of animal counts along a strip (e.g., because of bad weather) must also be noted. If a survey involves multispecies counts, each species and its corresponding counts must be clearly identified.

Navigational errors during strip sampling can be dealt with in two ways. Either the strip can be surveyed again, or the pilot can indicate the actual path of the aircraft on the flight map before surveying the next strip. This information can be used later to determine the actual path taken by the aircraft and the length of the transect traversed.

Investigators should make reconnaissance flights to identify the boundaries of the sample plots before they are sampled. The boundaries of blocks are easier to locate than those of quadrats, because they are clearly defined by natural landmarks. All crew members must agree on the sample plot boundaries before surveys commence. Several flight patterns can be used to cover all parts of the sample plot during the actual survey. For instance, an aircraft can fly along the boundaries of a plot and gradually spiral toward its center in smaller and smaller concentric circles. This flight pattern (spiraling) involves constant banking; observers can become airsick or be repeatedly blinded by the glare of the sun if it is low on the horizon. An aircraft can also take a zigzag route, which is the preferred flight pattern for surveying blocks. When blocks include mountainous areas, an aircraft can follow the contours of the terrain (contouring), beginning with contours at high elevations and gradually descending to the valley bottom. Helicopters are more suitable than fixed-wing aircraft for this type of work. Regardless of flight pattern, there can be no gaps in the areas searched.

The areas searched and the locations of animals seen must be accurately recorded on a map. Each location is assigned a serial number that is noted on a data sheet with the species, number of animals seen, and other information. When animals are located, the aircraft can deviate from the general flight path to allow for photographs and for closer observations. Overlapping photographs may be necessary to record all individuals in a large cluster. Such photographs must be taken while the aircraft flies in a straight line (rather than around the cluster) to be useful for counting the group (Norton-Griffiths 1978).

DATA ANALYSIS

The number of animals observed within a quadrat or block is divided by the size of the sample plot to estimate the density of the target species. The density estimates of several sample plots are used to obtain a mean density estimate with the maximum level of precision (see "Determining Sample Size," Chapter 3). The mean density estimate is multiplied by the size of the survey area to obtain the population size of the target species.

In strip sampling the actual width and length of the strips that were sampled should be calculated first. The actual strip width (w) is obtained as follows:

$$w = \frac{W}{H} \cdot h$$

where h is the average of the radar altimeter readings and W and H are predetermined strip width and flying height, respectively. The actual strip lengths are obtained from the flight map by measuring the true flight path of the aircraft. At this stage any navigational errors or temporary interruptions in observation are taken into account by modifying the strip length accordingly. The length and width of each transect are then multiplied to obtain area, and the areas of all strips are summed to obtain the total area surveyed.

The total number of animals seen within each strip is obtained from the visual counts and photographic records. Normally, visual counts are recorded for all animal sightings. Photographs are generally taken only of clusters exceeding a certain size (e.g., ≥ 5 animals). If all photographs provide clear prints, the number of animals seen within a strip is easily obtained by summing the visual counts of fewer than five individuals and the counts of individuals in all clusters in the photographs. It is likely, however, that some clusters will not have been photographed or that some prints will be of a quality too poor to use. In such instances, correction factors based on visual counts of a target species accompanied by clear photographic records are developed to enhance the accuracy of visual counts of groups. Suppose that four visual counts of 6, 10, 11, and 8 recorded within a strip are accompanied by photographs showing 7, 11, 14, and 10 animals, respectively. The correction factor (or counting bias) is obtained from the ratio of the sum of the visual counts (35) to the sum of the counts from the corresponding photographs (42). Each visual count of five or more individuals without an accompanying photographic record is divided by this ratio (0.83). Thus, a visual count of 10 without a photographic record would be adjusted to 12. The adjusted count(s), photographic counts, and visual counts of fewer than five individuals are summed to obtain the total number of animals seen within the strip. This process is repeated for all strips, and data from all strips are summed to estimate the total number of animals located within the area surveyed. This number is divided by the total area surveyed to provide a density estimate for the target species. Population size is obtained by multiplying the density estimate by the size of the survey area, which can be measured from a map (see "Measuring the Size of a Sample Unit," Chapter 5). More detailed analyses of aerial survey data are possible and are explained in Chapter 10 (see "Incomplete Counts"; see also Caughley 1977 and Norton-Griffiths 1978).

PERSONNEL AND MATERIALS

When selecting an aircraft, one must consider factors such as cost, range, visibility, safety, avionics, and space for the survey crew. High-wing aircraft with good lateral visibility are suitable for strip, block, and quadrat sampling. For strip sampling the aircraft should also have wing struts for attaching streamers and a radar altimeter for measuring altitude accurately. Aircraft, such as bombers and helicopters, with maximal visibility forward or down along the flight path are essential for line-transect sampling.

The survey crew should include a pilot and at least one observer. The pilot must be able to navigate the aircraft at relatively low altitudes, using only a map of the survey area and small ground features. Because crosswinds are frequent at low altitudes, the pilot must be experienced in dealing with aircraft drift and banking. The pilot is also responsible for checking the fuel supply and refueling.

Four-seat aircraft can carry the pilot, two observers, and a fourth person to assist the pilot in reading the flight map and the observers in changing and storing film or recording data. With one observer counting animals on each side of the aircraft, the area surveyed per unit time is doubled. Observers must either have experience in aerial surveying or be given intensive training prior to a survey, especially in counting animals rapidly and accurately. Such training can be accomplished by projecting slides of the target species taken from the air and having the trainees count the animals within a few seconds. Their counts are immediately compared with the actual numbers on the slide to provide instant feedback on counting accuracy. As counting accuracy improves, the counting rate is gradually increased by reducing the length of time during which slides are projected.

Fast color film (ASA 400) is best for aerial photography, but black and white film can be used for dark-colored species such as elephant, wildebeest, and buffalo. If tape recorders are used to record data, their batteries, microphones, and tapes must be checked regularly and replaced when necessary. Unexposed film, audiotapes, and batteries must be carried on the survey aircraft.

SPECIAL CONSIDERATIONS

All survey aircraft should carry safety equipment, including first-aid and survival kits, flares, signaling mirrors, and a radio for air-ground communication. Before takeoff, a detailed flight plan with estimated times of arrival at different locations within the survey area and back at base must be filed with the nearest air traffic control tower. The position of the aircraft during the survey should be reported regularly to the control tower. A second aircraft should always be available on the ground for search and rescue operations.

Marine Mammal Surveys

Surveying marine mammals presents a special problem, because animals can remain submerged out of sight of an observer for extended periods. This problem is particularly acute when surveying marine mammals from relatively fast-moving mobile platforms. Some investigators have taken advantage of particular behaviors to develop species-specific survey techniques. For instance, some whales, such as the gray and the bowhead, swim close to shore when migrating between their calving grounds and feeding areas. These species have been counted from shore or ice-based sites placed as near as possible to the path of migration; data taken from two observation sites provided an estimate of abundance (Braham 1982). Similarly, sea lions periodically come on land in large numbers to give birth, to mate, or to molt. At these times they can easily be photographed from an aircraft for a total count (Braham et al. 1980). Such counts are affected to different degrees by variation in animal detectability and should be considered minimal estimates of the population. Density cannot be estimated because the area from which the population is drawn is unknown.

Call Playbacks

Species-specific calls that cause animals to react in predictable ways may be broadcast in an area as a means of increasing the visibility of the animals for counting. In surveys of the burrow-dwelling thirteen-lined ground squirrel, adult alarm calls were played from various locations at the periphery of a population (Lishak 1977, 1982). Individuals responded to these calls by returning to or emerging from their home burrows, assuming an erect posture, and emitting vocalizations similar to the calls played. Lactating females were enticed from their burrows during the period of offspring production with distress calls of young ground squirrels. Animals responding to playbacks were rapidly counted. Counting was facilitated by open conditions in the short-grass habitat of this squirrel. Playbacks were repeated, and the highest response tally

was taken as the size of the population. The best results were obtained during early morning hours when individuals were actively foraging outside their burrows. Because the average litter size of the thirteen-lined ground squirrel is known (McCarley 1966), it was also possible to predict what the total population size would be at the time of emergence of the young.

In another study, recordings of spotted hyena vocalizations were used to attract conspecifics from as far away as 3 km (Kruuk 1972). Sounds of hyenas squabbling over food, struggling with another clan, or contesting lions over a kill elicited the greatest responses. Prior to broadcast, several hyenas were immobilized and marked with distinctive ear notches. The proportion of marked to unmarked animals responding to playbacks was used to estimate population size. Density could not be determined because the area occupied by the population was unknown.

Night Surveys

An inventory is not complete until the presence and abundance of nocturnal animals are determined. Most techniques for surveying nocturnal mammals depend on capture or animal sign, but line-transect counts can also be used. In general, line-transect counts are appropriate for surveys of terrestrial species inhabiting fairly open habitats (Chapman and Wilner 1986; Drew et al. 1988) and of medium-size to large nocturnal, arboreal mammals.

A nocturnal line-transect count is conducted by traversing a well-defined straight line on foot or by vehicle and counting animals seen along the line with the aid of a spotlight. The technique is based on the same theory and assumptions discussed earlier, but problems of animal detectability are more acute at night. Therefore, changes in research design and data collection methods are necessary to ensure reliable estimates of abundance and density. One assumption of nocturnal counts is that the spotlight reveals all animals along the transect line. Thus, the spotlight must have a fairly powerful beam and should be used systematically to scan all areas ahead and above and on either side of the transect line. The rate of scanning should be slow enough to allow the observer(s) to detect all animals along the transect. Spotlighting works best with species that have a bright eyeshine and do not look away from the beam of light. Knowledge of the size, color, and brightness of the eyeshine of the target species is a prerequisite to successful spotlighting, and the investigator must be able to distinguish between the eyeshines of different species in the habitat. Body form, vocalizations, or movement may also serve to identify animals in a spotlight beam.

Once an animal is detected, the investigator measures the sighting distance and sighting angle. Sighting angles can be measured easily using a compass with a luminous dial, but accurate distance measurements can be a problem because range finders are of little use in the dark. Sometimes sighting distances may be paced by the observer or measured with a tape without disturbing the animal. In most situations, however (as in dense habitats), the best approach is to assign sighting records to predetermined distance interval classes. Such data can then be used as frequency scores to estimate the density of the target species.

CONTRIBUTORS: THOMAS H. FRITTS, HELENE MARSH, DALE G. MIQUELLE, AND THOMAS J. O'SHEA

Chapter 7

Observational Techniques for Bats

Thomas H. Kunz, Donald W. Thomas, Gregory C. Richards,
Christopher R. Tidemann, Elizabeth D. Pierson, and Paul A. Racey

Introduction

The principal objective of surveying and censusing bats using observational methods is to determine community composition, species richness, and abundance. The methods selected and the kinds of information obtained will vary with the size and mobility of the animals being investigated, observer access to the animals, the available instruments for extending the sensory ability of the observer, and the number of animals present. Before beginning a survey, an investigator should have some knowledge of roosting habits, nightly foraging activities, seasonal movements, and effects of environmental factors—such as local topography, temperature, humidity, light intensity, and habitat structure—on abundance patterns. Knowledge of historical and temporal-spatial patterns associated with a particular species or population may also be important. Observers should be thoroughly familiar with the operation of binoculars, night-vision scopes, video cameras, ultrasonic detectors, or other instruments used to extend their sensory capabilities and should understand their detection limits. Because roost sites of some bat species are relatively easy to locate and often house large aggregations of individuals, they offer considerable potential for assessing population characteristics. Consequently, most efforts to estimate relative or absolute numbers of bats have focused on captures or observations made at or near roosts (e.g., Barclay and Bell 1988; Thomas and LaVal 1988).

Roosting bats

Generally, four observational methods can be used to survey bats in roost situations: direct roost counts, disturbance counts, nightly dispersal counts, and nightly emergence counts. Access to some roost sites may require specialized equipment, knowledge, and skills. Some roost sites may pose safety risks to the observers. The location of some roosts may severely restrict opportunities for observing bats, because of factors such as foliage density or height above the ground or because the roost site may be physically inaccessible (e.g., caves and mines with openings too small for human ingress and egress). To conduct surveys and censuses of bats that occupy caves and mines may require climbing experience, including rope work. A novice should not enter a complex cave or mine unless accompanied by an experienced caver. In situations in which direct access to the interior of a cave or building roost is limited, nightly emergence, dispersal, and disturbance counts offer the best alternatives for estimating colony size.

Estimates of absolute numbers present at roost sites are seldom feasible except for some relatively small, gregarious species. For other species, indices of abundance may be the only type of data that can be obtained. For some gregarious microchiropterans whose roosts are not accessible to observers, nightly emergence counts provide the most reliable estimates of colony size. For some large, gregarious pteropodids, estimates can be obtained by making roost counts, dispersal counts, or disturbance counts.

Direct Roost Counts

RESEARCH DESIGN

Direct roost counts are carried out by groups of observers stationed at designated positions relative to a colony or survey area. Observers should be positioned to ensure unobstructed views of the roosting bats (Wiles 1987), which are then enumerated directly, with or without the aid of binoculars or spotting scopes. Observers are assigned areas and count only those bats in the designated area. The reliability of the estimate depends on the experience of the observers and the visibility of the bats. Photography, videography, and radar can be used to augment direct observations of some species.

The most consistent and reliable counts of large, tree-roosting megachiropterans are made soon after sunrise and during the late afternoon, when wind velocity is low and cloud cover is minimal (Wiles et al. 1989). Under such conditions it may be possible to count the bats in a colony from distances up to 1,000 m from the roost (Wiles et al. 1989). The reliability of a roost count generally decreases with increasing numbers of bats, the inaccessibility of the roost to observers, and the extent to which direct observations are limited by dense foliage (Wiles 1987). In addition, only some individuals of a population may aggregate and be observed.

DATA ANALYSIS

Total numbers of roosting bats are obtained by summing the individual counts made by each observer. The density of bats is determined by dividing the total count by the area surveyed. Population estimates of species that form large aggregations can sometimes be made if all known sites are censused on the same day (Martin 1987). Richards (1990) used a network of interested individuals in Australia, who reported the presence or absence of *Pteropus* at traditional roosts each month by telephone. He was able to survey a study area that exceeded 250,000 km^2 rapidly and frequently. Although this approach may not yield reliable colony counts, it can target areas that warrant further investigation by qualified individuals. When trained observers are available, the total number of bats present can be estimated by counting the number of bats roosting in "average" trees and then extrapolating this number to the number of trees occupied by bats in a given region. Alternatively, the number of bats present can be estimated by counting bats in "patches" or groups and then multiplying the average of these values by the number of patches in each roost tree. These two approaches were used to estimate populations of *Eidolon helvum*. Although estimates differed by as much as 33%, the trends were in general agreement (Mutere 1980; Baranga and Kiregyera 1982).

Bats can be completely censused only in well-defined geographic areas, such as small oceanic islands, or at roost sites. Boats have been used to locate and count the number of trees occupied by pteropodids roosting on small islands (Tidemann 1985; Wiles et al. 1989), although direct visual counts of individual bats may not be possible. Counts of bats occupying dense foliage may underestimate actual numbers by as much as 20% (Wiles 1987).

Populations of some gregarious species may be difficult to assess if their roost sites extend over large areas, and species that roost singly or in small groups often cannot be surveyed or censused effectively. It is not uncommon to find large colonies of pteropodids spread over areas more than 1 km long, especially along water courses. Large colonies of *Pteropus* spp. in eastern Australia can exceed 100,000 individuals and cover areas up to 5 km long and 50 m wide (Nelson 1965b). Counts made at single locations are likely to underestimate total colony size unless observers can account for colony members that roost at alternate sites (Martin 1987; Wiles et al. 1989, 1991). For example, Tidemann (1985, 1987) found that only about half the population of *Pteropus melanotus* on Christmas Island actually formed colonies; other individuals were widely dispersed.

Disturbance Counts at Roosts

For disturbance counts at roost sites, bats are stimulated to take flight during the day and are counted as they become airborne (Racey 1979). Distur-

bance counts can be used when roosting megachiropterans are difficult to observe because of dense foliage or other visual obstructions.

At least two observers are required for disturbance counts. One person enters the roost area and makes loud noises, and the other individual counts bats. Often several "beaters" operating in unison are needed to encourage bats to take flight, although such efforts are only successful over a limited area. Sharp metallic sounds, such as those made by hitting metal pipes or tent pegs together, will cause some species of pteropodids to take flight (C. R. Tidemann, unpubl. data). The vocalizations of animals in very large roosting aggregations may mask the sounds made by investigators, however, making it difficult to use this method effectively. Shouting usually does not induce pteropodids to take flight (Racey 1979), although the noise from a discharged shotgun may do so (Ratcliffe 1931).

Once bats become airborne, they may be counted directly or photographed with a camera fitted with a wide-angle lens. Colony size can be estimated by projecting 35 mm transparencies on a screen and counting the number of bats, with corrections for overlapping topographic features. The success of a disturbance count depends on the sensitivity of bats to disturbance, the skill of the individuals causing the disturbance, and the position of the photographer relative to the flying bats. Obviously, this method will underestimate the number of bats present if some fail to fly (e.g., flightless young) or if others become disturbed prematurely. Videography has considerable potential for estimating colony size, if all bats take flight following disturbance, because the images of bats can be downloaded into a computer and the particles (= bats) counted frame by frame in the laboratory (G. C. Richards, unpubl. data).

Nightly Dispersal Counts

This technique involves counting bats as they disperse nightly from diurnal roost sites. Such counts are most effective if flying bats are silhouetted against the sky or open ocean and if there are several observers occupying different vantage points (Parry-Jones and Augee 1992). Observers should be at their stations at least one hour before nightfall and should count only those bats that depart within a preassigned arc surrounding the roost. Although decreasing light levels at the time of nightly dispersal may reduce visibility, use of light-gathering binoculars can facilitate counting (Racey 1979; Nicoll and Racey 1981).

Radar can also be used to census night-dispersing pteropodids (e.g., Tidemann 1985; Parry-Jones and Martin 1987; G. C. Richards, unpubl. data). An important advantage of radar over direct visual observation is that it can be used equally well both day and night. In addition, individual animals can be detected at distances up to 4 km (groups at much greater distances), a 360° area can be scanned at one time, and the data can be stored on video for later analysis (C. R. Tidemann, unpubl. data). In areas where the view of dispersing bats is uninterrupted, radar can be used to determine dispersal directions. The relatively low portability of the equipment and the relatively high acquisition and operational costs, however, are important disadvantages of radar.

Dispersal counts, like direct roost counts and disturbance counts, provide minimum estimates, because flightless young may remain in the roost when their mothers depart or cling to them when they fly off (Wiles et al. 1989). Bats should be counted only once during a dispersal flight, but many individuals circle the roost area before dispersing.

Roost Counts at Maternity Colonies

Many species of Microchiroptera form large roosting aggregations in caves, mines, buildings, and similar structures, sometimes numbering several million individuals. Because most species are highly susceptible to disturbance during the maternity period, efforts to estimate numbers of bats should be designed to minimize disturbance. This can best be accomplished by making infrequent visits to a colony after adults have departed and before they return from feeding.

Even in the most carefully designed study, it may not be possible to account for all bats present. For example, as the size and complexity of the roost site increase, the probability that all bats will be observed and counted directly decreases. Some species may roost in multiple layers, especially where the roost substrate is irregular (Tuttle 1975). Before undertaking a roost census, investigators should assemble background information on the roosting habits of the target species. Such information should include likely reproductive condition and physiological state of the roosting individuals, times of nightly emergence and return, effects of temperature and precipitation on roosting behavior, and nightly foraging behavior. This knowledge can be used to design a sampling protocol that will reduce disturbance to the bats and minimize sample bias.

For many species, an investigator will need to use a combination of observational and capture methods to obtain an accurate estimate of colony size and composition. An observer can sometimes estimate the number of lactating females present at maternity roosts by counting nonvolant pups in the roost following the nightly departure of adults. During the maternity period, it may be possible to use pup counts and the ratio of lactating bats to pregnant bats or to postlactating bats captured in traps or nets at the cave entrance to estimate the number of females present at a given maternity roost (Tuttle 1975, 1979). Using this method, the number of lactating females should equal the number of young present divided by litter size.

Some bat species leave stains from skin oils and urine on the ceilings and walls of caves and mines where they have roosted for many years, and sometimes one can derive population estimates using this information. Tuttle (1979) estimated the density of *Myotis grisescens* per unit of roost surface and multiplied this value times the ceiling area that was covered by stains to estimate colony size. Colony size also may be estimated by measuring the area on the floor of a cave covered by fresh guano deposits, and multiplying this area by the cluster density of roosting bats (Tuttle 1979; Thomas and LaVal 1988). Estimating cluster density and spatial coverage of bats in maternity roosts can be highly disruptive, however, because of the sensitivity of bats to disturbance during this period (Tuttle 1976a, 1979; Kingsley et al. 1991).

Nightly Emergence Counts

Nightly emergence counts made as bats depart from traditional day roosts in caves, mines, tree cavities, and buildings can be one of the most effective ways to estimate the number of bats that roost in inaccessible places or in mines and caves that are unsafe to enter, or that are disturbed by entry of observers into the roosting site. Ideally, such counts should be made repeatedly over several weeks to establish intracolony variation in the number of bats present. Nightly emergence counts are nondisruptive, and they can provide important baseline data for comparing results from different sampling methods (e.g., number of bats captured in the roost vs. emergence counts).

Nightly emergence counts are most effective when departing bats are silhouetted against a clear sky (Kunz and Anthony in press). As with nightly dispersal counts, decreasing light levels at the time of emergence can reduce bat visibility and lead to biased estimates. Use of light-gathering binoculars or night-vision goggles may improve visibility of bats in some situations.

The number of observers required to conduct nightly emergence counts will depend on the size and configuration of the roost site and the number of openings used by the departing bats. Observers should be assigned specific exits or fields of view and should be present at their stations before the onset of nightly emergence to ensure that the earliest departing bats are counted. It is important to avoid counting bats more than once, especially if bats reenter a roost during the emergence period. The number of reentering bats should be subtracted from the total number of departing ones to derive the actual count. As with direct roost counts, nightly emergence counts may underestimate the number of bats present, especially if flightless young are left in the roost when their mothers depart.

Still photography, cinematography, videography, and electronic or mechanical counting devices can be used to count emerging bats (e.g., Daan 1973; Altenbach et al. 1979; Mitchell-Jones 1987; Rodriguez-Duran and Lewis 1987; Thomas and LaVal 1988; Speakman et al. 1992). Such devices allow bats to be censused in the absence of an observer and can provide multiple records over extended periods. Before using such devices, it is important to validate the signals (mechanical or electronic) with visual counts. An advantage of "remote" devices is that recordings of emerging bats can be "viewed" and analyzed independently by different observers. In order to obtain consistent results, however, electronic devices must be maintained in excellent working order, and the power supply must be reliable.

It may not be possible to make reliable visual counts of emerging bats at roosts of exceptionally large aggregations. In these situations, sequentially timed still photographs of emerging bats may provide important data (e.g., Humphrey 1971; Rodriguez-Duran and Lewis 1987). Emerging bats can sometimes be photographed at fixed intervals against an open sky or a white backdrop placed near the cave or mine entrance. Colony size can be estimated by counting the number of bats per photograph and integrating these counts with estimates of flight speed and the duration of the emergence period. This method assumes that the number of bats in each photograph is independent of the time the photograph was taken. The photographs are treated as stratified random samples of the bat column, from which the colony size can be estimated. This method is used most effectively at roost sites occupied by single species.

Counts of Foliage- and Cavity-Roosting Bats

Censusing foliage-roosting micro- and mega-chiropterans poses special challenges to the observer, especially if the bats are rare, solitary, or widely dispersed. Methods are usually labor intensive and limited to searching for roosting animals at potential roosting sites or observing their activities at feeding or night roosts. Knowledge of general roosting habits can facilitate the location of bats once the observer has formed a search image based on roost characteristics. Tent-making bats (e.g., Timm 1987; Kunz et al. 1994) and other foliage-roosting species (Constantine 1966; Findley and Wilson 1974; Brosset 1976) often can be found in this manner.

Finding small bats roosting in tree hollows, tents, natural crevices, and foliage involves location and systematic observation of all possible roost sites in a designated area. Because bats are often cryptic, however, other methods may be more fruitful. For example, individuals or groups of foliage-roosting and crevice-dwelling bats can often be located by attaching radio transmitters to individuals captured in foraging or commuting areas and then following those individuals to their roosts. This approach has been used successfully to locate bats that roost in unmodified foliage (Thomas and Fenton 1978; Morrison 1980; Barclay 1989; Spencer and Fleming 1989), in leaf tents (Charles-Dominique 1993), in tree hollows (e.g., Fenton et al. 1985; Luney et al. 1985; Fenton and Rautenbach 1986; Tidemann and Flavel 1987;

Kurta et al. 1993), and beneath exfoliating bark (Kurta et al. 1993).

Counts of Hibernating Bats

Ideally, hibernating bats should be counted in midwinter, when the population is at its peak. Human safety considerations and the size and complexity of the hibernaculum will often dictate the number of personnel needed to conduct such a census. To minimize disturbance to hibernating bats, normally no more than two or three observers should participate in a winter census, and each site should be censused no more than once per winter. If bat populations are threatened or endangered, the census should be conducted only every two years.

Because many hibernating bats roost in large, tightly packed clusters, counting all individuals may be difficult or impractical. Instead, population size can be estimated by determining the cluster densities at selected sites and extrapolating those to the total area of the cave covered by the hibernating bats. Because cluster density can vary with species, season, and characteristics of the roost substrate (e.g., Tuttle 1975), estimates of average cluster density should be based on several different clusters at a given site. In addition, because cluster density may show considerable intercave variation, samples should be taken from several different clusters and hibernacula before extrapolating to entire species (Thomas and LaVal 1988).

Survey personnel can estimate cluster density by placing a frame of known dimensions over a cluster and counting each bat within the frame (LaVal and LaVal 1980) or by photographing the frame and cluster and then counting the bats from the photograph. In this way the total ceiling or wall area covered by bats may be estimated from photographs. The geometric properties of clustered bats, however, may introduce sources of error when estimating the total area covered. If the scale of photographs is not known precisely, the camera-to-cluster distance can only be approximated. The walls and ceilings of hibernacula are often uneven, and a projection of hibernating bats onto a flat plane will have a smaller surface area than the actual colony. If the film plane is not parallel to the colony, the projected surface will be smaller than the actual surface. Stereophotography overcomes these limitations and makes it possible to determine the camera-colony distance and the actual surface of the bat cluster, independent of the irregularities of the cave substrate and the angle from which the photograph was taken (Palmeirim and Rodrigues 1989).

Observers who census hibernating bats should be alert to potential adverse effects of their presence on the bats, because frequent winter arousals can significantly reduce critical fat reserves and, thus, survival (Tuttle 1979; Speakman et al. 1992). Intense or heat-generating light sources (e.g., floodlights, carbide and kerosene lanterns) should be avoided. Roost temperatures and other environmental variables (e.g., light, wind velocity, humidity) should be recorded at the time census data are taken, but prolonged exploration and extensive mapping should be minimized during the hibernation period. Extensive roost mapping should be carried out during the summer when bats are absent from the cave.

Flying bats

Assessing numbers of flying bats poses different challenges to observers. Because most bats commute to and from roost sites and forage well into the night, opportunities for direct visual observations are usually limited. Many small bats are difficult to identify in flight at night, and in forested regions identification is nearly impossible. At high latitudes, with prolonged periods of twilight, and in open desert regions, it is possible to observe flying bats without visual aids. In such circumstances, trained observers may be able to identify different species by their unique flight characteristics.

Night-vision devices, low-light-level videography, motion detectors, and ultrasonic detectors may aid an observer when assessing the relative numbers of flying bats. Even with such devices, however, it is not always possible to distinguish among different species or to obtain accurate counts. Ultrasonic bat detectors may be valuable for assessing species composition in some habitats, but they cannot be used to determine the numbers of bats present, because they do not distinguish echolocation calls made by one bat from those made by several individuals. Moreover, bat detectors do not detect megachiropterans and some microchiropterans (e.g., most phyllostomids). Although large megachiropterans may be observed flying during daylight or twilight hours, dense foliage and irregular topography often limit the ability of observers to assess numbers of bats reliably.

Counts with Motion Detectors

Indirect (remote) census methods minimize disturbance to sensitive colonies and allow for regular and frequent assessment of bat activity with minimum expenditure of time. Such approaches may be used to assess either seasonal or nightly changes in relative numbers of bats. Infrared motion detectors can be used to monitor both winter and summer roosts inside caves and other similar structures (Pierson et al. 1991).

As bats fly through the field of detection, individuals are recorded as a "pass." Infrared detectors do not distinguish among individuals or species, but they are useful for monitoring roost sites occupied by single species. When motion detectors are used, the motion counts must always be validated with direct observation using night-vision devices.

Ultrasonic Bat Detection

Ultrasonic bat detectors can sometimes be used to identify species of microchiropterans and to estimate their relative abundances in areas where they commute or forage (e.g., Crome and Richards 1988; Richards 1989; Ahlen 1990; Rydell 1990). A few species have distinctive echolocation calls and are relatively easy to identify; other species are more difficult to identify (Ahlen 1990). Learning to distinguish the echolocation calls of different bat species based on species-specific features such as frequency composition, changes in frequency with time, and duration of pulse repetition rate requires considerable practice, good acoustic memory, and much patience (e.g., Fenton et al. 1983; Miller and Andersen 1983; Pye 1983; Crome and Richards 1988; Richards 1989; Ahlen 1990; Rydell 1990).

When echolocating bats navigate on the wing and search for prey, many emit "cruising" pulses at repetition rates of approximately 2 to 10 pulses per second. Most of the calls are highly structured signals with frequencies that range from 20 kHz to 200 kHz (Pye 1983; Fenton 1988; Ahlen 1990). Calls of some species are frequency modulated and dominated by sweeps from high to low frequencies over a span ranging from 2 to 15 msec. Calls of other species are dominated by a constant-frequency component in which most of the signal remains steadily at a fixed frequency, usually for more than 10 msec. When an echolocating bat detects a potential prey item, it typically increases its repetition rate to about 100 pulses per second as it closes in for capture, terminating in what is commonly called a "feeding buzz." Some species of bats, notably those that glean arthropod prey from surfaces, may not emit a detectable feeding buzz (Fenton and Bell 1979), and species that produce low-intensity calls cannot be detected with most bat detectors.

INSTRUMENTATION AND RESEARCH DESIGN

Several types of bat detectors are available, and the reader should consult the literature (Downes 1982; Fenton 1988; Ahlen 1990) for details. Bat detectors can have one or more types of sound-detecting systems to convert ultrasounds into audible sounds: heterodyning, frequency division, and time expansion with digital memory. A heterodyning system has the highest signal-to-noise ratio and potentially the greatest sensitivity of the three detection systems. It is the only system present in the least expensive, and less sensitive, detectors (e.g., Ultra Sound Advice Mini-2, Petersson Electronik D-100, and Flan and Skye), but it is also included in more expensive models (Ultra Sound Advice S-25 and Petersson D-940 and D-980). Inexpensive heterodyne detectors can be tuned manually to monitor a selected 10-kHz frequency range. The Petersson D-940 and D-980 are unique in having a heterodyne system that continuously scans a user-defined frequency range and locks in a digital frequency display when a signal that exceeds an adjustable threshold is detected. Tape recordings of the audio output from either tunable or scanning heterodyne detectors can serve as event recorders, but they do not retain frequency or other information from the original sound.

In contrast to the narrow band and high sensitivity provided by heterodyning systems, frequency division systems offer lower sensitivity but broad band detection. The frequency of incoming signals is divided 10 to 20 times (into the audible range). Some bat detectors (Petersson D-940 and D-980), retain the amplitude structure of the original call, but in all systems harmonics are lost. Tape recordings of stepped-down calls do permit analysis of time-frequency structure for use in subsequent species identification (Thomas and West 1989). Some detectors (Westec and Titley Anabat II) only have frequency division modes, whereas others (Ultra Sound Advice S-25 and Pettersson D-940 and D-980) have both frequency division and heterodyne features.

Bat detectors with time-expansion systems digitally sample the incoming high-frequency sounds (250–300 kHz) at a high sampling rate for a few seconds and then play back the stored sample slowed down 10 to 20 times, making it possible for the observer to hear and record the call (Ahlen 1990). Harmonics are retained when the signal is slowed, and resolutions comparable to those obtained with large, expensive tape recorders can be achieved, while permitting field signal storage on a moderately priced cassette recorder (Fenton 1988; Ahlen 1990). The principal disadvantages of time-expansion systems are their decreased sensitivity to weak signals that can be detected by heterodyne detectors of comparable quality and their discontinuous signal (3 sec of capture followed by 30 sec of output). Time-expanded analog signals are easily analyzed with audio frequency hardware and software used for studies of bird song or human speech.

SPECIAL CONSIDERATIONS

It is important that researchers using bat detectors to conduct field surveys validate the sounds produced by each species. Investigators should capture and record representative individuals in the field and establish a reference library of unique frequencies that can be consulted when calls are being analyzed (e.g., Ahlen 1980, 1981, 1990; Fenton and Bell 1981; Woodside and Taylor 1985; Thomas et al. 1987; Thomas 1988). Cruising echolocation calls of most genera and many (but by no means all!) species in a given bat community differ in frequency span, duration, or shape. Once call characteristics are identified, it may be possible to determine activity levels of certain species within or between habitats. Ultrasonic bat detectors have also been used with varied success to determine the presence and relative abundances of bats in several different communities, including those in savanna woodlands in Africa (Aldridge and Rautenbach 1987), tropical wet forests in Australia (Crome and Richards 1988; Richards 1989), old-growth forests in the Pacific Northwest of the United States (Thomas 1988; Thomas and West 1989), rural areas in northern Europe (Limpens et al.

1989; Rydell 1990), and agricultural areas and forest reserves in eastern Europe (Gaisler and Kolibac 1992).

Bat detectors should be positioned at specific locations for both survey and census work, because station counts are more effective than transects at detecting rare taxa (Edwards et al. 1981; Crome and Richards 1988; Thomas 1988; Thomas and West 1989). Use of several detectors at or near ground level and suspended in foliage is appropriate for most survey work. Placement will be dictated by the height and density of the foliage. Bat detectors will be needed for each stratum in highly stratified forest. Fewer detectors are needed to cover the same area in desert regions.

LIMITATIONS

Ultrasonic bat detectors have several important limitations with respect to census and survey work. First, the investigator employing the devices must have a comprehensive knowledge of echolocation, bioacoustics, and methods of sound analysis, and at least a minimal knowledge of electronics. In addition, although some species of bats emit clearly recognizable calls, the call characteristics of others overlap considerably. For example, using broad-band, frequency-division detectors, Thomas (1988) was able to distinguish only eight of 12 species occurring in the Pacific Northwest of the United States. He was not able to distinguish the *Myotis* species based on their echolocation calls (Thomas 1988; Thomas and West 1989). In a study of a bat community in the rain forest of Australia, however, Crome and Richards (1988) were able to distinguish all 12 species present using a QMC S-100. Gaisler and Kolibac (1992) were able to distinguish only one of five species known to occur in an agricultural area of southern Moravia using QMC minidetectors. Obviously, the ability to discriminate between closely related taxa varies with the experience and skill of the observer.

Another problem with bat detectors is their unequal sensitivity to the echolocation calls of different species (Fenton 1988; Fenton et al. 1992). Bats of several families (e.g., Vespertilionidae, Molossidae, and Rhinolophidae) emit echolocation calls intense enough to be detected with most available bat detectors. In contrast, echolocation calls produced by other bat species (notably members of the Phyllostomidae) cannot be detected except at very close range (Forbes and Newhook 1990; Fenton et al. 1992). This means that some species cannot be censused with bat detectors. Moreover, because dense vegetation and high humidity reduce the propagation of high-frequency sounds, detection limits of bat detectors may vary by habitat and time of night (Griffin 1971; Lawrence and Simmons 1982). The consequence of atmospheric attenuation in the field is a perceived loss of the highest frequencies in calls at increasing distances from the source (Thomas et al. 1987). Because of these numerous limitations, echolocation surveys should not be considered equivalent to songbird surveys.

Richards et al. (1992, unpubl. data) showed that up to 40% of the bats in some communities can be detected but never captured. If the echolocation calls can be distinguished to species unambiguously, the calls can be used to estimate relative levels of "traffic" or "activity" in different habitats. In areas with several species of bats whose echolocation call characteristics overlap, relative activity can be used to identify habitats where mist nets and traps can be used most effectively. Data from bat detectors should be integrated with data from direct observations and captured animals to characterize the composition of bat communities fully.

Chapter 8

Capturing Mammals

Clyde Jones, William J. McShea, Michael J. Conroy, and Thomas H. Kunz

Introduction

Most methods for inventorying mammals and many methods for estimating their abundance require that animals be captured. Capture is the only means by which to obtain voucher specimens (see "Voucher Specimens," Chapter 4, and Appendix 3) beyond haphazard collection of carcasses. Likewise, information on the body mass and reproductive condition of many species (Appendix 5) can be obtained only after an animal is in hand. Estimating population abundance often involves marking animals (Chapter 10), usually after capture. Although capture is not always practical, fre-

quently it is an integral part of a successful inventory and monitoring program. Nevertheless, before selecting specific capture techniques, investigators must have a clear idea of the purpose for which they will capture animals and collect other inventory and monitoring data and a well-designed sampling scheme (Chapter 3).

There are over 4,600 species of mammals, and it is impossible to explain how to capture each species in every habitat. In addition, many researchers working on the same species in similar habitats have used different techniques, with comparable effectiveness. In this chapter we recommend specific techniques that have proved effective for typical mammals of particular sizes and habitats. One trap size and one trap spacing cannot be effective for all species. In particular, estimates of popula-

---

The authors listed above are responsible for all sections in this chapter except "Small Volant Mammals."

tion abundance should depend on trapping techniques tailored for specific species or size classes of animals within the habitat, based on experience obtained through preliminary trapping. First attempts to inventory mammals in a habitat, however, should follow general approaches, which are rooted in the size and habitat of the mammal.

Small terrestrial mammals are often sampled in preliminary inventories and, with the exception of tropical bats, usually have the greatest species richness. For each life-form, we describe our choice of (1) capture devices, (2) baits, (3) trap arrays, (4) trapping intervals, and (5) methods of handling animals. In many cases a method that can be used for one life-form is also appropriate for another. When we have described a particular method in detail for one life-form (e.g., small terrestrial mammals), later description (e.g., for medium-size mammals) will be brief. Capture generally becomes more difficult as the size of the animal increases. Therefore, observational techniques (see Chapters 6 and 7) and indirect evidence (see Chapter 9) are sometimes better for obtaining both inventory and density information.

Small terrestrial mammals

Capture Devices

Numerous kinds of traps can be used to capture small (<50 g), terrestrial mammals. The most appropriate kind depends on the size of the mammal as well as the purpose of capture. Snap traps kill the mammal (removal), and box traps capture the mammal unharmed for later release (capture-recapture). Mammals captured in live traps can of course be killed, if appropriate.

SNAP TRAPS

In a snap trap, a metal bale powered by a spring is released when the animal contacts a pan containing bait, killing the animal (usually by cervical dislocation). The most effective snap traps

for small terrestrial mammals are Museum Special mouse traps and rat traps (Victor, Four-way: Woodstream Corporation; see Appendix 9) used in domestic rodent control (Smith et al. 1971; Wiener and Smith 1972). Although mouse traps for domestic use are readily available, they are too small for effective capture of most small terrestrial mammals and are useful for inventory only when used in conjunction with larger traps. Snap traps must be of a size and power sufficient to kill the animal on impact.

BOX TRAPS

Mammal box traps (e.g., those manufactured by Sherman, Longworth, Allcock, and Tomahawk; see Appendix 9) are the most effective means for capturing small terrestrial mammals unharmed. A typical box trap is rectangular and open at one or both ends and contains a trip pan with bait inside (Figs. 8 and 9). The animal is captured when it contacts the trip pan, releasing spring-loaded doors, which close. Mammals may be retained in live traps for a few hours, if bait and bedding are provided. Wire-mesh structures can be placed around traps to prevent disturbance by other animals (Layne 1987).

The size of the trap (Maly and Cranford 1985; Slade et al. 1993), trap type (O'Farrell et al. 1994), ambient conditions at the trap site (Mengak and Guynn 1987), and body size of the species to be captured (Rose et al. 1977) all influence trap effectiveness. Such issues can only be addressed through preliminary study of particular species at the survey site.

PITFALL TRAPS

Pitfall traps provide the most effective means of capturing the smallest (<10 g) terrestrial mammals, such as shrews (McComb et al. 1991; Kalko and Handley 1992). A pitfall trap is a container placed in the ground so that its open end is flush with the surface. Animals are captured when they fall through the opening into the container below. Pitfall traps can be constructed

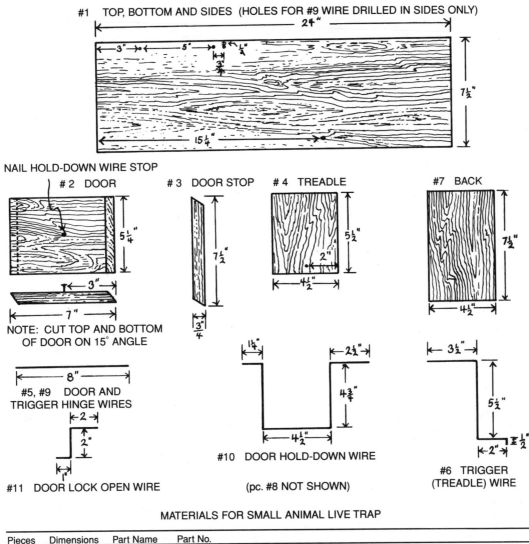

#1　TOP, BOTTOM AND SIDES　(HOLES FOR #9 WIRE DRILLED IN SIDES ONLY)

#2　DOOR

NAIL HOLD-DOWN WIRE STOP

NOTE:　CUT TOP AND BOTTOM
OF DOOR ON 15° ANGLE

#3　DOOR STOP

#4　TREADLE

#7　BACK

#5, #9　DOOR AND
TRIGGER HINGE WIRES

#11　DOOR LOCK OPEN WIRE

#10　DOOR HOLD-DOWN WIRE

(pc. #8 NOT SHOWN)

#6　TRIGGER
(TREADLE) WIRE

MATERIALS FOR SMALL ANIMAL LIVE TRAP

Pieces	Dimensions	Part Name	Part No.
	ONE INCH DRESSED WOOD		
4 pcs.	$7^{1/2}$" x 24"	Bottom, top, sides	#1
1 pc.	$5^{1/4}$" x 7"	Door	#2
2 pcs.	1" x $7^{1/2}$"	Door stops	#3
1 pc.	$4^{1/2}$" x $5^{1/2}$"	Trigger treadle	#4
1 pc.	$4^{1/2}$" x $7^{1/2}$"	Back of trap	#7
	HARDWARE CLOTH OR WELDED WIRE		
1 pc.	$7^{1/2}$" x 9"	Back of trap	#8

Pieces	Dimensions	Part Name	Part No.
	FROM 9 GAUGE SMOOTH WIRE		
2 pcs.	8"	Door and treadle hinges	#5, # 9
1 pc.	11"	Treadle trigger	#6
1 pc.	$17^{3/4}$"	Door hold down wire	#10
1 pc.	5"	Door lock wire	#11

HARDWARE

18	6d box nails	4	214 eye screws
12	$3/4$" staples	5	2d box nails

Figure 8. Diagram for construction of a small-mammal live trap, including a parts list. Redrawn with permission from Mosby (1955).

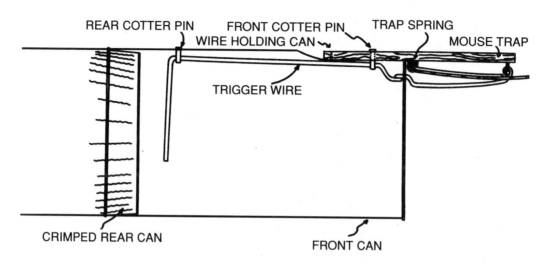

Figure 9. Scheffer small-mammal trap. The door at the right swings down to propel the animal into the trap. The animal is removed from the trap by separating the front can from the rear can. Redrawn after Davis (1956) from Day et al. (1980) with permission of The Wildlife Society.

from PVC pipe, buckets, cans, or drums and should be 40 to 50 cm deep and 20 to 40 cm in diameter. Containers for pitfall traps can be cylindrical or conical (e.g., Nellis et al. 1974). It may be advisable to cover the top of the trap, to keep the container from filling with water or litter. Covers can be constructed of wood, metal, or plastic and should be at least 5 cm in diameter greater than the opening to minimize runoff into the trap. The pitfall trap can be opened by elevating or removing the cover and closed by lowering the cover.

Capture rates of most species of small terrestrial mammals are enhanced greatly if pitfall traps are operated in conjunction with a drift fence that crosses the open pits (Kirkland and Sheppard 1994). A drift fence is a barrier designed to "corral" small mammals and direct them into traps. It is usually 20 to 30 cm high and constructed of aluminum, tin, fiberglass, metal flywire, or wood. Pitfall trap and drift fence arrays vary in length from 2 m to 20 m, usually with at least one pitfall trap per 5 m of drift fence (Handley and Kalko 1993).

Pitfall traps designed to kill are partially filled with water in order to drown the trapped ani-

mals. Pitfalls designed for live capture must be at least 40 cm deep, because some small mammals are excellent jumpers and can escape from shallower containers (Fig. 10). As most insectivores have voracious appetites, live pitfalls must be checked frequently (at least twice a day) to prevent animals from starving or consuming each other.

OTHER DEVICES

Some kinds of terrestrial small mammals can be effectively caught by hand (Shepherd et al. 1978). Examples include bats in roosts; mammals in nests, runways, and burrows; and some mammals (e.g., *Dipodomys*) when spotlighted. It is also possible to establish an array of nestboxes that serve to "trap" small mammals for examination (Goundie and Vessey 1986; Kaufman and Kaufman 1989). Kirsch and Waller (1979) suggested that shrew opossums can be captured with wire traps baited with meat.

Bait

Not all traps need to be baited. Pitfall traps, in conjunction with drift fences, are effective with-

Figure 10. Small-mammal pitfall trap.

out bait, as are live traps along runways. Soiled traps (i.e., those containing urine, feces, or another attractant) can be used to capture "curious" individuals. Bait can, however, increase the probability of capture. Most mammals that respond to bait are granivores (seed-eaters) or generalists that consume primarily insects but are attracted to other foods. A bait composed of small seeds or grains mixed with a sticky substance such as peanut butter or fruit is usually effective (Getz and Prather 1975; Anderson and Ohmart 1977). As with all our recommendations, however, a preliminary study should be carried out in this case to determine the most effective bait for the target species in the survey area.

Bait can be applied to either kill or live traps; if applied in sufficient quantity, it can withstand the onslaught of insects long enough to enhance trapping success. In tropical habitats, ants quickly remove bait, so bait must be reapplied every time traps are checked. For live traps, bait should be sufficient to feed a small mammal at least until the traps are next checked. One tablespoon of bait is usually sufficient, although more should be supplied as the trapping interval lengthens. Caution should be used when changing bait types between habitats or seasons. Baits can vary in attractiveness and can alter habitat use (Manville et al. 1992), making comparisons based on capture rates problematic for most estimation methods. Often, a single bait is not the most effective one for all species of rodents at a site. Use of multiple traps at a station, each containing a different bait, is appropriate in such situations (Szaro et al. 1988; Lomolino 1994).

Scents can serve as positive or negative stimuli for small mammals, depending on species. Feces, urine, or blood on traps attracts some mammals while repelling others (Drickamer

1984). To increase consistency in trapping success, we recommend cleaning all traps with soap and water after each trapping session. If that is not possible, then large quantities of water alone (e.g., traps can be left in a stream overnight and rinsed the next morning) may be sufficient. In dry environments, it may be necessary to bring traps out of the field between trapping sessions to be cleaned.

Trap Arrays

The type of trap and the type of bait used are the same whether the purpose of a study is a species list (inventory) or an abundance estimate. The way the traps are arrayed in the habitat, however, depends on the question being asked and the estimation methods used. For inventories, accurate estimates of abundance (total numbers of animals) or density (numbers per unit area) are not necessary: the primary concerns are ensuring an inventory of all species and the sampling of sufficient space.

The easiest way to array traps is along a transect, and we recommend this sampling technique for inventory of small terrestrial mammals. Traps are placed at equal intervals along a line, which is located randomly within a habitat type. Additional lines placed within the same habitat type are also spaced at equal intervals. Spacing distances should be a function of habitat complexity, with lines and traps more closely placed in more complex habitats. Size of the target species is also a consideration, because smaller mammals tend to travel shorter distances than larger mammals. A general rule is to space traps at a distance no greater than the radius of a circle having an area equal to that of the average home range (if known) of the target species. Such spacing provides at least one trap per individual home range. For inventory of a rodent community, trap spacing should be determined by the species with the smallest home range. As a starting point, we recommend that a trap transect be at least 150 m long, with traps placed every 10 to 15 m.

Whatever the spacing, traps should be placed at habitat features (e.g., log, tree, runway, burrow) as long as they lie within 2 m of the point. We recommend placing two traps at every station to avoid the saturation of traps with "trap-happy" individuals or species, that is, with individuals that are readily captured. Such practice increases the opportunity for animals that are less active or less attracted to traps to be caught (Drickamer 1987).

Trapping effort is the product of the number of traps used and the time over which those traps are monitored (see "Time Interval," below). The number of traps multiplied by the number of daily trapping periods (e.g., sunset to sunrise) gives the number of "trap-nights" for a particular study. We recommend a minimum of 500 trap-nights for the preliminary inventory of a habitat. The trapping effort actually required to complete the inventory can be determined with a species accumulation curve, a plot of cumulative number of species captured versus cumulative trapping effort. When this curve reaches a plateau, or when capture of species or individuals no longer increases with additional effort, the trapping effort may be adequate. If a species accumulation curve continues to increase with additional trapping, however, then more trapping effort is required to inventory the terrestrial small-mammal fauna at the site. If removal (or kill) sampling or closed population models are used to estimate density or abundance, trapping duration should be limited to avoid influx of new animals into the sampling area or loss of animals through death or emigration, which would violate assumptions of closure (see Chapter 10). Data from studies with similar trapping effort can sometimes be compared using relatively simple models (see "Capture Indices" under "Abundance Indices," Chapter 10).

Trap transects are not suitable for density estimation under most circumstances. Density by definition is abundance divided by area, so the "effective area" (area from which animals are

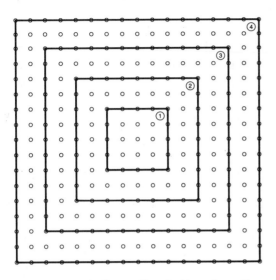

Figure 11. Trap-grid diagram. Nested grids are denoted by bold lines and numbers. Redrawn with permission from White et al. (1982).

sampled) of the trap array must be known. The effective area of a transect or quadrat can be estimated using assessment lines, which are trap lines that bisect the transect or quadrat and extend sufficiently far beyond its borders to encompass the likely range of the target species (O'Farrell et al. 1977; Kennedy et al. 1986). Animals marked on the transect or quadrat and subsequently captured at varying distances along the assessment lines provide information on the effective sampling area of the transect or quadrat. Transects may provide valid density estimations through use of assessment lines, but far fewer assumptions are required if traps are arrayed in a square grid (Fig. 11; White et al. 1982) or a circular web (Fig. 12; Anderson et al. 1983). In a square grid, the spacing between traps remains the same (10–15 m) as in a transect, but the spacing between the trap lines is equated to the inter-trap interval. We recommend that each grid be composed of at least a 10 × 10 array of points, with two traps at every point. In a circular web (Fig. 12), traps are equally spaced along each line radiating from the center. The distance between traps along ad-

jacent lines is variable, however, with the traps being spaced at greater distances from each other as one moves away from the web center. Trap webs may have an advantage over square or rectangular grids for density estimation; however, strong assumptions about trapability near the center are required before these methods can be used (see "Trapping Web," Chapter 10).

If abundance (numbers) rather than density (numbers/area) is of interest, we recommend square or rectangular grids and the collection of mark-recapture data (Nichols and Pollack 1983). Density estimates may also be obtained using these methods. Typically, however, many traps and large numbers of captures are required, because the data for several subgrids nested within the overall trapping grid must be analyzed separately to calculate effective trapping area. We urge investigators to read Chapter 10 carefully for an understanding of the sampling effort required to obtain useful estimates of either density or abundance.

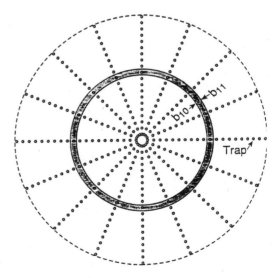

Figure 12. Trap-web diagram. Dots correspond to trap locations; additional traps may be placed at the center of the web. The points along each line that fall halfway between successive traps are denoted as b_i, $i = 0, 1, \ldots, T$, where b_0 is the center of the web and b_T is the boundary of the web just beyond the last trap. Captures in the 11th ring of traps are included in the shaded ring, with the area $C_{11} - C_{10}$, where $C_i = \pi(b_i)^2$. Redrawn with permission from Anderson et al. (1983).

Time Interval

The timing of the operation of traps significantly influences capture success for small terrestrial mammals, because the frequency and duration of their activities are highly variable (Madison 1985). Some species, such as small shrews, have multiple periods of activity during a single 24-hour interval, whereas others have only one. For the latter, traps may have to be left in place for 5 or 6 days; for the former, a shorter trapping period should adequately sample the population. Ideally, traps should be operated on a schedule that coincides with as many as five or six activity periods for a species.

The interval between trap checks can be lengthened if covers are used to protect traps (and the mammals therein) from the elements (sun, snow, and rain), and if live traps are provisioned with sufficient bedding materials (cotton or other plant fibers) and food (i.e., bait). The duration of a trapping session can be shortened and trapping effectiveness increased if animals are exposed to the traps for several days before actual trapping begins. For baited traps, we recommend a preliminary 1- to 3-day period, during which the trap is baited but locked open.

Handling

The goal when handling all mammals is to obtain the necessary information quickly, without pain or injury to either the animal or the researcher. Injury or pain to the animal should be avoided both as a humane action and to reduce the "trap-shyness" of released animals in mark-recapture studies. Trap-shyness refers to the tendency for capture probabilities of animals caught once to be lower than those of animals that have not yet been caught. Proper euthanasia of voucher specimens is treated elsewhere (Appendix 3), as is the risk of disease transmission from wildlife species (Appendix 2). Both treatments should be read before a capture program is initiated.

Small terrestrial mammals can be held in plastic or cloth bags until processed. The end of the trap is inserted into the opening of the bag, the door is opened, and the trap is shaken to drop the animal into the bag. Investigators should wear gloves to avoid being bitten by animals. Some small mammals, such as many desert rodents, can be picked up by the body, but we recommend grasping most small mammals and all unfamiliar species by the nape of the neck. The animal is initially grasped through the bag and then the bag is peeled back. After placing the animal on a flat surface, the investigator positions his or her thumb and forefinger on each side of the neck, against the back of the skull, squeezes, and pulls back, so that the fingers close only on the skin. Firmly grasping all the loose skin across the upper back, especially the skin behind the neck, restricts the movement of the animal's head and allows the researcher to lift the animal and turn it to view the ventral surface for determination of sex and reproductive condition (Appendix 5). The animal is marked, if necessary, and finally weighed. The bag can be rearranged to facilitate access to the animal for attachment of a spring balance; alternatively, the animal and bag can be weighed together and the bag weight subtracted. For animals with an abundance of loose skin, or other features that would prevent a firm grasp, we recommend using a handling cone (see "Handling" under "Medium-Size Terrestrial Herbivores," below).

Small volant mammals

THOMAS H. KUNZ, CHRISTOPHER R. TIDEMANN, AND GREGORY C. RICHARDS

No single capture method is suitable for all species of bats. Mist nets and harp traps are used most commonly, because they are easily deployed and suitable in a variety of situations (see reviews in Greenhall and Paradiso 1968; Tuttle

1976b; Nagorsen and Peterson 1980; Helman and Churchill 1986; Finnemore and Richardson 1987; Kunz and Kurta 1988). Several other capture methods, however, enhance capture success in special situations (e.g., Borell 1937; Nyholm 1965; Gaisler et al. 1979; Cotterill and Fergusson 1993).

When capturing bats for census and survey work, investigators should consider weather conditions, habitat, moonlight, and other variables that may influence capture success. Knowledge of roosting habits, daily and nightly activity patterns, seasonal movements, and colony size should also be considered in the design of capture protocols. If data from different surveys, inventories, and censuses are to be compared, types, sizes, numbers, and configurations of capture devices must be standardized. Moreover, investigators should be prepared to adjust their capture protocols if use of a particular capture technique alters the behavior of individuals or populations. For example, repeated capture efforts at a single site may cause some species to abandon an area or allow some individuals to learn to avoid capture. On the other hand, methods and capture designs that yield the highest capture success should be used with caution so as not to cause excessive disturbance of the bat community.

In the following sections we review various methods for capturing roosting and flying bats. We discuss net and trap deployment and describe techniques for handling bats following capture.

Capturing Bats in Roosts

The appropriate time for capturing bats in their roosts will depend on the objectives of the study, the season, and the age and life-history stages of the target subjects. Sampling protocols should always take into account the frequency of the capture effort; excessive capture attempts may cause bats to abandon roost sites, and too few attempts may mean that not all life-history stages of a particular species will be taken. Other factors to consider in designing capture protocols at roosts include the nutritional status of the bats prior to capture and the need for special handling once individuals have been captured. Typically, roosting bats are in a better physiological condition (hydrated and nourished) if they are captured soon after they have returned from feeding (e.g., Kunz 1974; Anthony and Kunz 1977). However, bats captured in late afternoon and early evening, immediately before emerging to feed, should yield the most reliable and least variable data on body mass, because the guts of bats at this time do not contain food from a previous feeding. When animals are processed following capture (i.e., marked, weighed, measured, or held for the collection of feces, parasites, blood, and other tissue samples), mortality can be minimized if individuals are provisioned with food and/or water.

When designing protocols for capturing bats at roost sites, special attention should be given to the nature of the roost (e.g., maternity, hibernation, transient), the time of capture (night vs. day), the reproductive stage (e.g., pregnancy, lactation), and the sensitivity of bats to investigator disturbance. Some species and individuals at certain reproductive stages are more sensitive to (or conversely tolerant of) capture than others. Such sensitivity should be established before a repeated-measure sampling design is initiated. Hibernating bats are especially intolerant of a repeated-capture design, because frequent captures and arousals can deplete critical fat reserves (see "Counts of Hibernating Bats," Chapter 7).

Repeated captures should also be minimized during the reproductive season. Females of most species are highly sensitive to being captured during late pregnancy, and some species may abandon maternity roosts if they are subjected to single-capture efforts at this time. In general, investigators should avoid capturing females during late pregnancy and during early lactation; disturbance at these times may increase mortality caused by premature births and falling young. Capturing bats at maternity roosts in the postweaning period, at transient roosts during migration, and at night or feeding roosts should cause less overall disturbance.

HAND CAPTURE

Bats that roost in caves, human-made structures, and foliage can often be captured by hand (Kunz and Kurta 1988). Hand capture is especially effective for individuals roosting in small groups. For example, bats roosting on the inner surfaces of furled leaves of *Heliconia, Musa,* and *Strelitza* may be captured by grasping and closing the open end of the leaf and quickly bending the leaf downward. With gentle squeezing and prodding, bats can be prompted to move to the open end of the leaf, from which they can be removed by hand or transferred directly to a holding bag. Long forceps can be used to extract roosting bats from narrow crevices or cavities in trees or rock surfaces, and from human-made structures. If instruments are used, special care must be taken to avoid injury to delicate wings.

HOOP NETS

Hoop nets are invaluable devices for capturing bats that roost in foliage, on walls and ceilings in caves and mines, in buildings, and inside large tree cavities. The hoop should be small enough, the net bag deep enough, and the handle long enough for the investigator to manipulate the net effectively in a given situation and to prevent captured bats from escaping (Kunz and Kurta 1988). If the inside of the net is lined with a flap of plastic, bats are less prone to escape after capture, as long as the net is relatively deep. Commercially available hoop nets, such as those used for capturing insects, may be suitable for capturing small bats, although similar types of nets can be assembled from readily available materials (e.g., heavy flexible wire, lightweight

cloth). It is preferable to use nets with the deepest available bags (~0.5 m). Almost any type of net handle (bamboo, aluminum, wood) will suffice, but aluminum extension poles improve maneuverability of the net under a variety of situations. Finally, capture success is greatly improved if the net position relative to the pole and pole length can be adjusted, and if a drawstring is incorporated into the net bag (Kunz and Kurta 1988).

To capture most foliage-roosting bats (including tent-making bats) with a hoop net, the investigator should move quietly and quickly toward the bat, making an upward sweeping motion with the net on the final approach to the roost. Although a long pole may be useful in some situations, unless it is collapsible or adjustable, it may be too unwieldy to use effectively. Bats that have been captured with hoop nets on previous occasions are often more adept at avoiding subsequent capture (Kunz and Kurta 1988).

BUCKET TRAPS

A bucket trap is simply a bucket that can be placed over a cluster of roosting bats on the ceiling of a building or the roof of a cave. Bucket traps can be fabricated from a variety of materials, including plastic or metal waste containers. Such containers may be used alone or they can be attached to an extension pole (Kunz and Kurta 1988). A useful type of bucket trap is made by cutting out the bottom of a cylindrical wastebasket or similar container and replacing it with a wire basket made from hardware cloth. The basket is attached to the container using wire or similar material woven through small holes drilled into the sides of the container. The opening of the container should be small enough to prevent captured bats from escaping. The attached wire basket should be deep enough to provide adequate ventilation and suitable places for bats to hang. The smooth sides of the plastic container will usually prevent bats from crawling out. However, it may be necessary to cover the opening of a bucket trap to prevent small species that are capable of hovering or flying in confined spaces from escaping.

Bucket traps are ideal for capturing groups of bats roosting in solution cavities in the ceilings of caves or similar roosting sites. For example, entire harem groups of roosting *Phyllostomus hastatus* have been captured with bucket traps (McCracken and Bradbury 1981; Kunz et al. 1983). These and similar groups of roosting bats are dislodged with a flexible wire spatula (a bent wire loop attached to a wooden handle) inserted over the upper edge of the bucket. To ensure a complete census of such roosting groups, care must be taken to dislodge all bats before the bucket is removed from the cave ceiling. If the bucket trap has a wire bottom, the person holding the pole and/or bucket can see through the bottom of the wire basket to determine if all bats have been dislodged. Bucket traps are most effective if the open end of the bucket fits flush with the roost substrate. Soft rubber foam may be attached along the top edge of the bucket trap to compensate for an uneven ceiling. Bucket traps work best if one person holds the bucket and/or an attached pole and another person manipulates the spatula to dislodge the roosting bats.

At times it may be necessary to attach a long extension pole to a bucket trap to capture bats that roost high on cave ceilings or high in foliage. Small groups of *Artibeus jamaicensis* that roosted in solution cavities at heights up to 4 m above the cave floor were captured using a bucket trap attached to a long pole (Kunz et al. 1983). A second pole was inserted through a small hole cut into the bottom of the wire basket and used to dislodge the roosting bats. Ideally, poles should be adjustable or removable so that when the trap is lowered from a capture position the bats cannot escape.

FUNNEL-AND-BAG TRAPS

A funnel-and-bag trap consists of a tube or funnel that directs a bat downward into a bag (or

hopper) that serves as a holding area for captured animals. Funnel-and-bag traps have been used successfully to capture bats as they crawl through small openings and crevices in human-made structures and trees at the time of their nightly departure from their roost and open their wings to take flight (Greenhall and Paradiso 1968; Gaisler et al. 1979; Kunz and Kurta 1988). Funnel-and-bag traps can be assembled from various materials, including closely woven fabric or polyethylene plastic sheeting. The open end of the trap is supported by heavy-gauge wire (e.g., wire from a clothes hanger), and the plastic sheeting is attached with duct tape. The frame of a funnel trap is then mounted over the opening used by bats. Captured bats are unable to climb on the polyethylene sheeting and escape. A funnel (or "laundry chute") may be incorporated into a trap to direct bats to ground level, where they can be removed (Griffin 1940). Funnel traps work best if alternative exits from roosts are blocked. Modified versions of the "Griffin" hopper trap, which incorporate hard transparent plastic or several wire guides to deflect departing bats into the trap, have also been used successfully (Davis et al. 1962; Gaisler et al. 1979). Bats should be removed immediately from such traps to reduce intraspecific fighting and the possibility of suffocation.

Capturing Flying Bats

Flying bats can often be caught as they depart from or return to roosting sites and when flying in enclosed spaces. If roost sites are not known, trapping efforts should be directed toward expected or potential commuting, foraging, and drinking sites. Capture methods will vary with the body size and flight characteristics of the bats being censused. Previous assessment of local topography, habitat structure (foliage density), and visual or acoustic surveys (see "Ultrasonic Bat Detection," Chapter 7) can often aid the investigator in determining where to deploy particular capture devices.

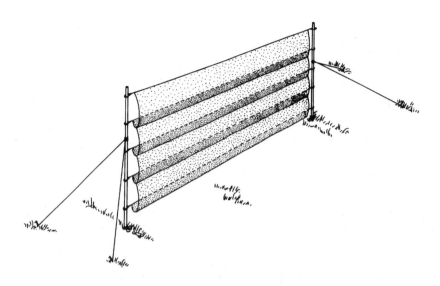

Figure 13. Typical ground-level mist net set for capturing bats. Guy lines run from the poles to stakes in the ground or to nearby vegetation. Horizontal strings are adjusted so that the amount of bag in the net is sufficient to trap bats that encounter the net. Redrawn with permission from *The Australian Bird Bander's Manual* (Lowe 1989); original artwork by P. de Rebeira.

HOOP NETS

If hoop nets are used to capture flying bats, they should be swung from behind the bat, in the plane and direction of the flight path, and they should not be swung too vigorously. This approach reduces the chances that a bat will avoid the net visually or acoustically. It also minimizes the possibility of a collision between the flying bat and the rim of the hoop net, which can break or damage wings or cause other serious injury.

MIST NETS

Mist nets are by far the most effective devices for capturing flying bats. They can be deployed in a variety of situations (e.g., Nelson 1965a; Handley 1968) and have increasingly been used in novel ways (Helman and Churchill 1986; Kunz and Kurta 1988). Advantages of mist nets are their ease in deployment, relatively low cost, and portability. Disadvantages of mist nets are that they must be tended constantly and that captured bats, when they become entangled,

must be removed individually. For this reason, mist nets should not be used at roosts with large numbers of bats, where large numbers of individuals are expected to emerge or return over a specified capture period. Too many bats may be captured at one time, and it may not be possible to remove them in a timely fashion.

Mist nets are constructed from a mesh of fine synthetic fibers (monofilament nylon, braided nylon, or braided Dacron polyester) supported by a rectangular framework of braided nylon or Dacron and a variable number of trammels or horizontal shelf cords. The net frame and trammels are tied so that when the net is properly set, it forms a capture area perpendicular to the ground with four or five long horizontal pockets made from fine netting (Fig. 13). Nylon or cotton loops are attached to each trammel and to the top and bottom cords so that the net can be secured to poles or other supporting structures. Bats are captured when they fly into the netting and fall into the net pocket, from which they are usually unable to escape.

Mist nets are relatively lightweight, compact, easily transported, and readily deployed in the field under a variety of conditions. Several practical guides to capturing bats are available, and these should be consulted for additional detail (e.g., Nelson 1965a; Handley 1968; Tuttle 1976b; Nagorsen and Peterson 1980; Helman and Churchill 1986; Finnemore and Richardson 1987; Kunz and Kurta 1988). The type and number of nets used and the manner in which they are deployed can greatly influence overall capture success (e.g., Helman and Churchill 1986; Kunz and Kurta 1988). The number of individuals of a given species captured is expressed as the number of bats per net hour to facilitate comparison of capture efforts among different studies and at different sites. For example, the capture of 100 bats in two 12-m nets during a 12-hour night would be expressed as 4.17 bats per 12-m-net hour. Because mist nets come in different lengths, it is important to present data adjusted to some standard length.

NET TYPES.   Two basic types of mist net are currently available: monofilament nylon and braided nylon or Dacron polyester. Nets made from monofilament fibers may be preferred for the capture of small bats, because the fibers are finer than those made of braided nylon or Dacron. Helman and Churchill (1986) reported high capture success with monofilament nets, but some workers find such nets cumbersome to use. Bats can be difficult to remove from monofilament nets, and the nets are more easily damaged and more difficult to repair than are braided nets. For most applications braided nylon or Dacron nets are effective for capturing bats. Both types of nets are available in the same linear dimensions, mesh sizes, and colors, but we are unaware of reports comparing their success at capturing bats.

Braided nylon or Dacron mist nets are currently available in different mesh size and denier. Mesh size is defined as the distance between two diagonal corners in the mesh of a stretched net. Denier refers to the number of grams of nylon or Dacron in 9,000 meters of fiber and reflects the "weight" or thickness of the fiber. Ply refers to the number of fibers that make up a braided strand. For example, a 50 denier/two-ply nylon net is made of two strands of nylon fiber (9,000 meters of which weighs 50 g) braided together to form the thread. A two-ply, braided net with a 36-mm mesh of either 50-denier nylon or 50- or 70-denier Dacron can be used successfully to catch most species of bats. Nets of smaller (30-mm) mesh size are available, but they are not suitable for capturing bats. Nylon nets with a denier of 50 are comparable to Dacron nets with a denier of 70, and both are suitable for capturing both microchiropterans and most small megachiropterans. Heavier-denier nets may be more durable than lighter ones but are easier for bats to detect both visually and acoustically. Tethering nets (i.e., tying the netting to the upper and/or lower trammels at regular intervals) prevents them from becoming bunched at one end during windy conditions.

Mist nets generally are available in lengths of 5.5 m (18 ft), 9.2 m (30 ft), 12.9 m (42 ft), and 18.5 m (60 ft), and they range from approximately 2.1 m to 2.4 m high when set. Large nets up to 60 m long and 6 m high also are available (Rautenbach 1985). Nets of almost any size can be fabricated from rolls of stock netting. Investigators should have several sizes of nets to provide maximum flexibility in net deployment. Short nets (5.5 m or 9.2 m) are most versatile in cluttered habitats, and they are easier for one person to set and dismantle. Long nets are usually better for spanning large areas, such as gap openings, ponds, and streams, although a large area can be spanned by butting several small nets end to end. In setting canopy nets, several nets may be stacked on top of one another or the trammels (horizontal shelf cords) from a single net may be removed and restrung perpendicular to the long axis to create a tall, narrow net (Munn 1991).

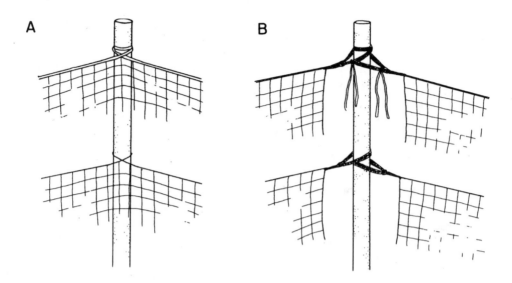

Figure 14. Methods of attaching mist nets to poles. A. Attachment of one net to a pivotal pole to form a V-configuration. B. Attachment of two nets to a single pole to extend the length of a net array. Reprinted with permission from Kunz and Kurta (1988).

PREPARATION OF NETS FOR FIELD USE.   A new net should be carefully unfolded to expose the loops at the ends of each trammel, and loops in each set should be arranged in sequential order. This is done by placing each loop sequentially over the fingers of one hand. The top loop of some nets is color-coded or may be identified by its double (and tethered) shelf string. Once the loops are gathered, the top loop should be threaded through the others and secured with an overhand knot. Arranging net loops in the proper order before a net is used and each time thereafter makes deployment easier and faster. Additional techniques for preparing mist nets for field use are given elsewhere (Handley 1968; Helman and Churchill 1986; Kunz and Kurta 1988; Lowe 1989).

NET POLES.   Net poles can be fabricated from almost any lightweight, sturdy material. However, investigators should use good judgment, considering local laws and customs and seeking permission from landowners, before cutting any vegetation (e.g., small trees and bamboo culm)

for poles. Relatively inexpensive poles can also be made from electrical conduit or aluminum tent poles. Two shorter lengths (1.5 m) of tubing assembled in the field may be more convenient to transport and handle than one long pole, particularly in rugged terrain. Short sections can be joined by inserting the compressed end of one pole into the expanded end of another pole. An alternative is to rivet a female conduit coupler to the end of one pole. Aluminum net poles are available commercially, and these often are sold with carrying cases for ease of transport. It may be necessary to sharpen the end of a pole so that it can be inserted into the ground. Metal poles can be sharpened by flattening one end with a hammer; poles made from small saplings can be sharpened with a machete.

Generally, two poles, each about 3 m long, are needed for each net, although the same pole may be used to attach the ends of two nets (Fig. 14). When the substrate is soft (e.g., sand or mud), poles usually can be inserted deep enough into the ground to support the net and remain stable. A pole can also be stabilized by placing large stones at its base or by connecting it to nearby objects with guy lines (Figs. 13, 15, and 16).

When poles are not available or desirable, ropes suspended from tree branches can be substituted. If no suitable branches are available, as on palm trees, nets can be secured to a rope suspended from another rope that has been strung between trees. Alternatively, net loops can be hooked over nails driven into trees. Nails may be used effectively when a site is to be sampled repeatedly, but nets must be removed between sampling efforts.

DEPLOYMENT OF NETS. Mist nets can be deployed successfully at almost any site where bats are expected to fly. Often the most successful capture sites are near roosts, at water holes and feeding sites, and along flyways such as animal- or human-made trails, natural forest gaps, and mountain ridges. Productive netting sites can sometimes be identified in advance of netting with acoustic surveys of echolocating bats (see "Ultrasonic Bat Detection," Chapter 7). Potential capture sites for Old and New World fruit bats can be identified by listening for sounds made by flying and feeding individuals (Handley et al. 1991). For survey and census work, placing mist nets randomly along transects may provide the least biased estimates of species richness and relative abundance. The configuration in which mist nets are deployed will depend largely on the type of habitat (Helman and Churchill 1986; Kunz and Kurta 1988). Combinations of high and low nets and T-, V-, and Z-configurations can often improve capture success (Figs. 15, 16, and 17). Nets set along ridge tops and above waterfalls also may be effective at capturing flying bats (Fig. 18), because individuals often fly close to the ground or near the water in these situations. When mist nets are deployed over water, it is important to adjust the lower shelf cord so that the lowest bag of the net does not touch the water. Nets set over streams are best placed where the water is calm and, if possible, where overhanging branches direct the flight path of the bats downward. A small rubber raft or flat-bottom boat may be required for monitoring mist nets stretched over deep water (Kunz and Kurta 1988).

Several rigging systems for suspending mist nets in the forest canopy have been described (e.g., Greenlaw and Swinebroad 1967; Handley 1967; Humphrey et al. 1968; Kunz and Kurta 1988). Generally, a combination of ropes and pulleys (Fig. 19) is used so that the entire system can be raised or lowered by one or two persons and the bats removed before they can damage the nets or themselves. The area beneath a canopy net should be cleared and covered with a tarp to prevent the net from becoming entangled in ground vegetation and litter once it is lowered to the ground.

Three or four mist nets can be stacked one above the other by tying the horizontal nylon cords together. Appropriate spacing and attachment of the net loops are maintained by tying vertical spacing lines to carabiners (or shower curtain rings). Such a system allows nets to be raised as high as 30 m above the forest floor. An alternative method for deploying mist nets in the forest canopy was described by Greenlaw and Swinebroad (1967) and modified by Munn (1991). This method involves tying a 3-m-long bamboo pole to each end of a 9-m or 12-m net, removing the three middle shelf strings, and restringing these perpendicular to the long axis (Fig. 19). When the shelf strings are restrung, they are tied so as to ensure adequate net bag for capturing bats. The restrung net is suspended in the forest canopy and is ideal for netting bats in narrow gaps in the vegetation.

Canopy nets are generally suspended from ropes positioned over high branches. The ropes are positioned by tying a lead fishing weight to a monofilament fishing line and shooting it over a tree branch using a bow and arrow (Greenlaw and Swinebroad 1967), crossbow, line-shooting gun, small slingshot (Kunz and Kurta 1988;

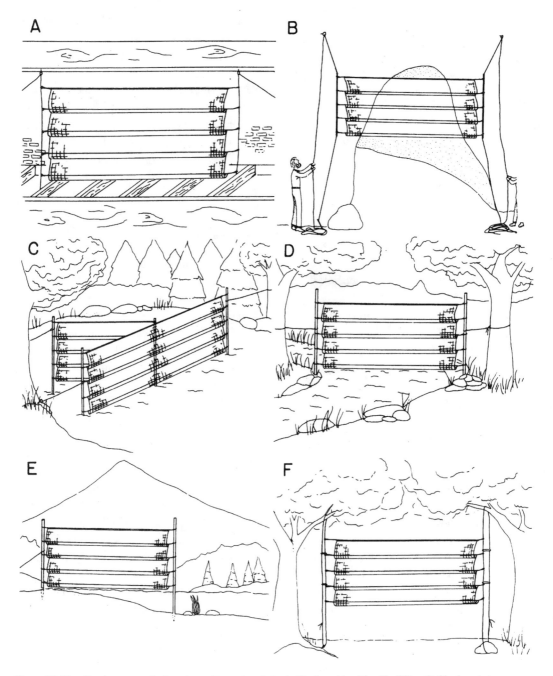

Figure 15. Sites for placement of mist nets used to capture bats. A. Single net in attic of building. B. Single net at cave entrance. C. T-net configuration over pond. D. Single net over stream. E. Single net at edge of lake. F. Single net along forest trail. Reprinted with permission from Kunz and Kurta (1988).

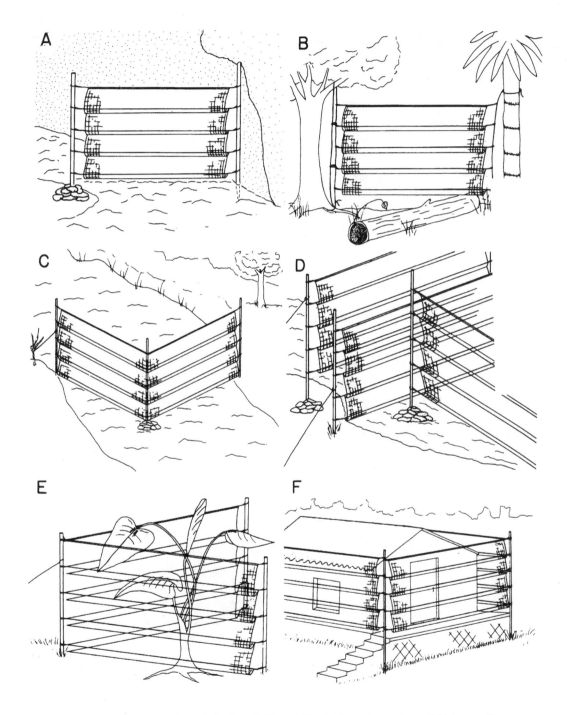

Figure 16. Deployment of mist nets for capturing bats. A. Net stabilized by large stones at base. B. Net secured to ropes suspended from tree branches. C. V-net over stream. D. T-net configuration over stream. E. V-net configuration around foliage. F. Net array around a building. Reprinted with permission from Kunz and Kurta (1988).

Figure 17. Multiple-net configurations over bodies of water. A. Y-net used to funnel bats toward a main net. B. Z-net configuration used to funnel bats into corners of net. Redrawn with permission of *Macroderma* from Helman and Churchill.

Nadkarni 1988; Munn 1991), or other home-made or commercially available device. Once the line is in the desired position, it is used to hoist successively heavier cords and rigging over the branch. To prevent the monofilament line from becoming entangled as the lead weight was projected over the desired tree branch, Munn (1991) fed the line from an open-faced

spinning reel capable of holding up to 200 m of 5-lb to 10-lb nylon line. If a field investigator uses a slingshot, he or she should wear an impact-resistant face shield and a thick leather glove on the hand that holds the slingshot.

Figure 18. Placement of a mist net over the edge of a water-fall or along a mountain ridge. Bats usually fly low over such sites and often can be readily captured. Redrawn with permission of *Macroderma* from Helman and Churchill (1986).

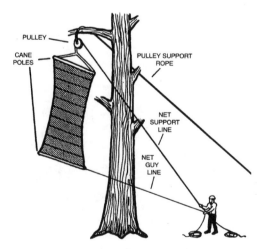

Figure 19. Simple rigging for suspending a vertically oriented mist net in a forest subcanopy. Reprinted with permission from Munn (1991).

Figure 20. Typical rigging for freestanding net poles used to suspend multiple nets in the subcanopy of forests. Reprinted with permission of *Bat Research News* from Gardner et al. (1989).

Leather gloves also should be worn when handling the braided nylon rope used to suspend the net. Nets to be transported should be wrapped around the poles and covered with the folded tarp. More complex permanent net rigging may require access to the tree canopy. Discussion of appropriate climbing techniques and practical suggestions for deploying rope rigging can be found in Perry (1978), Daunt-Mergens (1981), Perry and Williams (1981), Whitacre (1981), and Wheelock and Robbins (1988).

Nets can also be raised to considerable heights using freestanding poles (e.g., Humphrey et al. 1977; LaVal and LaVal 1980; Rautenbach 1985; Kunz and Kurta 1988; Gardner et al. 1989). With freestanding poles an investigator can place nets wherever desired, although the additional bulk and relative lack of portability of such a system of poles may be a problem. Freestanding net poles up to 9 m high can be assembled by fitting together sections of aluminum tubing (at least 2.5 cm in diameter, with a wall at least 1.6 mm thick). Poles can be joined together by inserting

solid rods of a smaller diameter into the ends of the abutting poles, or by adding a metal sleeve at the ends of the abutting poles to increase the overall rigidity. A pair of three-section, 9-m poles should be rigid enough to support four standard-size mist nets stacked one above the other. Each pole is fitted with a pulley system that is used to elevate and lower the stacked nets.

Rautenbach (1985) described an elaborate freestanding pole and rigging system to support a large 12-shelf mist net (6 m high × 30 m wide). Each pole was made from 6.2-m lengths of aluminum sailboat mast, ropes, and stainless steel pulleys. Each end of the net was attached to the posts by a series of curtain-rail runners, designed to slide freely up and down within the sail track of the mast. Each pole was secured with four guy lines anchored to the ground with large tent pegs.

Freestanding net poles taller than 3 m should always be stabilized with two to four guy lines (Fig. 20). Guy lines can be secured to the top and middle of each pole with eyebolts or with quick-release swivels, which preclude the inconvenience of having to tie and untie knots. Mechanical pulleys or O-rings can be used to suspend ropes from the tops of poles (Kunz and Kurta 1988; Gardner et al. 1989). Freestanding poles greater than 10 m in height when assembled are impractical, because they tend to bend and are difficult to stabilize.

Placing mist nets at moderate heights above the ground can greatly enhance capture success and, therefore, the potential information obtained on species richness and relative abundance of bats. For example, 85% of the bats that Rautenbach (1985) captured were caught in net shelves located approximately 3 to 6 m above the ground. Most bats were caught in the highest shelves, located 4 to 5 m above the ground. By contrast, more than half of the bats captured by Gardner et al. (1989), representing 11 species, were captured in nets at ground level; the remaining bats were captured in nets located 2.1 to 4 m above the ground or water. Sixty-five percent of the *Myotis sodalis* captured were captured in nets set above ground level. Without canopy nets, *M. sodalis* would have been captured at only 14 of 28 sites.

In most instances investigators will need to reposition mist nets at regular intervals, sometimes even nightly, to prevent bats from learning net locations and thus avoiding capture (Kunz and Brock 1975). Mist nets should be tended regularly so that bats will not become unnecessarily tangled, escape, or damage the net. The muzzles and tongues of small macroglossine bats are occasionally injured when they become tangled in nets. If bats enter deep torpor before they can be removed from a net, they should be rewarmed before release to reduce their vulnerability to predators (G. C. Richards, unpubl. data).

Normally, mist nets are opened at sunset and dismantled or folded before sunrise to avoid capture of diurnal birds. In fact, nets should be opened well in advance of sunset to ensure the capture of the earliest-flying bats. To capture bats that rely mostly on vision for orientation, however, it may be necessary to set nets well after dark (see Walton and Trowbridge 1983). Whichever method is employed, diligent attendance of nets is important, especially in the early evening hours when bat activity is highest.

NET FLICKING.   Net flicking involves moving a net quickly from a slanted to a vertical position in order to capture flying bats. Typically, a 6-m or 9-m net attached to two poles is used (Fig. 21). Net loops are secured with elastic bands or other materials so that the loops do not slide as the net is maneuvered. This ensures that the amount of bag between shelf strings is sufficient to trap a bat (Finnemore and Richardson 1987).

With single-pole flicking, one pole is securely anchored to the substrate. The investigator holds the other pole slightly above the horizontal and then moves it quickly as the bat comes within range of the net (Fig. 21A). Single-pole flicking can be used successfully along flyways and in feeding areas. Usually a second person is needed to remove bats from the net. A disadvantage of single-pole flicking is that the net cannot be conveniently moved from one area to another. Double-pole flicking requires two individuals, one holding each pole in order to keep the net in a taut, horizontal position (Fig. 21B). As a bat comes within range, the net is flicked upward to capture the bat. It is preferable to have a third person available for extracting bats from the net.

Flying bats may be detected either visually or with the aid of a tuned bat detector placed on the ground in front of the net. Sometimes, echolocating bats can be attracted by throwing a small pebble into the air. This technique requires practice and coordination by the two individuals maneuvering the net poles (Finnemore and Richardson 1987). Net flicking is especially effective for capturing flying bats in open areas.

REMOVING BATS FROM MIST NETS.   With training and patience, almost anyone can remove a

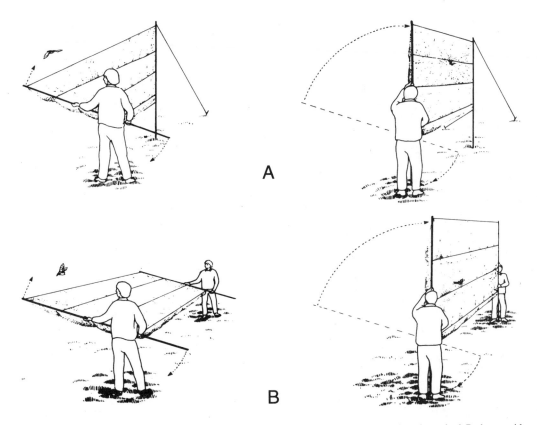

Figure 21. Net-flicking methods used to capture flying bats. A. Single-pole method. B. Double-pole method. Redrawn with permission from Finnemore and Richardson (1987); original illustration by T. P. McOwat/English Nature.

bat from a net within a few minutes. If the bat enters a net above a normal working height, the net can be lowered or pulled down within reach of the investigator. Some workers grasp the bat with one hand and use the other hand to free the netting (Fig. 22). They frequently wear a glove on the hand holding the bat.

The first step for effectively removing a bat from a mist net is to determine the direction from which it entered the net. Bats are always removed from the side of entry. As a general rule, the feet or a wing of the bat should be removed first, because these parts usually are the last to enter the net. Special care must be taken to avoid breaking the fragile wing bones or tearing the wing membranes. Wings must be extracted one at a time, and each wing may have to be opened to remove the netting (Finnemore and Richard-

son 1987). Sometimes, tension placed on the net helps to free the body, head, and other wing. Sometimes it is helpful to allow the bat to bite a loose part of a glove or a cloth bag, as this may prevent the bat from biting the investigator or from chewing on the net. A crochet hook or forceps can be useful for removing net strands from around the face and other body parts (Heideman and Erickson 1992). The longer a bat is left in a net the more entangled it becomes. If a bat is badly tangled, it may be necessary to cut some strands of the net, although this should be done only as a last resort (Finnemore and Richardson 1987; Kunz and Kurta 1988).

FURLING AND DISMANTLING MIST NETS. If an investigator wishes to stop catching bats, but intends to resume netting at a later time at the

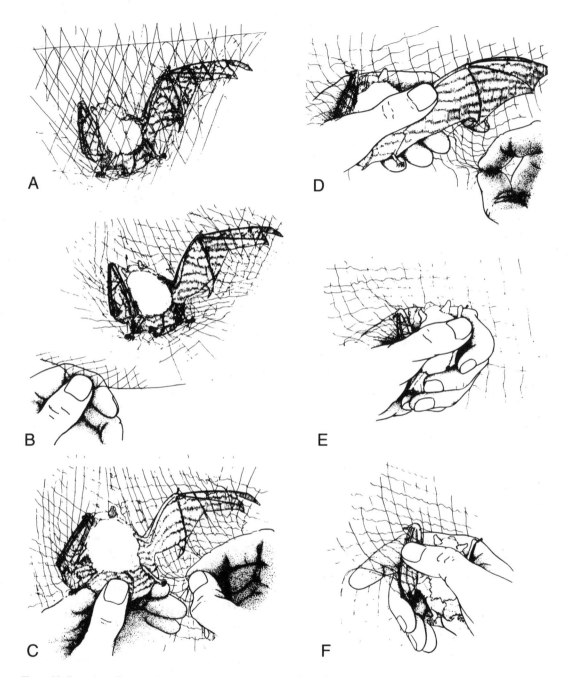

Figure 22. Procedures for removing a bat from a mist net. A. Bat entrapped in mist net bag. B. Net bag opened to expose the captured bat. C. Tail and hindfeet of bat cleared from the net. D. Body of bat firmly held, and net removed from an extended wing. E. Body of bat and one free wing grasped to prevent reentanglement. F. Remaining wing removed from the net. Modified with permission from Finnemore and Richardson (1987); original illustration by T. P. McOwat/English Nature.

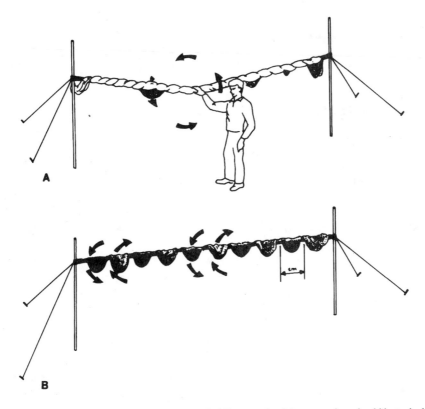

Figure 23. Methods for furling mist nets. A. Spinning. B. Draping. Loose ends of the net pockets should be tucked into shelf strings. Redrawn with permission from *The Australian Bird Bander's Manual* (Lowe 1989); original artwork by P. de Rebeira.

same site, it is common to furl rather than dismantle the nets (Lowe 1989). Before nets are furled (or dismantled), all debris, such as leaves and twigs, must be removed from the net (Kunz and Kurta 1988). The next step is to gather all of the loops at the top of each pole. The net can then be furled by spinning or repeatedly draping the net around the gathered shelf strings or trammels. Spinning is an easy way to furl a net, but nets closed in this manner may take longer to reopen (Fig. 23). Because furled mist nets tend to unwind during windy conditions, it may be helpful to tie strips of cloth around the net at approximately 1-m intervals.

To dismantle an erected mist net completely, the loops are gathered at the top of each pole and tied together in sequence, using the top loop or a separate tie-string (Fig. 24A). The loops are removed from one pole and held, maintaining tension on the net. The net is then folded by successively grasping areas of net at about 1-m intervals and doubling the net back upon itself (Fig. 24B). The folding procedure is repeated until the opposite end of the net is reached. After the net is removed from both poles, it is folded again several times to form a small bundle and stored in a labeled bag (Fig. 24B). An alternative to folding a net is to stuff it into a storage bag as one walks from one pole to the other. When opening a net stored in this manner, the loops located at the bag opening are attached to a pole, and the net is fed out as the investigator walks to the other pole. For temporary storage, net holding bags made of cotton cloth are preferable to

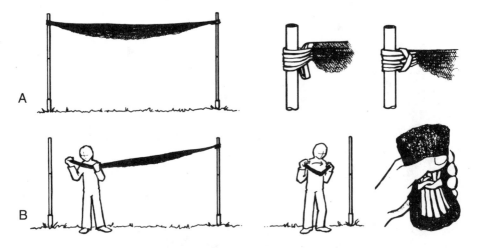

Figure 24. Procedure for dismantling and folding ground-level mist nets. A. Net gathered and secured at top of poles. B. Sequence of folding and gathering net into bundle. Reprinted with permission of *Macroderma* from Helman and Churchill (1986).

those made of plastic. Mist nets should always be thoroughly dry before they are stored for prolonged periods.

HARP TRAPS

The original harp trap was designed by Constantine (1958) to capture Mexican free-tailed bats (*Tadarida brasiliensis*). This trap consisted of a large rectangular frame crossed by a series of vertical wires. To maintain tension, the wires were attached at each end to spring coils, spaced about 2.5 cm apart. When a bat hit the bank of wires, it usually fell into the bag beneath the trap, from which it could be easily removed. This design had an advantage over that of mist nets, because it eliminated the tedious task of extracting each bat separately. Constantine's trap and other harp traps work on the principle that the wires cannot be easily detected visually or acoustically by approaching bats. In addition, bats that fly in familiar areas often navigate by spatial memory and may not listen or pay attention to their acoustic or visual input. Consequently, harp traps set at or near the openings to caves, buildings, or tree hollows, where bats roost, will often capture large numbers of individuals depending on the time of year. Harp traps

designed since Constantine's original model are more portable (Tuttle 1974a; Tidemann and Woodside 1978) and efficient (Francis 1989), and some can be used to capture large flying foxes (Tidemann and Loughland 1993).

DOUBLE- AND MULTIPLE-FRAME HARP TRAPS. Most commonly-used harp traps consist of two or more rectangular frames, usually made of aluminum tubing (Tuttle 1974a; Tidemann and Woodside 1978; Francis 1989). They work on the same principle as the original Constantine trap, except that bats that pass through the first frame of wires become caught in the second, third, or fourth bank of wires. Capture success appears to be higher in traps with more than two banks of wires. Bats "trapped" between banks of wires usually drop or flutter into the canvas trap bag located below the frame. The bag is lined along the opening with clear polyethylene, which discourages bats from crawling out.

Harp traps may be made in almost any dimension, although portability is an important consideration. The harp trap described by Tuttle (1974a) is approximately 2 m high by 1.8 m wide. The two frames are connected with four threaded rods, a design that allows the distance

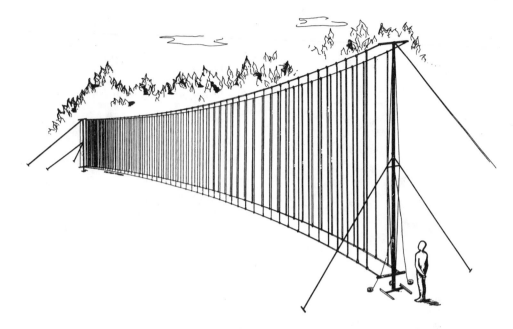

Figure 25. Harp trap for capturing large megachiropterans. Two upright supports are assembled from sections of yacht masts with sail tracks and stabilized by wire cables anchored to the ground. Reprinted with permission from Tidemann and Loughland (1993).

between frames to be adjusted. Traps used at ground level are supported by four adjustable, tubular legs, each attached to an upright section of the frame with two long bolts. Legs are not needed if the trap is suspended by a rope. In Tuttle's (1974a) original double-frame design, each frame was strung with vertical steel wires. Most workers now prefer to use 6-lb- or 8-lb-test, monofilament fishing line instead. Monofilament is more readily replaced if broken, and both ends are easily attached to horizontal tension bars with fishing swivels, or inserted through holes drilled into the bars, and tied. The recommended spacing for vertical wires in a harp trap is 2.5 cm. The distance between frames can vary from 7 to 10 cm without an apparent reduction in trap success. Drainage holes should be provided in the bottom of the trap bag (grommets are ideal for this purpose) so that the trap will automatically drain during periods of rain.

Most assembled harp traps are easily carried short distances by one or two individuals. For longer distances, they are best transported on a luggage rack or canoe carrier firmly attached to the top of an automobile. The collapsible trap designed by Tidemann and Woodside (1978) can be easily carried by one person. An advantage of small traps is that once they are assembled and deployed, only minimal maintenance and attendance are required (Kunz and Kurta 1988). Although a harp-trap bag can hold several bats, we suggest that bats be removed as soon as possible after capture, because predators may be attracted to sounds or odors produced by the captured bats, and larger species sometimes harass smaller species in the bag.

Recently, a large double-frame harp trap (15 m high and 17 m wide) has been used successfully to capture large flying foxes (Fig. 25; Tidemann and Loughland 1993). More than 200 individuals of *Pteropus poliocephalus, P. scapu-*

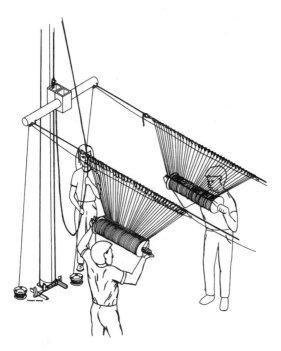

Figure 26. Two upper string assemblies from a harp trap for capturing large megachiropterans. Upper string assemblies (consisting of 60 strings per band) are attached with split rings to horizontal wire cables at the top of the trap. The string assemblies are expanded for deployment or bunched together and rolled on a hand-held drum for storage and transport. Bottom string sections vary in length to accommodate the parabola formed by the horizontal, supporting cables under tension. Upper and lower string assemblies are clipped together and opened for deployment. Reprinted with permission from Tidemann and Loughland (1993).

latus, and *P. alecto* were captured near colony sites each day. The trap consists of two banks of vertically strung, plastic-covered, stainless steel wires, each attached to horizontal steel cables in a design similar to that of a suspension bridge. With the frame in place, the bank of vertical wires can be deployed in a way similar to the opening and closing of household drapes (Fig. 26). No trap bag is used, because large pteropodids tend to injure one another if they are

confined in such bags. Thus, the trap must be tended continuously once it is set, and the bats are removed as soon as possible after capture. Typically, it takes three people almost three hours to set such a trap and about one hour to dismantle it.

TRAP PLACEMENT.   As with mist nets, harp traps are best deployed in natural flyways such as along trails, along ridges, over streams and small ponds, between trees and rock faces, in cave passages and attics of buildings, and at access points used by bats going to and from their roosts (Fig. 27). Setting two harp traps perpendicular to one another adjacent to access points takes advantage of the circling behavior of some species before they enter their roosts. Harp traps set at roost sites should not entirely block the opening, especially if large numbers of bats are present. As with mist nets, harp traps are most effective if they are partially concealed by adjacent or overhanging vegetation. If more bats are captured than can be conveniently removed and processed, a harp trap can be turned perpendicular to the flow of bats or the trap bag can be removed (Kunz and Kurta 1988). For maximum capture success, traps that are placed in doorways or at cave and mine entrances should be positioned so that the characteristics of the entrance, as perceived by the bats, are not altered. Thus, if bats are to be captured as they depart, the frame of the trap should be placed outside the opening. Conversely, if bats are to be captured as they return, the trap should be placed inside.

TRAP EFFECTIVENESS AND POTENTIAL BIASES. Tension and spacing of wires are two variables that can affect the capture success of harp traps. Tension on monofilament wires can be adjusted

Figure 27. Deployment of double-frame harp traps. A. Trap suspended in canyon. B. Trap set on forest trail. C. Trap set in shallow pond and used as an anchor for mist nets. D. Two traps arranged in an L-configuration in front of a closed door (opposite crevice opening). E. Trap suspended beneath ridge pole inside attic of building. F. Trap in doorway. G. Trap set at cave opening. H. Trap suspended outside building near roof line (opposite crevice opening). Reprinted with permission from Kunz and Kurta (1988).

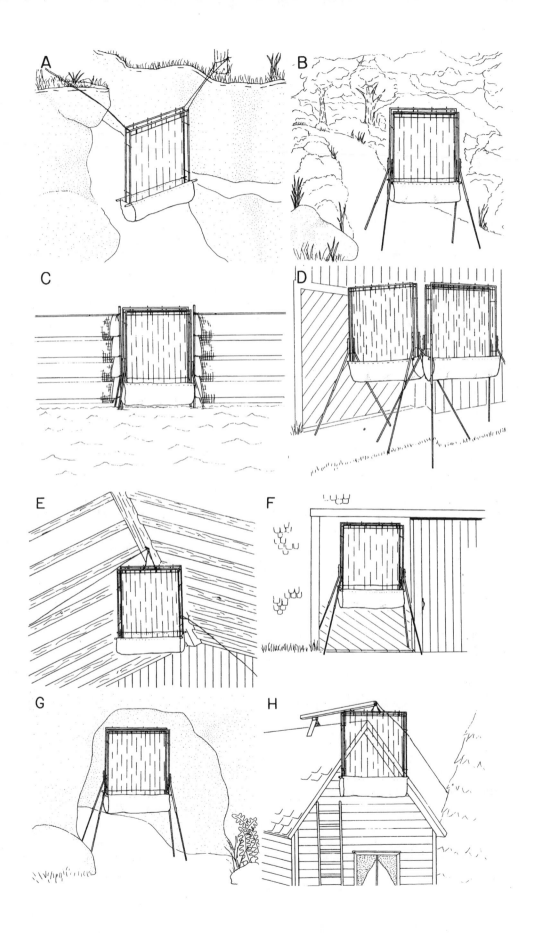

using wing nuts on the threaded rods that support the tension bars on a harp trap. In principle, line tension should be proportional to the flight speed of the bat (Tuttle 1974b). In most situations, however, wire tension can be adjusted only generally, because several species may be present in a given community. These adjustments are usually made empirically. As a general rule, if bats tend to bounce off the banks of wires, the tension on the vertical lines should be decreased. Conversely, if bats pass through the bank of wires too easily, the tension should be increased.

MIST NETS VERSUS HARP TRAPS

Both mist nets and harp traps are effective for capturing bats at ground level, at different heights in the subcanopy, and high in the forest canopy (e.g., Handley 1967; Bonaccorso 1979; Kurta 1982; Gardner et al. 1989; Heideman and Heaney 1989; Handley et al. 1991), but one or the other may be more effective for certain taxa or sizes of bats. Mist nets have been used most effectively to capture bats weighing less than 150 g as they fly within the forest subcanopy (Heideman and Heaney 1989; Ingle 1990; Francis 1994; R. C. B. Utzurrum, pers. comm.). However, mist nets and standard-size, double-frame harp traps are not very effective for capturing large pteropodids (>200 g), especially those that fly above the forest canopy. In one study, double-frame harp traps were 10 times more efficient than mist nets when both number and diversity of insectivorous bats were considered (Tidemann and Woodside 1978). Harp traps with multiple banks or frames appear to be more effective in capturing microchiropterans and small megachiropterans than those with two banks (Francis 1989). Small megachiropterans (10–80 g) have been captured at similar rates in mist nets and double-frame harp traps, but four-frame harp traps were 2 to 6 times more effective than two-bank traps (Francis 1989). Harp traps are also superior to mist nets in light rain (LaVal and Fitch 1977).

Capture rates of bats and species composition of bat captures in mist nets may differ markedly depending on the placement of the nets. Capture rates of nets placed in the subcanopy may be up to 10 times higher than those of nets set at ground level, and bats captured in the canopy are strongly biased toward plant-visiting species (Francis 1994). In the Philippines megachiropterans accounted for 95% of all captures in subcanopy nets but only 25% of the captures in ground-level nets (Ingle 1990). Up to 50 times as many bats were captured in canopy nets set along aerial walkways as were captured in ground-level nets set in forested areas of Papua New Guinea and Sulawesi, Indonesia (Gaskell 1984). Palmeirim and Etheridge (1985) found that mist nets placed along movement corridors or at foraging sites yielded higher capture rates than those placed randomly in a forest.

MISCELLANEOUS AND ANCILLARY METHODS FOR CAPTURING BATS

Cotterill and Fergusson (1993) described a specialized mobile trap for capturing bats in caves as they emerge from roost crevices or vertical rock faces at heights exceeding 60 m. The trap consists of a deflector and bag supported by a rectangular aluminum frame to which four bicycle wheels are fastened. A single polyethylene (plastic) sheet is folded and joined to form a deflector and trap bag. Two people are needed to manipulate the nylon ropes and pulleys in order to move the assembled trap into position below a roost, from which the bats are captured as they depart.

Small harem groups of emballonurids roosting in buttress cavities can be successfully captured with hand nets after the cavity is enclosed with nylon netting (Bradbury and Emmons 1974; Bradbury and Vehrencamp 1976). Bats in similar situations can also be captured by using a mist net stretched between two hand-carried poles (Bradbury and Emmons 1974). Mist nets and other types of netting can also be draped

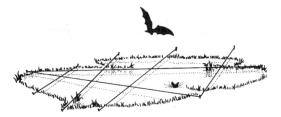

Figure 28. Deployment of trip lines over a large body of water. Redrawn with permission of *Macroderma* from Helman and Churchill (1986) as modified from Borell (1937).

over cave entrances or a frame of poles attached to a hollow tree to capture bats as they emerge from roost sites (e.g., Turner 1975; Tuttle 1976b; McCracken and Bradbury 1981; Gardner et al. 1989). Flying bats can sometimes be captured by stretching trip lines over bodies of water that are too large and difficult to net using conventional methods (Fig. 28).

Use of attractants at potential capture sites can sometimes improve capture success. Rieger and Jakob (1988) increased capture success of neotropical fruit-eating phyllostomids by "baiting" mist nets with ripe fruit. Some researchers have taken advantage of the propensity of insectivorous bats to feed on insects that have been attracted to lights to capture or shoot the bats (Youngson and McKenzie 1977; Kunz and Kurta 1988). Portable ultraviolet lights placed near bat traps and mist nets can attract clouds of insects and thus increase the rates of bat capture (G. C. Richards, unpubl. data).

Some insectivorous bats respond to the sounds made by insects (Nyholm 1965; Buchler and Childs 1981; Bell 1985) or to the sounds of other bats feeding on insects (Fenton and Morris 1976; Barclay 1982). Thus, it may be possible to capture bats in mist nets or harp traps by playing recordings of insect sounds or feeding bats adjacent to capture devices (G. Wilkinson, pers. comm.). Similarly, some bat species respond to distress calls emitted by captured bats held in cloth bags near traps and nets or to playbacks of such calls (Handley et al. 1991). Other bats may

be lured by odors produced by captured bats (Handley et al. 1991).

Preparation for Field Study

It is important to assemble all necessary capture equipment well in advance of field studies. Materials include nets of different sizes, poles, ropes, and in the case of canopy nets, devices (e.g., pulleys, carabiners, O-rings, slingshots) for positioning and hoisting nets into the subcanopy. Investigators using harp traps should have extra parts (e.g., wing nuts, thumbscrews, monofilament line, fishing swivels) and hand tools available for making emergency repairs. Extra lengths of nylon shelf cord (from a previously damaged net) for repairing mist nets that have been chewed by bats, nylon cord for making guy lines, and duct tape and wire for repairing and/or positioning capture devices in novel situations are also useful. Light- and heavyweight leather gloves, cloth or net holding bags, a pair of scissors (or fingernail clippers), and a crochet needle are invaluable aids for removing bats from mist nets and traps.

Handling Bats

Small bats can be held loosely in the palm of the hand with fingers gently but firmly wrapped around the body (Finnemore and Richardson 1987). Depending on the part of the body to be examined, the bat can be oriented so the head protrudes between the thumb and forefinger (Fig. 29A) or extends beyond the fingers in the other direction (Fig. 29B). With the bat cupped in the palm of the hand, the humerus can be held out to extend the wing (Fig. 29D). Bats can also be grasped by holding the forearms over the back with the thumb and index finger (Fig. 29C). When holding a bat for banding or taking forearm measurements, one wing should be extended and stabilized until the procedure is completed. Both hands may be required for a

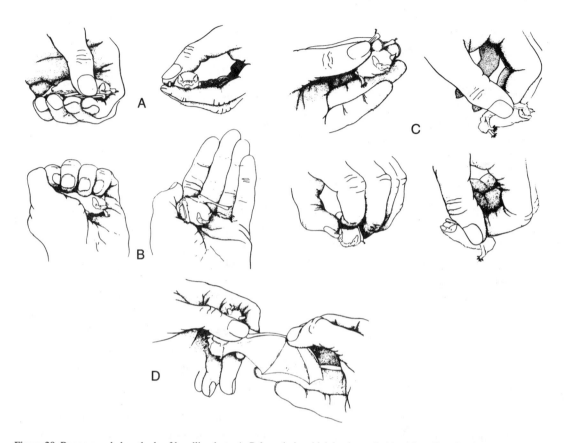

Figure 29. Recommended methods of handling bats. A. Palm grip in which bat is cradled in palm of hand and restrained with the thumb and fingers. B. Variant of palm grip in which the bat is restrained only with the thumb. C. Forearm grip in which both forearms are grasped and held over the back of the bat (care must be taken not to strain the forearms or flight muscles). D. Recommended method for holding a bat while examining the wing, attaching a band, or reading a band number. Redrawn with permission from Racey (1987); original illustration by T. P. McOwat/English Nature.

firm grasp on the body and wings of some large pteropodids. It may be necessary to grasp the bat around the neck to stabilize its head and prevent it from biting.

Fieldworkers handling and removing bats from mist nets and traps should wear gloves that are impervious to the long, sharp bat canines. For most microchiropterans, lightweight leather garden gloves will suffice. Some investigators have found it convenient to wear loose-fitting gloves so that the bats can bite into the extra "loose" material. For very small species, ultra-light weight gloves, such as those used by golf-ers, baseball batters, or cricket and handball players are useful. Handling larger megachirop-terans requires the use of thick, heavyweight leather or welder's gloves.

Holding Devices for Bats

When bats are removed from mist nets and traps, they should be placed in temporary holding de-vices. Different types of holding devices are dis-cussed in Kunz and Kurta (1988). If cloth bags are used, they should be made of soft, open-weave fabric, to reduce injury to bat wrists, el-bows, and knees and to prevent bats from overheating and suffocating (Kunz and Kurta

1988; Handley et al. 1991). Each bag should be equipped with a tie-string or barrel lock to prevent bats from escaping. Bags with drawstrings and a barrel lock can be closed more quickly than those with ties. Depending upon the size of the bag, several small individuals of a given species may be temporarily held together. In some species, individuals (especially males) may bite one another if held in the same bag. It is important not to place different species in the same bag. Large pteropodids should be bagged individually. Bats should be processed as quickly as possible, because prolonged restraint may cause unnecessary stress or depletion of energy reserves and body water. Allowing small groups of the same species to cluster in the same bag may minimize stress and reduce energy and water loss.

Examining Bats and Keeping Records

Species, sex, age (young or adult), reproductive condition, body mass, and forearm length as well as information on position in net, time of capture, and evidence of recent feeding (determined visually or by abdominal palpation) should be recorded for bats that have been captured for later release. Body masses of small bats should be recorded at least to the nearest 0.1 g, or for species weighing less than 5 g, to the nearest 0.01 g. Portable spring scales (e.g., those manufactured by Pesola) are seldom accurate to the nearest 0.1 g, but some portable electronic balances are accurate to the nearest 0.01 g. Bats weighing more than 50 g should be weighed at least to the nearest gram. Bats weighed in the field are placed in preweighed cloth or plastic bags; body mass is determined by subtracting the mass of the bag from the combined mass of the bat and the bag. If the same bag is used for weighing different bats, it should be reweighed periodically, because bats defecate and urinate in bags; the accumulation of such matter may change bag weight.

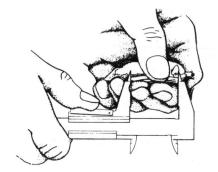

Figure 30. Recommended method of holding a bat to measure forearm length. Redrawn with permission from Racey (1987); original illustration by T. P. McOwat/English Nature.

The forearm length of small bats (<150 mm) should be measured to the nearest 0.1 mm with vernier or dial calipers (Fig. 30). A metric ruler will suffice for larger species, but the level of precision is greatly reduced. The abdomen of female bats should be palpated for evidence of pregnancy, and mammary glands should be palpated to detect the presence of milk (see Appendix 5). Males should be examined for the presence and developmental state of chest or other skin glands, and the length and width of the testes should be measured with calipers. Teeth should be examined and scored for relative wear, or measured for absolute wear. Pelage condition should be recorded, including whether the head or other body parts carry pollen. Pollen and hair samples can be taken at the time of capture and stored in glassine envelopes or attached to transparent sticky tape for later study.

Records should also be kept of food (fruits, flower parts, leaves) carried by the bats at the time of capture. Feces and urine may be collected, preferably while the bats are in the net or when they are being measured. Discarded items (insect wings, fruit parts, leaves), fecal splats (often containing seeds), and fecal pellets deposited beneath feeding roosts often provide important dietary information. It may also be possible to identify the species of bat based on dental or palatine imprints left on pellets discarded by plant-visiting bats.

Investigators should prepare maps of each capture locality, including the location of trails, streams, roads, and other important landmarks. Observation and trapping areas should be drawn to scale, indicating the location of each sample site and, if possible, the placement and configuration of nets and/or traps. The importance of recording environmental variables is highlighted by the findings that the variance in species diversity of bats is best explained by environmental correlates (Richards 1987, 1990). The collection of environmental data should be supplemented with photographs taken of roost trees, roosting groups, individual bats, and sexually dimorphic characters (e.g., epaulets, skin glands).

It is also important to obtain precise latitudinal and longitudinal data for all capture localities. Information from satellite-based global positioning systems can be used in combination with geographic information systems to map habitat characteristics and species distributions. Using this technology, Richards et al. (1992) were able to predict distributions, map centers of high diversity, and identify key conservation areas for bats on the Cape York Peninsula in Australia.

Medium-size terrestrial carnivores

Capture Devices

TRAPS

Steel traps, such as small leg-hold traps with offset, padded jaws, as well as conibear traps (spring-powered kill traps) can be used effectively to trap medium-size (50 g–5 kg) carnivores (Palmisano and Dupuie 1975; Storm et al. 1976). Leg-hold traps ordinarily are designed to capture animals alive and thus must be checked frequently (at least once per day). They should be padded to prevent injury to captives (Linhart et al. 1986; Olsen et al. 1986; Onderka et al. 1990), and they must be of the appropriate size and strength for the target species. Animals will

escape from traps that are too small or too weak and will be injured by traps that are too large or too strong. Conibear traps are specifically designed to kill and must be of sufficient strength and size to do so swiftly and humanely (Linhart and Dasch 1992; Proulx et al. 1994). Traps of either type must be anchored securely in place in order to avoid being dragged away by a captured mammal. Leg-hold traps usually must be washed, coated with wax, and handled with rubber gloves, to prevent human scent from contaminating the trap-set (Howard and Kelley 1977).

Box traps (e.g., those manufactured by Sherman, Allcock, Tomahawk; see Appendix 9) of appropriate size are also effective for the capture of medium-size carnivores, especially if properly baited (see "Bait," below), but they require more effort to transport and set up than leg-hold or conibear traps. Pitfall and snap traps ordinarily will not be effective for carnivores because of their size.

SNARES

Snares can be effective at capturing some kinds of carnivores (Hawkins et al. 1967a; Nellis 1968). They are constructed from a length of rope or wire, which is formed into an appropriate-size, movable loop at one end and anchored firmly at the other. Snares are placed over entrances to burrows, in crawlways, and across slides and trails. Carnivores (e.g., canids) may be captured alive using foot-hold snares, but these must be set with care to avoid injury. Snares designed to kill do so by strangulation or cervical dislocation and must be properly set to be effective and humane. We recommend the use of more humane kill or capture methods, unless the user is experienced or trained in the use of snares.

Bait

Use of baits with appropriate traps (steel traps, box traps) is as described in the section "Bait" under "Small Terrestrial Mammals" earlier in this chapter. For carnivores, pieces of flesh (mammal remains, poultry, fish) and scents (urine, gland secretions, fish oil, rotten eggs) should be used as attractants (Dobie 1961; Linhart and Knowlton 1975; Lindzey et al. 1977). Carnivores can also be easily repelled by some scents, especially human odors (Howard and Kelley 1977). Therefore, we recommend that traps for carnivores be cleaned thoroughly after each use and handled as little as possible. In some cases, live bait, especially if the bait-animals vocalize, is effective and practical for attracting carnivores to traps. Box traps can be fitted with live-boxes or wells (compartments inside traps for holding live animals) for domestic chickens or other prey. The prey require food and water and regular checking to assure their survival. Tape recordings of prey calls can also be used to attract some carnivores.

Trap Arrays

Traps set for carnivores should be obscured with a light covering of vegetation and litter; leg-hold traps are often placed in holes in the ground. Because of the large body size and home range of these mammals, we recommend spacing of 100 m or more between traps. Placement of traps along transects is appropriate when inventory is the study objective. Random placement of traps is also appropriate for inventory. The habitat is divided into squares, and traps are placed in randomly selected squares in association with some habitat attribute (e.g., burrow, slide, trail, rock, log, tree). If possible, traps should be moved daily to new, randomly selected locations. We recommend grid or web layouts, with spacing as given previously, for estimates of abundance and density. Grids can be delineated on maps of the study area and specific trap sites selected opportunistically near each grid node.

Time Interval

Traps for medium-size terrestrial mammals should be checked at least every 12 hours. We recommend that traps be operated constantly for a period of at least 7 days. As with small terrestrial mammals, a species accumulation curve can provide an effective indication of the success of an inventory.

Handling

Considerable care should be exercised when handling medium-size carnivores because of their size and strength. Gloves must be worn, and the animal should be grasped firmly at the base of the neck in order to restrict movement of the head. The other hand is used to hold the animal by the hips. Hands and arms should be kept above the back of the mammal, to avoid the claws. Many carnivores are vectors of diseases (including rabies) communicable to humans, and appropriate precautions must be taken to avoid infection (Appendix 2).

Other Capture Methods

Many carnivores are difficult to capture in conventional traps. Alternate methods that are often effective are discussed in this section.

NETS

Nets include any fixed or movable device (except snares, considered above) for entangling animals. For medium-size carnivores, drop nets or cannon nets are probably most effective. Animals are attracted to a site with baits or auditory or visual decoys. Drop nets are attached to a frame and raised above ground over the bait site. They are triggered remotely, using a trigger cord or radio-controlled release device (Garrott and Hayes 1984). Cannon nets are attached to projectiles that are launched by powder charges, which are remotely detonated using a blasting machine or auto battery (Hawkins et al. 1968). Cannon nets are rolled or folded and lightly concealed with straw or litter. Bait is placed directly in front of the net, and an observer concealed in a blind detonates the charges when the target animal(s) are in the area covered by the outstretched net. The outgoing net should cover but should not strike the animal, because the force of projection can cause death or serious injury. Personnel should also exercise extreme caution to avoid serious injury or death in case of premature detonation. In particular, workers should *never* load a net that is armed with live charges, or stand in front of it, unless certain that the electric detonator is disconnected and that the wires on the detonator end are twisted together or grounded, to preclude detonation by static electricity.

GUNS

If animals are not to be released, small-caliber firearms can be used to obtain specimens. Selection of the appropriate firearm and ammunition is important to ensure that the target animal is brought down quickly. Small-bore (0.22- to 0.30-caliber) rifles or shotguns with no. 6 to BB-size shot are appropriate, depending on the size of the animal. Animals can be shot opportunistically, or they can be attracted to bait or lures near a blind where the collector is hiding. If live specimens are needed, then capture guns (Hawkins et al. 1968) that fire darts or pellets containing sedative drugs can be substituted for conventional firearms. Use of such devices is discussed later in this chapter (see "Guns" under "Large Terrestrial Mammals").

DRUGGED BAIT

Baits laced with either lethal or sedative drugs may be used in cases in which other techniques are impractical (see Table 6). Caution must be used, especially with lethal drugs, to avoid nontarget animals, especially humans or domestic stock. Nonlethal drugs must be quick-acting and should be used only in areas where immediate retrieval of dosed animals is possible.

Medium-size terrestrial herbivores

Capture Devices

Some medium-size (50 g–5 kg) herbivorous mammals (e.g., rabbits) can be captured with large snap traps (standard rat traps); most herbivores, however, will be too large. Both steel traps (such as small leg-hold traps with offset, padded jaws) and conibear traps are effective if securely anchored so that the captured mammal does not drag them away. Box traps (e.g., those manufactured by Sherman, Allcock, Tomahawk; see Appendix 9) of appropriate size can also be used, but pitfall traps are not effective for mammals in this size group. Snares are appropriate for capturing some kinds of herbivores. Snares are constructed and placed as already described for medium-size terrestrial carnivores. Snare size is determined by the size of the animal and by the type of capture intended (i.e., live [the animal is caught by the foot or leg], or kill [the animal is caught by the neck]). Snares are most effective as kill traps, but for humane reasons we do not recommend their use by other than trained personnel.

It is possible to drive some medium-size mammals into capture chutes or pens for inventory or marking. This technique involves large numbers of workers that are spaced close enough (ca. every 5 m) to "push" animals before them. The workers drive the animals down a natural funnel that eventually becomes a wall or fence and channels animals into a holding pen. Within the holding pen animals are grabbed by hand or directed into a smaller funnel one at a time. The smaller funnel constricts until the animal cannot turn around and can be grabbed. Drive-trapping is effective for secretive medium-size mammals that cannot escape below ground or into trees and that live within terrain that can be traversed yet provides natural funnels (Spillett and Zobell 1967; Kattel and Alldredge 1991; Sullivan et al. 1991).

Table 6. Drugs and Dosages for Capturing and Handling Mammals[a]

Genus	Common name	Dosage	Reference
Succinylcholine chloride (Anectine and Sucostrin)			
Alces	Moose	15–17 mg/adult	Houston 1969
Antilcapra	Pronghorn	0.13–2.20 mg/kg	Beale and Smith 1967
Cervus	Elk	27.0 mg/adult (male)	Varland 1976
Mirounga	Elephant seal	2.5 mg/kg	Ling et al. 1967
Odocoileus	White-tailed deer	0.07–0.10 mg/kg	Allen 1970
	Black-tailed deer	0.15–0.31 mg/kg	Miller 1968
Ovis	Bighorn sheep	0.44–1.45 mg/kg	Stelfox and Robertson 1976
Etorphine hydrochloride (M-99)			
Aepyeros	Impala	0.5 mg (male)	Pienaar 1975
Alces	Moose	4–5 mg/moose	Roussel and Patenaude 1975
Cervus	Elk	7 mg/adult (female)	Coggins 1975
Dama	Fallow deer	2 mg/100kg	Harrington and Wilson 1974
Loxodonta	African elephant	9.0 mg	Alford et al. 1974
Odocoileus	White-tailed deer	2 mg/deer	Day 1974
Ovis	Bighorn sheep	2–2.5 mg/bighorn	Wilson et al. 1973
Taurotragus	Eland	4–6.5 mg/adult	Pienaar 1975
Ursus	Black bear	1.6 mg/100 g	Beeman et al. 1974

[a]From Day et al. (1980).

Bait

A mash of seeds, water, flour, and salt is often effective as bait. A similar mash can be made by adding water to commercial food pellets made for domesticated herbivores (e.g., Purina Rat Chow) or to pellets composed primarily of alfalfa hay (e.g., Purina Rabbit Chow). We recommend prebaiting, or placing bait in the area of the trap for a night or two before trapping. Traps or snares placed on runways do not need to be baited.

Trap Arrays

We recommend placement of traps for medium-size mammals at 50- to 100-m intervals along transects for inventory of a specific habitat. Exact spacing will depend on home range size (see "Small Terrestrial Mammals," above).

Traps should be placed along trails or at burrow entrances, watering spots, or other locations that animals are likely to visit. A random-trap-placement grid can be used to estimate abundance. The habitat is divided into 50- to 100-m squares, and traps are placed in randomly selected squares in association with a burrow, slide, trail, rock, log, tree, or other attribute. Traps should be moved daily, if possible, to new, randomly selected locations. As the number of traps increases, the array approaches the grid configuration used for small mammals.

Time Interval

We recommend that traps be operated constantly for at least 7 days. A species accumulation curve (see "Time Interval" under "Small Terrestrial Mammals," above) can help determine if capture effort is sufficient for an inventory. Traps should be checked at least every 12 hours; traps in direct sun should be checked every hour. The time required to check traps for medium-size mammals will be greater than that required to check traps for small mammals because of greater trap spacing. Therefore, more personnel will be required to rotate on trap checks or alternate between periods when traps are opened and closed.

Handling

Gloves should be used, especially with larger rodents such as capybara, whose bites can be severe. Animals should be grasped firmly at the base of the neck with one hand and held by the hips with the other hand. The handler should also keep his or her hands and arms above the back of the mammal, to avoid being clawed. Investigators capturing smaller mammals (e.g., rabbits) in box traps should use a handling cone, a tapered cone of wire or stiff plastic fitted with a cloth "skirt" that fits over the end of the trap (Fig. 31). The end of the trap inside the cone is opened, and the animal

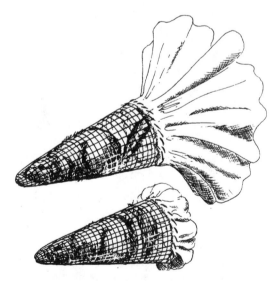

Figure 31. Handling cone for holding small or medium-size mammals. Reprinted with permission of The Wildlife Society from Day et al. (1980).

moves down the cone until it cannot turn. The skirt is then folded into the cone behind the animal, and weight, body measurements, marks, and other data are recorded. Most of these mammals do not have to be sedated unless handling lasts more than 15 minutes. Placing a mask over the eyes of small cervids, such as musk deer, reduces the stress of handling.

Other Capture Methods

These methods are similar to those described under "Other Capture Methods" for medium-size carnivores, above. Animals can be baited or otherwise attracted to a site and captured with drop nets or cannon nets (Conner et al. 1987). Personnel using cannon nets should be extremely careful. Small-caliber firearms can be used to collect specimens, also as described previously for carnivores. Capture guns using darts or pellets containing sedative drugs are used to obtain live specimens (Table 6). As with carnivores, baits containing lethal or sedative drugs can be used to capture medium-size herbivores (Table 6). Extreme caution must be exercised, especially when using lethal drugs.

Other medium-size mammals

Scansorial Species

Although scansorial species are semiarboreal, they usually forage close to the ground, and extensive arboreal trapping is not necessary. Nevertheless, because the degree of arboreality of the species may be unknown, we recommend placing 10% to 15% of the traps in trees. Traps can be fastened directly to a tree or placed on a trap platform, a shelf constructed of wood and bracing that is nailed or wired to the trunk or lower branches of a tree. Traps can also be suspended from a harness in the canopy (see "Deployment of Nets" under "Mist Nets" in

"Capturing Flying Bats," above). Any bait used for medium-size terrestrial mammals (both herbivore and carnivore) is appropriate for scansorial mammals.

A trap array based on random cells within a grid, with 10 to 15% of the traps placed off the ground, is appropriate for these mammals. Traps on the ground should be placed next to trees, with the opening oriented toward the tree trunk. Trapping should be conducted over a 7-day period, with traps checked at least every 12 hours. Traps exposed to full sun should be checked every hour. Captured mammals will be less stressed if the trap is covered to provide shelter and concealment. Some scansorial mammals (e.g., gray, red, and Albert's squirrels) appear to be more sensitive to handling than other medium-size mammals. Investigators should consider sedating these animals upon capture. Use of a handling cone (see "Handling" under "Medium-Size Terrestrial Herbivores," above) will reduce handling stress.

Arboreal Species

Arboreal mammals are difficult to capture. All traps must be placed in trees, with 10% to 15% placed in the canopy, using a method outlined earlier (see "Deployment of Nets" under "Mist Nets" in "Capturing Flying Bats," above). Tree trunks may be scaled with tree-climbing gear, but use requires prior training. Use of trap platforms and a 7-day prebaiting period increase the probability of capture. For inventory work, observation or use of a gun is probably a more effective means of sampling than trapping, but many population estimation methods require marking of some individuals, and, therefore, capture.

Semifossorial and Semiaquatic Species

The easiest time to capture semifossorial and semiaquatic mammals is when they are active above ground or terrestrially. Traps should be oriented to burrow entrances or emergence sites,

because activities in both groups are usually focused around these areas. Snares placed at burrow entrances are effective for trapping, but the user must be experienced in the proper use of these devices. For semiaquatic mammals the use of snares and leg-hold traps almost always results in mortality, unless the traps are placed on less-used trails away from water. We do not recommend excavation of burrows except as a last resort, because semifossorial animals can usually burrow faster than a researcher can dig, and because burrows of semiaquatic mammals have several exit holes, often under water. Some mammals (e.g., muskrats) build dams that can be effectively excavated to capture animals.

Large terrestrial mammals

Large terrestrial mammals (>5 kg) are usually captured live, although kill-sampling with a gun is possible.

Capture Devices

Steel leg-hold traps of appropriate size often are effective, especially for carnivores. As with medium-size mammals, leg-hold traps ordinarily are designed to capture animals live and must be checked at regular intervals. Traps must be anchored securely so that the captured mammal will not drag them away. Large box or culvert (Fig. 32) traps (Erickson 1957; Black 1958; Troyer et al. 1962) are effective, if properly baited (see "Bait"), but require much more effort than leg-hold traps. Ordinarily box and culvert traps for large mammals must be designed and constructed with a particular species and field situation in mind; many designs are possible. Snares (see "Capture Devices" under "Medium-Size Terrestrial Carnivores," above) are effective for capturing some kinds of large mammals, but again we caution against their use except by trained and experienced persons. Leg snares em-

Figure 32. Culvert trap for capturing large terrestrial mammals. Photo: Mark Jecker (Arizona Game and Fish Department).

ployed for medium-size mammals also work effectively on most large carnivores. Drive-trapping, as described earlier for medium-size herbivores (see "Capture Devices" under "Medium-Size Terrestrial Herbivores," above) is effective for some large herbivores (e.g., ungulates). In open terrain, aircraft can be used to herd or drive large herbivores into traps or corrals.

Bait

Carnivores are often attracted by pieces of flesh and by scents; we recommend prebaiting. Live bait is also effective. For herbivores, whole or mashed grains, commercial feed pellets, and salt blocks are appropriate (Hawkins et al. 1967a; Ramsey 1968; Mattfeld et al. 1972).

Trap Arrays

Traps set for large mammals should be camouflaged if possible; steel traps should be obscured with a light covering of vegetation and litter or placed in ground cavities. Traps should be spaced at 100 to 1000-m intervals, depending on home range size (see "Trap Arrays" under "Small Terrestrial Mammals," above). Placement of traps along transects is appropriate for inventory purposes. We recommend a grid or web array, with spacing as indicated previously, for estimates of abundance and density. Grid lines can be delineated on a map and trap sites within grid squares selected on the basis of logistical feasibility (e.g., proximity to roads).

Time Interval

Large terrestrial-mammal traps should be checked at 12-hour or shorter intervals. We recommend that traps be operated constantly for a period of at least 14 days. A species accumulation curve may indicate whether trapping effort is sufficient for inventory of the mammalian fauna of an area (see "Time Interval" under "Small Terrestrial Mammals," above).

Handling

Considerable care should be exercised when handling large mammals because of their size and strength. It is often necessary to drug such animals, especially large carnivores, before removing them from traps (Seidensticker et al. 1974). Nooses, choke sticks, and other physical restraining devices can also be used (Hawkins et al. 1968). Handling large mammals is a demanding activity, requiring physical strength and experience. Two or more people should be involved, and at least one individual should be experienced in the proper restraint and handling of the species being captured. Many large mammals are vectors for diseases, including rabies, that are communicable to humans; these animals can also inflict serious injury through biting, kicking, and goring. Appropriate precautions must be taken (Appendix 2). Personnel should be trained to administer first aid in case of injury.

Other Capture Methods

NETS

Drop nets (Clover 1954; Sparrowe and Springer 1970; Conner et al. 1987) or cannon nets (see "Nets" in "Other Capture Methods" in "Medium-Size Terrestrial Carnivores," above) can be effective for capturing large mammals, especially herbivorous ones. Animals are attracted to nets at fixed sites with baits or lures. Aircraft can be useful in netting large mammals, especially ungulates in open country; in some cases helicopters can approach animals close enough to drop nets from above.

GUNS

Shooting skill and selection of appropriate firearms and ammunition are critical to the efficient and humane dispatch of large mammals. For collection of large mammals we recommend high-powered rifles equipped with telescopic sights and, for nocturnal mammals, use of spotlights or infrared scopes. Shots should be fired only when conditions allow for a quick death, to avoid crippling animals or lengthy pursuits of wounded individuals. We recommend a head shot unless an intact skull is required, in which case a neck shot will usually suffice.

Guns using darts or pellets containing sedative drugs can be employed to obtain live specimens (Crockford et al. 1958). Recommended drug dosages must be followed (Table 6), and the time taken to immobilize the target species must be known. If induction time is relatively long and allows a darted animal to move away from the darting site, prior arrangements for personnel and/or logistic support sufficient to follow the animal must be made. The needles on darts should be no longer than necessary to inject the drug. Long-needled darts can cause extensive damage to muscle tissue and may also puncture the internal organs of an animal. In some types of guns the power used to deliver a dart can be adjusted to reflect distance from the target animal. Effective ranges of capture guns are much shorter than those of conventional firearms, but power should always be kept to a minimum to prevent injury to the darted animal. The condition of the darted animal should be monitored closely during the time it remains immobilized.

Investigators must wear protective gloves when handling potentially lethal drugs such as Immobilon and M-99, and they should take precautions against spilling a drug during transfer from the vial to the dart. When a drug-filled dart is inserted into the barrel of a gun, the safety mechanism should always be locked. Darting should not be attempted in dense vegetation or under conditions of heavy winds or low ambient light if other alternatives are available, because tracking drugged animals for retrieval can be difficult and dangerous. Under such circumstances, tracking dogs are helpful, if not essential. The investigator should aim to strike the flank or rump of the target animal, after ensuring that the path of the dart is clear. Darts that miss their targets should be retrieved, especially if they contain lethal drugs. Radio transmitters can be attached to darts to facilitate either their retrieval or the location of the darted animal. If darts without transmitters are used, the darted animal must be followed so that it can be recovered and processed as soon as possible.

DRUGGED BAIT

Large mammals can be captured using baits laced with either lethal or sedative drugs (e.g., Austin and Peoples 1967; Table 6). Nonlethal drugs must be quick-acting and must be used only when immediate retrieval of the animal is possible. Again, care must be taken to prevent nontarget animals and people from coming into contact with the bait.

Fossorial mammals

Capture Devices

Fossorial mammals (e.g., moles, golden moles, pocket gophers, pine voles, tuco-tucos, mole rats) are highly adapted for subterranean existence and for the most part are not readily captured by traditional methods. Exceptions include the large species of tuco-tucos, which can be caught by small steel traps (nos. 0–1) set at the entrance to their burrows. Some other fossorial mammals, such as pine voles, can be taken in conventional snap traps set on the surface, especially in moist habitats and dense vegetation.

Most fossorial mammals construct shallow subsurface tunnels as well as deep burrows. Subsurface tunnels are usually indicated by surface ridges of soil and/or by cone-shaped mounds of earth. The nature and extent of such tunnel systems differ greatly among the various fossorial species. Many of the fossorial mammals are trapped most effectively within the subsurface tunnels by means of special devices (Hawkins et al. 1968), such as mole traps (choker-loop, harpoon, scissors-jaw) and gopher traps (Macabee, pincher). A mole trap is pressed into the soil of a subsurface tunnel so that the trigger lies within the soil blockade or rests on top of it. Any disturbance of the soil releases the trap mechanism. A gopher trap is designed for placement inside an excavated tunnel. Several kinds of homemade devices are also available for trapping these mammals (Schemnitz 1980). It is possible to modify these traps to allow live capture, but the traps must be checked several times a day to prevent mortality (Howard 1952; Jensen 1982).

Trap Use

Fossorial mammals that can be caught in conventional traps set on the surface respond to traditional bait for small terrestrial mammals. Specialized traps for capturing animals in subsurface tunnels are designed to be used without bait.

The specialized traps are used most effectively in association with either subsurface tunnels or mounds of earth. Spacing depends on home range size (see "Trap Arrays" under "Small Terrestrial Mammals," above). Placement of traps along transects is appropriate for inventory. We recommend grid or web layouts for estimates of abundance and density. Distributions of fossorial mammals are generally patchy and largely dictated by local soil conditions. Information on soils should be used to determine the location of traps for stratified sampling (see "Stratification," Chapter 3).

Traps should be checked at intervals of 12 hours or less. They can be left in place for several days or until an animal is captured. The smaller fossorial mammals can be handled in a manner similar to that already described for small terrestrial mammals. For the larger forms, we recommend that gloves be worn for protection against biting and clawing.

Unusual mammals

Several groups of mammals are sufficiently unusual to warrant special consideration. Extra-large terrestrial mammals (e.g., elephants, rhinoceroses, hippopotamuses, giraffes), in particular, present special problems. Only qualified veterinarians and other trained personnel should attempt to capture those mammals. Many such animals are, in fact, sufficiently conspicuous because of size or habits to preclude the need for trapping. Information required for either inventory or abundance estimation can be obtained more efficiently by other methods (see Chapters 6, 7, and 10).

Aquatic mammals by nature are difficult to capture and handle, and we do not recommend capture for inventory or abundance-monitoring purposes. We refer readers to Chapter 6 for a discussion of survey techniques dealing with aquatic mammals.

CONTRIBUTORS: KYLE R. BARBEHENN, STUART C. CAIRNS, TIMOTHY F. CLANCY, MARTIN DENNY, MICHAEL E. DORCAS, LOUISE H. EMMONS, SARAH B. GEORGE, J. EDWARD KAUTZ, GORDON W. KIRKLAND, JR., ELIZABETH D. PIERSON, WILLIAM E. RAINEY, AND GALEN B. RATHBUN

Chapter 9

Mammalian Sign

Christen Wemmer, Thomas H. Kunz, Geoffrey Lundie-Jenkins, and William J. McShea

Introduction

The natural history literature is replete with anecdotes about astute naturalists and noble savages who, like Sherlock Holmes, could decipher dramas from traces in the dust. As they showed, even the most secretive mammals leave a record of their presence and activities in a variety of subtle ways. The recognition of those signs—which may be visual, olfactory, or auditory—requires perceptiveness, hours of field experience, and serendipity, all of which figure in the background of celebrated naturalists (Seton 1958; Tinbergen 1965; Corbet 1978).

The concept of a "search image," or the idea that a wide range of animals develop expecta-tions of stimuli that are associated with certain predators and prey and that are conditioned by reinforcement, is relevant (Uexkull and Kriszat 1934; Tinbergen 1960). Naturalists also develop "search images" based on their experiences, but they have often relied on intuition and guess-work rather than verification when attributing sign to a particular species. To be reliable as an indicator of the presence and activities of a given species, sign must have been previously vali-dated—that is, that species must have been iden-tified in the act of producing the sign. In the following sections we discuss different catego-ries of sign and provide guidelines for the use of sign in mammalian inventory and monitoring programs.

Visually detected sign

Structures and Habitat Features

A number of animals create structures for protection and for rearing young that are easily detected visually (von Frisch 1974). Examples include leaf and grass nests of tree squirrels, chimpanzees, and dormice; hole nests of scansorial rodents; earthen mounds of mole rats, gophers, and moles; earthen dens of aardvarks, foxes, and tuco-tucos; stick and reed lodges of beavers, muskrats, and nutria; tents made from living leaves modified by bats; and grass dens of civets (Fig. 33A).

Equally distinctive are signs produced by the feeding activities of many herbivorous mammals. Some rodents (e.g., meadow mice, voles, rats) leave droppings and grass cuttings in their runways. Others cut stems and branches (e.g., beaver, rabbits, mice, mountain beavers, pocket gophers, wood rats, squirrels, and porcupines) or debud shrubs and trees (e.g., squirrels and chipmunks) (Schemnitz 1980). Deer and antelope nip buds and leaves in a characteristic way and may produce "browse lines" on trees that often resemble the handiwork of a gardener with a hedge clipper. Dinerstein (1992) reported the distinctive browsing effects of the great one-horned rhinoceros on the growth form of the kutmiro tree in Nepal. Many carnivores, rodents, and pigs leave behind characteristic diggings after foraging for invertebrates, small vertebrates, and roots (Fig. 33B). Crystallized urea develops curious forms in some caves and human-made dwellings used by wood rats and colonially roosting bats (Kunz 1982). Wallows of certain species (sambar, pigs, buffalo) can be detected by tracks, impressions of the user's body, and shed hair. Desert rodents produce dust wallows, as did the American bison. Bison wallows persisted for over a century after the species disappeared from the American West, and some are still apparent today.

Many mammals, including rodents, bats, and small marsupials, use rock crevices, caves, recesses, tree cavities, existing burrows, and a variety of artificial structures as permanent and temporary resting places. In most such cases the presence of mammals can be detected by visual signs of feces and urine, prey items, and sounds and odors (Kunz 1982).

Tracks

Opportunities to observe mammals in the field are often limited because most are small, nocturnal, and secretive. Even many large diurnal mammals are secretive and cannot be observed directly. Learning how to identify, interpret, and preserve tracks and other signs left by mammals can provide information about their habits that cannot be obtained in any other way. These signs also provide an obvious record of the presence of the species in the survey area. Mammal tracks can often be found in wet or muddy areas near lakes, ponds, and streams where animals come to feed or drink, and along trails used to move between different habitats. Sometimes it is possible to create areas of moist soil or sand in habitats where animals regularly travel so that tracks will be left behind. Sandy substrates are less than ideal, but also record tracks with sufficient detail for identification. In cold regions there are often opportunities for observing and recording mammal tracks left in the snow. Not all snow is suitable for tracking, however; the best snow is slightly moist and only a few centimeters deep.

Mammal tracks may provide clues about behavior, age, social status, mode of locomotion, and foraging behavior as well as the identity of the animal. For example, two or more different track patterns may indicate that an animal was traveling in a group; foot size may hint at the animal's age and sex; and the distance between prints and the footfall pattern may indicate the animal's gait. Associated sign—such as partially

Figure 33. Animal sign. A. Den of the large Indian civet (*Viverra zibetha*). B. Extensive disturbance to riverine grassland caused by the rooting of wild pigs (*Sus scrofa*). Royal Chitwan National Park, Nepal. Photos: C. Wemmer.

eaten carcasses, fur, feathers, scats, and tracks left by other animals—may provide evidence of predatory habits.

PHOTOGRAPHING MAMMAL TRACKS

Photography is one of the simplest and most effective ways to obtain a permanent record of animal tracks. It can provide the observer with information about movement patterns as well as details of individual footprints. Photography may also allow an observer to record related events or activities of the animal that might have contributed to the formation of the track impression in the first place. An object of known size should be included in each photograph for scale, in addition to a label indicating the date, locality, name, and catalogue number of the specimen. Information on scale is especially important if two or more species leave footprints of similar shape but of different size (e.g., small and large canids, felids, and cervids).

CASTING TRACKS

At times it is useful to make an actual physical record or cast of a track. Track-casting should be used only as a supplement to photography, which quickly provides information about the tracks and the habitat. The two most commonly used media for casting animal footprints are plaster of Paris and rosin-paraffin moulage. Choice of one or the other may be dictated by the availability and cost of supplies, preparation time, location of the track, type of substrate, ambient temperature, and quality of the imprint.

PLASTER OF PARIS METHOD.   Plaster of Paris is the most commonly used medium for casting animal tracks (Leutscher 1960; Murie 1974). It is universally available, requires minimal preparation time, sets quickly, and stores easily. It cannot, however, be used to cast tracks in snow, because it can be relatively dense when wet and because it generates heat as it reacts with water. Materials needed to prepare track casts from

plaster of Paris include powdered plaster of Paris (stored in a waterproof container), mixing container, stirrer/pouring stick, forceps or tweezers, casting frame, water, salt or vinegar, talcum power, and soda straw. These materials are available in most urban areas.

When an investigator locates a suitable track, he or she cleans all debris from the track with small forceps or tweezers and then blows a fine dusting of talcum powder into the track with a soda straw. Dusting is important, because it prevents soil particles from adhering to the plaster of Paris. The next step is to position a casting frame around the track, gently pushing the edges of the frame into the soil while being careful not to distort the track. If the soil is dry or frozen, it may be necessary to build up a small amount of soil around the outside of the frame for additional support. The casting frame should be large enough to encompass the track and a surrounding space 3 to 4 cm wide.

Casting frames can be assembled from almost any kind of material that will contain the liquid plaster of Paris as it is poured. Sections of half-gallon or quart paper milk cartons 3 to 4 cm wide provide simple, inexpensive frames that can be prepared in advance, folded, and easily transported in a pocket or backpack. However, most paper casting frames can only be used once. Metal or plastic pastry molds (of various sizes) can also be used as casting frames, but the inner surface of the mold must be dusted with talcum power to prevent the plaster of Paris from sticking to the sides. Such molds can be used repeatedly, but they are more expensive and take up more storage space, and there is a risk of breaking the track cast when it is removed from the mold.

When a track is ready for casting, plaster of Paris should be added to water in a mixing container until the mixture has the consistency of pancake batter or is thick but pourable. A small amount of salt or vinegar may be added to the mixture to increase setting time. The resultant mixture should be poured into the track immediately. For large tracks, it may be advantageous to reinforce the cast by adding sticks or wire to the plaster before it has set. The cast and the frame should be gently lifted from the ground after approximately 30 minutes. After at least 1 hour, the frame can be removed and the edges of the cast trimmed with a sharp knife. Remaining soil and debris should be washed from the track surface of the cast by immersing the cast in water.

ROSIN-PARAFFIN MOULAGE METHOD. The rosin-paraffin moulage method is more versatile than the plaster of Paris method. This substance also has better casting quality than plaster of Paris (Kent et al. 1985), and it can be used to cast tracks in snow and ice, as well as in various types of soil. A field kit for using rosin-paraffin moulage should contain the following items or their equivalents: several boat-shaped aluminum molds, each containing prepared rosin-paraffin moulage; candle; candle base (e.g., plastic lid of coffee can); chimney stand (e.g., half-gallon paper milk carton); casting frame; forceps or tweezers; and waterproof matches.

The rosin-paraffin moulage is prepared in advance of fieldwork by melting 1 part (by weight) rosin and 3 parts paraffin in a double boiler. Canning paraffin is available in hardware or grocery stores, and rock rosin (also used by gymnasts and violinists) is usually available at specialty stores. A double boiler can be fashioned by placing a 1-lb coffee can into (but not touching the bottom of) a larger container filled with water that can be heated. Investigators should be careful when using an open flame, because both paraffin and rosin are flammable. The paraffin is melted first. Then the mixture is stirred as a preweighed amount of crushed or powdered rosin is added. After the melted paraffin and rosin are thoroughly mixed, the moulage is poured into each aluminum mold and allowed to harden at ambient temperature. Boat-shaped

molds are made from pieces of heavy-duty aluminum foil. The foil should be folded so that melted moulage cannot seep from the mold. Each mold should contain enough moulage (approximately 150 ml) for a single, moderate-size (8 × 8 cm) mammal track. Tracks of other sizes may require more or less moulage.

In the field, the moulage can be remelted over a candle, stove, or similar heat source. Candles should be supported upright in the soil or in a candle holder, made by cutting an X into the center of a plastic coffee-can lid. Either a half-gallon cardboard milk carton or a 2-l metal can opened at both ends can be used as a chimney. The chimney supports the mold containing the rosin-paraffin moulage above the candle flame and prevents the candle from being extinguished by wind. Paper milk cartons take up less space (when stored flat) than metal containers and can be trimmed to support a foil mold at an optimum height above the candle flame. If the mold is too close to the flame, the moulage may burn or the candle may be extinguished; if it is too far away, the moulage will not melt.

While the moulage is melting, the investigator should prepare the track and place a casting frame around it, as described previously (see "Plaster of Paris Method"), except that the track should *not* be dusted with talcum power. If the track is in snow, the track should be sprayed with a fine mist of water from an atomizer (perfume sprayers or pump-action sprayers are ideal). This procedure will coat the track with a thin layer of ice, making it possible to pour the liquid moulage without it melting and distorting the track. This step can be omitted if the track is already frozen and the air temperature is below −10° C. Melted moulage should be allowed to cool for a few minutes (until the edges become opaque) before it is poured into the casting frame.

One side of the mold is held close to the casting frame. The ends of the mold are then lifted upward and toward each other so that the mold folds in the middle to form a spout that facilitates pouring. The moulage should be poured carefully into a small groove running into one side of the track. Enough moulage should be poured into the casting frame to make a cast 1 cm to 1.5 cm thick. The cast can be removed from the frame when the moulage is completely opaque and firm to the touch.

LABELING TRACK CASTS. Each track cast should be labeled with the species name (if known), date of collection, location, and collector's name and catalogue number. These data should be written in permanent ink on a waterproof paper label (museum skin tag) that is inserted into the edge of the plaster of Paris or rosin-paraffin moulage before it hardens. Once the cast has hardened, the collector's name and catalogue number, the date, the locality of collection, and the species identification should be permanently inscribed on the underside of the cast using a sharp object.

HANDLING AND STORING TRACK CASTS. Plaster of Paris takes about 24 hours to "cure" fully. Rosin-paraffin moulage usually sets within 1 hour. Fully hardened casts should be stored in a cool, dry place. Cast detail may be distorted by softening if rosin-paraffin casts are exposed to temperatures above 40° C. Before transport, each cast should be wrapped separately in tissue paper and packed carefully to avoid breakage. Plaster of Paris casts generally are more fragile than casts made from rosin-paraffin moulage.

TRACK BOARDS AND FLUORESCENT POWDER

The presence of mammals can be recorded from tracks left on boards covered with smoked kymographic paper or smoked aluminum plates (Raphael et al. 1986; Carey and Witt 1991). This technique has the same limitations as camera censuses using trip-plates (see "Remote-Trip Cameras," below): the animal must step in the right place, and the step-plate (i.e., the smoked

paper or aluminum) cannot be exposed to rain. Track boards made of a mixture of clay and petroleum jelly can survive a hard rain, but prints are usually marred beyond recognition. Under ideal conditions, track boards provide evidence of the presence of particular species. However, in a study of northern flying squirrels (*Glaucomys sabrinus*), Carey and Witt (1991) found little correlation between track board counts and abundance determined through trapping. In addition, the technique provides no information on the movements of individuals.

The development of tracking with fluorescent powder (Lemen and Freeman 1985) has provided a dynamic means to measure the movements of animals without the expense of radiotelemetry, and in some cases a better means than track boards to record presence. Powder can be mixed with bait for small mammals and tracks around bait stations used to identify species (M. V. McDonald, pers. comm.). These bait stations can be placed in the same array as track boards. Fluorescent powders are inexpensive (U.S. $10/lb), and a pound is sufficient to mark 100 animals. Handheld ultraviolet lights cost between U.S. $50 and $150 (see Appendix 9).

SCENT STATIONS

The use of scent stations to determine the presence of carnivores within survey areas has become increasingly popular (Johnson and Pelton 1981). Each station consists of a circle of soft earth, about 1 m in diameter, with a centrally placed synthetic attractant (Turkowski et al. 1983) or natural lures such as fermented egg or bobcat urine (Linhart and Knowlton 1975; Conner et al. 1983). Animals visiting the scent station are identified by the presence of their tracks on the soft earth. Scent stations are usually located from 0.3 to 0.5 km apart along several transect lines.

Conner et al. (1983) activated scent stations in the evening and checked them the following morning, sometimes on several consecutive days. They defined a visit as one or more tracks of a species per station and expressed visitation rates as the percentage of stations visited by a species per transect-night (Conner et al. 1983). This value was used as an index of relative abundance.

The technique was originally developed to determine the relative abundance of red and gray foxes (Wood 1959). It was later applied to coyotes (Linhart and Knowlton 1975), black bears (Lindzey and Meslow 1977), bobcats, raccoons, and opossums (Conner et al. 1983) and was also adapted for river otter and mink (Humphrey and Zinn 1982). Conner et al. (1983) compared scent station indices from bobcats, raccoons, gray foxes, and opossums with their population abundances as estimated through trapping, radiotelemetry, and radioisotope tagging techniques. They found that scent station indices accurately reflected the population abundances of bobcats, raccoons, and gray foxes, but not of opossums. Scent station indices, therefore, may not be related to population abundance in some species. In addition, between-species comparisons of abundance cannot be made because of species-specific visitation rates.

We recommend the following procedures for standardization. Scent stations should be operated when target species are most active (Roughton and Sweeny 1982). The stations should be placed at equal distances along transect lines and distributed proportionately among the major habitat types of the survey area. The tracking surface, the attractant, and the method of presentation of the attractant should be standardized (Roughton and Sweeny 1982; Conner et al. 1983). Finally, multiple-night sampling should be considered for species that occur at low densities.

IDENTIFYING AND INTERPRETING MAMMAL TRACKS

Tracks can be identified through reference to field guides (e.g., Perkins 1954; Seton 1958;

Leutscher 1960; Lawrence and Brown 1967; Murie 1974; Twigg 1975; Aranda Sánchez 1981; Brown 1983; Headstrom 1983; Stokes and Stokes 1986), although, ideally, the identity of the animal making the track should be confirmed by direct observation. In many groups identification to genus is feasible, and with practice it may be possible to identify tracks of some mammals to species. Tracks made by medium- to large-size species are often easier to identify than tracks of smaller ones. Photographs and written records of the habitat where tracks are observed should be made, including evidence of ancillary sign (e.g., hair, fecal pellets, partially eaten or rejected food items).

Stored Food and Food Remains

Specific behavioral adaptations for the storage of food may create unique signs of the presence of a species. The most remarkable examples of food storage are the larder hoards of many rodents, although these collections of seeds and grasses tend to be concealed under vegetation, soil, or rock piles. Traces of food are often visible where the rodent prepares seeds and cones for hoarding. Undigestible remains of prey also reveal the activities of predators, and wings of moths and birds, beetle elytra, fruit husks, and pulp castings often litter the feeding places of bats. The remains of large mammalian prey such as antelope and deer often accumulate near carnivore dens or resting areas, and tooth markings on bones may be used to identify the predator. Owl pellets invariably contain the hair and bones of mammalian prey, usually well prepared for scientific identification.

Scats

Feces, or scats, of mammals may be found almost anywhere, but many species, particularly carnivores, leave their feces in predictable places on trails or elevated objects, or beside prominent landmarks (Fig. 34). A few species ha-

Figure 34. Mammal feces used as a record of species in an area. A. Feces deposited on a wooden bridge in the Nepalese terai. The two hair-filled scats (left and center) were made by an unknown carnivore; the black scat (right) is from a sloth bear (*Melursus ursinus*). B. Old tiger (*Panthera tigris*) feces packed with sambar (*Cervus unicolor*) hair. Royal Chitwan National Park, Nepal. Photos: C. Wemmer.

bitually defecate in a particular place, or latrine, which may be used by a single individual (e.g., in shrews, porcupines, wood rats, and antelopes) or several individuals (e.g., in bats, spotted hyenas, aardwolves, and brow-antlered deer) (Walther et al. 1983; Putman 1984). Scats disintegrate rapidly in tropical areas, where they also are often consumed by coprophagous insects (Wiles 1980).

A typical survey of pellet groups involves searching a sample of randomly or systematically selected circular or rectangular plots. Often, permanently marked plots, which are periodically cleared of pellets, are used, but temporary plots where a scat deposition period can be determined (e.g., time since autumn leaf-fall) can also be used. An assumed daily rate of defecation is then used to derive an estimate of density. Eberhardt and Van Etten (1956), for example, estimated density of white-tailed deer as follows:

$$\text{Deer/plot} = \frac{\text{average number of pellet groups per plot}}{12.7 \text{ pellet groups/deer/day} \cdot (\text{days since leaf-fall})}$$

Numbers of deer per plot were then converted to animals per unit area (e.g., km^2) based on plot dimensions.

The quantity, size, shape, and consistency of feces depend to a large extent on diet, which often shows seasonal changes even in the humid tropics. Species identification of scats requires strong circumstantial evidence (footprints, hair, or camera-trap photos) or, preferably, direct validation of the animal "in the act" to collect an identifiable voucher. Scat from certain sympatric mammalian herbivores as well as some carnivores can be distinguished by pH (Howard 1967; Hansen 1978).

Olfactory sign

Olfaction is a prominent sensory mode of most mammals. Specialized gland secretions and excretory and fecal products with characteristic odors are the primary carriers of olfactory information. Scent serves a variety of social and anti-predator functions; it may be passively dispersed or intentionally deposited in specific places (scent marks). Scent marks are often associated with specific environmental features such as tufts of grass, twigs, or visually prominent objects. The zone of detectable scent has been called the "active space" by chemical ecologists, and the longevity of a scent depends upon the molecular characteristics of the substance (Bossert and Wilson 1963). Strong scent often emanates from mammalian habitations and even from animals themselves, and it thus can serve as a secondary cue to the location of the animal or its den. The cuscuses (*Phalanger* spp.) and porcupines (*Coendou* spp.) are particularly well known for their pungent odors, which may allow the biologist to locate a tree occupied by an animal. Colonial roosting bats may also be detected from a distance by the odors of skin gland secretions and accumulated guano. The recent presence of many large ungulates and carnivores (mustelids in particular) can be detected by odor, and scents of some species, such as skunks, maned wolf, and bush dog, may persist for days or weeks. Scents of some mammals are not detected by the human nose, and these species may be detected visually more easily than by smell.

Remote-trip cameras

Remote-trip cameras, or "camera traps"— cameras in which the animal itself triggers the shutter—have been used by wildlife photographers since the early 1900s (Nesbit 1926; Champion 1928; Shiras 1936; Gregory 1939). The advent of automatic film advance and electronic flash, however, has greatly increased the application of remote-trip photography for wildlife studies. Remote-trip cameras are ideal for identifying the species living in a particular area (Fig. 35), for monitoring relative and absolute

Figure 35. Mammals photographed with remote-trip cameras. A. Javan rhinoceros (*Rhinoceros sondaicus*). B. Leopard (*Panthera pardus*). Ujung Kulon National Park, Java, Indonesia. Photos: M. Griffiths (World Wildlife Fund, Indonesia).

abundance of species, and for studying activity patterns (Seydack 1984; Rappole et al. 1985; McShea and Rappole 1992). These devices have also been used to address a surprising variety of ecological and conservation-related questions (Pearson 1959; Carthew and Slater 1991; Griffiths and van Schaik 1993; Leimgruber et al. 1994).

Remote-trip cameras have several advantages over other inventory methods. The cameras are relatively nonintrusive: An area can be monitored with minimal human disturbance, and the animals do not have to be captured. Large areas can be surveyed by only a few people (Seydack 1984; Rappole et al. 1985), and investigators do not have to be in constant attendance. The cameras are ideal for detection of cryptic terrestrial species that are difficult to capture but that use established trails, feeding sites, or dens. Individual animals can be studied if they have distinctive markings, scars, or other marks (Seydack 1984; Rappole et al. 1985), and if the animals are trapped and marked, remote-trip photography can greatly augment data on abundance, movement, and activity (Griffiths and van Schaik 1993; Karanth 1995). The disadvantages of the method are the relatively high costs of equipment and film, the risk of equipment theft, and the tendency of many trigger devices to produce size-biased (and therefore, species-biased) records. Remote-trip cameras are also difficult to repair in the field. In addition, information easily obtained from captured animals, such as mass and reproductive condition, cannot be recorded with cameras, although shoulder height can be measured accurately by reference to a meter stick fixed within the camera frame.

Camera Equipment

A remote-trip camera must have a battery-operated film advance and an electronic flash. Many 35 mm cameras now on the market have these features, and the electronic shutter-release

Figure 36. Camera in a protective housing made of plywood and Plexiglas and painted green. A plastic canopy protects the unit from rain. Lore Lindu National Park, Sulawesi, Indonesia. Photo: C. Wemmer.

feature allows easy connection to a variety of trip mechanisms (Figs. 36 and 37). With a little ingenuity, older camera models, including Polaroid and small-frame cine cameras, can also be modified as "camera traps" (Abbott and Coombs 1964; Joslin 1977, 1986; Smythe 1978; Goetz 1981; Mims 1982; Plage and Plage 1985; Picman 1987; Savidge and Seibert 1988). However, retrofitting older models is an activity best left to the committed mechanic and tinkerer.

Remote-trip cameras using an active infrared sensor (see "Trigger Mechanisms," below) should be fitted with a wide-angle autofocus lens. All major camera manufacturers now produce such lenses that are suitable for use in camera traps. A quartz crystal data-back will automatically record time or date information on each photograph. Unfortunately, cameras generally do not

Figure 37. Camera arrangement used to identify tigers. The pressure pad is in the right foreground. Royal Chitwan National Park, Nepal. Photo: Charles McDougal.

stamp the photo with both date and time. When using fixed-location sensors (trip-plate or photic beam), a battery-operated clock can be positioned within the focal plane. A single-frame camera with trip-plate can be constructed quite inexpensively (Picman 1987), but an automatic unit with all the above features costs U.S. $375 to $500 (e.g., Camtrak or Trailmaster models; see Appendix 9).

Mammals photographed with Kodak Pan-X and Tri-Pan high-contrast black and white film are often individually identifiable from negatives, but color film gives a more realistic rendering of visual details. Electronic flash allows the researcher to overcome problems of changing ambient light levels and limited depth of field. Many electronic flash units are calibrated for indoor photography where walls are highly reflective; when used in the forest, they produce slightly underexposed pictures. Test trials allow the user to determine proper f-stop adjustment. Usually the aperture needs to be opened one-half

to one full f-stop beyond that suggested by the manufacturer. Relatively inexpensive "sureshot" cameras with auto-focus and built-in flash units avoid the vexing problems of older-model cameras, which need an external flash unit and rely on a fixed focus (Goetz 1981; Seydack 1984; Plage and Plage 1985; Carthew and Slater 1991). Sureshot cameras are the models that are modified commercially as camera traps.

Shutter noise and light from the flash may frighten or attract a passing animal (Fig. 35). This can affect sampling and may lead to equipment damage. Tigers in Nepal avoided a camera trap for weeks after experiencing camera noise, although tracks indicated that the animals were in the area (Charles McDougal, pers. comm.). Wildlife cinematographers use commercially made "blimps," or mufflers, to absorb camera sound, but these are costly accessories that can be easily made from cloth and cotton batting. One or two mufflers are placed over a camera after it is positioned on the tripod. Alternatively, a camera may be placed in a solid housing lined with a sound-absorbent material such as cotton batting or the plastic bubble sheets used as packing material. It is also possible to place a camera some distance from the trail and use a longer-focal-length lens. With the flash positioned close to the trail excellent pictures can be obtained. Commercially produced remote-trip cameras are equipped with an adjustable delay circuit, which helps prevent the exposure of an entire roll of film when a single curious animal investigates the unit. For protection against animals and the elements, camera units should be enclosed in a protective housing (Fig. 36).

Trigger Mechanisms

There are mechanical, electric, and electronic, light-sensitive trip mechanisms. Proficiency in using various trigger mechanisms is achieved only through experimentation and practice.

Mechanical trips have been used successfully by Joslin (1977, 1986), who attached meat baits to strings. Up to three baited strings were connected directly to the shutter release of a Polaroid camera, and as many as three species were recorded in a single night. A flexible steel wire or splinter of bamboo can also be used as a "whisker" trip along narrow trails. By tripping the "whisker," the passing animal completes an electric circuit and takes its own picture.

Trip-plates are the simplest means of activating a shutter (Fig. 37). These can be constructed using two pieces of wood hinged together in such a way that pressure completes an electric circuit. A small spring is used to maintain the circuit in an open position. Commercially made waterproof, pressure-sensitive pads (the kind used to trigger automatic doors in stores) are a workable alternative. The animal must tread directly on the trip-plate for it to function properly (Seydack 1984). The plate is placed at a constriction in a trail, usually under a sheet of plas-

tic, and camouflaged with soil and leaf litter (Seydack 1984). It also may be placed beneath a baited site (Goetz 1981). Homemade trip-plates are sensitive to weather conditions and should be sealed in plastic bags; special precautions should be taken to ensure drainage and protection from dew (Goetz 1981).

Photic cells and camera units have been used to record the passage of animals through a beam (Pearson 1959; Wemmer and Watling 1986; Savidge and Seibert 1988; Carthew and Slater 1991). Proper alignment of the beam is critical, often necessitating use of a few frames of film to fix the position of transmitter and receiver (Carthew and Slater 1991). In addition, falling leaves and flying insects, especially large night-flying moths, may break the beam and trigger the camera. This problem can be overcome by using a *pulsed-infrared beam,* in which the receiver monitors the pulse rate rather than the magnitude of the beam; such a beam also eliminates the problem of ambient light triggering the camera (Carthew and Slater 1991). By adjusting the height of the beam, animals to be photographed can be selected on the basis of size.

Infrared beams (Rappole et al. 1985; Carthew and Slater 1991) and *motion sensors* (Rappole et al. 1985; McShea and Rappole 1992) offer alternative methods of triggering a shutter. Some investigators have used *active infrared sensors* in place of infrared beams (Rappole et al. 1985; Leimgruber et al. 1994). Because these sensors monitor a "heat field," it is not necessary to align the transmitter and receiver. The Trailmaster active infrared trail monitor consists of a unit that generates a wide beam and a receiver that detects a small (3/8-in.-wide) segment of that beam. The investigator places the detection beam at the chest height of the desired species and sets the length of time estimated for the animal to pass through it. Because infrared sensors are triggered by sudden changes in ambient heat, some models are prone to "false triggering" if sunlight moves across the sensor window. This problem can be minimized by placing the camera out of direct sunlight or by using a "daylight" or photo-electric switch, which activates the camera only at night.

Passive infrared sensors are not selective and record any warm-blooded animal passing through a broad wedge-shaped detection field. Power is needed to drive the transmitters for both the infrared beam and the infrared sensor systems. A rechargeable motorcycle battery (12 volt, 24 amp) is sufficient to power the beam and sensor systems for at least 20 days. Because the sensor is triggered by heat, detection of reptiles and amphibians is not assured unless they cast a shadow across the sensor.

Video Cameras

Time-lapse video cameras have been used to monitor birds, either continuously or intermittently when triggered by a sensor (Capen 1978; Montalbano et al. 1985; Mudge et al. 1987). One tape can monitor a 24-hour period in the time-lapse mode and can be viewed in less than 2 hours on a video player or recorder. Video cameras can also be linked to intervalometers, which take single-frame exposures at specified intervals. But as video units retail for over U.S. $2,000, costs can be prohibitive.

A 35 mm camera can be modified to trigger at regular intervals and to simulate the qualities of a video camera (Rutnagur et al. 1990). Commercial interval cameras are also available (see Appendix 9). However, the large quantities of film needed for such cameras can be expensive. Nevertheless, video or interval cameras are desirable when monitoring species visitation to highly frequented sites such as bat roosts and the fallen fruit beneath tropical trees.

Camera Placement

Proper positioning of remote-trip cameras is critical and often time consuming. Game trails

are logical places to place cameras because they are used by many species (Seydack 1984; Wemmer and Watling 1986; Griffiths and van Schaik 1993). Ideally, a trail should funnel animal traffic in front of the camera topographically, but strategically placed sharp stones and thorny vegetation can also be used. Channeling an animal's movement is especially critical when using pressure pads to trigger the camera. The optimal camera position should provide a full-body shot of the desired species, or of the largest species to be photographed in the study area. A tripod is useful for positioning a camera, because it can be easily moved and adjusted; it can, however, be knocked over and destroyed by large mammals. Commercially available camera traps come with a Velcro belt that can be strapped to any available tree.

Baits can also be used to attract predators (Joslin 1977, 1986; Rappole et al. 1985; Wemmer and Watling 1986) and frugivores (Smythe 1978) to a camera trap. Scent stations are another potential means of attracting mammals. To our knowledge, no one has yet used a drift fence to funnel animals past a camera trap, but this technique could prove useful for small terrestrial mammals.

Finally, a camera trap can be placed at a natural site that is known from sign to be regularly used by a particular species. Watering holes, salt licks, burrows, latrines, and fruiting and flowering trees are examples. Such natural and simulated features have been used to attract mammals within range of sensors by investigators studying nest predation (Savidge and Seibert 1988; Leimgruber et al. 1994) and pollination (Carthew and Slater 1991).

Sampling

Statistical inference used with camera-trap methodology is yet to be perfected, but Karanth applied capture-recapture models to camera-trap data "for estimating objectively parameters such as size, density, survival, and recruitment for populations of tigers and other secretive animal species" (Karanth 1995:333). Although more work is needed in this promising area, we offer some guidelines and generalizations about sampling.

First, triggering devices are not "objective detectors" of all passing wildlife. There are limitations to using the method to inventory a wide range of wildlife (e.g., large homeotherms to small poikilotherms) or to sample species of differing habits (e.g., terrestrial vs. arboreal rodents). Because trigger mechanisms differ in species sensitivity, the fieldworker must validate the characteristics of the chosen system. Not all animals may readily pass through the picture window framed by photic beams, and thus density will likely be underestimated. The detection distance of active infrared sensors depends on the size of the animal and the contrast it makes with background temperatures. Large animals passing by on a cold night are more likely to trigger the camera than are small animals passing by during the day. Comparisons between species that differ in size or activity period may not be valid unless detection distances are standardized. In areas of heavy understory, for example, detection distance can be standardized if all patches of vegetation cleared are of the same size. Angling the camera down from a high vantage point will also reduce the importance of animal size in detection. For all of these reasons, it may prove wise when beginning a study to survey a single species or several similar species at one time. In the process, differences in detectability that are due to differences in size and behavior may become apparent.

Second, remote-trip cameras cannot provide a measure of individuals per unit area *unless* individual animals can be identified and the size of the area being sampled by the camera is known. Individuals of many species can be identified by distinctive markings and scars (J. L. D. Smith et al. 1989), but long-lasting artificial markers

(e.g., colored and numbered ear tags, dyes for light-colored animals, freeze brands for dark-colored animals) can also be used (Hill and Clayton 1985). Determining the size of the area sampled depends upon a knowledge of the home range of the species under study, and this requires validation by fieldwork. In long-lived animals, combining an annual capture/marking period with monthly camera censuses may provide a great deal of information. When individual animals cannot be identified, density cannot be estimated.

Third, cameras placed along trails or at feeding sites do not randomly sample a habitat, and the data are subject to the problems inherent in biased line transects (Burnham et al. 1980). We recommend that investigators use a stratified placement of cameras to sample large herbivores, such as deer and antelope, and large carnivores. One camera should be placed within each 1-km^2 block, along an obvious trail or at a watering or feeding area. The camera site may be baited for small carnivores, but the effects of differential attractiveness of baits must be considered. The cameras should be set for a fixed period, which will depend upon the frequency of trail visitation. Intensive examination of the relationship between camera-trap sampling and the movement of radio-collared animals within known home ranges is needed.

Target organisms and habitats

Abundance estimates based on counts of sign are particularly useful in situations in which animals are difficult to see and to count consistently. Cryptic species and species occupying habitats in which visibility is poor can often be more easily surveyed by counting sign. Quite clearly, however, these techniques can be applied only to species that produce conspicuous sign that can be easily and reliably distinguished from sign of other species.

Sign counts are not appropriate for broad-scale survey work; they are best used for monitoring individual species or a small number of target species. Population size and habitat use of North American Cervidae have been investigated using fecal pellet indices (Bennett et al. 1940; Neff 1968). In Australia, sign counts have been used predominantly to assess differential use of habitats by medium-size to large mammals, particularly macropods, by estimating relative abundance of species across different habitats (Caughley 1964; Hill 1981, 1982).

Few investigations have attempted to use sign other than fecal pellets to estimate animal abundance, and in nearly all cases only relative abundance was considered. Pocket gophers are an exception; it is possible to obtain total counts of populations in the nonbreeding season by spatial sampling (see "Spatial Sampling," Chapter 10) for fresh mounds. Other signs used to monitor populations include trails (Newsome et al. 1975; Johnson 1977; Coulson 1979; Statham 1983), holes in fences (Coulson 1979), nests and diggings (Christensen 1980), scent stations (Lindzey et al. 1977; Randall 1979; Conner et al. 1983), and tracks (Justice 1961; Downing et al. 1965; Lord et al. 1970; Sarrazin and Bider 1973; Boonstra et al. 1992).

All mammals leave some record of their presence in an environment (e.g., fecal pellets, tracks, nests, or diggings). Such animal signs can be surveyed using the same techniques used to count animals directly, and the results can be analyzed with the same statistical treatments. It is important to note that counts of sign are simply that, and they can be translated into animals only by calculating from a known ratio of signs per animal (Putman 1984). Counts of sign need not, however, be considered an index of animal abundance and can often be examined simply to document occurrence, depending on the aims of the study. As a means of analyzing comparative densities or population trends, a density index is often useful (Hill 1981).

Many authors have expressed reservations about the use of sign to assess abundance and have stressed the potential sources of error (Hill 1981; Putman 1984; Johnson and Jarman 1987; Southwell 1989). Sources of variation in counts of sign include the following:

1. *Environmental heterogeneity.* Differences in vegetation or topography may result in nonrandom use of the population range and thus result in nonrandom distributions of animal sign.
2. *Nonregular production of sign.* Animals may have "preferred" times of day for producing sign, preferred areas within their range, or preferred vegetation types. Animals of different sex or different age may produce sign at different rates.
3. *Mobility of animals.* The patterns and rates of movement of animals in a population may vary by time of day, season, or vegetation type, resulting in irregular distribution of sign in the survey area, even if the amount of sign produced per unit time is constant.
4. *Detectability of sign.* Detectability of sign may vary in different habitats and on different substrates.
5. *Decay rates.* The rates at which sign becomes undetectable may vary according to habitat, substrate, time of day, season, and weather conditions.

Most of these problems can be avoided with the use of proper sampling protocols. It is important, therefore, that these limitations be addressed in the design of any survey based on sign counts. Sampling can be designed to detect and evaluate the extent of the variability in these parameters, especially environmental heterogeneity and animal mobility. It is also essential that the number of signs produced per animal per unit time be determined independently for each population when populations are to be compared. This number is used to relate counts of sign to animal numbers (see "Scat and Other Sign Counts," Chapter 10).

Although there are many potential sources of error in estimates of animal abundance based on sign counts, a number of detailed studies report a good fit between estimates of absolute population size based on counts of sign and those derived by other methods (Johnson 1977; Christensen 1980; Johnson and Jarman 1987). A similar good fit is reported in studies of habitat use by a number of macropod species in which estimates from sign counts were compared with those based on more traditional survey techniques (Coulson 1979; Hill 1982). Johnson (1977), Coulson (1979), and Stewart (1982) repeatedly counted fecal pellets produced by populations of known size to evaluate the accuracy of sign counts in estimating absolute abundance.

Research design

For surveys of animal signs, as with direct counts of animals, the investigator must decide upon the type of sampling unit to be used, the placement of these units in the survey area, and whether to stratify the survey area (Southwell 1989). For practical reasons it usually is impossible to count all animal sign in an area. The only total counts reported in the literature were carried out inside animal enclosures (Stewart 1982). Samples of some sort must therefore be drawn from the population to arrive at an index or estimate of abundance.

One of the first decisions to be made in conducting a survey based on counts of animal sign is what sign is to be used and whether all such sign is to be included. Some investigators have attempted to age signs and to record only those determined to have been recently deposited or created. Age classes of fecal pellets have often been based on external appearance, whereas classes of signs such as nests or diggings have been based on evidence of recent use. Such clas-

sifications are best reached by reference to sign of known age and origin or through controlled experiments on the decay rates of sign.

In general, two methods are available for estimating population abundance from counts of animal sign. Most previous studies have involved assessment of "standing crop" of animal sign in random quadrats or transects. More recent studies have examined the rate of accumulation of sign in fixed sample plots that are regularly cleared. Both methods have been developed largely from studies involving counts of fecal pellets, but the methods are equally applicable to other types of animal sign. In the following discussion we provide a general outline of each method and the sampling procedures employed. More detailed reviews of techniques applied specifically to fecal pellet counts are given in Johnson (1977) and Southwell (1989).

Standing Crop Method

With the standing crop method, the population size in an area is calculated by dividing the accumulation of some animal sign over a known period (i.e., standing crop) in an area by the known rate of production of that sign for the species in question (Putman 1984). Thus, this method requires, in addition to counts of sign, determination of the time period over which the sign has accumulated and determination of a production rate for the sign. The first can be established, in theory, by studies of the rate at which sign becomes undetectable (decay rate), and the second from long-term observations of animals. The need to derive estimates of decay and production rates for sign can be overcome to some degree by counting only sign of a particular age or origin as determined by reference samples of known age.

The rates at which signs become undetectable will vary widely with substrate, season, and climatic conditions. Thus, extremely detailed study of the decay patterns for the sign being investigated is required if sign standing crop is to be used to estimate the absolute abundance of a population. The technique is also limited by the general criticisms noted earlier regarding nonrandom patterns of sign production, both temporal and spatial. Because of these problems, standing crop counts are best for estimating the mean relative abundance of animals in an area. This method has been successfully applied to a number of macropod species (Floyd 1980; Hill 1982).

Accumulation of Sign Method

The more certain method of measuring the rate of accumulation of sign in an area is to establish permanently marked plots or transects, clear them of sign, and then count the sign on them at a later date. The counts should be corrected to account for unfound sign or sign that may have disappeared in the interval between deposition and counting. Rates of disappearance (decay rates) can be determined by counting the proportion of sign lost from fixed control plots during a fixed time interval due to agents such as decay and natural or artificial disturbance. This method resolves many of the problems inherent in the standing crop method and provides an unbiased estimate of sign deposition rates. Although clearing of fixed plots or transects may cause animals to be attracted or repelled by the area, this method has quantitative advantages over the standing crop method (Putman 1984). It may not always be logistically possible to use such techniques, however, and many studies must be based on standing crop sampling.

Sign-Marking Methods

Sign marking provides a basis for calibrating sign counts using estimates of sign production from marked individuals within a population (Skalski 1991). Examples of this approach include embedding materials (visual markers) that outlast the scat in food, injecting dyes that later

appear in fecal pellets into subjects, and toe-clipping animals to distinguish tracks left by different individuals (Justice 1961; Pelton and Marcum 1977). The interpretation and analysis of sign-marking data have often been handled in a fashion similar to that used with mark-recapture data. Skalski (1991) provided a statistical model and variance measures that permit abundance and density estimation.

Size, Shape, and Placement of Samples

The next factor of research design to consider is the size, shape, and placement of sampling units within the survey area. The relative merits of different configurations of sampling units have been considerably debated (Johnson 1977; Southwell 1989). Circular plots that are systematically or randomly placed within the survey area have been used most commonly (Southwell 1989). This approach effectively combines the advantages of quadrats and transects. To date there is no single preferred option, and choices of sampling units will be determined by the habitat characteristics of the survey area and the type of animal sign being surveyed. In general, a decision on the size and shape of sampling units should attempt to maximize precision and optimize efficiency over time. A large number of small plots will normally provide better precision than fewer but larger plots that cover the same sample area (Neff 1968). In other words, less total sample area will be required to achieve the same statistical precision if small plots rather than large plots are used. Small plots are also more time-efficient (Johnson 1977). Different size and shape combinations can be tested to arrive at an appropriate design to suit the aims of a particular biodiversity inventory and monitoring study. Comparability can be achieved through the statistical treatment of the results.

Sampling intensity, or the proportion of a study area that is sampled, affects the degree of statistical precision of the estimate of sign density. A pilot study should be conducted to determine the optimum sampling intensity, using the following formula (Neff 1968):

$$N = \frac{t^2 s^2}{(a \cdot x)^2}$$

where

N = number of plots required
t = Student's t-value for a selected probability level
s^2 = variance of the data from the pilot study
a = selected risk error (e.g., 0.20, if the estimate is required to be within 20% of the mean on 95% of occasions)
x = mean sign density from the pilot study.

Alternative equations for determining sampling intensity have been derived by Cochran (1977) and Overton and Davis (1969).

Many surveys of animal sign have employed a stratified sampling design (Caughley 1964; Floyd 1980). In general, this design has been used not to improve the accuracy of population estimates but as a means of examining differential use of habitat within a survey area.

If long-term, repeated monitoring of a population is planned, the timing and frequency of sampling should be planned to ensure that short-term seasonal variations do not obscure any long-term differences that are being investigated. Regardless of the experimental design chosen, at some point early in the study the data and field methods should be evaluated and methods modified as appropriate.

Field methods

Procedures for counting animal signs are straightforward, irrespective of the sampling technique selected. Quadrats or transects are systematically searched, and the quantity of animal sign is recorded according to any predetermined categories. Information for each species being sur-

veyed is recorded separately. Depending on the extent and desired precision of the survey, it may be necessary to establish a permanent transect or grid system prior to the initial count. Fixed marker points enable areas or transects to be resurveyed and facilitate the selection of random sampling units.

Data handling, analysis, and interpretation

Data Handling

The type and amount of data collected will depend on the specific goals of the study. The simplest record is a tally of the animal sign recorded within each sampling unit. When the sign is categorized, or when more than one species is sampled, each sign type is recorded separately. The observer may also record microhabitat information such as substrate type, dominant vegetation cover, or distance from ecotone. A detailed description of all habitats sampled should accompany any sign-count survey.

Analysis and Interpretation

Counts of animal signs are used as an alternative to direct counts of animals for the purpose of monitoring abundance. The results of such surveys provide the investigator with an index of abundance for the target species in the survey area over the sampling period. The manner in which the data are then extrapolated depends upon the aims of the study. The index can be used directly to represent a measure of relative abundance using values for the production and decay rates of sign calculated during the study. It is important to note that, whether dealing with relative or absolute abundance, estimates produced using counts of sign are estimates of mean population density over the period during which the recorded sign has accumulated in the study area. These abundance estimates, however, may still be treated in much the same way as estimates derived from other direct or indirect survey techniques (see "Estimation of Mammal Abundance," Chapter 10).

In cases in which microhabitat data have been incorporated into the survey design, additional information can be obtained in relation to the distribution of sign (and, therefore, the population within the survey area) in both space and time. Nonparametric statistics can be used to analyze differences in the abundance of sign or animals on the basis of microhabitat features such as substrate and habitat type. Such analysis provides useful information on the study population and also an opportunity to examine the validity of any assumptions or limitations identified for the technique.

When areas are surveyed more than once during a sampling period by different observers, it is desirable to evaluate interobserver variation in results, using analysis of variance. This type of analysis is especially important when both experienced and inexperienced observers are used.

Special Considerations

The primary practical limitation to using counts of animal sign for monitoring abundance is that counts are generally time-consuming and therefore expensive to apply over any but small areas (Southwell 1989). For this reason the technique is most effective for monitoring small, localized populations and for monitoring the use of small areas of habitat that are difficult to survey with more conventional techniques.

The accuracy of sign counts in estimating absolute abundance is largely influenced by the accuracy of the decay and production rates determined for the animal sign being used. Both parameters can be highly variable in time and space and can result in potentially large errors. Effective training of observers prior to a survey can obviate many of these problems.

Personnel and materials

The number of persons required to conduct a survey using sign counts depends on the size of the area to be sampled, the sampling intensity, and the characteristics of the sign and habitat being surveyed. Where possible, the number of persons involved in conducting a sign-based survey should be kept to a minimum to reduce interobserver bias. The best results from sign counts are achieved by a single person sampling all units. If two or more observers are used, they should be rotated among sampling areas and times so that each individual samples all areas. Presampling training can also significantly reduce individual differences.

Few materials are required for counts of animal signs: data sheets, pens with permanent ink, a map of the sampling area(s), a measuring tape, and a compass. Stakes and flagging tape are required if transects or grids are to be marked permanently. When observers are inexperienced, training and reference collections or photos of sign may increase the accuracy of identifications. For this reason, when sign counts are to be used to estimate absolute abundance, fixed control plots must be incorporated into the survey design.

CONTRIBUTOR: THOMAS GRANT

Chapter 10

Techniques for Estimating Abundance and Species Richness

We are seldom able to enumerate all of the individual animals of a particular species in any study area. It is also frequently difficult to count all of the mammalian species in an area of interest. In this chapter, we present methods for indexing and, more importantly, for estimating abundance of single species and species richness from various kinds of count statistics. We refer readers wanting a more detailed treatment of abundance estimation to Seber (1982). The literature on methods for estimating species richness is not nearly so well developed, and we are aware of no general source for details of these methods.

Estimation of mammal abundance

Introduction

JAMES D. NICHOLS AND MICHAEL J. CONROY

Knowledge of biodiversity requires knowledge of the number of species (or other taxa of interest) and the respective abundances of these species. All diversity indices can be constructed with this information. In this section, we focus primarily on methods for estimating the abundance of animals. Resulting abundance estimates can then be used to develop species abundance distributions or associated diversity indices.

CONCEPTUAL FRAMEWORK

Methods of abundance estimation appear diverse and are typically presented as unrelated entries in methods "cookbooks." In this section, we emphasize the underlying concepts that are shared by virtually all estimation methods (see Lancia et al. 1994). The biologist interested in estimating animal abundance faces problems related to (1) observability of those animals and (2) sampling of the space that they occupy (see Chapter 3). All methods of abundance estimation represent solutions to problems in one or both of these classes.

OBSERVABILITY.   All abundance estimates are based on count statistics. Number of animals counted in aerial surveys, number of animals counted on ground transects, and number of animals caught in traps are examples of count statistics used in abundance estimation. Count statistics usually represent unknown fractions of target populations and are of little use by themselves for abundance estimation. Some knowledge of the fraction of animals represented by a count statistic is needed in order to use the statistic to draw reasonable inferences about the population. The central problem of abundance estimation thus involves estimation of the observability (observation probability) associated with a count statistic.

If we let C denote a count statistic, N population size (the quantity of interest), and β the observation probability, then we can write the following relationship:

$$E(C) = \beta N \qquad (1)$$

where $E(C)$ denotes the expected value of the count statistic (i.e., the average value of the count statistic if a large number of samples could be taken). In probabilistic terms, β is the probability that an individual animal in the population (N) will appear in the count statistic (C). If we can figure out some way to estimate β, then we can estimate N as follows:

$$\hat{N} = \frac{C}{\hat{\beta}} \qquad (2)$$

where hats denote estimates. For example, say that we catch 20 deer mice in traps ($C = 20$) and that we estimate that about 25% of the animals in the sampled area were caught ($\hat{\beta} = 0.25$). Then we can estimate the number of mice in the area as $\hat{N} = 20/0.25 = 80$.

SPATIAL SAMPLING.   A second problem is that limitations of time and money typically preclude obtaining count statistics over the entire area of interest. In many instances, the investigator must select sample areas that represent some fraction, α, of the total area of interest. Unlike the fraction associated with observability (β), the fraction of area sampled is frequently known with reasonable accuracy. If we can estimate the number of animals for a representative sample area as $\hat{N}'$, then we can estimate number of animals in the total area ($\hat{N}$) as follows:

$$\hat{N} = \frac{\hat{N}'}{\alpha} \qquad (3)$$

If α is not known, but is estimated, then we replace α in equation 3 with the estimate, $\hat{\alpha}$. If we estimate that 32 white-tailed deer ($\hat{N}'$) occupy sample areas representing 10% of some larger area of interest ($\alpha = 0.10$), then we estimate the number of deer in the total area as $\hat{N} = 32/0.10 = 320$.

CANONICAL ESTIMATOR.   We can combine concerns about observability and spatial sampling into a single canonical estimator:

$$\hat{N} = \frac{C}{\alpha\hat{\beta}} \qquad (4)$$

where C is a count statistic representing a fraction, β, of the animals present in sample areas that represent fraction α of the total area of interest. Virtually all estimators of animal abundance can be expressed in the form given by equation 4. Throughout this chapter, we note the relationship of different abundance estimators to this equation as a means of emphasizing the common conceptual basis underlying all of the methods we discuss.

Whenever we estimate abundance, we would like to estimate the sampling variance of our estimate as a means of assessing how close our estimate is likely to be to the true abundance (see "Sampling Variability," Chapter 3). The general variance for the canonical estimator is provided by Lancia et al. (1994) as follows:

$$\text{var}(\hat{N}) = \left[\left(\frac{\text{var}(C)}{C^2}\right)(1-\alpha) + \left(\frac{\text{var}(\hat{\beta})}{\beta^2}\right)\right] \qquad (5)$$

where var denotes sampling variance. The size of the component associated with $\text{var}(\hat{\beta})$ depends on the method used to estimate $\hat{\beta}$, and it can be large. The component associated with $\text{var}(C)$ depends on the variation in true abundance over the area of interest and on the method of sampling the space; it is best estimated from replicate samples. The variance of $\hat{N}$ becomes smaller as α becomes larger, corresponding to the intuitive idea that variance should decrease as the proportion of the total area that is sampled becomes large. The variance of $\hat{N}$ also decreases as β increases, corresponding to the intuitive idea that the population estimate becomes less variable as the proportion of the population counted becomes larger. Knowledge of how these different components contribute to the total sampling variance of $\hat{N}$ is important in designing a study to estimate animal abundance (Skalski and Robson 1992).

Abundance Indices

MICHAEL J. CONROY

An index of density (number of animals per unit area) can be defined as "any measurable correlative of density" (Caughley 1977:12). An index is usually a count statistic that is obtained in the field and that carries information about a population. In most cases, an index increases directly when the population increases in size and decreases when the population decreases. An investigator might choose to use an index rather than some more direct measure of population status for several reasons. In many situations, however, an index is not appropriate.

One reason for using an index is that individuals of the species in question may be difficult to observe and count because of size, habits (e.g., nocturnal, fossorial), or some other attribute. The animals also may be difficult to capture and tag, rendering capture-recapture techniques infeasible. In addition, the investigator may only need a measure of relative abundance or change in abundance, and for this goal a formal abundance estimate may be unnecessarily expensive or time-consuming.

Decisions about the use of an index to address questions of biodiversity and selection of an index for such use will depend on whether one is interested in measuring primarily abundance, species richness, or both. Here I discuss the use of indices to assess changes in relative abundance across time and space, with some remarks about interspecific comparisons. The use of indices to address questions about species richness will be considered in the section "Indices of Species Richness" later in this chapter.

RELATIONSHIP OF INDICES OF ABUNDANCE TO ACTUAL ABUNDANCE

By definition, an index must be associated in some manner with actual abundance (Caughley 1977). This association may be positive or negative, but one would hope for an index whose association with abundance is positive and monotonic. By this, I mean that as population size or density increases, so does the value of the index (Fig. 38A). One could use an index that is negatively related to abundance, but such indices would be rare unless the index were expressed in units that are naturally larger at small population densities (e.g., hectares of home range per animal). For certain ranges of abundance, an index might increase with increasing abundance, whereas over a different range of abundance the index might remain constant or even decrease (Fig. 38B). For example, if an index were based on some behavioral response of individual animals, which in turn was inhibited by high densities, then, when density was low or moderate, one might expect a more or less linear increase in the index. For high densities, however, the index might show little change or might actually decrease.

If an index is positively and monotonically related to abundance, then one can conclude from observing sequential indices that differences in the index over time or between habitats represent at least ordinal or relative differences in abundance. On the other hand, if the relationship between the index and abundance is not monotonic (Fig. 38B), one cannot be sure, for example, whether an increase in the index from one year to the next reflects an increase in abundance, or whether a decrease reflects a population decline (see Caughley 1977: fig. 4.2). Indices whose relationships with abundance are nonmonotonic probably are of little value for comparing abundances, unless the exact relationship between the index and abundance is known (in which case an index probably would not be needed).

For some purposes, a simple rank association between an index and abundance may suffice (i.e., the index values take on the same rank as abundance). In such cases, any positive, monotonic index will do. Suppose, for example, that

we wish to know only that populations are highest in year 1, lowest in year 2, and intermediate in year 3, but we do not care about the size of the differences. If an index has a simple rank association with abundance, then the index will provide the same information about the ordering of density over time as would an absolute abundance estimator. Generally, however, we like to know if a 10% decrease over time in an index represents a 10% decrease in abundance; if a value of 0.4 for an index in Habitat a and a value of 0.1 for the same index in Habitat b represent a proportional difference of four times between the two habitats; or if values of 0.1, 0.1, and 0.1 for three species of ungulates in a particular park really mean that the species are equally abundant. In order to draw such inferences from indices, an investigator must either know or, more commonly, assume that a particular type of mathematical relationship exists between the index and abundance. Such relationships may be linear with a constant slope, linear with a variable slope, or nonlinear.

LINEAR RELATIONSHIP, CONSTANT SLOPE. The relationship between an index and abundance is most commonly assumed to be *linear,* in which case the index is associated with abundance by the general equation

$$Y = \beta_0 + \beta_1 N \qquad (6)$$

where Y is the value of the index, β_0 and β_1 represent the intercept and slope, respectively, of the linear relationship, and N is the size or density of the animal population. For zero values of the index, one might expect $N > 0$ if, at very low densities, the index is too small to be observed or its value is too small to be measured. In such situations N_0 represents the "threshold" density at which the index can be observed (Fig. 39A). Below N_0, the species will not be detected, and values of the index will be 0 (theoretically the Y-intercept, β_0, is a negative value for the index

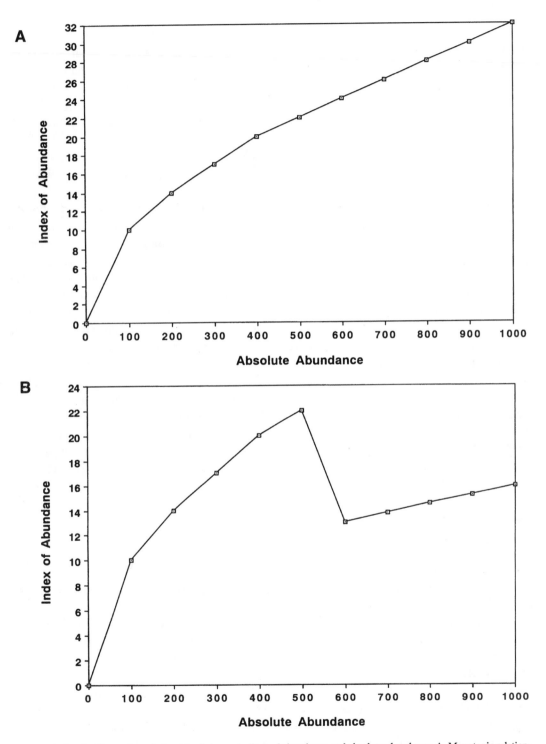

Figure 38. Graphs illustrating relationships between indices of abundance and absolute abundance. A. Monotonic relationship. B. Nonmonotonic relationship.

in this situation). In practice $Y = 0$ for $0 < N < N_0$). If one can assume, however, that population abundance is zero when the value of the index is zero, then equation 6 reduces to the simple relationship

$$Y = \beta_1 N \qquad (7)$$

where β_1 is a constant proportion or multiple of abundance (Fig. 39B). When Y is a count statistic, then β_1 is equivalent to β, the observation probability or fraction of all animals represented by the count statistic (see equation 1 and "Conceptual Framework," above).

The critical fact here is that β_1, the constant of proportionality relating N and Y, remains constant for all relevant comparisons. For example, if one is comparing the value of Y, the index, between two different habitat types, between two species, or at two points in time, one must assume that β_1 is constant across habitats, between species, or over time. This is a very strong assumption, and one that often is not justified. If the assumption does not hold, comparisons based on the index alone may yield misleading inferences about the populations.

LINEAR RELATIONSHIP, VARIABLE SLOPE. Depending on the index statistic, β_1 may depend on behavioral characteristics of a species, on habitat circumstances, or on factors such as weather. For example, an index based on sighting of animals may depend on their activity, which in turn depends on weather conditions. Index values measured on two days with different weather may differ not because of changes in abundance, but because activity was less on one day. Similarly, animals may be more visible in an open forest habitat than in a dense shrub thicket. For any constant set of weather conditions in the same habitat, the index may relate linearly to abundance by a constant, β_1, whereas under a different set of conditions the relationship would be expressed by β'_1, a different slope.

One can imagine a situation in which an index exhibits a linear relationship with abundance, but with variable slope dependent on the habitat in which the index is obtained (Fig. 40).

NONLINEAR RELATIONSHIP. There are a number of situations in which one might expect an index to exhibit a nonlinear relationship with abundance. In some cases an index may depend on a behavioral mechanism that changes in intensity in a density-dependent fashion. For example, subdominant animals may visit scent posts or other tracking stations less frequently at high densities than at low densities because of behavioral inhibition. Thus, the index may reach a "saturation point," beyond which it is influenced little by additional increments in population size. Caughley (1977:20) pointed out that many nonratio measures, such as frequencies, ordinal classes of abundance, and group size, are probably related in a nonlinear fashion to absolute abundance. For example, an index based on the proportion p of n sample units in which a positive index value occurs is theoretically related to the mean density (animals per unit area, $x = N/A$, where x is mean density and N is animal abundance in an area of size A) by a nonlinear relationship:

$$p = 1 - e^{-x} \qquad (8)$$

where e is the base of natural logarithms (Fig. 41). Intuitively this makes sense: whereas x can theoretically take on infinitely large values, p (a proportion) is necessarily bounded above by 1. It is not possible to observe the index as occurring on more than n of n plots, no matter how large the population is! Similarly, the "largest" of n index values of the type given in equation 8 has meaning only in a relative (ordinal) sense; it cannot be translated into a proportional or other ratio relationship for abundance, because another set of n index values for a population varying by 10 times as much could provide the same

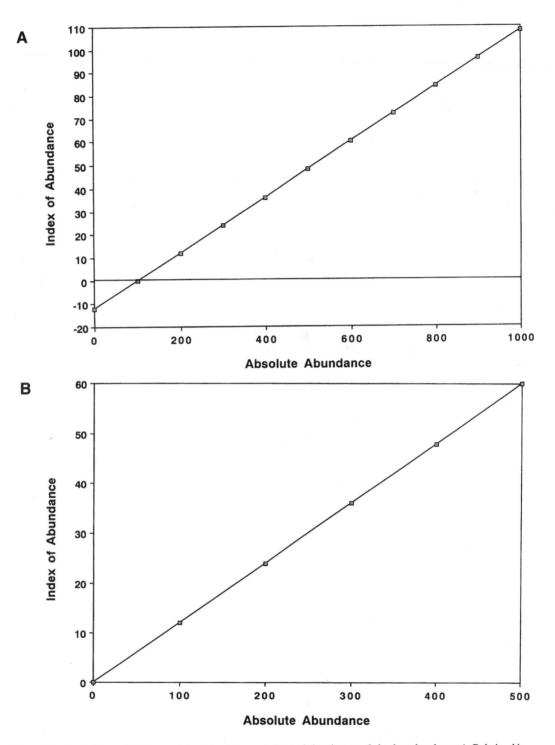

Figure 39. Graphs illustrating linear relationships between indices of abundance and absolute abundance. A. Relationship with a nonzero intercept. B. Relationship with a zero intercept.

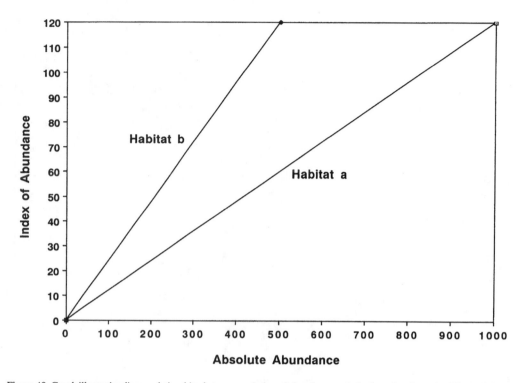

Figure 40. Graph illustrating linear relationships between an index of abundance and absolute abundance in different habitats.

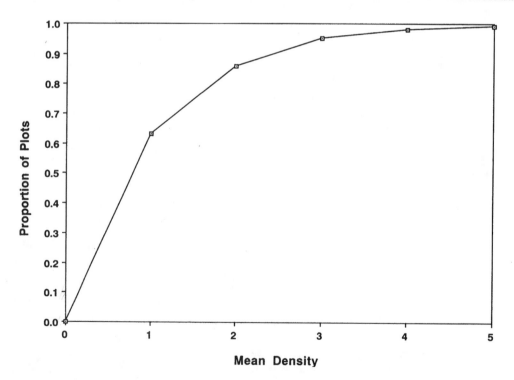

Figure 41. Theoretical relationship between mean density and proportion of plots where individuals are present.

rank statistics. For example, the index values for years 1 through 5 of 2, 4, 5, 1, 3, respectively, could have come from a population having a 5-year trajectory of 200, 400, 500, 100, and 300, or 220, 240, 250, 210, and 230.

DIRECT INDICES TO MAMMALIAN ABUNDANCE

Direct indices are those based on direct observations of animals, either visually or through capture or harvest (Seber 1982:52). Typically *n* animals are observed in a sample, and these individuals constitute some unknown proportion of the total population. For the ideal case of a constant, proportional, direct index, $n = \beta_1 N$, and the proportionality constant β_1 is simply the observability fraction, β (equation 1).

INCOMPLETE COUNTS. An *incomplete count* is obtained when animals are counted as they occur in nature (rather than after they are captured or harvested) in a manner that does not include every animal in the sampled area; at least some animals are missed. This type of measure is common for many large, conspicuous mammals. Examples of techniques that can yield incomplete counts include aerial surveys, road counts, spotlight surveys, drives, and line-transect surveys. These techniques are discussed in detail in Chapters 6 and 7.

If the incomplete count can be assumed to be a constant proportion of abundance, then a simple relationship (e.g., Fig. 39) applies, and comparisons over space and time are straightforward. Several studies suggest, however, that incomplete counts are often influenced by factors other than abundance. For example, Samuel et al. (1987) evaluated aerial surveys of elk (*Cervus elaphus*) in Idaho and found that group size and vegetative cover significantly influenced visibility of elk from aircraft. Experimental surveys indicated that visibility rates could be less than 10% for single animals in dense cover but would approach 100% for groups of 20 ani-

mals or more in sparse cover. Similarly, the observability of manatees (*Trichechus manatus*) from the air was influenced by several factors, including habitat type, manatee distribution, weather conditions, and observer experience (Packard et al. 1985). Finally, habitat and animal behavior greatly influenced the results of a spotlight count of white-tailed deer (*Odocoileus virginianus*) in Michigan (McCullough 1982). Therefore, the assumption of counts representing a constant proportion of abundance frequently does not hold, reducing the ability to use incomplete count indices for the purpose of comparing abundance.

Limited attempts have been made to validate the relationship between incomplete counts and population density. Allen and Sargeant (1975) evaluated roadside counts of red foxes in North Dakota and found a positive association ($r = 0.96$) between the roadside index and aerial estimates of fox density. This correlation, although suggesting a relationship between the index and density, was not a validation, because the aerial surveys were themselves of unknown validity.

Besides missing some animals in incomplete counts, it is possible to count individuals more than once, for example bats that reenter caves during emergence counts and then exit again (Barclay et al. 1980; Thomas and LaVal 1988). If multiple counting of individuals occurs sufficiently frequently, then the slope parameter, β, relating the index to abundance may be >1. However, β must still be constant in order to use the count index for comparative purposes.

CAPTURE INDICES. A *capture index* is a direct index that is based on the number of individual animals captured per unit of time and effort (e.g., trap-night) for a "standardized" study design (see "Use of Count Statistics in Comparative Studies," Chapter 3). Various standardized methods of trapping (Wood 1959; Wood and Odum 1964; Gaisler et al. 1979) and mist netting (Tuttle 1974a) have been described. For small mammals, a standard trap grid (e.g., 10×10 traps) might be run for several nights (e.g., 1 week), and the results recorded as the numbers of animals captured per trap-night (e.g., 70 animals caught in 700 trap-nights would yield an index of 0.1 animals caught per trap-night).

Capture indices implicitly assume that given some constant trapping effort, changes in captures of animals over time represent proportional changes in abundance (i.e., equation 2). The assumption that capture probabilities are constant is critical. Smith et al. (1984) found strong evidence of variable capture probabilities among years for all species in a study of furbearers using trapline captures. This evidence suggested that the trapline method was of dubious value for making year-to-year comparisons. Capture probabilities also differ among species, making capture frequencies invalid as indicators of relative abundance among species in a given year (Nichols 1986). Even at confined sites such as bat roosts, some unknown proportion of the population is typically missed (Thomas and LaVal 1988), and there is no assurance that the proportion missed is consistent over time or space. I recommend that if animals are captured they be marked and released. Data resulting from such sampling can be used to estimate abundance with formal methods that do not require unrealistic assumptions about the constancy of capture probabilities (see "Capture-Recapture Methods," below).

HARVEST INDICES.   A *harvest index* is an index of abundance computed using the number of animals harvested for a specified amount of effort or period of time. Numbers of animals harvested can be used as a constant-proportion index if harvest rates are constant, but ordinarily it is not reasonable to assume constancy. If harvest rate and size of the harvest can be estimated, then population size can be estimated using capture-recapture or removal methods (see "Removal Methods" and "Capture-Recapture Meth-

ods," below). Alternatively, correlatives of harvest effort (e.g., license sales) can be used to compute a harvest index (harvest per unit effort). Determining the proportionality of such a correlative index to abundance, however, requires that the investigator make strong assumptions about the population dynamics and the harvest-effort relationship, which are not always justified (Seber 1982).

INDIRECT INDICES TO MAMMALIAN ABUNDANCE

Indirect indices (Seber 1982:54) are based on indirect evidence of an animal's presence (i.e., evidence other than direct observation of the animal). Presumably, such evidence is absent (the index takes a value of 0) when the animal is absent, but increases with increasing population size. Examples of indirect indices include those based on counts of tracks, animal calls or other sounds, scent-station visits, structures, scats, and sizes of territories. In general, indirect indices are assumed to have positive, monotonic, and (ideally) linear relationships with abundance (equations 6 and 7). For indirect indices, however, β_1 need not fall between 0 and 1 as with typical count statistics used as direct indices; β_1 can be greater than 1 when an individual animal produces multiple indices (e.g., scats, sets of tracks, sounds).

TRACK COUNTS.   Track counts have been used as indices of abundance for species that are difficult to observe or capture directly but that have tracks that are easily seen and identified (see "Tracks," Chapter 9). There is some support for the assumption that track counts are related to population size in a linear, or at least a monotonic, fashion. Tyson (1959) compared track counts to drive censuses in a white-tailed deer population and concluded that an approximate 1:1 ratio existed between the counts (the index) and population size. Tyson (1959) and others (e.g., Downing et al. 1965) believed, however, that this relationship probably was not constant over years or areas. Similarly, Van Dyke et al. (1986) suggested that track counts of mountain lions (*Felis concolor*) are "potentially" directly related to abundance under "ideal tracking conditions," but that otherwise the relationship is uncertain. In recent work track counts have been used in conjunction with probability sampling methods, to provide actual estimates of mammal abundance (Becker 1991; Van Sickle and Lindzey 1991).

SCENT STATION SURVEYS. Scent station surveys are used for mammal species that scent-mark. Typically, stations are established at fixed intervals along roads or trails. A small (e.g., 1-m diameter) area is cleared and covered with a tracking medium (e.g., lime or sand), and a scent attractant (e.g., urine or a synthetic odor) is placed in the center of the station. Stations are subsequently visited, and the presence or absence of evidence of visitation by the target species (e.g., tracks) is noted for each. The proportion of stations visited is computed as an index to abundance. Detailed descriptions of scent station methodologies are provided in Chapter 9 (see "Scent Stations").

Implicit in the scent station methodology is the assumption that the visitation rate is associated with population density. As noted by Caughley (1977; also see discussion of indices with "Nonlinear Relationship," above), the theoretical relationship between any such index and mean abundance is likely to be nonlinear. For nonextreme densities, however, the relationship can be approximately linear, and transformations (e.g., arcsine) can be used to "linearize" the relationship. More important is the establishment of the empirical relationship between scent-station indices and abundance. Conner et al. (1983) compared scent station indices to abundance estimates obtained from mark-recapture and radio telemetric studies of bobcats (*Felis rufus*), raccoons (*Procyon lotor*), gray

foxes (*Urocyon cinereoargenteus*), and opossums (*Didelphis virginianus*); they concluded that the indices accurately reflected trends in population estimates of all species except opossums. Conner et al. (1983) did not formally estimate the relationship between the indices and population sizes, however, and pointed out that visitation rates (i.e., β_1) almost certainly vary among habitats, over time, and between species, making spatial, temporal, and interspecific comparisons of dubious validity. Diefenbach et al. (1994) observed changes in the frequency of scent station visits as abundance of bobcats increased on an island to which bobcats were introduced in stages over a 2-year period. Although the general association between scent station visits and abundance was positive, the index varied greatly, suggesting that results based on a single sample of stations (as is typical) are of limited value in making inferences about population trends (Diefenbach et al. 1994).

AUDITORY INDICES. An *auditory index* is an index based on counts of animal sounds, such as territorial calls of wolves (*Canis lupus*) and coyotes (*Canis latrans*). Recorded calls of conspecifics can be used to elicit vocalization and/or approach by wolves, coyotes, foxes, and other predators (Morse and Balser 1961; Harrington and Mech 1982; Fuller and Sampson 1988) (see "Call Playbacks," Chapter 6). Ultrasonically detected bat calls (e.g., Ahlen 1981) have also been used as an index to abundance (see "Ultrasonic Bat Detection," Chapter 7). Harrington and Mech (1982) provided detailed guidance for the design of howling-response surveys for gray wolves, including optimal times of day, time of year, weather conditions, sequence of recorded calls, and numbers of trials.

The key assumption for auditory indices is that the index (numbers of calls heard or elicited) bears the same relationship to abundance at each sampling location or occasion.

Work by Harrington and Mech (1982), Fuller and Sampson (1988), and Laundré (1981), among others, suggested that behavioral and environmental factors (e.g., daily and seasonal social dynamics, animal movement, weather and lighting conditions, techniques used by observers to hear or elicit calls) influence call rates of wolves and coyotes. In addition, variations in topography and vegetative cover can greatly affect the radius within which either animals or observers respond to or detect calls (Harrington and Mech 1982). Similarly, bat motion detectors may not provide useful indices of bat abundance, because the association between number of bat passes and number of individuals present is weak (Thomas and LaVal 1988). I strongly encourage investigators considering the use of either active or passive auditory indices to examine the assumption that the relationship between the index and abundance is constant on their study sites. Ideally, investigators should compare the index directly to animals known to be present (e.g., through radio telemetry; Harrison and Mech 1982; Fuller and Sampson 1988). At a minimum, field trials using recorded calls should be used to determine a radius of observer detectability of calls, and efforts should be made to conduct surveys under optimal and comparable conditions (Harrington and Mech 1982). Use of line-transect and variable-circular-plot methods (Roeder et al. 1987; Buckland et al. 1993) for estimating density should also be considered, if distances to calling animals can be estimated (see "Line-Transect Sampling," below).

STRUCTURE SURVEYS.   A number of species construct nests, lodges, food caches, or other conspicuous structures that may serve as indices to population abundance (see "Structures and Habitat Features," Chapter 9). These structures can be surveyed according to a predetermined design, for example, stratified random or systematic quadrats, and an index computed for the sampled area (e.g., number of beaver lodges/

km^2). The index is then assumed to be proportional to the abundance of the animals present in the study area.

A key assumption of structure indices is that the number of animals per structure is the same between surveys. Uhlig (1956) found a constant relationship between the number of squirrel nests and squirrel density in one study area. Hay (1958) and Easter-Pilcher (1990) both found a poor correlation between numbers of beaver lodges and beaver density, but Easter-Pilcher (1990) noted a good association between cache size and beaver colony size. Repeated counts of muskrat houses at the same time of year may provide a good index to population trends (Proulx and Gilbert 1984). I strongly encourage investigators to verify the relationship between animal structures and abundance for their own studies before using counts of structures to index abundance (see "Designing and Implementing an Index Survey," below).

SCAT AND OTHER SIGN COUNTS.   Counts of scat have been used as indirect indices to the abundances of large mammals, especially ungulates that deposit groups of highly visible pellets (Bennett et al. 1940; Eberhardt and Van Etten 1956; Neff 1968). Procedures for the proper statistical analysis of pellet-group data to detect population changes are described by White and Eberhardt (1980).

The critical assumption made in scat surveys is that the relationship between observed amount of scat and animal abundance is constant. Attempts were made to validate that assumption in two studies of white-tailed deer in enclosures (Eberhardt and Van Etten 1956; Ryel 1959). Replication was limited in both studies, but an examination of Ryel's (1959) data (in Neff 1968) shows a nearly inverse rank correlation between pellet counts and known population sizes. The relationship between pellet count and known deer abundance varied over years in both of the enclosures studied by Eberhardt and Van Etten

(1956). Fuller (1991) compared pellet counts with estimates of deer density from aerial surveys over a 5-year study and found poor correlations of the index with the survey estimates. White (1992) questioned the basis for this inference, however, and noted that the statistical power of Fuller's (1991) test was very low. Krebs et al. (1987) found evidence of a linear relationship between fecal pellet counts of snowshoe hares (*Lepus americanus*) and estimates of hare abundance based on capture-recapture models.

Caution should be exercised in making inferences about habitat use or preference based on pellet distributions, because the observability of pellets can differ in different habitats, and because the connection between defecation rates and other animal activities is not always known (Neff 1968). Similarly, staining or fecal accumulation rates at bat roost sites vary because of physical characteristics of cave or mine ceilings and floors (Thomas and LaVal 1988), greatly limiting the utility of comparisons based on fecal indices.

TERRITORY AND HOME RANGE SIZE. Home range or territory size has been used as an index of population density. Home range size is believed to be negatively correlated to density, under the assumption that animals "pack" into a given habitat by reducing their home range sizes, or at least the defended portions thereof ("territory"; Burt 1943; Overton and Davis 1969:417). For example, a decrease in average territory size from 2.0 ha in one year to 0.5 ha in a subsequent year suggests a four-fold increase in animal density.

The use of home range or territory size as an index to density depends on restrictive and ordinarily unrealistic assumptions about animal behavior and population dynamics, for example, that home ranges do not overlap and that individuals use space in the same way, regardless of age, sex, dominance, habitat quality, or time of

year. In addition, approaches used to estimate home range size are inconsistent, probably rendering many published estimates of home range size of doubtful value for comparative purposes (e.g., Laundré and Keller 1984). Because considerable effort is required to estimate territory or home range size accurately, and because few investigators have attempted to define the relationship between home range size and density rigorously, I do not recommend the use of territory or home range size measures as indices to abundance.

DESIGNING AND IMPLEMENTING AN INDEX SURVEY

It is important for users of indices to adhere to proper sampling procedures and to be aware of statistical estimation issues (Chapter 3). If use of a sampling design is invalid, the resulting index data may apply to a target population different from the one the investigator intended. For example, track counts made along a road that bisects only one of four habitat types present in a study area will not permit valid extension of results to the entire study area. Similarly, observability and sampling variability must be considered for correct interpretation of index results. Trap indices uncorrected for variation in trapping effort (and other sources of variation in capture probability) over time or between habitats will not provide valid indications of relative abundance. Unless sampling variability is considered in such comparisons, differences due to sampling error may be interpreted as "real" population differences.

Before initiating a survey, the investigator must have a clear idea of the goals of the investigation. The goal may be to describe population abundance (e.g., with an estimate or an index), or it may be to detect a change in abundance over time or space (e.g., a trend). The investigator must also specify the degree of reliability she or he desires in the resulting data, for example,

"Is detection of a 50% population decrease, but not a 10% decrease, adequate?" Many times such considerations will determine the type of index that is most appropriate or will suggest an alternative methodology. The investigator must then clearly specify the target population; make appropriate decisions regarding stratification; specify the size, shape, and number of units to be sampled; and provide a scheme for selecting them. These decisions in turn require some kind of mapping or other delineation of the study area and stratum boundaries, and preliminary estimates of sampling variability, based on a pilot sample, a previous study, or an educated guess.

Many of these points were covered in Chapter 3 and are the same in principle, regardless of whether one is using an index or some other method such as a formal abundance estimator. Here I concentrate on features of study design that are particularly relevant to indices to abundance.

INDEX-ABUNDANCE RELATIONSHIP.   If an investigator uses an index to, rather than a formal estimate of, abundance, then he or she must first consider the relationship of the index to abundance. If the relationship is highly predictable (e.g., a high linear correlation), then the index may be useable as a surrogate for direct measures of abundance. The investigator must obtain as much information as possible about this relationship before investing significant time and resources in an elaborate sampling design using the index. General, qualitative information on the relationship of the index to abundance may be available from the literature. In rare cases such relationships have been validated with some generality, so that one may be reasonably comfortable simply adapting a previous investigator's results to the current survey. More frequently, it is best to conduct a pilot survey using the index in an area where abundance is known or can be estimated well.

SAMPLING VARIABILITY OF AN INDEX.   Regardless of how well an index predicts abundance, index values from different samples within a population will vary, both because of the sampling process (see "Sampling Variability," Chapter 3) and because of imprecision in the true relationship of the index to abundance. In most cases these two sources of variability will not be separable, and only a "sampling error" that includes variability from both sources can be estimated. Preliminary estimates of sampling error from a pilot study will be useful in establishing guidelines for sampling effort and in allocating samples to strata.

CALIBRATING AND UPDATING THE INDEX.   It is common for indices to be used for years with no adjustment or modification in either the index method or the sampling design. Unless the investigator is sure that the index-abundance relationship does not vary over time, it is advantageous to calibrate the index periodically (Overton and Davis 1969). Index calibration refers to the estimation of the relationship between the index and abundance (Eberhardt and Simmons 1987) and serves three purposes: (1) it tests the critical assumption underlying the use of indices (that the relationship between the index and abundance remains constant for the comparisons for which the index is to be used); (2) it improves overall precision and index sensitivity by estimating the index-abundance relationship more precisely; and (3) if the calibration is sufficiently accurate, it allows the investigator to convert the index into an estimate of abundance.

CONSTANT-PROPORTION INDICES.   In some cases, a high correlation between an index and abundance may justify the direct conversion of the index to an estimate of abundance. For example, Allen and Sargeant (1975) converted roadside survey counts made by rural mail carriers (RMC) to abundance estimates of red fox because of the high correlation between RMC counts and aerial estimates. Their aerial survey estimates were of unknown validity, however, and in general it is unusual for the relationship between an index and abundance to be fixed. In the rare case in which a known constant-proportion relationship exists, an estimate of N (population size) can be obtained by solving equation 7 to yield equation 2:

$$\hat{N} = \frac{Y}{\hat{\beta}_1}$$

where $\hat{\beta}_1$ is an estimate of the proportionality constant (observability), provided, for example, by a comparison of the index to known levels of population abundance. More generally, if a known, linear relationship exists between an index and abundance, then the solution of equation 6 provides an estimate of N:

$$\hat{N} = \frac{Y - \hat{\beta}_0}{\hat{\beta}_1}$$

where $\hat{\beta}_0$ and $\hat{\beta}_1$ are estimates from the regression of Y on $\hat{N}$, again obtained from a calibration study.

VARIABLE-PROPORTION INDICES. It is probably more common for the relationship between an index and population size to vary over time, among observers, across habitats, and so forth, than for it to remain constant (e.g., Conner et al. 1983; Packard et al. 1985; Samuel et al. 1987). Thus, a single calibration of an index with respect to abundance may not be valid for future use of the index, particularly if conditions (e.g., habitats, weather, observers) of future use differ from those of the calibration study. Two approaches are possible for dealing with this problem. First, one may conduct a study in an attempt (this may not be possible) to identify all signifi-

cant factors (besides abundance) related to the index and specify their relationship to the index. In future studies, such factors can be measured and used to adjust the index value, given some predictive model (e.g., Samuel et al. 1987). One can also simply control for as many nuisance factors as possible by standardizing methods (e.g., always surveying with the same observers, under the same weather conditions, and so forth). The second approach does not assume that the relationship of the index to abundance or other variables is the same at each sample, but instead reevaluates this relationship at each sample. In this approach, a double sample is taken at each occasion (see "Double Sampling," below). In the first sample (consisting of n' sampling units), the investigator obtains both an index statistic and a count or estimate of the true number of animals for each unit (or for some subgroup of animals in each unit). The ratio of the index statistic to the true number of animals in the primary sample is used to estimate the relationship between the index and N. In the second sample ($n > n'$ sampling units), only the index (along with variables such as weather, habitats, and observers) is measured. Regression or ratio estimators are then used to estimate N from the combined sample (Cochran 1977; Eberhardt and Simmons 1987).

REDUCING INDEX VARIABILITY. As suggested earlier, many variables besides abundance can influence index values for a particular sample. Even if it is not possible to predict abundance from the index, for example because it is never possible to estimate abundance directly, it still may be possible to use auxiliary variables such as weather conditions, observers, and so forth to reduce index variability (e.g., via multiple regression) (Overton and Davis 1969). The resulting adjusted index values are not estimates of abundance, but they are more comparable and have less sampling variability than raw index values, perhaps enabling more accurate and powerful comparisons over time or space.

ALTERNATIVES TO INDICES

I believe that directly estimating an abundance parameter (e.g., population size or density) is generally better than employing an index to abundance. That is an opinion with which some eminent colleagues might disagree. For example, Caughley (1977:13) stated that "estimates of absolute density have been used in many studies where density indices would have provided equally enlightening information." If appropriate indices are available, and if their relationships to abundance are known to be invariant over time and survey conditions, then there probably is no need to estimate absolute density, if one simply needs to compare relative abundance over time or space.

In my experience, however, such ideal conditions almost never hold true, with potentially disastrous consequences for even relative comparisons. Furthermore, in many (most?) cases, data collected with the objective of estimating absolute abundance also can be used to compute an index. For example, if one designs a mark-recapture study to estimate population size, one can still compute and use a capture index (e.g., if the estimation procedure fails). The converse often is not true; that is, if one designs a study with the objective of obtaining an index, typically one is left with only one option, that of using the index and accepting all the assumptions required for its use.

I have already alluded to estimates of abundance, obtainable by calibrating index data, for special cases when the relationship between the index and abundance is known to be constant or can be estimated by concurrent (double) sampling. Virtually all sample population or density estimators are based on the principle of collecting sufficient data to estimate the proportion of the population that has been sampled (observability). These methods employ statistical

estimation procedures that explicitly incorporate these observation probabilities, enabling comparison of estimates obtained under different sampling conditions. Before deciding on any method, an investigator should consult appropriate chapters in this manual for a more detailed treatment of each method (more than one may be appropriate).

Complete Counts

PETER JARMAN, ANDREW P. SMITH, AND COLIN SOUTHWELL

METHOD

In a complete count, or census, the investigator determines population size (N) in a survey area directly from the number of animals counted (C), without correction for sampling (α) or observation (β) probabilities (i.e., $N = C$). There is no need to estimate and correct for sampling and observation probabilities, and this feature distinguishes a *census* from an *estimate*.

The assumptions required for a census are as follows:

1. The sampling fraction is 1 ($\alpha = 1$ in equation 4).
2. The observability is 1 ($\beta = 1$ in equation 4).
3. The population is closed during the census.

Assumption 1 requires that the entire survey area, not just sample plots within it, be searched. Assumption 2 requires that all animals in the survey area be detected (typically by visual means) and counted, and that none be counted twice. The animals must therefore be easily detected, either because of their size or behavior or because they are counted in a manner that makes them easily detectable. For assumption 3 to be fulfilled the census must be conducted over a short period so that no immigrations, emigrations, births, or deaths through natural processes occur, and in a way that ensures that no animal will evade the observer and leave the area before it is counted.

FIELD APPLICATION

A well-defined area within which a sedentary population of a readily visible species is dispersed, in which the individuals are unaffected by the presence of observers, and in which all individuals can be located and distinguished, is ideal for a census. The observer can count the animals either by scanning from a single location or by moving through the area. In practice, such ideal situations seldom exist. No mammals are sufficiently visible for one observer to count all individuals present over a usefully large area from one location. In addition, most mammals respond to the presence of an observer with evasive behavior, which can lead to some animals being missed, to other animals being counted more than once, or to animals leaving the study area before being counted. Even if animals do not respond to an observer, the assumption that the observation probability is equal to 1 (i.e., that all animals are seen) is difficult to fulfill.

It is sometimes possible to overcome these problems by making use of special characteristics of the target animal and/or the survey area, or by employing a sufficiently large number of observers to conduct a census. Under such circumstances, three methods may be used to census mammals in the field: drive counts, counts based on individual identification, and counts of animals emerging from dens and burrows. These methods, their assumptions, and their limitations are discussed in detail in Chapter 6 (see "Drives for Total Counts," "Individual Identification," and "Observation of Emergence from Dens and Burrows"). Circumstances appropriate for censuses are rarely encountered in real field situations, however, and total counts cannot, therefore, be generally applied.

Estimation of Population Size and Density When Counts Are Incomplete

COLIN SOUTHWELL

In most situations counts of animals in sample areas are incomplete and represent unknown fractions of the animals present. The probability of detecting an animal that is present in an area being surveyed is variously termed observability, visibility, sightability, and detectability. In this section, I use these terms interchangeably.

The most important factors influencing observability (from Graham and Bell 1969, Caughley 1974, and Morgan 1986) can be classified as follows:

1. Characteristics of the animal (size, color, and behavior).
2. Medium between the observer and the animal (atmosphere, vegetation, and topography).
3. Visual background noise.
4. Temporal factors (time of day and season).
5. Relative spatial positions of observer and animal (both vertical and horizontal).
6. Characteristics of the observer (ability, experience, and fatigue).
7. Observer's rate of travel.

Variation in these factors can lead to variation in sightability within or between surveys. For example, Bayliss and Giles (1985) conducted aerial surveys of kangaroos in the same area on consecutive days at 3-month intervals over 7 years (Fig. 42). During each survey, kangaroos were counted in 100-m-wide transects by observers on each side of an aircraft flying at a constant speed (167 km/hr) and height (90 m). The counts showed an overall long-term increase, realistically reflecting increasing population levels, but the noise about this general trend, indicating between-day and between-season variation in counts, was large. This short-term variation cannot be attributed to varying popula-

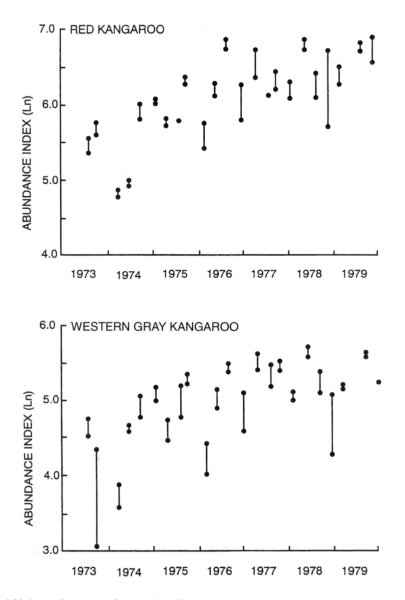

Figure 42. Variable incomplete counts of two species of kangaroos made from the air at different seasons. Vertical lines represent consecutive survey days. Redrawn with permission of The Wildlife Society from Bayliss and Giles (1985).

tion levels but rather must reflect incomplete counting, which varied in extent. The investigators showed that much of the short-term variation was related to ambient temperature at the time of the survey; at warmer temperatures kangaroos lay down to rest or sought shelter under shrubs, thus becoming less visible. Variable sightability can ultimately be attributed to temporal (temperature varies with both time of day and season) and animal (behavior) factors.

In addition to these dynamic effects on sightability, a set of static effects is likely to have resulted in a constant, or near-constant, number of kangaroos in the sample strips being over-

looked. It is extremely unlikely that all kangaroos in a 100-m strip could be seen by observers
flying at a height of 90 m and a speed of
167 km/hr, because of the distances between and
the relative positions of the observers and the
kangaroos, the large area of ground to search per
unit time, and the rate of travel of the observers.

To estimate population size and density accurately, the investigator must correct for incom-
pleteness in counts. Several methods that have
been developed for this purpose are outlined
below.

DOUBLE SAMPLING

METHOD. With double sampling, a complete
count is made in a subsample of a larger sample
of areas to which incomplete counting is applied. The observation probability (β) in the incomplete count is then estimated as the ratio of
the mean incomplete count ($\bar{y}$) to the mean complete count ($\bar{x}$) from those areas sampled by both
methods, as follows:

$$\hat{\beta} = \frac{\bar{y}}{\bar{x}} \qquad (9)$$

The number of animals present in the sampled
area is estimated as the number counted (n) in
the incomplete count divided by the estimated
observability:

$$\hat{N} = \frac{n}{\hat{\beta}} \qquad (10)$$

(e.g., see equation 2). Population size for the
survey area is then obtained through division by
the sampling fraction α:

$$\hat{N} = \frac{n}{\alpha\beta} \qquad (11)$$

yielding the canonical estimator (equation 4)
discussed in the "Conceptual Framework" sec-

tion, above. The practical rationale underlying
double sampling is that the complete counts are
accurate yet expensive, whereas the incomplete
counts are inaccurate but typically inexpensive.
Double sampling attempts to combine the advantages of cost-efficiency and accuracy from
the two complementary methods.

For unbiased estimation of population size
and density from the double-sampling method,
the following conditions must be met:

1. Complete count is accurate.
2. Population is closed to additions (births and
   immigrants) and losses (deaths and emi-
   grants) between application of complete and
   incomplete counts.
3. Complete and incomplete counts are independent.

FIELD APPLICATION. In most applications of
double sampling, some form of ground count is
used as the complete counting method for correction of incomplete aerial counts (e.g., Short
and Bayliss 1985; Eberhardt and Simmons 1987;
Short and Hone 1988). Typically, a plane flies
along predetermined transect lines with observers counting animals seen in strips of specified
width on each side of the plane; ground counts
are made in subsampled sections of the surveyed
strip. Double sampling thus uses the accurate yet
expensive ground counts over small areas to estimate observability and to correct the less expensive, incomplete counts that cover the larger
area.

Aerial photography can also be used as the
complete counting method. This method is most
successful when the target animals are conspicuous and form large, dense aggregations. Mammals that fit these requirements include large,
herding, savanna species (Sinclair 1972) and
pinnipeds that haul out of water onto ice or land
to breed (Eberhardt et al. 1979). In Africa photography has been used extensively to estimate
observability for aerial survey of large ungu-

lates, but usually in a restricted capacity. Generally, groups larger than a certain size (say, 10), for which total enumeration is difficult, are photographed (Sinclair 1972; Norton-Griffiths 1974). This approach allows for correction of missed animals in groups but not for correction of missed groups. Aerial survey could be used for both incomplete and complete counts, with fixed-wing aerial strip transects as the extensive incomplete count and intensive searching in subsample areas treated as the "complete count." Similarly, ground survey alone could be used, with extensive counts from a vehicle corrected with intensive counts in subsample areas on foot. The double-sampling principle can also be used to convert index data into estimates of population size (e.g., Ryel 1971; Perry and Braysher 1986; see "Abundance Indices," above).

The double-sampling method can be illustrated with data from Short and Hone (1988). They employed this method to estimate detection probability for aerial surveys of kangaroos. The methodology for broad-scale aerial surveys is standardized across Australia: counts are made in 200-m-wide strips from a fixed-wing plane flying at a constant speed (185 km/hr) and height (76 m). Short and Hone (1988) employed the standard aerial-survey methodology to count kangaroos in a small (12 km^2) area where barriers to kangaroo movement (fences and a lake) would also facilitate a total count from the ground. The open shrubland habitat of this area is typical of much larger areas surveyed from the air. The site was first sampled from the air to obtain an estimate of observed density, $\bar{y}$, and subsequently a drive count was conducted to determine true density, $\bar{x}$ (note that we retain the $\bar{y}$ and $\bar{x}$ notations of equation 9 despite the use of a single site, rather than multiple sites, in this specific example). For the red kangaroo, observed density was $\bar{y} = 5.6/\text{km}^2$ and true density, $\bar{x} = 14.7/\text{km}^2$; detection probability was estimated, therefore, as $\hat{\beta} = 5.6/14.7 = 0.38$. For a second kangaroo species in the area, the western

gray kangaroo, detection probability was estimated as $\hat{\beta} = 3.8/24.7 = 0.15$. The higher detection probability for red kangaroos compared to western gray kangaroos was attributed to their more conspicuous pelage and their preference for more open microhabitats. These results have subsequently been used to correct for observation bias in broad-scale surveys over similar habitat. In this study the special conditions required for the drive count restricted double sampling to only one subsample area. The design thus does not adhere strictly to the general one described previously, but the same basic principles were employed.

POSSIBLE PROBLEMS IN FIELD APPLICATION. The assumption of double sampling most likely to be violated is that the counts for the subsample are complete. Situations in which a truly complete count can be achieved are rare and usually depend on certain special characteristics of the target animal or the survey area. Often ground counts are assumed to be accurate, but in practice they may not be, because they too are based on assumptions that are often violated. Because opportunities for complete counts are so rare, investigators sometimes use bias-corrected counts in subsample areas in lieu of a complete count (i.e., observability is estimated for the counting method used in the subsamples and used with equation 2 to estimate abundance in the subsample areas).

Application of complete and incomplete counts must be carefully planned to ensure their independence (assumption 3). For example, intensive searching on foot or by aircraft to obtain a complete count could cause animals to vacate the subsample unit temporarily. Until the animals have returned, any counts in the area will yield a negatively biased estimate of observation probability. In Short and Hone's (1988) drive count, more than 80% of the kangaroos were forced to vacate the site. In practice, incomplete counts are less disruptive than complete counts and so should be conducted first. The time between counts should be as short as possible to avoid violation of the second assumption (additions to or losses from the population).

MARKED SUBSAMPLE

METHOD. This marked-subsample technique is an adaptation of the capture-recapture method in which a sample of animals is captured and marked in such a way that animals are recognizable as marked or unmarked in subsequent observational surveys. In the two-sample capture-recapture method a sample of n_1 animals is captured, marked, and returned to the population. After allowing time for marked and unmarked animals to mix, a second sample of n_2 animals is captured or observed, and the number (m_2) of those animals that are marked is recorded.

A natural estimator for the probability, β, of capture or sighting in the second sample is

$$\hat{\beta} = \frac{m_2}{n_1} \qquad (12)$$

From equation 2, one can estimate population size by dividing the number of animals seen in the second sample by the estimated probability of sighting:

$$\hat{N} = \frac{n_2}{\hat{\beta}} \qquad (13)$$

Substituting equation 12 into equation 13 gives

$$\hat{N} = \frac{n_2}{(m_2/n_1)} = \frac{n_1 n_2}{m_2} \qquad (14)$$

This is the unmodified Lincoln-Petersen estimator (Seber 1982; see "Capture-Recapture Methods," below). In practice, I recommend use of the bias-adjusted modification of this estimator (Chapman 1951):

$$\hat{N} = \frac{(n_1 + 1)\,(n_2 + 1)}{m_2 + 1} - 1 \qquad (15)$$

Equations 14 and 15 apply to the situation in which a single survey is used to reobserve marked animals; repeated surveys permit more precise estimates (Minta and Mangel 1989; Arnason et al. 1991; Neal et al. 1993).

In the context of the marked-subsample method, the assumptions of the model on which the Lincoln-Petersen estimator is based can be written as follows:

1. The population is closed between marking and surveying.
2. All animals have the same probability of being caught in the first sample or seen in the survey.
3. The marking of an animal does not affect the probability of its being sighted in the survey.
4. Animals do not lose their marks between capture and the survey.

FIELD APPLICATION. The marked subsample method has been applied to both terrestrial and aquatic mammals in conjunction with both aerial and ground surveys (Table 7). The first and most difficult aspect of application is the capture and marking of animals. A substantial portion of the population should be marked within as short a period as possible. Marked animals should be evenly distributed throughout the population. The marks, usually comprising conspicuous collars (e.g., Bear et al. 1989) or radio transmitters, must allow accurate determination of whether an animal seen during the survey is marked or unmarked. Marks must be sturdily constructed and securely attached to the animals to ensure that none is lost or becomes ineffective before the survey is completed.

The design of the survey should ensure that the area is searched as thoroughly as possible, so that all animals have an equal chance of being

Table 7. Examples of Studies Using Marked Subsamples to Estimate Mammal Population Size

Species	Type of survey	Mark	Number of repeated surveys	Approximate percent of population marked	Reference
Manatee	Aerial	Radio	5	—	Packard et al. (1985)
Mule deer	Aerial	Collar	12	35	Bartmann et al. (1987)
White-tailed deer	Aerial	Collar	5	10	Rice and Harder (1977)
Roe deer	Ground	Collar	[a]	77	Strandgaard (1967)
Gray kangaroo	Ground	Collar	5	—	Arnold and Maller (1987)
Elk	Aerial	Radio and collar	2–5[b]	5–19	Bear et al. (1989)

[a]Observation effort not subdivided into discrete surveys; no specific number of surveys run.

[b]Range of values from five sites to which method was applied.

sighted (assumption 2) but duplicate sightings are avoided. Sighted animals are classified as marked or unmarked. Multiple surveys should be conducted to provide replication.

The amounts of effort allocated to the stages of marking and surveying are interdependent. For example, Rice and Harder (1977) showed that to obtain a population estimate of a specified precision, investigators must mark 13% of a population if 40% of animals can be observed on each of two surveys; if ten surveys are undertaken, however, only 3% of the population need be marked. Bear et al. (1989) found that five separate flights with 10% of the population marked and approximately 50% sighted resulted in coefficients of variation of 10% or less. In practice, investigators usually have more control over the sampling intensity (which determines the observation probability) of the survey than the proportion of animals that can be marked, so it is prudent to set the sampling intensity of the survey based on the expected or realized capture success rather than vice versa.

The proportion of animals that can be marked for a given amount of effort will be inversely related to the size of the survey area. The success of Bartmann et al. (1987), for example, in mark-

ing a high (35%) proportion of their study population was undoubtedly facilitated by the small size of their survey area (2.6 km²). Because of this relationship, the marked-subsample method is not suitable for estimating populations in large areas (survey areas for the studies listed in Table 7 are all <100 km²).

Bartmann et al. (1987) marked 8 deer with radio transmitters in an enclosure of less than 1 km² and subsequently attempted 12 complete aerial searches from a helicopter using standardized searching procedures. After completion of the surveys, the exact number of deer was determined with a complete count, in this case possible only because the area was enclosed. On the first aerial survey, 9 deer were sighted, of which 5 were marked. Sighting probability was estimated as:

$$\hat{\beta} = \frac{m_2}{n_1} = \frac{5}{8} = 0.63$$

and population size as:

$$\hat{N} = \frac{(8+1)(9+1)}{(5+1)} - 1 = 14$$

They repeated this computation for each of the remaining surveys. Averaging across the 12 estimates, they estimated population size to be 12.5, close to the true size of 13.

POSSIBLE PROBLEMS IN FIELD APPLICATION. The most important consideration in application of the marked-subsample method is the nature of the mark. It must be conspicuous enough to be identifiable during the survey but, to fulfill assumption 3, not so conspicuous that it draws attention to the animal, making it more visible than an unmarked individual. If animals are radio-collared, collars need not be conspicuous, but the radiotelemetry equipment must be able to identify accurately whether a sighted animal carries a radio.

Bear et al. (1989) concluded that collar conspicuousness had little effect on the probability of sighting elk, because the overriding cue to sighting was movement. Of greater concern to them was incorrect classification of marked animals as unmarked. This is a violation of assumption 4 in which the marks are visually, rather than physically, lost (i.e., they are overlooked). This problem can occur for several reasons. Bear et al. (1989) suggested that congenitally defective color vision may be a contributing factor when marks are colored collars. The color vision of observers should be considered in the design of marked subsample studies.

Equal catchability of animals for marking (assumption 2) may be difficult to achieve in the field because of problems in distributing the capture effort uniformly throughout the population (Bear et al. 1989), or because of age- or sex-related bias in capture vulnerability (Garrott and White 1982). It may also be difficult to capture and mark a sufficiently large number of animals in a relatively short period. If marking is carried out over an extended period of several weeks or months, then marked animals may die or emigrate from the survey area before the survey is conducted, thus violating assumption 1.

Radiotelemetry can be invaluable in this instance, because it allows the investigator to determine the number of marked animals that are present and alive in the survey area immediately before or after the survey (Eberhardt 1990). In this case n_1 from equation 15 is the number of marked animals known to be present at the time of the survey and not the (possibly greater) total number of animals that was marked. Similarly, m_2 is the number of these "known to be present" animals that are seen from the air.

MULTIPLE INDEPENDENT OBSERVERS

METHOD. This is another adaptation of the capture-recapture concept. Two observers count animals simultaneously and independently in the same area in a manner that allows partitioning of sightings into those seen by each observer and those seen by both observers.

Considering the method in the standard capture-recapture framework outlined previously, animals seen by one observer (n_1) are those "captured," "marked," and returned to the population in the first sample, animals seen by the other observer (n_2) are those "captured" in the second sample, and those seen by both observers (m_2) are those "marked" in the first sample and "recaptured" in the second. Population size is again estimated from the bias-adjusted Lincoln-Petersen estimator (equation 15).

The capture-recapture assumptions can be reframed into the multiple-independent-observer context as follows:

1. The population is closed between counts by the two observers.
2. All animals are equally likely to be seen by a given observer.
3. Sighting of an animal by one observer does not affect the probability of its sighting by the other observer.
4. Animals seen by both observers can be accurately identified.

Table 8. Examples of Studies Using Multiple Independent Observers to Estimate Mammal Population Size

Species	Type of survey	Method for determining duplicate sightings	Reference
Sea otter	Aerial/ground	Mapping	Geibel and Miller (1984)
Sea otter	Ground	Mapping	Estes and Jameson (1988)
Feral stock	Aerial	Transect unit	Bayliss and Yeomans (1989)
Horse, donkey	Aerial	Immediate recording	Graham and Bell (1989)
Dugong	Aerial	Immediate recording	Marsh and Sinclair (1989b)

FIELD APPLICATION. The multiple-independent-observer method has been used most frequently to correct for detection bias in broad-scale aerial surveys of mammal populations (Table 8). Usually two observers seated on the same side of an aircraft simultaneously and independently count animals within a strip of specified width. Observers in a plane and on the ground have also conducted simultaneous counts, as have multiple ground observers.

Partitioning of sightings into those seen by one observer, those seen by the other observer, and those seen by both observers can be achieved in several ways. One method for aerial surveys is to divide each strip transect into short lengths or units such that no more than one animal or cluster of animals is likely to be present in each unit (Caughley and Grice 1982; Bayliss and Yeomans 1989). Then m_2 can be obtained by tallying those units that had nonzero counts by both observers. This method can be refined by additional division of the transect into substrips, with m_2 obtained by tallying those substrips with nonzero counts by both observers. Each observer can record sightings and the exact time of each sighting, permitting sightings by both observers to be identified by the relative times of their reporting (Graham and Bell 1989; Marsh and Sinclair 1989a).

An alternative method most applicable to immobile objects such as burrows or nests is to mark each sighting precisely on a map of the survey area and to compare map locations between observers later. If the target objects are mobile, then mapping by the independent teams must occur simultaneously for reasonable confidence in animal identification. If animals aggregate into well-defined groups or clusters, then determination of which individuals were seen by each observer may be difficult. In this case the cluster should be treated as the target object to be sighted, with auxiliary information on cluster size recorded at each sighting.

Marsh and Sinclair (1989b) carried out an aerial survey of dugongs in the Great Barrier Reef Marine Park off northern Australia. Two observers seated in the front and rear seats on the same side of a small fixed-wing aircraft simultaneously counted dugong groups within 200-m-wide strip transects. Observers wore headphones and were separated by a curtain in order to isolate them from each other both visually and acoustically. Sightings were recorded immediately onto a two-track tape recorder, and animals seen by both observers, m_2, were determined from the relative times of reported sightings. In one sample area the observer in the front seat saw 94 clusters (n_1), the observer in the back seat saw 76 clusters (n_2), and both observers saw 58 clusters (m_2). The sighting probability for dugong clusters available for counting in the sample transects, $\hat{\beta}$, is estimated as

$$\hat{\beta} = \frac{58}{94} = 0.62$$

and the number of dugong clusters, $\hat{N}_c$, is estimated using equation 15 as

$$\hat{N}_c = \frac{(94+1)\,(76+1)}{(58+1)} - 1 = 123$$

Dugong population size in the sample transects was then calculated as the product of the number of clusters and mean cluster size.

POSSIBLE PROBLEMS IN FIELD APPLICATION. Investigators are assured of meeting assumption 1 of the multiple-observer method, that is, the requirement for a closed population, if observers count simultaneously or if the target objects are immobile. On the other hand, the assumption that all animals have an equal chance of being seen (assumption 2) is under the least control of the investigator and is most likely to be violated. Heterogeneity in sighting probabilities will produce negative bias in the Lincoln-Petersen estimator, resulting in underestimation of population size. An extreme case of heterogeneity occurs in surveys of aquatic animals where animals below the water surface have no chance of detection ($\beta = 0$) and are completely unavailable for counting, whereas animals at the surface have positive detection probabilities ($\beta > 0$). Marsh and Sinclair (1989b) were careful to emphasize for their dugong survey that $\hat{N}_c$ was an estimate of the number of clusters "available" to observers and not necessarily the total number of clusters in the population (so N_c may exceed $\hat{N}_c$). Their estimate of N_c was based on the assumption that all available clusters were equally likely to be seen. Converting estimates of available clusters to estimates of actual clusters requires information on the proportion of time spent by dugongs at the sea surface. Fulfillment of the other assumptions can usually be ensured with appropriate field procedures.

The third assumption requires that observers count independently of each other. Independence is assured if observers count in two separate surveys (e.g., two aerial surveys or corresponding air and ground surveys), but this is usually possible only when surveying immobile objects (but see Estes and Jameson 1988 for application to mobile objects). Observers surveying mobile objects are usually in close proximity (e.g., in the front and back seats of a small plane), and the possibility of dependence between observers is high. Lack of independence can occur if some activity of one observer—such as speaking into a tape recorder, writing on a map, or even looking more intently than usual—alerts the other observer to the possibility of an animal being nearby. In these situations the investigator must devise some way of isolating observers from each other when they are counting; the earphones and curtain used by Marsh and Sinclair (1989b) and described previously are an effective strategy. Lack of independence between observers will lead to a positive bias (overestimation) in the estimation of sighting probability and consequent negative bias (underestimation) in the estimation of population size.

The investigator must employ data collection procedures that allow accurate identification of animals seen by both observers (assumption 4). Such procedures include partitioning the area counted into units or substrips and immediately recording duplicate sightings, as outlined earlier, but the success of these methods depends on the degree of their refinement and the density of the population. Division of a transect into short units (e.g., the 5-km lengths by Caughley and Grice 1982 and Bayliss and Yeomans 1989) reduces the density of the target objects (for Caughley and Grice, to <0.02/km^2) so that occurrence of more than one object in each unit is rare. This practice can reduce the chance of misidentified sightings considerably, but not entirely. Marsh and Sinclair (1989b) reduced the chance of misidentifying sightings of dugong clusters to an absolute practical minimum by dividing the strip transect into units and substrips and simultaneously recording observers' sightings on a two-track tape recorder. These procedures were adequate for the range of dugong cluster densities encountered (mean = $0.17/km^2$). For mammals that form large, loosely structured, high-density aggregations, even this highly refined procedure may break down, because of density-dependent saturation in the recording of sightings and the difficulty in defining a cluster of animals.

MULTIPLE DEPENDENT OBSERVERS

METHOD. This method is an adaptation of the two-sample removal model (Seber 1982; also see "Removal Methods," below). The method requires that two observers, designated as "primary" and "secondary," count animals such that the number seen by the primary observer and the additional number seen by the secondary observer (i.e., additional to those seen by the primary observer) are determined.

The number of animals seen by the first observer is denoted n_1, and the additional number located by the second observer is denoted n_2. If the two observers expend equal effort, and if $n_1 \geq n_2$, then population size is estimated as

$$\hat{N} = \frac{n_1^2}{(n_1 - n_2)} \qquad (16)$$

(see Cook and Jacobson 1979 and Pollock and Kendall 1987). If we consider the multiple-dependent-observer method in the two-sample removal framework, animals seen by the primary observer (n_1) are those "captured" and "removed" from the population in the first sample, and additional animals seen by the secondary observer (n_2) are those "captured" in the second sample.

The assumptions required by the removal model underlying the multiple-dependent-observer method are as follows:

1. The population is closed between counts by the two observers.
2. All animals are equally likely to be seen by a given observer.
3. The observers have equal sighting abilities and expend equal effort.

FIELD APPLICATION. Unlike the independent-observer method, this procedure requires the observers to communicate in order to determine n_1 and n_2. The dependent-observer method is most suited to aerial survey but could also be applied to ground or shipboard surveys.

Pollock and Kendall (1987) considered the independent-observer method preferable to the dependent-observer method because it is simpler to understand, is more precise, and allows for differing sighting abilities by observers. For these reasons the dependent-observer method has received relatively little use compared to other available methods.

POSSIBLE PROBLEMS IN FIELD APPLICATION. The assumption most likely to be violated in the field is that the two observers have equal sighting abilities (e.g., Caughley et al. 1976; Short and Bayliss 1985). This problem can be overcome if observers switch roles halfway through the survey (Cook and Jacobson 1979), in which case the only assumption is that the sighting ability of each observer does not change with his or her role as primary or secondary observer. Estimators for this method that are more complex than equation 16 are given by Cook and Jacobson (1979) and Pollock and Kendall (1987).

The dependent-observer method is subject to the same problems of heterogeneous sighting probability (assumption 2) as outlined for the independent-observer method. Heterogeneity will again lead to negative bias in estimates of population size.

LINE-TRANSECT SAMPLING

METHOD. In the line-transect method an observer traverses one or more straight lines within the survey area and counts all animals seen. In some instances a maximum perpendicular distance w from the line is established beyond which sighted animals are ignored (data are said to be "truncated"). If data are not truncated, then w is taken to be the maximum perpendicular distance at which an animal was observed. The investigator records the number of animals counted and the perpendicular distance from the transect line to each animal, or the radial distance from the observer to the animal and the sighting angle between the line of sight to the animal and the transect line at the moment of detection (Fig. 43). The data from a line-transect survey thus comprise the transect length (L), the number of animals or clusters of animals sighted (n), and a set of perpendicular distances ($x_1, \ldots, x_n$) or of radial distances ($r_1, \ldots, r_n$) and sighting angles ($\theta_1, \ldots, \theta_n$).

If all animals present in the strip of width $2w$ are counted, it is a case of complete counts on sample plots (see "Complete Counts," above) where the plot is rectangular. In this situation population size in the surveyed strip (N_w) is simply the number counted:

$$\hat{N}_w = n \qquad (17)$$

and population density is the number counted divided by the area of the strip,

$$\hat{D}_w = \frac{n}{2Lw} \qquad (18)$$

This special case of line-transect methodology is usually referred to as a strip transect.

In practice, however, observers often fail to count some animals within distance w from the line. The probability of detection in the strip (β_w) is then less than 1, and equations 17 and 18 will provide negatively biased estimates of population size and density. One can correct for observability less than 1 using equation 2 and dividing the count statistic (n) by the estimated sighting probability ($\hat{\beta}_w$), as follows:

$$\hat{N}_w = \frac{n}{\hat{\beta}_w} \qquad (19)$$

and

$$\hat{D}_w = \frac{n}{2Lw\hat{\beta}_w} \qquad (20)$$

The line-transect method addresses the problem of detection bias by using perpendicular-distance data ($x_1, \ldots, x_n$) or radial distance ($r_1, \ldots, r_n$) and sighting-angle ($\theta_1, \ldots, \theta_n$) data to estimate sighting probability $\hat{\beta}_w$. I recommend that perpendicular-distance data be used to estimate $\hat{\beta}_w$, whenever possible, because use of sighting-distance and angle data requires unrealistic assumptions about the detection process that are not required by perpendicular-distance methods (Burnham et al. 1980; Buckland et al. 1993). In the remaining discussion, therefore, I deal only with methods based on perpendicular distances. Readers interested in radial-distance and sighting-angle methods should consult

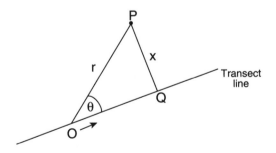

Figure 43. Line-transect parameters. The observer at position O sights an animal at position P. Q is the point on the transect line that is perpendicular to the animal. The radial distance is r, the sighting angle is θ, and the perpendicular distance from the object to the line is x. The observer travels from O to Q.

Burnham et al. (1980), Hayes and Buckland (1983), and Buckland et al. (1993).

Estimation of the sighting probability, $\hat{\beta}_w$, from perpendicular data is based on the following assumptions, listed from most to least critical:

1. Animals directly on the transect line are never missed.
2. Animals are detected at their initial location, prior to any movement in response to the observer, and are not counted twice.
3. Distances and angles are measured accurately.
4. Detections are independent events.

Note that a requirement for complete counting is inherent in the line-transect method but relates only to animals directly on the line. This is far less restrictive and less likely to be violated than the analogous requirement for complete counting on plots.

The basic concept underlying the estimation of sighting probability from perpendicular-distance data is that the probability of detecting an animal decreases as its perpendicular distance from the line increases. Mathematically this is represented by a detection function, or curve, denoted $g(x)$, which describes the probability of detecting an object given its perpendicular distance, x, from the line. The detection function can be approximated by drawing a smooth curve through a histogram of detections grouped by perpendicular-distance intervals (for example, 0–10, 11–20, 21–30, . . ., m). Typically, the detection function has an initial slow decline in detection frequency (this part of the curve is called the "shoulder"), followed by a more rapid decline at intermediate sighting distances, and a final slow decline to the maximum observed perpendicular distance (Fig. 44).

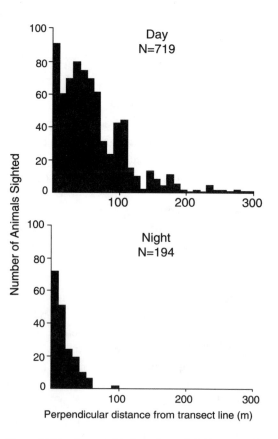

Figure 44. Line-transect data from day and night surveys of the same kangaroo population.

Burnham et al. (1980) showed that the detection probability β_w can be estimated from perpendicular-distance data as follows:

$$\hat{\beta}_w = \frac{1}{w\hat{f}(0)} \tag{21}$$

where $f(x)$ is the probability-density function of the perpendicular-distance data and can be thought of as the underlying probability distribution from which the observed distance data were generated. If the detection function, $g(x)$, is rescaled so as to integrate to 1, we obtain the relative probability-density function $f(x)$. The parameter estimate $\hat{f}(0)$ in equation 21 is obtained by evaluating $f(x)$ at $x = 0$ (i.e., where it cuts the vertical axis or, conceptually, on the line where detection probability is 1). By substituting equation 21 into equations 19 and 20, we have

$$\hat{N}_w = \frac{n}{[1/w\hat{f}(0)]}$$

$$= nw\hat{f}(0) \qquad (22)$$

and

$$\hat{D}_w = \frac{n}{2Lw[1/w\hat{f}(0)]}$$

$$= \frac{n\hat{f}(0)}{2L} \qquad (23)$$

These are the general equations for estimation of population size and density from perpendicular-distance data from line transects. Note that w has dropped out of equation 23; estimation of population density (but not population size) thus allows generalization of the sampling situation to transects with no predetermined width.

The values of n, w, and L needed for equations 22 and 23 are known from the survey; $f(0)$ must be estimated. The main problem of the line-transect method lies in developing an appropriate model for $f(x)$ and, subsequently, an estimator for $f(0)$. Many estimators have been described in the literature, and choosing among them can be very confusing for the biologist using line transects. To assist in this task Burnham et al. (1980) proposed four criteria against which to judge an estimator:

1. Flexibility and capability of fitting detection functions of varying shape.
2. Good performance despite variations in detection probability for fixed distance x.
3. A "shoulder" near $x = 0$.
4. Estimation of $f(0)$ with the smallest possible variance.

In their more recent discussion of model selection for line-transect data, Buckland et al. (1993) referred to criteria 1 and 2 as model robustness, criterion 3 as the shape criterion, and criterion 4 as efficiency. Details of model selection and model fitting are beyond the scope of this chapter, but the general approach recommended by Buckland et al. (1993) is (1) to select a key function as a starting point based on visual inspection of the distance data, and then (2) to select a flexible series expansion to adjust the key function to fit the distance data better.

FIELD APPLICATION AND DATA ANALYSIS. The line-transect method has been used to estimate mammal densities in a variety of different sampling situations (Table 9). In its classical application an observer traverses a transect line on foot searching for animals. Field considerations relevant to defining and following the transect lines are presented in Chapter 6 (see "Line Transects"). When an animal is sighted, its perpendicular distance from the transect line is measured. In practice it is often more convenient to measure the radial distance (r) and sighting angle (θ), and to calculate perpendicular distance (x) from the relationship $x = r(\sin(\theta))$, than to measure the distance directly in the field.

If the target animals aggregate into well-defined clusters, then distance and angle measurements are made to the geometric center of the cluster, and the size of each cluster is recorded. Otherwise (i.e., if the distance to each clustered individual is recorded), sightings of individuals are not independent, as required in assumption 4. Cluster density is then estimated using standard line-transect methodology, and individual density is estimated as the product of cluster density and mean cluster size. In situations in which exact measurement is not possible, sightings should be classified into distance-class intervals (for example, 0–10, 11–20, 21–30, . . ., m), and $\hat{\beta}_w$ should be estimated

Table 9. Examples of Studies Using the Line-Transect Method to Estimate Mammal Population Size

Species	Type of survey	Data collection		Reference
		Grouping	Truncation	
Minke whale	Ship	Ungrouped	Untruncated	Butterworth and Best (1982)
Fallow deer	Ground	Grouped	Truncated	Gogan et al. (1986)
Gray kangaroo	Ground	Ungrouped	Untruncated	Coulson and Raines (1985)
Bowhead whale	Ground	Ungrouped	Untruncated	Hammond (1984)
Whales	Aerial	Grouped	Untruncated	Hay (1982)
Pig	Aerial	Grouped	Truncated	Hone (1988)
Jackrabbit	Ground	Ungrouped	Untruncated	Smith and Nydegger (1985)
Mule deer	Aerial	Grouped	Truncated	White et al. (1989)

from frequency data (such data are said to be *grouped*).

Historically, two comprehensive computer programs have been available for analysis of line-transect data: TRANSECT (Laake et al. 1979) and LINETRAN (Gates 1980). TRANSECT has now been superseded by DISTANCE (Laake et al. 1991; Buckland et al. 1993). These programs provide several estimators for computing population density both from perpendicular-distance data and from radial-distance and sighting-angle data. DISTANCE follows the two-step, modeling approach of selecting a key function and then a series expansion (Buckland et al. 1993) and handles the necessary computations. Population size is estimated as the product of the population density estimate and the size of the survey area.

If distances are measured to clusters instead of individuals, then the investigator should test whether detection probability for fixed distance, *x*, depends on cluster size (a relationship termed size bias). Significant size bias can lead to bias in the estimation of mean cluster size and ultimately to bias in the estimation of population density. A program called SIZTRAN (Drummer and McDonald 1987) is available to test for size bias. Drummer and McDonald (1987), Otto and

Pollock (1990), and Quang (1991) discussed models for density estimation when size bias occurs.

DISTANCE (and TRANSECT and LINE-TRAN) will analyze data in both ungrouped and grouped form. When data are collected in ungrouped form, either analytical option is possible. Ungrouped analysis is preferable if distance and angle measurements are accurate, but if measurement inaccuracies result in accumulation of measurements at certain obvious values (termed "heaping"; see "Possible Problems in Field Application," below) then postsurvey grouping of the data may be used as a "smoothing" technique (Burnham et al. 1980; Hayes and Buckland 1983; Buckland et al. 1993).

Truncation can be used in the analytical stage as well as during data collection. Sometimes spurious peaks at extreme distances (outliers) occur in the observed detection function. These can detrimentally affect the performance of models in estimation of $f(0)$ and should be excluded from analysis. Burnham et al. (1980) recommended routine truncation of 1 to 3% of sightings.

I will illustrate line-transect estimation of population density with data from two surveys of the same population of kangaroos. The surveys were identical in all respects except time of day (one carried out during daylight hours and the other at night) (C. Southwell, unpubl. data; also see Southwell 1994). The total number of animals counted over a 21.8-km transect during the day (719) was 3.7 times higher than the number counted at night (194) (Fig. 44). Because population density was known to be constant at 198 kangaroos/km^2 during the surveys, the difference in numbers must reflect varying visibility. By using TRANSECT to fit the Fourier series model to these untruncated detection frequencies, $f(0)$ was estimated to be 0.01122 for daytime data and 0.04142 for night data. Densities for day and night surveys are estimated (using equation 23) as follows:

$$\hat{D} = \frac{1,000 \times 719 \times 0.01122}{2 \times 21.8} = 185.0/\text{km}^2$$

and

$$\hat{D} = \frac{1,000 \times 194 \times 0.04142}{2 \times 21.8} = 184.3/\text{km}^2$$

respectively. The factor of 1,000 is included in the numerator to convert perpendicular distance measurements, made in the field in meters, to kilometers. With distance units for x and L both in kilometers the resultant density estimate has a unit of animals/km^2.

If we assume that the data were truncated at 80 m during the survey, then the number counted in the $2w = 160$-m-wide strip would be 533 and 192 for day and night, respectively. The estimates of $f(0)$ for truncated data using the Fourier series model would then be 0.01642 and 0.04207, and from equation 21, respective day and night detection probabilities would be the following:

$$\beta_{80} = \frac{1}{(80 \times 0.01642)} = 0.7613$$

and

$$\beta_{80} = \frac{1}{(80 \times 0.04207)} = 0.2971$$

The probability of detecting a kangaroo within 80 m of the transect line is thus 2.5 higher during the day than at night. Respective day and night densities are estimated as

$$\hat{D}_{80} = \frac{1,000 \times 543 \times 0.01642}{2 \times 21.8} = 204.5/\text{km}^2$$

and

$$\hat{D}_{80} = \frac{1,000 \times 192 \times 0.04207}{2 \times 21.8} = 185.3/\text{km}^2$$

Thus despite the large difference in detection probabilities between day and night, estimates of density differ little from each other (whether based on truncated or unbounded data) or from the true density.

POSSIBLE PROBLEMS IN FIELD APPLICATION. The most important assumption of the line-transect method is that all animals on the line will be counted, i.e., $g(0) = 1$. Field sampling situations most likely to be associated with violation of this assumption were discussed in Chapter 6 (see "Research Design" under "Line Transects"). Violation of this assumption will lead to overestimation of $g(0)$ and consequent underestimation of population density and size, unless additional means of estimating $g(0)$ are employed. Pollock and Kendall (1987) suggested using multiple observers to overcome this problem on aerial line-transect surveys. Alldredge and Gates (1985) proposed a general model and estimators for this problem of left-truncated distributions of distance data. Elaborate experiments have been designed to estimate $g(0)$ for shipboard surveys of aquatic mammals (e.g., Kishino 1986; Butterworth and Borchers 1988). Leatherwood et al. (1982) provided modified line-transect models to account for obstructed visibility of the line in aerial surveys.

Most mammals react to an approaching observer. Generally, they attempt to hide or move away, although some approach the observer to investigate. Consequently, assumption 2 will often be violated in line-transect surveys of mammals. The extent to which such violations lead to bias in population estimates depends largely on the relative speeds of the observer and reacting animals and the behavior of the animal. Foot surveys of swiftly moving mammals may produce highly biased estimates if sighted animals flee along the line and flush other animals ahead and out of sight of the observer. This behavioral response is less of a problem for surveys from a vehicle and negligible for surveys from aircraft, because the observer is moving too fast for animals to flush ahead. Because detection of animals often depends on conspicuous movement behavior, an animal is often moving when first sighted, and its original position (before it detected the observer) may be difficult to locate. This effect is greatest for animals that move inconspicuously for some distance before flushing, and least for animals that flush immediately and conspicuously.

Observers can minimize inaccuracy in distance and angle measurements with careful field procedures. The most frequent problem in distance and angle measurement is "heaping" of data at certain obvious values (for example, on the line or at the truncation distance). Grouping of data (Hayes and Buckland 1983) or application of sophisticated smearing techniques (Butterworth 1982) can be used to lessen the effects of heaping, but such procedures are no substitute for accuracy.

SIGHTING-PROBABILITY MODELS

METHOD.   In all of the methods I have described so far for estimating population size from in-

complete counts, sighting probability is estimated from data collected concurrently with the incomplete counts. An alternative approach is to estimate sighting probability and to collect incomplete count data in separate surveys.

The first step in this approach is to conduct experimental surveys in order to identify those variables (e.g., vegetation, animal group size) that influence sighting probability and to develop a model for predicting sighting probability. During subsequent operational surveys, incomplete counts are made and data are collected on the variables in the model. These data are used with the model to predict a sighting probability for the conditions encountered during the survey, and these predictions are used with the incomplete counts to estimate population size.

The success of the sighting-probability-model approach for converting incomplete counts to accurate estimates of population size and density depends on the design of and results from the experimental survey. In particular, (1) the variables having the greatest influence on sighting probability should be identified; (2) the number of important variables should be manageably small; (3) conditions encountered in the operational survey must be similar to those in the experimental survey; and (4) important variables should be easily measured in operational surveys.

FIELD APPLICATION.   Sighting-probability models have been used primarily with aerial surveys (Table 10), but such models could also be used with ground or shipboard surveys. The most critical aspect of applying sighting-probability models is the choice of variables to consider as possibly influential. Omission of an important variable(s) can reduce the effectiveness of the model. Selection of variables is largely an intuitive process, which is greatly enhanced if the researcher has direct experience with the animals and habitat to be surveyed. Data collected on operational surveys include counts of the an-

Table 10. Examples of Studies Using Sighting-Probability Models to Estimate Mammal Population Size

Species	Type of survey	Variables in the model	Reference
Red kangaroo	Aerial	Speed, height, transect width, vegetation cover	Caughley et al. (1976)
Cattle	Aerial	Group size, vegetation cover	Newsome et al. (1981)
Red and gray kangaroos	Aerial	Temperature	Bayliss and Giles (1985)
Elk	Aerial	Group size	Samuel et al. (1987)

imals (or clusters of animals) and measurements of the variables included in the model for each sighting. Operational surveys should be carried out under conditions in which measurements of variables will fall within or close to the range of measurements obtained during the experimental survey.

Samuel et al. (1987) used radio-marked elk to estimate sighting probability during winter helicopter surveys. Radio-marking permitted determination of the exact location of the elk and, thus, permitted determination of whether or not they were sighted during the helicopter survey. At each sighting, investigators recorded data on variables thought to be important to sightability (group size, vegetation cover, snow cover, behavior, search rate, study area, and observer). They found group size and vegetation cover to be the most important determinants of sightability and developed a model to predict a sighting probability for each group of elk seen:

$$\text{Sighting probability} = \frac{\exp(u)}{1 + \exp(u)}$$

where exp is the exponential function and $u = 1.22 + 1.55 \ln(\text{group size}) - 0.05(\% \text{ vegetation cover})$. The coefficients of the model indicate the pattern of influence of each variable on sightability. The positive coefficient for group size reflects increasing sightability with increasing group size, and the negative coefficient for

vegetation cover indicates decreasing sightability with increasing vegetation cover.

To estimate population size, the number of groups of a given size in a given vegetation cover class is first divided by the sighting probability predicted by the model. The total number of animals in the sampled area is computed as the sum of these corrected values, and this sum is divided by the fraction of the survey area sampled. Once again, the basic form of the estimator is that of equation 4.

POSSIBLE PROBLEMS IN FIELD APPLICATION. Most problems in applying this method are likely to be related to development of the model. Incorporation of the most important variables influencing sightability into the model is critical, but such variables can be overlooked. For example Caughley et al. (1976) developed a sighting-probability model for red kangaroos on the a priori assumption that speed and height of the airplane, transect strip width, and vegetation cover were the survey variables most important in determining sightability. Subsequent work by Bayliss and Giles (1985) showed, however, that environmental variables, particularly ambient temperature and cloud cover, were also important.

Problems can occur if conditions encountered in the operational survey are outside the experimental range of conditions from which the model was developed. Caughley et al. (1976) developed a sighting model for a single species (red kangaroo) in two vegetation cover types (open and medium woodland). In subsequent broad-scale surveys (Caughley et al. 1977), however, two species (red and gray kangaroo) were sighted in vegetation cover ranging across and beyond the experimental range of percent vegetation cover. Caughley et al. (1977) addressed the problem by interpolation and extrapolation of the model's predicted detection probabilities for vegetation cover and by assuming that detection probabilities were the same for red and gray kangaroos. More recent work has revealed that this approach produces markedly inaccurate estimates of population size (Short and Bayliss 1985; Short and Hone 1988).

Removal Methods

RICHARD A. LANCIA AND JOHN W. BISHIR

Removal methods of estimating animal abundance are based on an accurate recording of the number of animals removed from or added to a population, in conjunction with observations of the population before, after, or during the removal process. These methods are attractive because removal data can often be collected by individuals other than the investigator (e.g., hunters), so such data are often inexpensive to obtain.

Two common removal methods are the change-in-ratio (CIR) method and the catch-per-unit-effort (C/E) method. CIR methods can be applied when the removals are "selective" and involve a particular class of animals. Hunters, for example, might remove only antlered deer from a population. Change-in-ratio estimators can be used when the proportions of subclasses of animals (e.g., sex, age, antler, or size subclasses) in the removals are substantially different from the proportions of the subclasses in the preremoval population. In addition, species in a community can be thought of as subclasses of

the community, and "removals" can be reductions or additions of individuals from particular species.

If removals are not selective, then either C/E models or fixed-effort removal models can be used to estimate population size. The fixed-effort models assume that equal effort is expended on catching or removing animals at each sampling occasion, whereas C/E models can be applied when sampling effort varies among sampling periods but is known or can be estimated.

CHANGE-IN-RATIO METHOD

The CIR method uses changes in the proportions of subclasses in a population or community to estimate the size of the subclasses. The method is predicated on the idea that the proportions of the subclasses in the population change over time because of disproportionate removals from or additions to the subclasses. In the most general CIR models, individuals from one or more subclasses can be removed. The removals occur during two or more distinct periods, where number of periods is denoted as r. A sample of the proportions of the subclasses in the population is taken before the first removal. An additional sample is taken after each subsequent removal for a total of $s = r + 1$ samples (Udevitz 1989; Udevitz and Pollock 1991). Most field applications of CIR with mammal populations will probably be limited to only a few subclasses, removals, and sample periods. In the following discussion, we illustrate the method for a situation involving two subclasses, one removal, and two sample periods. For extensions to different situations, see Udevitz (1989) and Udevitz and Pollock (1991).

METHOD.   The equations used to estimate population size with the CIR technique are essentially algebraic solutions to simultaneous equations relating population sizes of subclasses of animals before and after known removals. For more detail on derivation and a complete description of the technique see Paulik and Robson (1969), Seber (1982), Udevitz (1989), and Udevitz and Pollock (1991).

The basic CIR method assumes a closed population with two subclasses of animals, x-type and y-type. These could be males and females, antlered and antlerless deer, adults and juveniles, large and small individuals, or two different species. We can designate the proportions of x-type animals in the preremoval and postremoval populations as

$$P_1 = \frac{X_1}{N_1}$$

and

$$P_2 = \frac{X_1 - R_x}{N_1 - R} \qquad (24)$$

where

P_1 = proportion of x-type animals before the removal
X_1 = number of x-type animals in the initial (pre-removal) population
N_1 = total population size before the removal
P_2 = proportion of x-type animals after the removal
$R = R_x + R_y$ = total number of animals removed (known), where R_x and R_y are the numbers of x-type and y-type animals removed, respectively.

Substituting $X_1 = P_1 N_1$ in equation 24 yields

$$P_2 = \frac{P_1 N_1 - R_x}{N_1 - R}$$

Solving for N_1 yields the following estimator of total population size before the removal:

$$\hat{N}_1 = \frac{(R_x - R P_2)}{(P_1 - P_2)} \qquad (25)$$

Proportions P_1 and P_2 are unknown and must be estimated by some sampling scheme, such as

road counts or aerial surveys. These estimates, $\hat{P}_1$ and $\hat{P}_2$, together with the numbers removed, are substituted into equation 25 to estimate the preremoval population size. The number of x-type animals in the initial (preremoval) population is estimated by

$$\hat{X}_1 = \hat{P}_1 \hat{N}_1 \qquad (26)$$

From the estimates of N_1 and X_1 and the removals, the following estimates can be calculated: $\hat{Y}_1 = \hat{N}_1 - \hat{X}_1$; $\hat{X}_2 = \hat{X}_1 - R_x$; $\hat{Y}_2 = \hat{Y}_1 - R_y$; and $\hat{N}_2 = \hat{X}_2 + \hat{Y}_2$ or $\hat{N}_2 = \hat{N}_1 - R$. If estimates of P_1 and P_2 are assumed to be independent, variance estimates for $\hat{N}_1$ and $\hat{X}_1$ can be calculated as recommended by Seber (1982).

ASSUMPTIONS. The assumptions of the basic CIR method are as follows:

1. Observed proportions of x-type and y-type animals are unbiased estimates of the true proportions in the population.
2. The population is closed except for the removals.
3. Numbers of removals of x-type and y-type animals are known.
4. The proportion of the x-types removed is different from their proportion in the population.

Assumption 1 simply states that the x-type and y-type animals have equal probabilities of being sampled, that is, they are equally observable. The CIR method can be applied, however, when only a single type of animal is removed (e.g., only x-types being removed in an antlered-bucks-only hunt). In such a special case, the population estimate of the type removed (bucks) is appropriate regardless of whether or not x-types and y-types are equally sightable (Seber 1982).

Assumption 2 can best be met if the removal period and the time between the two estimates of the P ratios are short. Although the basic CIR method requires that numbers of removals of each type be known (assumption 3), the method can still be applied if unknown removals can be estimated (see Paulik and Robson 1969 and Seber 1982).

If the x-types are removed in the same proportion in which they occur in a population, then the P ratios are the same before and after the removal, that is, $P_1 = P_2$, and assumption 4 is not met. In this situation, the denominator in equation 25 is zero, and the method fails. This problem can be circumvented in more general applications of the CIR method in which there is more than one removal period, with different types of animals removed in the different periods (Pollock et al. 1985; Udevitz 1989; Udevitz and Pollock 1991). For example, in a two-period design only x-type animals might be removed during period 1 and only y-type animals during period 2. Estimates for these more general applications of CIR typically must be obtained iteratively (Pollock et al. 1985; Udevitz 1989; Udevitz and Pollock 1991), and therefore, we cannot provide estimators.

Several authors have investigated the effects of size and variability of samples used to estimate the P ratios on the accuracy and precision of CIR estimates of population size (Paulik and Robson 1969; Seber 1982; Pollock et al. 1985; Conner et al. 1986; Udevitz 1989; J. Bishir and R. Lancia, unpubl. data). The initial P ratio (P_1), the change in P ratios ($\Delta P = P_1 - P_2$), and the numbers of animals that are observed to estimate the P ratios all affect the accuracy and precision of CIR population estimates. In general, CIR estimates are most accurate when the change in ratio, ΔP, is large. Removals of large numbers of a single subclass, or removals of two or more subclasses widely disproportionate to their representation in the population, will produce large changes in ratios (large ΔPs). If the change in ratios (ΔP) is small, the method is likely to produce excessively large or small population estimates or model failures (i.e., negative popu-

lation estimates). Accuracy also improves as the sizes of samples used to estimate the P ratios increase. Using computer simulations, we showed that accurate population estimates were obtained with CIR when 70% or more of x-types of a single subclass were removed from populations of from 50 to 1,000 animals and large numbers of animals were sighted to estimate the P ratios (J. Bishir and R. Lancia, unpubl. data). Removals of this magnitude can occur, for example, when antlered deer only are removed during a harvest. Estimates based on smaller removals are more likely to be unreliable.

The foregoing considerations of accuracy and precision assume that the underlying assumptions of the method are met and that the observations used to estimate the P ratios are binomial random variables. In practice the assumption of equal observability (assumption 1) is the most difficult to meet.

Conner et al. (1986) used the CIR method to estimate population sizes of antlered and antlerless deer at Remington Farms in Maryland. During 54 prehunt road counts, 120 antlered (x-type) and 1,126 antlerless (y-type) deer were observed. Then 56 antlered (R_x) and 54 antlerless (R_y) deer were removed by hunters during a 1-week season. After the season, 43 x-type and 1,086 y-type deer were observed during 52 posthunt road counts. Therefore,

$$\hat{P}_1 = \frac{120}{1,246} = 0.0963$$

$$\hat{P}_2 = \frac{43}{1,129} = 0.0381$$

$$\hat{N}_1 = \frac{(R_x - R\hat{P}_2)}{(\hat{P}_1 - \hat{P}_2)}$$

$$= \frac{56 - 110(0.0381)}{(0.0963 - 0.0381)} = 890 \quad (\hat{SE} = 149)$$

$$\hat{X}_1 = \hat{P}_1\hat{N}_1 = 0.0963(890) = 86 \quad (\hat{SE} = 14)$$

In this example both *x*-types and *y*-types were removed. Because the *y*-type (antlerless) animals were probably more observable than the *x*-type animals, the equal-observability assumption was likely violated to some degree, and the population size estimates are likely to be positively biased.

GENERAL CIR MODELS AND ESTIMATORS. Udevitz (1989) provided a general framework for all CIR models and pointed out that, rather than requiring equal observability, the implicit assumption of the general method is that the relative observabilities of the subclasses are constant over time. The relative observability of two subclasses, λ, can be expressed as

$$\lambda = \frac{\text{observability of } y\text{-type animals}}{\text{observability of } x\text{-type animals}}$$

When both types are equally observable, then $\lambda = 1$, which is equivalent to the equal-observability assumption. In some of the more general applications of the CIR method, when there are >1 removal periods, λ can be estimated. For example, when observabilities of classes of animals differ, but the relative observabilities are constant, a two-removal, three-sample CIR analysis can be used to estimate population size.

Consider a hypothetical example based on white-tailed deer (Lancia et al. 1988). Antlered and antlerless deer are observed during roadside counts from vehicles to obtain estimates of the *P* ratios. Before the first hunt is held, $\hat{P}_1$ is estimated as 0.3333. Then 250 deer are removed during a 1-week antlered-deer-only hunting season. Subsequently, $\hat{P}_2$ is estimated as 0.2500. An additional 250 deer are removed during a 1-week antlerless-deer-only hunting season, and $\hat{P}_3$ is estimated at 0.3333. It follows then that

$$\hat{X}_1 = \frac{[R_x \hat{P}_1(1 - \hat{P}_2)]}{(\hat{P}_1 - \hat{P}_2)}$$

$$= \frac{[250(0.3333)(1 - 0.2500)]}{(0.3333 - 0.2500)} = 750$$

$$\hat{Y}_1 = \frac{[R_y \hat{P}_3(-\hat{P}_2)}{(\hat{P}_3 - \hat{P}_2)}$$

$$= \frac{[250(0.3333)(1 - 0.2500)]}{(0.3333 - 0.2500)} = 750$$

$$\hat{N}_1 = \hat{X}_1 + \hat{Y}_1 = 1{,}500$$

and

$$\hat{\lambda} = \frac{[R_x(1 - \hat{P}_1)(\hat{P}_3 - \hat{P}_2)]}{[R_y \hat{P}_3(\hat{P}_1 - \hat{P}_2)]}$$

$$= \frac{[250(1 - 0.3333)(0.3333 - 0.2500)}{[250(0.3333)(0.3333 - 0.2500)]} = 2.00$$

The large estimated value of $\hat{\lambda}$ (2.00) indicates that the antlerless deer were more observable than the antlered deer and that the two-removal CIR was a more appropriate method than the basic CIR. Calculation of standard errors for these multisample CIR estimators is described in Udevitz (1989).

CATCH-PER-UNIT-EFFORT METHOD

C/E estimators have been examined by many individuals, including Leslie and Davis (1939), Chapman (1954), Ricker (1958), Seber (1982), and Bishir and Lancia (unpubl. data). Overton and Davis (1969) provided a clear and simple description of the derivation of the basic C/E estimator.

C/E estimators are based on the premise that as more and more animals are removed from a population, fewer are available to be "caught," and catch per unit effort declines. For example, fewer animals can be seen per hour or fewer can be harvested per hunter-day as more animals are removed. Let

N = initial number of animals in the population

E_i = number of units of effort expended in the ith time period

C_i = number of animals "caught" during the ith period

$f_i = C_i/E_i$ = catch per unit effort in the ith period

x_i = total removal prior to the ith period (so $x_1 = 0$)

k = "catchability coefficient," the probability that a particular unit of effort catches an animal.

At the beginning of the ith period, $N - x_i$ animals remain in the population, and E_i units of effort are expended to catch these animals during the period, with each unit of effort having probability k of producing a "catch." From this, the expected number of catches in the ith period, $E(C_i)$, is $kE_i(N - x_i)$, and the expected catch per unit effort, z_i, is

$$z_i = E(f_i) = \frac{kE_i(N - x_i)}{E_i}$$

$$= kN - kx_i$$

If we write $A = kN$ and $B = -k$, then the relation becomes

$$z_i = A + Bx_i$$

a linear regression of z_i on x_i. Here A equals the y-intercept estimated by the regression equation, and B is the slope of the regression. Standard least-squares procedures produce the estimates, in which the observed values $f_i = C_i/E_i$ are inserted for z_i, as follows:

$$\hat{B} = \frac{\sum_i \frac{C_i}{E_i}(x_i - \bar{x})}{\sum_i (x_i - \bar{x})^2} \tag{27}$$

and

$$\hat{A} = \bar{z} - \hat{B}\bar{x}$$

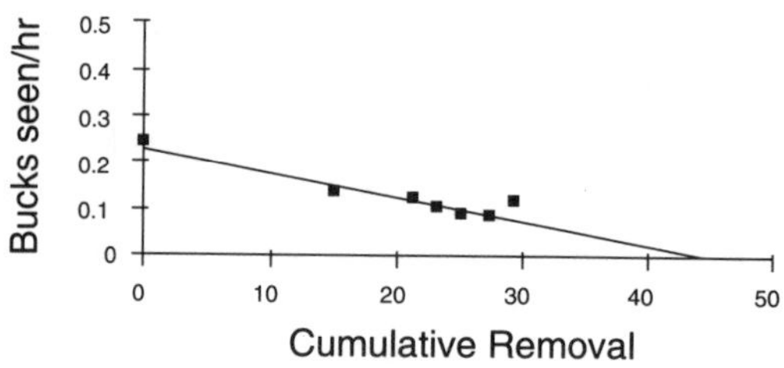

Figure 45. Catch per unit effort with effort expressed as the number of bucks seen per hour by hunters on Remington Farms, 1983. Data are from Table 11.

is inserted for B and A (where bars denote means), giving

$$\hat{k} = -\hat{B}$$

and

$$\hat{N} = \frac{\hat{A}}{\hat{k}} = \bar{x} + \frac{\bar{z}}{\hat{k}} \tag{28}$$

The estimated preremoval population size, $\hat{N}$, is the cumulative catch (population size) for which the expected catch per unit effort is zero ($y = 0$), that is, the x-intercept (Fig. 45).

Although maximum-likelihood (ML) and weighted-least-squares estimators for the C/E method have been published (e.g., Seber 1982:297; Dupont 1983; Pollock et al. 1984), no computer program for these estimation procedures is generally available. Dupont (1983) and Pollock et al. (1984), however, offer ML routines on request. Typically, the traditional regression method has been used to implement the C/E method for closed populations. More work is required to develop C/E computer programs analogous to the program CAPTURE (see "Closed-Population Models with K Samples" under "Capture-Recapture Methods," below). Novak et al. (1991) and Laake (1992) used open C/E models and ML estimates (Dupont 1983) to estimate the size of deer and elk populations, respectively.

The meanings of "removed" and "caught" with respect to removal methods require elaboration. Animals can be physically removed from a population (killed or live-trapped and relocated), or they can be figuratively "removed" by being marked (e.g., see the discussion of models M_b and M_{bh} in "Capture-Recapture Methods," below). In the latter case observations of marked animals are ignored in subsequent catches, and the statistics of interest are the numbers of new (unmarked) animals captured. Animals can be removed by any means; procedures used to obtain animals that are removed do not have to correspond to the effort used in the estimator. For example, removals can be hunter kills and catch per unit effort can be animals seen per day. All sources of removals, such as accidental roadkills, poached animals, or unretrieved hunting kills, are included in the cumulative total removed. Furthermore, animals "caught" during a C/E study can be shot, trapped, or sighted. They do not have to be physically taken or removed to be "caught."

Although the form of the C/E regression equation looks familiar, the expression may not be a typical regression. If kills are used as a measure of effort, then both y, the catch per unit effort, and x, the cumulative removals, depend on the same removals. This lack of independence makes calculation of variances and confidence intervals difficult. If sightings are used to measure effort, however, then standard variance calculations could be appropriate. Estimates of N do not necessarily follow a normal distribution, and standard expected value and variance equations (e.g., Seber 1982) may not be appropriate.

ASSUMPTIONS. The assumptions of the C/E method are similar to those of the CIR method:

1. The population is closed (except for the removals).
2. Each animal has an equal probability (k) of being caught by a particular unit of effort

for each sampling period (e.g., day, week), and k is constant over time.
3. All removals are known.

Assumption 1 can best be met by keeping the removal period as short as possible. Some models (e.g., Dupont 1983); however, relax this assumption.

Assumption 2 is the equal-catchability or equal-observability assumption. A qualitative test of this assumption is to examine the trajectory of the plot of catch per unit effort versus the cumulative removals. If it is not linear, then the equal-catchability assumption is probably violated. Caughley (1977) recommended that the technique be abandoned in such cases. The computer simulations of J. Bishir and R. Lancia (unpubl. data) suggest, however, that if a substantial proportion of the population is removed, then linearity probably is not a significant concern. The assumption of equal catchability can be relaxed if maximum likelihood or least-squares models are developed with additional parameters that allow k to vary, for example, with time (Dupont 1983; J. Bishir and R. Lancia, unpubl. data).

If the units of effort are constant, for example, as when the same number of traps are set each night or the same number of individuals hunt each day, then the catch per unit effort is equivalent to the number of animals caught on successive occasions. In this case constant-effort, capture-recapture models (models M_b, M_{bh}, and M_{th}; see "Closed-Population Models with K Samples" under "Capture-Recapture Methods," below) can be used to estimate population size. Model M_{bh} may be especially useful as it allows individual differences in catchability.

The likelihood of violating assumption 3 can be minimized by keeping the removal period short and by searching the study area for unreported removals.

ACCURACY AND PRECISION. We used computer simulation to examine the accuracy and

Table 11. Summary of Daily Hunter Reports, Remington Farms, 1983[a]

		\multicolumn{7}{c}{Day of the season}						
		1	2	3	4	5	6	7
Effort	Hunters	33	35	24	16	21	19	1
	Hours hunted	161.2	198.7	129.5	63.0	95.5	87.8	80.3
Catch	Bucks killed	15	6	2	2	2	2	5
	Bucks seen	40	26	15	6	8	7	9

[a]Data from E. C. Soutiere (pers. comm.).

precision of C/E estimates. The C/E method can fail if the slope of the regression line (B) is positive, resulting in a negative population estimate or negative x-intercept (Overton and Davis 1969). Similarly, excessively large estimates result if B is negative but very small. These occasional large values skew distributions of C/E estimates toward larger values. Both model failures and excessively large estimates are more likely to occur if a small proportion of the population is removed. C/E estimates are likely to be accurate and precise if more than 70% of the population is removed.

Removal methods of population estimation can often be used in conjunction with harvests of hunted species. For example, the number of antlered bucks in the Remington Farms deer herd was estimated from hunter diaries of the number

Table 12. Computations Related to the Catch-per-Unit-Effort Estimate of the Prehunt Number of Antlered Bucks at Remington Farms, 1983[a]

	\multicolumn{7}{c}{Day of the season}						
	1	2	3	4	5	6	7
Cumulative removal (x_i)	0	15	21	23	25	27	29
Catch per unit effort (z_i)	0.248	0.131	0.116	0.095	0.084	0.080	0.112

Summary of computations:

$$\bar{x} = 20 \qquad \bar{z} = 0.1237$$

$$\sum (x_i - \bar{x})^2 = 590 \qquad \sum z_i(x_i - \bar{x}) = 3.2293$$

$$\hat{k} = -\hat{B} = \frac{3.2293}{590} = 0.00547$$

$$\hat{A} = \bar{z} + k\bar{x} = 0.2331$$

$$\hat{N} = \frac{\hat{A}}{\hat{k}} = 42.6$$

[a]Effort = number of hours hunted; catch = number of bucks seen; removal = number of bucks killed. Based on data from Table 11.

of bucks seen per hour and check station data on the number of bucks killed per day during a 1-week hunting season in 1983 (Fig. 45, Table 11). The total kill was 34 bucks (Table 11), and the population estimate (Fig. 45, Table 12) was 42.6 ($\hat{SE}$ = 4.0). Apparently, a large proportion (ca. 0.8) of the population was removed, so this estimate is probably reasonably accurate.

CONTRIBUTORS: STUART C. CAIRNS AND TIMOTHY F. CLANCY

Capture-Recapture Methods

JAMES D. NICHOLS AND CHRIS R. DICKMAN

Capture-recapture models provide a means of estimating population size from data on numbers of animals captured and recaptured or resighted during trapping and observation efforts. The count statistic obtained is the number of animals caught or sighted in a sample period. Information derived from the initial captures and, more typically, the pattern of recaptures is then used to estimate capture or sighting probability (the observability fraction, β). Population size, N, is then estimated using equation 2 or, if spatial sampling is also involved, the canonical estimator given by equation 4.

TWO-SAMPLE ESTIMATOR

METHOD.   The Lincoln-Petersen estimator was the first capture-recapture estimator of animal abundance (Lincoln 1930). It is still used, and it provides an intuitive understanding of capture-recapture estimation that carries over to more complicated models. The basic method was described earlier (see "Marked Subsample" under "Estimation of Population Size and Density When Counts Are Incomplete," above); we discuss it here in more detail.

This estimator is based on a two-sample study. An initial sample of n_1 animals is caught at period 1. All of these animals are marked and released back into the population. The investigator then catches another sample of n_2 animals during a second capture period (e.g., the two samples might be taken on consecutive days). The investigator records the number (m_2) of animals in the second sample that are marked and, therefore, were known to have been caught in the first sample. The observability fraction or capture probability associated with the initial sample is defined as n_1/N. If certain assumptions hold true, then the proportion of marked animals in the second sample, m_2/n_2, should estimate both the proportion of marked animals in the population and the capture probability for sample 1:

$$\frac{m_2}{n_2} = \frac{n_1}{N} \tag{29}$$

We can rearrange equation 29 to obtain the Lincoln-Petersen estimator for population size, as follows:

$$\hat{N} = \frac{n_1 n_2}{m_2} \tag{30}$$

The estimator in equation 30 is of the same form as the canonical estimator (equation 4), in that it represents a count statistic, n_1, divided by an estimate of the observability fraction, m_2/n_2. Alternatively, we can view n_2 as the count statistic in sample 2 and m_2/n_1 as the corresponding capture probability estimate. In this case, equation 30 is again of the same form as equation 4.

Chapman (1951) provided a modified Lincoln-Petersen estimator with less bias, as follows:

$$\hat{N} = \frac{(n_1 + 1)(n_2 + 1)}{(m_2 + 1)} - 1 \tag{31}$$

Seber (1970) provided the following approximately unbiased estimate of the variance of $\hat{N}$:

$$\text{var}(\hat{N}) = \frac{(n_1 + 1)(n_2 + 1)(n_1 - m_2)(n_2 - m_2)}{(m_2 + 1)^2(m_2 + 2)} \quad (32)$$

We recommend the estimators of equations 31 and 32 for use in two-sample capture-recapture studies of mammalian abundance.

ASSUMPTIONS. Chapman's modified Lincoln-Petersen estimator (equation 31) is based on the following assumptions:

1. The sampled population is closed; no additions (through birth or immigration) or losses (through death or emigration) occur between the two sample periods.
2. All animals are equally likely to be captured in each sample.
3. Marks are neither lost nor incorrectly recorded.

Assumption 1 can be weakened. If only losses to the population occur between samples, then equation 31 still estimates the number of animals at the time of the first sample. If only additions occur between samples, then equation 31 estimates the population size at the time of sample 2.

Assumption 2 is important in all capture-recapture studies, and we generally think of two classes of violation of this assumption, heterogeneity and behavioral response. Heterogeneity refers to inherent variation in capture probability among individuals in a population. When such heterogeneity is associated with morphological or other characteristics that can be identified from a captured animal, then an investigator can deal with this problem via stratification. For example, if capture probability differs for male and female mice, then one can obtain separate estimates of abundance for each sex and add the estimates to get total abundance. Some variation in capture probability is always likely to occur, however, even among individuals that are mor-

phologically similar. A consequence of heterogeneity is that animals with relatively high capture probabilities will appear in both samples more frequently than other animals, so that m_2/n_2 will overestimate the proportion of marked animals in the population at the time of sample 2. This produces negative bias in $\hat{N}$; that is, on average, $\hat{N}$ will be smaller than the true population size.

Assumption 2 can also be violated when animals exhibit a behavioral response to being trapped. Many small mammals become "trap-happy," so that capture probability for marked animals is higher than that for unmarked animals. Other species (e.g., raccoons), however, tend to become wary or "trap-shy" and difficult to recapture. $\hat{N}$ tends to be too small in the case of a trap-happy response and too large in the situation of a trap-shy response. One methodological means of reducing both heterogeneity and trap response is to use different methods of catching animals in the two sample periods. Capture probabilities for the methods may differ, but that difference poses no problems for the use of the Lincoln-Petersen estimator.

If animals lose their marks (assumption 3), then m_2 will be too small, and $\hat{N}$ will tend to be too large. If loss of marks is anticipated to be a problem, then investigators should double-tag each animal so that they can estimate tag loss and use this estimate to correct the bias in $\hat{N}$ (Seber 1982).

As an example of the Lincoln-Petersen estimator, consider a study of Nuttall's cottontail (*Sylvilagus nuttalli*) in central Oregon in which 87 rabbits were live-trapped and released during August 1974 (Skalski et al. 1983). On 5 September 1974, a drive count was conducted; 7 of 14 animals counted were marked. The statistics $n_1 = 87$, $n_2 = 14$, and $m_2 = 7$ were used with Chapman's modified estimators (equations 31 and 32) to calculate population size, $\hat{N} = 164$, and estimate variance, $\text{var}(\hat{N}) = 1288.33$.

CLOSED-POPULATION MODELS
WITH K SAMPLES

The two-sample Lincoln-Petersen estimator performs well when its underlying assumptions are met. The capture-recapture data from a two-sample study cannot, however, be used to check whether those assumptions have been met. Information from studies with more than two sample periods can be used to test assumptions about population closure and sources of variation in capture probability. In addition, multisample studies yield more data and, therefore more precise estimates than two-sample studies and permit modeling of sources of variation in capture probability. When the population is open to both gains and losses between sample periods, the Lincoln-Petersen estimator does not perform well. When three or more sample periods are used, however, open-population models allowing for both gains and losses can be used. Data from K-sample capture-recapture studies are typically summarized as "capture histories" of individual animals. A capture history is simply a row of 1s and 0s, where 1 denotes capture (or observation in a resighting study) and 0 denotes no capture. The first entry in the row corresponds to the first sample period, the second entry to the second sample period, and so on. For example, a capture history of 1011 specifies an animal that was caught in the first sample period, was not caught during the second period, but was caught again in periods 3 and 4. A capture-recapture study yields a capture history for every animal caught. These data form the basis for all estimation in both closed- and open-population modeling.

STANDARD CAPTURE-RECAPTURE MODELS. K-sample capture-recapture models for closed populations are used in situations in which no gains or losses to the population are believed to occur between the sample periods of the study.

Such studies take place over relatively short periods (for example, a grid of traps might be run for 5 consecutive nights). The capture-history data arising from such studies are modeled in terms of capture probabilities. Let p_{ij} denote the probability that animal j is captured in sample period i. It is possible to make a number of different assumptions about capture probabilities and their sources of variation, and these different assumptions lead to different capture-recapture models for closed populations (see the excellent primer by White et al. 1982).

In the simplest model, M_0, capture probability is modeled as a constant over time periods and individuals ($p_{ij} = p$, for all i and all j). Model M_t assumes that capture probability varies with sample period but is the same for all individuals within each period ($p_{ij} = p_i$ for all j). This model is illustrated in Table 13, which shows the probabilities associated with each observable capture history in a three-sample study. Consider, for example, capture history 101. The probability associated with capture in period 1 is given by p_1, the probability of not being captured in period 2 is $(1 - p_2)$, and the probability of being caught in period 3 is p_3. The probability associated with the entire capture history is given by the product of these three probabilities.

Table 13. Possible Capture Histories and Associated Probabilities in a Three-Sample Study of a Closed Population under Model M_t

Capture history	Probability
111	$p_1 p_2 p_3$
110	$p_1 p_2 (1 - p_3)$
101	$p_1 (1 - p_2) p_3$
100	$p_1 (1 - p_2)(1 - p_3)$
011	$(1 - p_1) p_2 p_3$
010	$(1 - p_1) p_2 (1 - p_3)$
001	$(1 - p_1)(1 - p_2) p_3$

Model M_b incorporates a behavioral response of the animal to initial capture (trap-happiness or trap-shyness), such that all unmarked animals exhibit one capture probability, p, and all marked animals exhibit a different capture probability, c, but these probabilities do not vary otherwise by either time (sample period) or individual. Estimation of abundance using model M_b (and most behavioral-response models) is based only on initial captures of animals. Therefore, M_b can be used to estimate population size from removal trapping data (e.g., small mammals are often sampled by snap-traps that kill the animals and yield no releases or recaptures). M_b can be thought of as a special case of the C/E models in which effort is constant over time (see "Catch-per-Unit-Effort Method," above).

Model M_h permits heterogeneity, or variation in capture probability among the different individuals in the population. Each individual may have a capture probability different from those of all other individuals, but these probabilities do not change with time: $p_{ij} = p_j$ for all i.

In addition to these models incorporating single sources of variation in capture probabilities (time, behavioral response, heterogeneity), models have been developed to deal with more than one source of variation. Abundance estimation is possible with all of the two-source models, M_{bh}, M_{tb}, and M_{th}, but not under the most general model, M_{tbh}. Like M_b, model M_{bh} can be used to estimate abundance from removal data (see "Catch-per-Unit-Effort Method," above).

Unlike the Lincoln-Petersen estimator, these K-sample models for closed populations require a computer program to provide iterative solutions to the estimation equations. The program CAPTURE (Otis et al. 1978) was originally developed to provide estimates under these models. We recommend use of a newer version (Rexstad and Burnham 1991) for estimation of population size in capture-recapture studies of closed populations.

CAPTURE accepts input data summarized in the form of capture histories. The program computes estimates of population size and associated standard errors under any or all of the models described above. A test of the closure assumption is also computed by the program. CAPTURE also prints goodness-of-fit and between-model test statistics that can be used to judge which model (which hypothesis about variation in capture probabilities) best describes the data. An objective model selection procedure provides a recommended model for each data set (Rexstad and Burnham 1991).

Two assumptions are required by all models in CAPTURE: (1) that marks are neither lost nor incorrectly recorded and (2) that the population is closed. The closure assumption is testable and can frequently be met by limiting the duration of the study (populations are more likely to remain closed over short intervals). Each of the described models makes a different set of assumptions about capture probabilities. The investigator should attempt to reduce the number of sources of variation in capture probability. This will reduce the chances that the most general model, M_{tbh} (for which no abundance estimator currently exists), is required by the data. In addition, reductions in sources of variation in capture probability tend to yield increases in estimator precision (reduced variance).

As an example of the use of CAPTURE for estimating population size, consider capture-recapture data for meadow voles (*Microtus pennsylvanicus*) at Patuxent Wildlife Research Center in Laurel, Maryland (Pollock et al. 1990). A total of 102 animals was captured during a 5-day period, and the model-selection procedure of CAPTURE indicated the appropriateness of model M_h. The test of model M_0 versus M_h provided strong evidence of heterogeneity ($\chi_3^2 = 120.8$, $P < 0.01$), and the goodness-of-fit test for M_h was nonsignificant ($\chi_4^2 = 7.3$, $P = 0.12$). The statistics needed to compute population size estimates under this model are simply the capture frequencies, that is, the number of animals captured once, twice, three times, and so forth. For this data set, 29 animals were captured once, 15 twice, 15 three times, 16 four times, and 27 five times. The population size estimate from CAPTURE using M_h was $\hat{N} = 139$, with $\widehat{SE}(\hat{N}) = 10.85$ and an estimated average daily capture probability of $\bar{p} = 0.44$.

TRAPPING WEB. The trapping web can be used to estimate density (number of animals per unit area) of mammals based on capture or removal data (Anderson et al. 1983). The method differs fundamentally from the standard capture-recapture models described earlier, in that it involves no attempts to model capture-history data. Instead, the method makes use of concepts from distance sampling (e.g., see "Line-Transect Sampling," above).

In line-transect estimation methods, sighting or detection probabilities are assumed to decrease with increasing distance from the line. The modeling of detection probability as a function of distance forms the basis for estimation under these methods. The trapping web is a configuration of traps designed to yield a gradient of capture probabilities, decreasing with distance from the web center.

A trapping web consists of some number (Anderson et al. [1983] recommended 16) of equally spaced lines of equal length radiating from a randomly located center point (see Fig. 12, Chapter 8). Traps (e.g., 20 per line; Anderson et al. [1983] recommended a total of at least 250 traps) are placed along the radial lines at fixed distances from the center point of the web, with constant distance separating successive traps on a line. The web thus consists of a series of concentric rings of traps. Traps in the ring nearest to the web center are close together. Distances separating traps that form a particular ring increase with increasing distance of the ring from the web center. The web configuration produces a gradient in trap density and, therefore, in the probabil-

ity of detection or capture. Highest trap densities and capture probabilities occur at the web center.

Traps may be run for several (e.g., 3–5) consecutive days. Anderson et al. (1983) recommended that trapping be continued until at least 60 initial captures are obtained. The concentric ring (and thus, distance from the web center) of initial capture is recorded for each animal. Because only the initial capture is used in the analysis, the method is applicable to removal as well as recapture data.

The area trapped by a ring of traps is computed based on rings (circular zones) whose limits are defined by points halfway between adjacent traps along radial lines (see Fig. 12, Chapter 8). Information on area trapped and number of animals caught is available for each ring of traps and forms the basis for estimating the probability density function of the area sampled, $f(c)$. Animal density is then estimated as in the line-transect setting:

$$\hat{D} = M_{t+1} \hat{f}(0) \tag{33}$$

where D is density and M_{t+1} denotes the total number of animals caught during the web operation. The estimators from the program DISTANCE (Laake et al. 1991; Buckland et al. 1993) can be used to estimate density. The variance of this estimate, $\text{var}(\hat{D})$, can be estimated using DISTANCE output in conjunction with the estimator of Wilson and Anderson (1985).

Four assumptions underlie density estimation with the trapping web:

1. All animals located at the center of the web are caught with probability 1.0.
2. Animal movements are "stable" with respect to the web configuration (i.e., animals do not move preferentially toward or away from the web center).
3. Distances from the web center to each trap are measured accurately when the web is laid out.

4. Captures of different individuals are independent events (this assumption is necessary for variance estimation).

Assumption 1 is unlikely to be exactly true in real-world sampling situations but frequently may be met closely enough and often enough to provide reasonable density estimates. As capture probabilities of animals at the web center decrease, density estimates become increasingly negatively biased (i.e., underestimate density).

The trapping web seems to perform well with small mammals (Anderson et al. 1983; Jett and Nichols 1987). Link and Barker (1994) recently proposed a different, geometric approach to estimation, using the trapping web configuration. Although this new approach has been little tested, it appears to have some advantages over the distance-sampling approach to estimation.

OPEN-POPULATION MODELS WITH K SAMPLES

Open-population capture-recapture models are those that permit gains and losses of animals between sampling periods and are thus appropriate for long-term studies. For example, a small mammal study might involve trapping a grid once every 1 to 2 months for several years. Because losses to the population are permitted, models of capture history data require not only capture probabilities, p_i (the probability that an animal alive and in the sampled area in period i is captured then), but also survival probabilities, ϕ_i (the probability that an animal in the sampled area in period i is alive and still in the sampled area in period $i + 1$).

JOLLY-SEBER MODEL. The basic model for open populations (Jolly 1965; Seber 1965) includes separate capture and survival parameters for each sampling period (i). Probabilities associated with each possible capture history for animals captured in period 1 of a three-period

Table 14. Possible Capture Histories and Associated Probabilities for Animals Caught in Period 1 of a Three-Period Study of an Open Population under the Jolly-Seber Model

Capture history	Probability
111	$\phi_1 p_2 \phi_2 p_3$
110	$\phi_1 p_2 (1 - \phi_2 p_3)$
101	$\phi_1 (1 - p_2) \phi_2 p_3$
100	$(1 - \phi_1) + \phi_1 (1 - p_2)(1 - \phi_2 p_3)$

study are presented in Table 14. Capture history 111, for example, indicates that an animal released in period 1 was also captured in periods 2 and 3. Conditional on capture and release in period 1, this animal had to survive until period 2 (this event occurs with probability ϕ_1), be captured in period 2 (this occurs with probability p_2), survive until period 3 (ϕ_2), and be caught in period 3 (p_3). The probability associated with the entire capture history, conditional on release in period 1, is given by the product of the probabilities associated with each of these events, $\phi_1 p_2 \phi_2 p_3$.

Unlike those in many K-sample models, the estimators of quantities of interest under the Jolly-Seber model can be written in closed form. They are most easily written as functions of the following summary statistics obtained directly from the capture-history data:

m_i = number of marked animals caught in sampling period i

n_i = total number of animals (marked and unmarked) caught in period i

R_i = number of marked animals released into the population in period i

r_i = number of animals released at i that are ever captured again in subsequent sampling periods of the study

z_i = number of animals caught at some period before i and at some period after i, but not in sampling period i.

We can estimate population size at period i, N_i, by noting that the ratio of marked to total animals in the sample at i should approximately equal the ratio of marked to total animals in the population at that time:

$$\frac{m_i}{n_i} \approx \frac{M_i}{N_i}$$

where M_i is the total number of marked animals in the population at period i. This yields the following estimator of N_i:

$$\hat{N}_i = \frac{n_i \hat{M}_i}{m_i} \qquad (34)$$

This estimator for N_i is of the same form as our canonical estimator (equations 2 and 4) in that it consists of a count statistic, n_i, divided by the estimated capture probability, $m_i / \hat{M}_i$.

Our only problem with using equation 34 is that M_i is not a statistic obtained directly from capture-history data, but is instead a quantity that must be estimated (hence the "hat" notation, $\hat{M}_i$). In order to estimate M_i, the following ratios should be approximately equal:

$$\frac{z_i}{M_i - m_i} \approx \frac{r_i}{R_i} \qquad (35)$$

The denominator of the ratio on the left side of equation 35, $M_i - m_i$, is the number of marked animals that are in the population at i, but not caught at that time. The denominator of the right side of equation 35, R_i, is the number of animals caught at i and released with marks. The numerators of the two sides represent the numbers in each group of marked animals (z_i for those not caught at i; r_i for those caught at i) that are ever captured again during the remainder of the study ($>i$). Rearrangement of equation 35 yields the following estimator for M_i:

$$\hat{M}_1 = m_i + \frac{z_i R_i}{r_i} \qquad (36)$$

which is defined for periods $i = 2, \ldots, K - 1$. Thus, N_1 can be estimated only for periods 2 through $K - 1$.

The estimators defined in equations 34 and 36 exhibit some bias. To minimize this bias, Seber (1982) recommended the following modified estimators (where the tilde denotes correction for bias):

$$\tilde{M}_i = m_i + \frac{(R_i + 1)z_i}{r_i + 1} \qquad (37)$$

$$\tilde{N}_i = \frac{(n_i + 1)\tilde{M}_i}{m_i + 1} \qquad (38)$$

The approximate variance of $\hat{N}_i$ can also be written in closed form, but computation is tedious and is most easily accomplished using a computer program such as JOLLY (Pollock et al. 1990; Appendix 9) or POPAN-3 (Arnason and Schwarz 1987). Survival rate and number of recruits can also be estimated with the Jolly-Seber model. Because survival and recruitment are not subjects of this handbook, we do not treat them here but refer interested readers to Seber (1982) and Pollock et al. (1990).

The Jolly-Seber model is based on the following assumptions:

1. Every animal present in the population at sample period i has the same probability of capture, p_i.
2. Every marked animal present in the population immediately after sample period i has the same probability (ϕ_i) of surviving (and remaining in the population) until period $i + 1$.
3. Marks are neither lost nor overlooked.
4. All samples are instantaneous, and each release immediately follows sampling.

Stratification (by, e.g., size, age, sex, or any other characteristic likely to be associated with capture-probability differences) can be used to

Table 15. Capture-Recapture Statistics[a] and Jolly-Seber Estimates with Approximate Standard Errors of Population Size for Meadow Voles Trapped at Patuxent Wildlife Research Center, Laurel, Maryland, 1981

Period	Date	n_i	m_i	R_i	r_i	z_i	$\tilde{N}_i$	$\hat{SE}$
1	27 June–1 July	108		105	87			
2	1 August–5 August	127	84	121	76	5	138.4	4.14
3	29 August–2 September	102	73	101	68	8	118.1	4.41
4	3 October–7 October	103	73	102	63	3	109.4	2.93
5	31 October–4 November	102	61	100	84	5	111.2	3.13
6	4 December–8 December	149	89	148				

[a]Data from Pollock et al. (1990).

minimize heterogeneity of capture probabilities. When heterogeneity does exist, abundance estimates, $\hat{N}_i$, are negatively biased (i.e., underestimate abundance). A permanent behavioral response to trapping (animals caught at least once exhibit a different capture probability from unmarked animals) can also cause problems. A trap-happy response can produce a negative bias in $\hat{N}_i$, and a trap-shy response can produce a positive bias in $\hat{N}_i$ (Pollock et al. 1990).

Heterogeneous survival rates can produce either positive or negative bias in $\hat{N}_i$ depending upon the nature of the heterogeneity (Pollock et al. 1990). Loss of tags is much more likely to occur in studies of open populations because of the longer duration of such studies. Tag loss does not result in biased estimates of population size, although it does decrease precision of $\hat{N}_i$. Regarding assumption 4, instantaneous sampling is an ideal that is never actually achieved. The primary concern is that the duration of the sample period be short relative to the duration of the interval between sampling periods, so that little mortality occurs during the sampling period. The occurrence of mortality during the sampling period causes differences in probabilities of animals surviving until the next period (animals tagged early in the period will have lower chances of surviving to the next period

than animals tagged later), resulting in violation of assumption 2. Goodness-of-fit tests are computed by programs such as JOLLY and POPAN-3, providing an assessment of how well the model assumptions are met by the data.

To illustrate Jolly-Seber model computations, we consider a live-trapping study of meadow voles conducted at Patuxent Wildlife Research Center, Laurel, Maryland (Pollock et al. 1990). A 10 × 10 grid was trapped for 5 consecutive days each month from June through December 1981. For the Jolly-Seber analysis, we considered whether or not an animal was captured at least once during the 5 days. Summary statistics were compiled on this basis (Table 15) and used with JOLLY, which computes the estimator of equation 38 for population size. Estimated capture probabilities (p_i) were very high, averaging about 0.9, and this sampling intensity produced precise estimates of population size (Table 15). The goodness-of-fit statistic (a test statistic used to judge how well the model fit the capture history data) provided evidence of a lack of fit to the data, probably because of heterogeneous capture and survival probabilities (see Pollock et al. 1990). However, only a fairly small bias in $\hat{N}_i$ is expected to result from such heterogeneity when capture probabilities are so high. This was the case when the $\hat{N}_i$ values were compared with alternate estimates based on closed models (M_h) that permit heterogeneity (see Pollock et al. 1990).

OTHER MODELS. The Jolly-Seber model has served as the basis for development of a number of other capture-recapture models. These models can be categorized as more or less general than the Jolly-Seber model. The less-general models include both partially open models and reduced-parameter models. The more general models include multiage models, multistratum models, and trap-response models.

As their name suggests, the partially open models restrict the degree to which the sampled population is open to gains and losses (Jolly 1965; Seber 1982; Pollock et al. 1990). The birth-only model permits gains to the population between sampling periods, but not losses. This model is likely to be appropriate in only a few sampling situations. The estimator for population size is of the same form as for the general Jolly-Seber model, except that the M_i are now known (because there are no losses of marked or unmarked animals):

$$\hat{N}_i = \frac{M_i n_i}{m_i} \tag{39}$$

These estimates of population size and their associated sampling variances can be computed with POPAN-3 (Arnason and Schwarz 1987).

The deaths-only model, on the other hand, permits losses but not gains and may be appropriate for studies of relatively short duration during periods outside the breeding season. The bias-adjusted population size estimator for the deaths-only model is

$$\hat{N}_i = \frac{(R_i + 1)z_i'}{(r_i + 1)} \tag{40}$$

where z_i' is the number of animals not caught in sampling period i but caught during some later period. Population size estimates and associated variance estimates under the deaths-only model can be computed with JOLLY and POPAN-3.

Reduced-parameter models are those in which capture probabilities, survival probabilities, or both sets of parameters are assumed to be constant over time (over all sampling periods). Such models were developed because large numbers of parameters (as in the Jolly-Seber model) lead to reduced precision of associated estimates (i.e., estimates of abundance tend to have large variances). Reduced-parameter models can yield substantial increases in the precision of estimates and should be used when they adequately fit capture-recapture data.

Among the reduced-parameter models, model B assumes constant survival probabilities but permits time-specific capture probabilities. Model C assumes constant capture probabilities but permits time-specific survival probabilities. Model D assumes constant survival and capture probabilities, and thus has only two parameters. Estimates of population size under these models require iterative computation. JOLLY (Pollock et al. 1990) can be used to compute estimates under models B and D, whereas POPAN-3 (Arnason and Schwarz 1987) computes estimates under models B, C, and D.

A number of generalizations (models with more parameters) of the Jolly-Seber model have been developed. Most of these focus on the estimation of survival rates. A general model permitting age-specific variation in capture and survival probabilities was developed by Pollock (1981). Numbers of animals in all age classes except the very youngest can be estimated under this model. In the case of two age classes (e.g., young and adult), JOLLYAGE (Pollock et al. 1990; Appendix 9) can be used to compute estimates of the number of adults using Pollock's (1981) general model or reduced-parameter models in which age-specific survival rates or capture and survival rates are assumed to be constant over time (Brownie et al. 1986; Pollock et al. 1990).

Pollock (1975) presented estimators for population size when capture and survival probability depend on previous capture history (e.g., whether or not an animal has been captured before). His models are very general, but software to compute estimates of population size under these models is not currently available. Arnason (1973) considered the situation in which different areas are trapped and animals can move between areas. He developed estimators for numbers of animals in the different subareas, and estimates can be computed with POPAN-3 (Arnason and Schwarz 1987).

POLLOCK'S ROBUST DESIGN.   Pollock (1982) recommended a capture-recapture study design that included primary capture periods, between which the population was open to gains and losses, and secondary periods (over which the population was effectively closed) within each primary period. Initial studies using the robust design made use of closed-population estimators of population size. Recently, however, Kendall et al. (1995) introduced models that make use of capture-history data over both primary and secondary periods to provide more efficient estimates of population size. A new computer program, RBSURVIV (see Appendix 9), has been developed to implement these models.

Recommendations

MICHAEL J. CONROY AND JAMES D. NICHOLS

Investigators interested in the abundance of one or more species should use one of the methods for abundance estimation described in this section. To assist them in selecting among those methods, we have devised some simple guidelines (Table 16). For animals that are readily observed and counted, we direct the user to observation-based methods for complete and incomplete counts. For animals that are not easily

counted but are readily caught, capture-recapture models may be most useful. Removal models, catch-effort models, and change-in-ratio estimators may be useful for species that are harvested but are not easily counted in the wild or captured. If relative abundance is of primary interest, and if detectability is known to be constant for the comparison of interest, then a count statistic can be used to index abundance.

The guidelines for abundance estimation (Table 16) are based on characteristics of the sampling situation and the relative ease with which animals can be sampled using different techniques. All of the described methods provide unbiased estimates of abundance when underlying model assumptions are reasonably met. Thus a decision about which method to use should be based on the investigator's ability to obtain the requisite data in a manner that is consistent with model assumptions. Selection of methods for estimating abundance is also discussed in Pollock et al. (1990:65–68) and Lancia et al. (1994:247–250).

Estimation of species richness

JAMES D. NICHOLS AND MICHAEL J. CONROY

Attempts to enumerate all the mammalian species inhabiting some area of interest (i.e., species richness) will necessarily involve many of the field methods described in this handbook. No matter how thorough the efforts, however, it is likely that some species will be missed; that is, the number of species counted will be less than the true number of species in the area. The number of species that are enumerated can be thought of as a count statistic, with the investigator interested in estimating its associated observability fraction (see "Observability," Chapter 3). Although much effort has been devoted to methods for estimating numbers of indi-

Table 16. Guidelines for Selecting Methods for Estimating Abundance

ABUNDANCE ESTIMATORS

I. Animals readily observable and countable[a]

   A. Animals easily counted without error: animals large or conspicuous, usually diurnal, occupying open habitats, and visible from air or visible/audible from ground; groups not too large.
   Method: Complete counts (censuses), often conducted from aircraft.

   B. Some individuals missed in counting, some underestimation likely: animals diurnal, relatively conspicuous, but obscured by habitat features, or groups large.

   1. Every individual in a small subunit of the sampling unit can be counted. Underestimation errors likely during primary survey method (e.g., counts from aircraft), but alternative counting methods (e.g., intensive methods such as ground counts) can yield exact counts.
   Method: Double sampling to estimate observability.

   2. Not every individual in a small subunit of the sampling unit can be counted.

      a. Marked subsample of animals available or easily obtained.
      Method: Marked subsample.

      b. Distances between the investigator and each detected animal readily determined.
      Method: Distance sampling—line transect and variable circular plot estimation models.

      c. Distances between the investigator and each detected animal not readily determined. Marked subsample of animals not available.
      Methods: Multiple observers; sighting probability models.

II. Animals catchable[b]
   Animals readily captured, marked, released, and later recaptured and examined for marks (may not be suitable for large animals that are difficult to handle or for predators or other sparsely distributed animals that are difficult to capture, unless such animals can be identified by natural markings and are readily photographed using camera traps) (see Karanth 1995).

   A. Study short-term: trapping, marking, and recapturing-resighting accomplished over a short period (e.g., several days) during which population can be assumed to be closed (i.e., mortality, recruitment, immigration, and emigration negligible); usually appropriate for abundant, easily trapped mammals (e.g., microtines).
   Methods: Lincoln-Petersen estimates; multiple capture-recapture estimates (CAPTURE); trap-web estimates.

   B. Study long-term: trapping, marking, and recapturing-resighting accomplished over a long period (e.g., months or years) during which population cannot be assumed to be closed (i.e., appreciable mortality and recruitment); mortality and (sometimes) recruitment rates must be estimated at each sampling period, in addition to population size.
   Methods: Jolly-Seber and related open population models (POPAN, JOLLY, JOLLYAGE).

III. Animals harvested in known numbers[c]
   Animals removed from the population by sport or commercial harvest (e.g., permit-controlled sport harvest of deer; some commercial fisheries); ordinarily, methods require that a substantial portion of the population be removed and that the size of the harvest be well documented for reasonable performance.

   A. Animals include two or more identifiable classes (e.g., sexes) whose relative frequencies can be observed before and after harvest; numbers of either or both removed must be known.
   Method: Change-in-ratio method, but only if harvest is large enough to change ratio (otherwise see B).

   B. Animals do not include identifiable classes whose relative frequencies can be observed before and after harvest.

(Continued)

Table 16. (*Continued*)

1. Effort expended in harvest quantified; harvest effort measured in terms of time (e.g., hunter-days, trap-nights), money (e.g. expenditures in commercial fisheries), or other quantifiable indicator.
   Methods: Catch-per-unit-effort estimators.

2. Effort expended in harvest not quantified, but removal effort constant over time (investigator controls removal, or "harvest")
   Method: Removal models (e.g., models M_b and M_{bh}).

ABUNDANCE INDICES

IV. Determination of relative abundance when relationship of index to abundance known to be constant or to depend on known variables in a known way (validated by past experience or by experiments); situation very rare.
   Methods: Collect indices or count statistics (e.g., using standardized methods); inclusion of covariates likely to influence index variability (e.g., observer experience, weather) desirable.

V. Determination of relative abundance when relationship of index to abundance not known to be constant or to depend on known variables in a known way (past experience suggests that the index and population density are related, perhaps monotonically); situation common (e.g., track and pellet surveys for deer; coyote vocal responses to distress calls, sirens; scent station surveys of foxes, bobcats).

   A. Census or abundance estimation for use in index calibration possible.
      Method: Calibrate index with population censuses or with estimates based on methods having limited or known bias (e.g. mark-recapture), and then use a double-sampling design to estimate population density or abundance from index.

   B. Census or abundance estimation for use in index calibration not possible.
      Method: Use the index (with proper replication, stratification, and recording of nuisance variables, as in IV), recognizing that it may or may not be correlated with actual abundance. Option to be used with extreme caution, as it relies on untested assumptions.

[a]See also II.   [b]See also I, III, and IV.   [c]See also I, II, and IV.

viduals in populations, the related problem of estimating species richness has not been so well studied. In the following sections, we first discuss indices of species richness. We then present methods of estimating species richness that we believe to be most useful. Estimation methods can be classified by the type of sampling design (quadrat-based, multiple sampling occasions, empirical species-abundance distributions) that they use.

Indices of Species Richness

In general, the principles that apply to abundance indices also apply to species-richness indices: An index should be related to the parameter of interest (i.e., species richness) in some consistent (e.g., positive, linear) fashion, and that relationship should be known to be constant over time and space or should be estimable from other information. An index to richness is of no value if the relationship of the index to richness changes over time or among the communities studied. Such an index would not support meaningful comparative statements about diversity over time or among different communities.

It is difficult to offer general guidelines about using indices as correlatives of species richness. Many indices—for example, scat surveys, scent station surveys, and structure surveys—are oriented toward individual species and depend on behavioral and other unique attributes of the particular species. Other indices—such as incomplete counts, capture indices, and track counts—can be used to survey numerous species and perhaps to provide an index to richness, but with obvious limitations. Trapping, for example, may provide an index to species richness for small herbivorous or granivorous mammals, but it will provide no data or only limited data on larger mammals and on certain small mammal taxa that are difficult to capture without specialized trapping techniques (e.g., shrews). This example illustrates the problem with using indices for estimating species richness: There is no assurance that a particular species count based on any sampling method (e.g., trapping) fairly represents the true number of species in the target community. Some species may be rare or not well sampled by the selected methods, so the probability of encountering them in a sample is low. In addition, sampling (e.g., trapping) probabilities are likely to vary from species to species (e.g., Nichols 1986). Thus, a method (such as trapping) that may provide satisfactory results for temporal or spatial comparisons of abundance of a single species may do a very poor job of providing comparative information about species richness. In general, we do not recommend indices as a means of drawing inferences about species richness.

Quadrat-Based Sampling

When quadrat-based methods are used to estimate species richness, the area under study is divided into a number of quadrats of roughly equal size and shape. The entire study area of interest should be specified, so that potential quadrat samples cover the community of interest (i.e., the target population). A sample of these quadrats is randomly selected, and a mammalian species list is developed for each quadrat. The species lists may be obtained using any variety of methods selected by the investigator, but the

same methods should be used on each quadrat. The number of individual animals counted, caught, observed, or otherwise identified for each species is not required by most quadrat-based estimation methods but may be used to obtain an alternative estimate of species richness (see "Empirical Species-Abundance Distributions," below).

The same capture-recapture models used to estimate number of individuals in a sampled population, based on releases and recaptures or reobservations of marked individuals, can be useful in estimating species richness. Assume a simple situation in which an investigator samples two quadrats in an area of interest, obtaining a count of the different species in each one. The total number of species identified in quadrat 1 is denoted as n_1, the number identified in quadrat 2 as n_2, and the number identified in both quadrats as m. If all species in the area of interest have equal probabilities of being detected in the quadrats, then the ratio of species detected in quadrat 1 to the total number of species in the area (N) should approximately equal the ratio of species found in both quadrats to total species found in quadrat 2:

$$\frac{n_1}{N} \approx \frac{m}{n_2} \qquad (41)$$

Solving for N, one obtains the following estimator for species richness:

$$\hat{N} = \frac{n_1 n_2}{m} \qquad (42)$$

In the context of equation 2, n_1 of equation 42 can be viewed as the count statistic and m/n_2 as the estimate of β. Alternatively, n_2 can be viewed as the count statistic and m/n_1 as the estimate of β. The estimator of equation 42 is simply the Lincoln-Petersen estimator, developed for estimating population size. We recommend that investigators using this estimation approach with data from two quadrats substitute the bias-adjusted estimator of Chapman (1951) (equation 31) for equation 42.

The problem with using an estimator such as equation 42, or its multiquadrat analogues, for species richness is that the estimator assumes equal probabilities of detecting all species. Certainly, species detection probabilities (observability fractions, β) will always differ, with some species being readily detected and others being very difficult to detect. Detection probabilities will vary with the relative ease of catching or observing individuals of different species and with the relative abundances of species in the area of interest. For example, a species that is very rare in the area of interest will be less likely to occur in a particular quadrat than will an abundant species. Even if one or two individuals of a rare species are present in a quadrat at the time of sampling, typically a species represented by many individuals in the quadrat will be easier to detect. The estimator in equation 42 will be less biased than the naive count statistic ($n_1 + n_2 - m$; the total number of different species detected in the two quadrats), but unequal species detection probabilities will still produce a negative bias (i.e., $\hat{N}$ will tend to underestimate the true number of species).

Burnham and Overton (1978, 1979) developed a jackknife estimation procedure for estimating population size when the individuals in a population have different capture probabilities (model M_h). These authors noted that their estimator could also be used to estimate species richness (Burnham and Overton 1979). Heltshe and Forrester (1983) appear to have developed the same estimator independently; they discussed it specifically in the context of estimating species richness. The estimator is based on the seemingly weak assumption that the N detection probabilities corresponding to the N different species in the community represent a random sample from some unspecified distribution. The

estimator does require that the detection probability for a given species be the same in all sampled quadrats (hence the need for equal-size quadrats and similar sampling efforts in the different quadrats).

The estimator presented by Heltshe and Forrester (1983) is the following first-order jackknife estimator of Burnham and Overton (1978, 1979):

$$\hat{N} = S + \frac{(t-1)f_1}{t} \qquad (43)$$

where S is the number of species found as a result of sampling all of the t quadrats and f_1 is the number of species that were found on only one quadrat. Burnham and Overton (1978, 1979) and Heltshe and Forrester (1983) presented variance estimators, and Burnham and Overton (1978, 1979) considered the use of higher-order jackknife estimators.

Chao (1987) noted that the jackknife estimator of Burnham and Overton (1978, 1979) does not perform well when many species (or individuals) have very low detection probabilities (i.e., when many species are encountered in only one or two quadrats). She presented an alternative estimator for model M_h derived specifically for the situation of many species with small detection probabilities. We recommend use of the program CAPTURE (Rexstad and Burnham 1991) for computing estimates of species richness under M_h.

A bootstrap resampling approach can also be used to estimate species richness with quadrat sampling when detection probabilities among the different species are heterogeneous (Smith and van Belle 1984). A comparison of the bootstrap and first-order jackknife approaches led Smith and van Belle (1984) to conclude that the jackknife is generally more accurate unless the number of sampled quadrats is large. Mingoti and Meeden (1992) proposed an empirical Bayes estimator for species richness from quadrat data and concluded that it should perform reasonably well. We are not aware of available software for computing bootstrap estimates based on the procedure of Smith and van Belle (1984).

The estimators of Burnham and Overton (1978, 1979) and Chao (1987) perform well in most situations. Thus, we recommend the use of one of these or of the empirical Bayes estimator of Mingoti and Meeden (1992) for estimating richness from quadrat data at the present time. CAPTURE can be used to compute both the jackknife estimator of Burnham and Overton (1978, 1979) and the estimator of Chao (1987). The program includes an algorithm for objective selection of the appropriate order jackknife estimator for a given data set. CAPTURE also computes a goodness-of-fit statistic for the underlying model (M_h), as well as the variance estimate associated with the richness estimate.

An estimator proposed by Chao et al. (1992) in the capture-recapture context permits relaxation of a key assumption underlying model M_h. In their model, M_{th}, detection probability varies not only among species, but also among the quadrats (note that in the application described by Chao et al., t corresponds to sampling periods, whereas in the species-richness application, t denotes different sampled quadrats). Such a situation could easily occur if randomly selected quadrats sample different microhabitats. CAPTURE (Rexstad and Burnham 1991) can be used to compute the Chao et al. (1992) estimate of species richness and its variance under M_{th}. CAPTURE also contains an objective model selection procedure that provides guidance about the model that is appropriate for a given data set.

Because CAPTURE was developed for use in estimating population size from animal capture-recapture data, we briefly describe its use in estimating species richness with presence-absence data from quadrats. Data for estimating species richness can be summarized (and entered into the program) in the form of an "X-matrix"

Table 17. Sample X-Matrix for Quadrat Data to Be Used with CAPTURE (Rexstad and Burnham 1991) to Estimate Species Richness[a]

Species	Quadrat				
	1	2	3	4	5
Peromyscus leucopus	1	1	1	0	1
Microtus pinetorum	0	1	0	0	0
.		.	.	.	.
.		.		.	
.		.	.	.	.

[a]A 1 in a quadrat column signifies species presence in that quadrat; a 0 signifies species absence.

(see example in Table 17). This matrix contains one row for each species identified in any of the quadrats. The species name (analogous to an animal identification or tag number in the capture-recapture context) is followed by a string of 1s and 0s denoting that the species was (1) or was not (0) found on a particular quadrat. Again, the quadrats are analogous to the trapping occasions in the capture-recapture context. CAPTURE computes estimates under several models. We have discussed only the two models (M_h, M_{th}) that we expect to be most useful for estimating species richness from quadrat data, based on our belief that detection probabilities are likely to vary greatly among species. Other models implemented in CAPTURE may be useful with quadrat data if detection probabilities are similar among species. The objective model selection procedure of CAPTURE can aid the investigator in selecting the most reasonable model for a particular data set.

Multiple Sampling Occasions

SINGLE INVESTIGATOR

Instead of dividing the area of interest into quadrats and sampling a subset of these, an investigator may attempt to survey the entire area, but do so on multiple occasions. As with quadrat sampling, the surveys may involve a variety of sampling methods and should include all methods that may lead to the detection and identification of different species. These same methods should be used every day for, say, 5 to 10 consecutive days (or some other unit of sampling time). The time between the first and last sampling occasions should be sufficiently short that the community composition (number and identity of species) is not expected to change. The data can be summarized in an X-matrix similar to that of Table 17, but with columns representing sample periods rather than sample quadrats.

The model most likely to be useful for estimating species richness with such data is the generalized removal model, M_{bh} (Otis et al. 1978; Pollock and Otto 1983). The critical data for the model are the number of different species first detected in each successive sample period (e.g., 1, 2, . . ., K). The idea underlying the estimator for model M_{bh} is that the number of undetected species decreases over time, and the resulting change in number of detections of new species over time provides information about the number yet to be detected. Model M_{bh} permits variation in detection probabilities among different species but assumes similar sampling effort on each sampling occasion. Two estimators for this model, one developed by Otis et al. (1978) and one developed by Pollock and Otto (1983), are computed by CAPTURE. We suspect that the other heterogeneity models, M_h and M_{th}, will not be as useful for such data collected by a single observer. We believe it is likely that detection probability for a particular species will increase following initial detection, because the observer will know what sign to look for, what specific areas to search, and so forth, leading to the need for model M_{bh}.

MULTIPLE INVESTIGATORS

Another means of estimating species richness in the absence of quadrat sampling involves the use of multiple investigators for an area of interest. One biologist samples the area for one or two days using whatever methods he or she chooses and develops a species list (numbers of individuals per species are not necessary). Then a second biologist samples the area for a day or so and develops a species list using his or her own methods (not necessarily the same methods as those of the first biologist). If five biologists generate independent species lists in this manner, the resulting data can be used to estimate species richness. Data are recorded in the same form as in Table 17, with the columns of the X-matrix corresponding to biologists, instead of sample quadrats. The model most likely to be useful in this situation is M_{th} (Chao et al. 1992). This model permits variation in detection probability among different species and among different observers or biologists (so there is no assumption of equal sampling efforts by the different observers). Estimates can be computed with CAPTURE. If the biologists attempt to standardize their sampling by using the same methods and expending the same effort, then it may be possible to use the M_h estimators for such data.

Empirical Species-Abundance Distributions

Instead of sampling at different points in space or time, the investigator can develop a species list based on catches or observations over the entire area of interest without regard to time or sampling occasion. If it is possible to record the number of different individuals encountered for each species (i.e., this may be possible if animals can be removed or caught and marked for future recognition, or if the area can be traversed at one

time in such a manner as to ensure that no animals are counted twice), then the resulting data
(number of individuals encountered for each
species) can be viewed as an empirical species-
abundance distribution.

A variety of theoretical distributions have
been considered as possible models of species-
abundance distributions (e.g., Engen 1978). If
the "correct" theoretical distribution is known,
then it may be possible to estimate species richness from species-abundance data. The problem
is that the correct distribution is never known,
and it is often difficult to distinguish which
model, among a set of competing models, corresponds most closely to a particular data set. In
fact, it is common for several different models to
fit such data well but yield very different estimates of species richness (see Cormack 1979).
In addition, the empirical species-abundance
distribution will usually differ substantially from
the true species-abundance distribution because
of interspecific variation in detection probabilities. Thus, we recommend against the use of
theoretical species-abundance distributions for
estimating species richness or for comparing
richness in different communities.

Burnham and Overton (1979) suggested a
nonparametric approach to estimating species
richness from empirical species-abundance-
distribution data using a limiting form of their
jackknife estimator under model M_h. They noted
that some studies lack trapping occasions (or
sampling units). Thus, they considered the limiting value of their jackknife estimator as t (number of quadrats or sampling occasions) becomes
infinite. The data required to estimate species
richness using this estimator are the total number
of species encountered and the numbers of species for which 1, 2, 3, 4, and 5 individuals are
encountered. SPECRICH, which selects the appropriate order jackknife (after Burnham and
Overton 1979) and computes the resulting estimate of species richness, can be obtained from J.

E. Hines and J. D. Nichols (National Biological
Service, Patuxent Wildlife Research Center,
Laurel, MD 20708).

Field Application

We have not presented examples of the preceding general methods for estimating species richness because we do not know of any situations in
which these estimators have been used with
mammalian data. Historical investigations of
mammalian diversity have assumed that all species are detected and that individuals of different
species are sampled in proportion to their relative abundances in the community. These assumptions are not likely to be met in practice, so
we have recommended estimation methods for
species richness that permit different sampling
probabilities for different species.

An example of the use of model M_h for estimating population size from individual capture
history data is presented in the section on "Capture-Recapture Methods," above. That example,
in which CAPTURE is used to compute estimates, corresponds to the use of M_h with species
list data from different quadrats or from different
individuals sampling the same area. The limiting
form of M_h has been used to estimate species
richness from empirical abundance distributions
obtained using bird count (Derleth et al. 1989)
and capture (Karr et al. 1990) data.

We have not discussed specific field methods
because we believe that investigators interested
in species richness should select specific methods as dictated by such factors as their experience, prior knowledge of the fauna, and the
nature of the sampled habitats. In recognition of
the variety of potential field sampling methods,
we have selected general estimation methods
that require minimal assumptions. Under the
quadrat approach, different quadrats should be
sampled using the same general methods if possible, so that M_h may be used. If different biolo-

gists prepare species lists for the same general area, then use of similar methods will again increase the potential applicability of M_h. If sampling methods vary among different quadrats (or different biologists), however, then M_{th} can always be used. If the entire area is surveyed by a single biologist (or team of biologists) on different sampling occasions, then similar sampling methods and effort should be used on the different occasions in order for M_{bh} to be applicable. If the entire area is to be surveyed by a single biologist (or simultaneously by a team of biologists) to compile an empirical species abundance distribution (for use with the limiting form of the M_h estimator), then sampling methods must yield counts of individuals per species. Methods that could lead to counting the same individual more than once should be avoided.

Most important, other than equation 42, none of the described methods assumes that different species are encountered or detected with equal probability. This assumption will certainly not be met when many methods are used to observe or to capture different species. Even if we restrict our interest to species richness of a subset of mammals susceptible to the same sampling method (e.g., a "guild"), it is unlikely that different species will be sampled with equal probabilities (Nichols 1986). We hope that the described methods are sufficiently general and flexible to accommodate a variety of field sampling methods under a wide variety of field situations.

Chapter 11

The Geographic Information System for Storage and Analysis of Biodiversity Data

Peter August, Carol Baker, Charles LaBash, and Christopher Smith

Introduction

A geographic information system (GIS) is a computer tool that is used to enter, edit, store, analyze, and report spatial information. The results of GIS analyses take the form of maps, statistical summaries, or derived data sets that can be used in other tasks such as modeling or hypothesis testing (Tomlinson 1987; Star and Estes 1990; Maguire et al. 1991; Berry 1993a, 1993b). GISs are frequently integrated with other tools that are used to measure various aspects of the landscape. For example, global positioning systems are used to record the location of collection sites or habitat boundaries in the field, and these data can be automatically transferred to a GIS for subsequent analysis (Goos 1990; Puterski et al. 1990; Slonecker and Carter 1990). Satellite-image processing is used to determine the distribution and extent of many relevant features of the landscape, such as habitat and changes in vegetation cover (Roughgarden et al. 1991). A GIS, when integrated with remote sensing, is especially valuable for assessing land cover and habitat distribution in regions that have not been accurately or recently mapped (Ehlers et al. 1989, 1991). GISs can also be an important tool for collecting and processing telemetry data obtained from animals carrying radio transmitters (White and Garrott 1990).

The data that are used to assess biodiversity are almost exclusively spatial data; that is, they can be represented by points, lines, areas, or volumes on the landscape. Fundamental elements of a biodiversity database are species, their distributions, and their habitats; the patterns of environmental factors that are associated with ecological communities (e.g., soils, drainage, topography); the locations of factors that threaten biodiversity (e.g., roads, human settlements); and the temporal changes that occur in the distributions and extent of species, habitats, and land cover (Jenkins 1988; Davis et al. 1990; Lombard et al. 1992; Stoms 1992).

In studies of biodiversity, the scale of analysis can vary immensely. The microhabitats of rodents are examined at scales in which the minimum mapping unit is an individual plant. This is an example of a large-scale, highly detailed map. Analyses of continental patterns of species diversity are carried out at scales for which the minimum mapping unit can be as large as 100 km^2. This is an example of a small-scale, low-resolution map. The GIS is a tool that has been developed to manage and analyze spatial data and is equally relevant for studies at extremely large (rodent microhabitats) or small (continental patterns) scales. The technical procedures for handling spatial data are the same regardless of scale.

A GIS is preferable to manual methods for handling spatial data for many reasons. In a computerized system, spatial data can be easily compiled, edited, and stored. To carry out these fundamental operations using manual cartographic techniques is slow, expensive, and sometimes inaccurate. In addition, certain cartographic details, such as map scale and projection, can be easily transformed using GIS procedures and are, thus, not an impediment to conducting spatial analyses. These factors can create innumerable logistic problems for manual

analysis of spatial data. Basic measurements such as area, circumference, and length, which are essential to the study of ecological pattern, are difficult to obtain by manual methods but are easily and accurately calculated using a GIS. Many analytical procedures found in GISs—such as overlay functions, proximity analyses, and three-dimensional data modeling—are difficult or impossible to perform by manual methods. For these reasons, GISs are becoming a standard tool for the assessment of biodiversity (Davis et al. 1990).

GISs can contribute to many aspects of the assessment of biodiversity. We will focus on the use of GISs for (1) data logging and storage, (2) analysis of spatial pattern, and (3) production of maps and derivative data sets.

Spatial data

Representing Map Data in Analog and Digital Formats

Any kind of information that can be illustrated or presented on a map, regardless of scale, is appropriate for a GIS. The fundamental requirement is that geographically relevant coordinates be used to record the location of features of interest. The coordinate system can be one that is internationally applicable, such as latitude/ longitude or the Universal Transverse Mercator (UTM) system; one that is of regional significance, such as the state plane coordinate system for a specific state in the United States; or one that is unique to a particular study site, such as a location on a sampling grid or a position within a trail system.

Spatial data have traditionally been presented in the form of analog maps. A single map usually represents one or more land features (e.g., vegetation or soils) for a given snapshot in time. Different components of a feature (e.g., classes of vegetation or soil types) are distinguished

VECTOR

RASTER

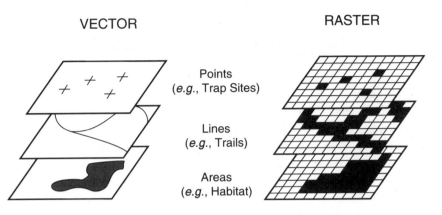

Points
(*e.g.*, Trap Sites)

Lines
(*e.g.*, Trails)

Areas
(*e.g.*, Habitat)

Figure 46. Description of landscape features in a geographic information system. In a vector-based GIS, features are represented by points, lines, or areas (polygons). In a raster-based GIS, grid cells or pixels are used to store map information.

using colors, patterns, textures, and symbols. The complex of features that together form a landscape can be depicted in two basic ways: vector representations and raster models (Fig. 46). In a vector system, a landscape is represented as points, lines, and areas. A point is the most fundamental unit of spatial data and consists of a location and one or more identifying characteristics or attributes. Examples of data represented by points are collection sites, locations of important food plants within an animal's home range, or burrow entrances. A line feature is a series of points strung together. A line consists of the geographic coordinates that identify its beginning, its end, and locations where it bends. Lines are characterized quantitatively by their length and direction, and they can have one or more descriptive attributes (e.g., road name, trail number, or water quality of stream segment). Examples of data represented by lines are trail systems, streams, and roads. An area (also referred to as a polygon) is the interior space created by one or more lines that close in on themselves. Polygons can be quantitatively described by their area and perimeter as well as by thematic attributes. Examples of data represented as polygons are species territories, the geographic range of a taxon, or the distribution of a certain habitat type.

Another way of describing a landscape is by dividing it into grid cells (also known as rasters or pixels) and defining each by the feature that dominates its area (Fig. 46). The value for the cell does not have to be a discrete or nominal variable but can be a continuous one—for example, the total number of species that occur within the cell. Grid cells can be of any size; the size is specified by the analyst or set by the instrument that is collecting the data (e.g., satellite sensor or raster scanner).

Topology, the location of one feature in relation to another, is a characteristic of spatial data that is essential to a GIS. For many analytical operations, it is critically important to know which features are contiguous with or near other map elements, that is, their *adjacency.* For example, ecotonal habitats can be identified by asking the GIS to map lines that bound the two habitats that form the ecotone. The degree to which habitats are connected, *connectivity,* is another characteristic of spatial data that requires explicit measurement of topology. Connectivity is a basic pattern of landscapes and has considerable ecological importance (Berry 1993b).

Two types of GIS have evolved over the past decade (Maffini 1987; Johnson 1990; Woodcock et al. 1990; Maguire et al. 1991; Berry 1993a). A *raster GIS* defines the landscape as grid cells. A *vector GIS* views the world as a collection of points, lines, and polygons. Raster systems are very efficient for mathematical manipulation of complex data, and we prefer them for modeling applications (Goodchild et al. 1993). Raster GISs, however, can be less accurate than vector systems in calculating areas, perimeters, and distances (although the level of accuracy also depends on the scale of the source data and the size of each grid cell). Maps created from raster GISs can be more abstract than maps created from vector systems because diagonal lines have a jagged, staircase appearance (Fig. 46). We prefer vector systems for cartographic production and analyses in which calculations of area and distance must be accurate. The two types of GIS systems complement each other, and the better GIS software packages have the capability of using either representation. Which representation is best depends on the questions being asked by the investigator and the questions that she or he might address in the future.

Attributes: What the Spatial Data Represent

The location of a feature on the landscape is only part of the data that are required to define that feature fully in a GIS. Attributes that describe the feature are also essential. Hard copy maps communicate attribute information with text, colors, and symbols. Such information is managed in a GIS in databases that are linked to the points, lines, polygons, and grid cells. The intimate linkage of attributes to spatial features is an essential component of a GIS. An investigator can use a GIS to ask complex questions of a database with answers output as maps: A researcher might ask, for example, "Where are the deciduous forest patches with areas of at least 5 ha that are within 2 km of standing water and lie within the home range of a specific study animal?"

Special Properties of Spatial Data: Cartography 101

Spatial data have many unique properties that must be considered when using GIS tools (August 1991, 1993). An investigator must choose carefully the geographic coordinate system in which to develop his or her database. Common coordinate systems in use are decimal degrees or decimal seconds of latitude and longitude, the UTM system, and the state plane system. In the United States, two geodetic reference systems in common use are the North American Datum of 1927 (NAD27) and the North American Datum of 1983 (NAD83). NAD83 is an internationally recognized reference point and is best for new mapping projects (Schwartz 1989). The United States mapping community is slowly converting from NAD27 to NAD83. It is important that all elements of a GIS database be expressed in the same datum. A geographic point given in NAD27 can be displaced by as much as 100 m if expressed in NAD83. For study areas that encompass large regions, all the maps in the database must be of the same cartographic projection (i.e., the manner in which a three-dimensional object, the surface of the earth, is portrayed on a two-dimensional map). Maps of different projection cannot be superimposed correctly (Snyder 1983; Monmonier and Schnell 1988). Most GIS software packages have the ability to transform data from one cartographic projection to another.

Map scale has the greatest effect on the resolution of a GIS database. Scale establishes the smallest feature that will appear in a data set (Fisher 1991), and it is one of several factors that determine the positional accuracy of a line or point in the database. Large-scale (small-area, high-detail) data sets may be appropriate for testing some hypotheses, but they may be inappropriate for addressing other questions. GIS data derived from large-scale sources may be accurate and detailed, but the data set may be excessively voluminous, unwieldy to use for large study areas, and possibly too expensive to acquire.

System design

Before an investigator designs a GIS database, he or she should know how the data will be used to assess biodiversity and what results the GIS will be expected to produce. A database designed to represent patterns of biodiversity in all of Costa Rica would be very different from a database representing the trails, canopy gaps, tree falls, flora, and fauna of La Selva Biological Station. Once an investigator has established the goals of the project, he or she can inventory available data to determine whether such data are sufficiently accurate to meet the project objectives (Lunetta et al. 1991). For example, a vegetation map digitized at a scale of 1:100,000 would be of little value in determining habitat preferences for radio-collared small mammals.

Investigators should address the following issues when assessing the accuracy of potential data for a project using a GIS:

1. Are the data sufficiently recent to meet the objectives of the study?
2. Are the source maps of acceptable quality and condition?
3. Are geographic coordinates and graticules (reference marks) clearly and accurately positioned on hard copy maps?
4. Are the data available at a scale that will provide sufficient resolution and detail for the intended project?
5. Were the data collected by a knowledgeable person?
6. Is documentation for the data sufficient to allow for correct interpretation of the information?
7. Have the data been converted from analog form (e.g., paper or mylar map) to the GIS accurately?
8. What errors are inherent in the data?

Developing a database

A project that uses GIS technology has two phases: (1) transformation of the data into a format that is accepted by the GIS hardware and software and (2) performance of the analytical procedures. August (1993) reviewed in detail the common methods of developing a GIS database. The process involves transferring information that exists in analog form (on maps or in notebooks) into computer files. We briefly describe below the standard methods of creating digital representations of spatial data.

Keyboard Entry of Coordinates

Geographic coordinates can be typed into a computer file, and the GIS will create a digital map from these points. Likewise, data that are recorded as a series of angles and distances, as might be done in surveying a transect line or recording the path that an animal has taken over a period of time, can be entered into the GIS as a simple file. The system will, using coordinate geometry, create a digital map of the track. Data sets created by manual encoding of geographic coordinates are only as accurate as the initial coordinates (and the typist!). Manual data entry can be very laborious for large volumes of data.

Digitizing

Cartographic data are most commonly entered into a computer with a digitizing tablet. The process is simple, albeit slow and tedious. It involves attaching a map manuscript to the sur-

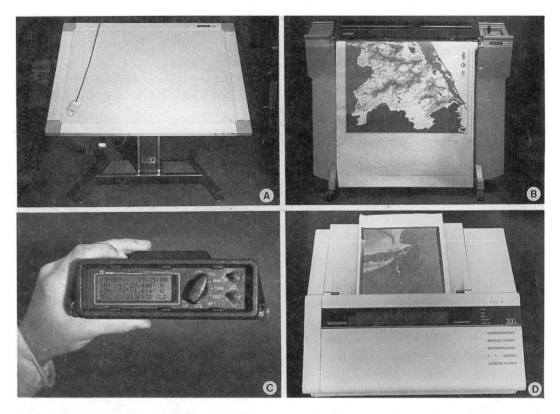

Figure 47. Devices that are used with a geographic information system. A. Digitizing tablet. B. Pen plotter. C. Global positioning system receiver. D. Color printer.

face of a digitizing tablet and tracing the features to be automated with an electronic tracing device (or puck; see Fig. 47). The digitizing tablet sends to the computer a stream of coordinates that correspond to the positions of the puck on the surface of the tablet. The computer converts the coordinates to geographic coordinates (e.g., UTM meters, decimal degrees, state plane feet). Once a map has been digitized, a proof plot is made at the same scale as the source document and superimposed on it. Any point or line that does not overlie its counterpart on the base map can be redigitized or edited. Attribute data that are associated with the points, lines, or polygons can be entered from the keyboard or digitizer or imported from other computer files.

Scanning

Source maps can be scanned into a computer with an electronic device called a *scanner*. The scanning process converts the map into a grid of extremely small cells (down to 50 microns on a side) and encodes each cell according to whether or not it encompasses a feature on the base map. The result is a raster representation of the map. The raster data set can be converted to vector structure using specialized software. A scanner can convert large amounts of data rapidly, but it requires very clean base maps. Any text, smudges, or symbols that appear on the analog map will also appear in the digital version and will have to be removed. Editing out unwanted information in a scanned manuscript sometimes requires more time than manually digitizing the data from the beginning. Scanning is an excellent way to convert aerial photography or other data into a form that can be used by the GIS. Scanning devices can be extremely expensive (>U.S. $50,000 for a large, accurate drum scanner) and require considerable expertise to operate properly. Less expensive (<U.S. $5,000) scanners can be purchased, but they usually do not have adequate resolution, the capacity to scan large maps, or software to edit scanned images.

Global Positioning Systems

The United States Air Force manages and maintains a constellation of 26 satellites orbiting the earth at an altitude of 20,000 km (Hurn 1989). The satellites transmit radio signals that can be received by small electronic devices called global positioning systems (GPSs; Fig. 47). These units process the satellite signals and calculate the geographic position of the receiver (Puterski et al. 1990). The coordinates of that position are displayed on a small screen and can be logged into the memory of the receiver. Coordinates recorded in the field can be transferred electronically to a GIS for storage and analysis. An inexpensive receiver (U.S. $1,000) can provide locations accurate to within 100 m. By applying certain correction factors and post-processing procedures (e.g., differential correction), GPS data obtained with inexpensive receivers can be corrected to the 2- to 6-m accuracy range (August et al. 1994). Survey-quality receivers (>U.S. $50,000) can provide locations accurate within centimeters.

GPS receivers should be standard equipment for field ecologists. Basic GPS receivers available today weigh less than a kilogram or two. The simplest of receivers permits in-the-field digitizing of points or lines, and this information can be easily transferred into a GIS. Examples of basic applications of GPSs include recording collection sites and digitizing habitat boundaries (Goos 1990; Slonecker and Carter 1990). The technology works best in open habitats where few obstructions protrude above 10° from the horizon in any direction (Wilkie 1989). Rough topography may cause satellite signals to reflect off the terrain, an artifact that can lead to distortion in reported locations. In some habitats, GPSs will not provide reliable results because satellite signals cannot penetrate dense canopy vegetation. Large artificial objects (e.g., buildings, storage tanks) or expanses of water (e.g., ponds, lakes) may reflect a satellite message in such a way that the GPS antenna will receive a delayed signal, and this may result in a positional error (Puterski et al. 1990).

Radiotelemetry

Much of the software that accompanies radiotelemetry systems will calculate geographic coordinates for estimated positions ("fixes"), and these data can be transferred into the GIS for storage and analysis. Telemetric systems, such as the one operated by Service Argos, that use satellites to receive location data and transmit it to processing centers on earth provide data in a format that is easily transferred into a GIS (Fancy et al. 1988; White and Garrott 1990).

Existing Digital Data

An immense volume of high-quality GIS data relevant to field ecologists is available at little cost. These data encompass all regions of the world and include themes such as land cover, coastlines, topography, hydrography, wetlands, soils, and climate. The best directory to public and private agencies from which to obtain existing data is the GIS Sourcebook (Parker 1991). Many of the bulletin boards and discussion lists (e.g., GIS-L, CONSGIS) found on the Internet provide listings of free data that are relevant to ecological applications (Mark and Zubrow 1993).

Data handling

It is important to establish data standards and quality control procedures when developing a GIS database (Campbell and Mortenson 1989). This practice is especially important if the data are to be distributed to other users. "Data standards" is a broad term with respect to GIS and covers a variety of issues. Each GIS software package requires data to be in a certain digital format and to meet certain technical specifications. These issues are discussed in detail in software manuals and general texts on GIS (Burrough 1986; Aronoff 1989; Star and Estes 1990; Antenucci et al. 1991).

GIS data sets must be carefully documented and should provide information on the source of the data, the accuracy of the information, and how the data are coded. The Federal Geographic Data Committee (1992) is in the process of developing a national standard for database documentation (i.e., metadata). Without such data documentation, an end-user cannot responsibly incorporate the information into a study.

All investigators who use computers have experienced the frustration of receiving data on a disk or tape or in a format that their particular system does not accept. Such problems also occur frequently in the exchange of GIS data, and data transfers among systems should be planned with care. Numerous formats can be used for the exchange of GIS data among software and hardware (Craig et al. 1991; Ramirez 1991). The United States Geological Survey, with assistance from many branches of the federal government, has developed a standard format for the exchange of spatial data (Fegeas et al. 1992). The Spatial Data Transfer Standard should emerge as the single acceptable format for transferring data among systems. GIS data sets can be extremely voluminous and can require substantial disk space (Calkins 1990; August 1993). Consequently, large GIS data sets are commonly exchanged on magnetic tapes or compact disks rather than floppy disks. For example, the land cover data set for an area as small as the state of Rhode Island (2,774 km^2) would require more than seventy 1.4-megabyte diskettes!

Data analysis

The analytical functions in a GIS distinguish it from computer drawing software, such as computer-aided design and/or drafting (CADD), systems. GIS software that would be used in studies of biodiversity is typically packaged as an analytical toolbox from which the user selects the procedures required for a particular task. The software has matured to the point that the toolbox is by now large and complex. The average biologist is likely to use only a small number of the functions. The analytical capabilities of GIS have been reviewed in detail by others (Johnson 1990, 1993; Tomlin 1990; Maguire et al. 1991; Berry 1993a, 1993b). We briefly review here those classes of functions that are clearly relevant to studies of biodiversity (Fig. 48).

Overlay Procedures

The ability to overlay different data sets on a single computer display or computer-generated plot is fundamental to any computer mapping system. The simultaneous viewing of different maps (e.g., 100 single-species distributions) can provide incredible insight into spatial pattern and spatial relationships among the data represented. An extension of the visual overlay process is digital overlay in which data from more than one map are combined to form new data

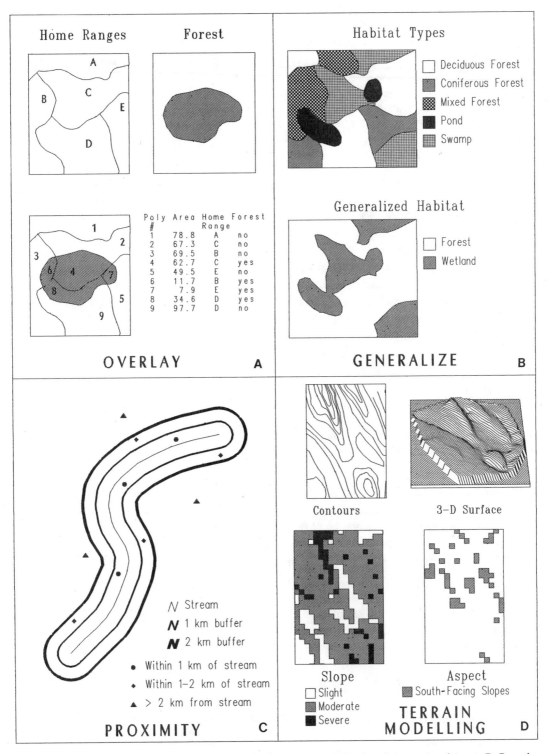

Figure 48. Basic analytical functions of a geographic information system. A. Overlay of two or more data sets. B. Generalization of a complex data set into a simpler representation. C. Analyses of proximity using buffering functions. D. Presentation and analysis of three-dimensional data.

(e.g., a map of changes that have occurred in the areal extent of a habitat between two time periods). When GIS information is merged digitally, the resulting data set retains the attributes and topological structures of the original data (Fig. 48). Subsequent analyses can be performed on the derivative data set.

Generalization

Many landscape characteristics are mapped in more detail than is required for an analysis. For example, soil maps are extremely detailed representations of landscape geomorphology that are typically used in a simplified, generalized manner. The ecologist or environmental scientist generally is less interested in the pattern of all soil types than in the distribution of a soil type that supports a particular type of vegetation. In a GIS environment, data generalization is a basic analytical process and one that is used frequently to simplify or extract subsets of complex data (Fig. 48).

Neighbor or Proximity Analysis

Questions about the proximity of one feature to another are fundamental to studies of biodiversity. The neighbor or proximity functions of GISs permit measurement of how near one set of features is to another set, and it is possible to calculate what areas fall within a specified distance of a feature (Fig. 48). This procedure permits an investigator to ask a question such as "Which of the rare species occur within 15 km of human-disturbed habitat?" Calculation of distances from a point or set of points to another set of points or lines is a basic function of most GISs and a process that is extremely laborious if done manually.

Analyses of Connectance

The networking functions of GISs allow a researcher to examine how features move along connected networks (e.g., cars along roads, water down rivers, or telephone calls through communication grids). These kinds of analyses take into account many variables such as direction (e.g., one-way or two-way), temporally changing impedances (e.g., traffic at rush hour versus that at midday), and random breaks in flow (e.g., a car crash stops traffic). Network analyses have significant relevance to ecological studies. Instead of cars and roads, the units of analysis might be individuals or species and networks of habitat on the landscape.

Modeling and Spatial Mathematics

A GIS is very much a predictive tool and can be integrated with predictive models (Berry 1993b). Raster GISs are extremely efficient in performing arithmetic and algebraic manipulations of spatial data. Many of the spatial data sets relevant to studies of biodiversity are characterized by continuous variables (e.g., number of species in an area, the probability that species "a" will occur in a forest patch, the population density within a certain habitat) that lend themselves to mathematical manipulation. Because the factors that affect species distributions vary over space, a GIS is the appropriate tool to account for this geographic variation in modeling applications. An especially powerful component of this approach is to include in the model the effects of factors in adjacent or nearby areas.

Terrain Analysis

The world is a three-dimensional space defined by location (x and y dimensions) and elevation (z dimension). Factors such as terrain, elevation, height, slope, aspect, texture, and volume of available space or habitat are fundamentally important to animal ecology. The z variable can represent factors other than elevation, such as number of species in a region, the variance in body proportions in an area, and the number of endemic taxa. Many GIS packages offer a rich suite of tools to handle such data. Some of the standard analytical tools include the ability to prepare contour and three-dimensional surface maps, determination of slope and aspect, cross-sectional profiling, calculation of volumetric characteristics, and measurement of true length of lines and areas on three-dimensional surfaces (Fig. 48).

Special considerations

GISs have powerful capabilities for producing maps. Indeed, the ability to enter data at one cartographic scale and produce an accurate map at another scale is a simple yet extremely useful function of computer mapping systems. Most major commercial map-making companies use GISs to assist in the production of atlas-quality maps. Many GIS software products have editing and transformation tools that can correct for relief displacement or distortion commonly found in data derived from aerial photography. These tools can also be used to correct mapping errors, resolve projection inconsistencies among data layers, and adjust cartographic data to fit a common base map. Many of the questions asked of a GIS are answered not with maps but with tabular summaries, statistical results, or digital data sets that can be transferred to other computer tools for further analysis. Almost every GIS can convert area, perimeter, distance, and other derived results into ASCII file format for export to other software packages.

Purchasing a system

Scores of GIS software products are available. Powerful software can be purchased for all classes of computers, from the personal computer (PC) to the large mainframe system. For small to modest databases (up to 200 megabytes), PC systems (DOS, Macintosh) are appro-

priate, especially because the same computer can be used for other purposes as well (e.g., word processing, database management). VGA-quality graphic monitors will suffice for most GIS applications on a PC. For large projects, or those that require management of large data sets (in the range of gigabytes), we prefer the UNIX-based workstation computers. When many users must simultaneously access a GIS and database, a network of workstations or minicomputer systems is a possible solution. PC and workstation computing technology is currently in a period of explosive advancement, so whatever recommendations we make with respect to computer specifications will soon be outdated.

The cost of GIS software runs from nothing, for packages in the public domain, to tens of thousands of dollars. The GIS Sourcebook (Parker 1991) provides a comprehensive review of GIS software packages, their costs, and their capabilities. A researcher should consider many variables before choosing a GIS. These variables are reviewed in Aronoff (1989), Woodcock et al. (1990), Antenucci et al. (1991), and Croswell (1991). Some of the critical points to consider are the following:

1. Size of the database required for the project and whether the software can handle such data sets.
2. Number of other users.
3. The GIS commonly used by colleagues in the investigator's profession or workplace.
4. Whether the product can carry out the analyses required.
5. Whether the software can read and write data in the formats required.
6. Company's reputation for service and technical support.
7. Whether the software documentation is easy to understand and use.
8. Cost of the software.
9. Cost of hardware maintenance contracts and annual software subscription fees.

10. Cost of software upgrades.
11. Whether the product accepts macro programs for automating repetitive processes.
12. Whether the software can handle three-dimensional analytical functions.

GIS software systems are notoriously user-unfriendly, and a sizable amount of time is often required to learn to operate them. Software vendors are, however, aggressively pursuing ways to make their products easier to learn and use. When researchers choose a system, they should consider how much time the users can devote to learning to use it.

The standard peripheral devices that are required for most GISs are a digitizing tablet, a plotter, and a printer. Digitizing tablets come in a variety of sizes. Small tablets with work surfaces on the order of 0.5 m × 0.5 m are good for training but are of little value for digitizing maps. Because the maps used for biodiversity studies can be large, the digitizing tablet should have an active work area of at least 1.0 m × 1.5 m (Fig. 47). The device should be able to distinguish points that are at least 0.1 mm apart. A backlit surface is useful for proofing digitized maps, but it is not essential. A base on which the elevation and angle of the work surface can be changed is necessary if people will be digitizing data for long periods. A high-quality digitizer can be obtained for U.S. $9,000.

Plotters are devices that are used to draw maps on paper or mylar. They come in many forms and sizes. An investigator must be able to generate maps of the size and scale required for a study using the plotter. If the project requires digitizing new information into the GIS, the plotter must be capable of making proof plots at the exact scale of the original base maps. Pen plotters are adequate for many applications (Fig. 47). Electrostatic plotters make atlas-quality images but are expensive (U.S. $50,000 for a high-quality device) and require considerable mainte-nance. Ink-jet and thermal wax printers are excellent for legal and letter-size maps. Laser printers are especially good for letter-size black and white maps. Before purchasing a particular plotter or printer, the researcher should be certain that the GIS software can communicate with the device.

Conclusions

Much of the business of inventory and assessment of mammalian biodiversity involves the management and analysis of spatial data (National Research Council 1993). A GIS is an appropriate, and sometimes necessary, tool to accomplish this. There is a developing literature on applications of GISs to studies of biodiversity. The reviews by Davis et al. (1990) and Johnson (1990, 1993) contain numerous examples of how GISs have been used to support the conservation of biological diversity. Studies by Johnston and Naiman (1990), Breininger et al. (1991), Pereira and Itami (1991), and Stoms (1992) illustrate how GISs can be used to evaluate associations between species and habitats. Scott et al. (1991), Lombard et al. (1992), and Beardsley and Stoms (1993) discussed how GISs can be used to identify lands that should be targeted for conservation. As the computer hardware and software used for spatial data analysis mature over the next few years, GISs will become as fundamental a resource to biologists as word processors are to writers and spreadsheets are to accountants.

Acknowledgments. Julie Forsyth prepared the photographs in Figure 47. We thank the "GIS-nerds" in the University of Rhode Island Environmental Data Center for providing the technical stimulation and conservation insight from which this chapter was developed. This is contribution number 2928 of the Rhode Island Agricultural Experiment Station.

Chapter 12

Conclusions and Recommendations

Attempts to standardize anything are unlikely to be accepted unconditionally. Anyone who has ever attempted to sample mammalian biodiversity, or who intends to do so in the future, will have an opinion on the best way to get a job done. Nevertheless, a major underlying theme of this series of books is that standardization will greatly facilitate comparison, both spatial and temporal. That theme has guided our presentation of methods for measuring and monitoring mammal diversity.

Although the advantages of standardization are adequately explained at the beginning of the volume, advocating standard methods is not without some risk. For one thing, it is a bit presumptuous for any subset of authors to specify which techniques are most appropriate for all situations at all times in all parts of the world. In addition, many of the techniques that we have suggested undoubtedly can be modified to improve their effectiveness in particular situations. Nevertheless, we believe that the benefits of standardization outweigh the costs in most circumstances.

We also realize that the study of mammalian diversity is a dynamic endeavor, and that new techniques will be developed steadily as investigators pursue new knowledge. We encourage that development and urge that such modifications or improvements be made available to the scientific community as rapidly as possible, and that appropriate analytical techniques be developed to allow comparisons of new data with those obtained in prior studies. We believe that a certain amount of variation in field techniques can be handled adequately with proper analytical and statistical tools, as discussed at length earlier in this volume.

Data Maintenance

We have devoted considerable space to the importance of collecting and maintaining voucher specimens and associated data. We are also concerned about the maintenance of data from the ever-increasing number of mammalian inventories being carried out. The 1992 United Nations Conference on Environment and Development and the Convention on Biological Diversity both focused heavily on the use of biodiversity for humans. The convention mandates the establishment of national repositories for biodiversity data in each signatory nation. If such data are to be useful in achieving both anthropocentric objectives and the conservation of species diversity for its inherent values, then they must be maintained over the long term, and they must be easily and inexpensively accessible.

We urge all practitioners to ensure that data resulting from biodiversity surveys and inventories are deposited in appropriate repositories and made available for future use. With all due respect for intellectual property rights, we also urge investigators to make their original data available to others as quickly as possible. The environmental problems facing the world are so severe and are increasing at such a rapid rate that data forming a sound scientific basis for decision making will be needed in as timely a fashion as possible.

What Next?

For investigators armed with an array of standard methods and appropriate analytical tools, what are the next steps in the study of global diversity of mammals? Obviously, many users of this volume will select the techniques and tools that they believe are most appropriate for a local sampling problem or monitoring program. We intended that the volume be used in just this way.

We would also argue, however, that such techniques and tools could form the basis of a much larger, better-organized, and coordinated program to study the world's mammalian diversity. Our knowledge of mammalian systematics, biogeography, and natural history could be advanced enormously if individual local sampling efforts could be expanded or combined into regional, national, and, eventually, global programs.

Perhaps the Convention on Biological Diversity could be used as a stepping-stone to such advances in our knowledge. If national biodiversity institutes are established, they could also serve as centers for the oversight and coordination of national sampling programs to provide baseline data on mammalian distributions. These data, in turn, would form the basis for both spatial comparisons across habitats, ecosystems, and landscapes and temporal comparisons through long-term monitoring programs.

Only such well-grounded, long-term sampling programs will provide data adequate for setting priorities among areas for protection, predicting and dealing with effects of habitat disturbance, and making wise land-use decisions. We urge the adoption of standardized sampling schemes for regional, national, and global studies of mammalian diversity. National and regional professional societies devoted to the study of mammals could play a leading role in developing such programs.

Mammals are among the most important groups of organisms on earth, regardless of the metric used to make such a determination. The fact that *Homo sapiens* belongs to the class Mammalia helps to ensure that humans will accord their fellow mammals the attention they deserve. The phylogenetic position of mammals, their adaptations to all but the most extreme and harsh environments on the earth, and their sensitivity to environmental perturbations argue for their continued position as a flagship group for conservation efforts.

The degree to which we will be able to deal with conservation needs will depend greatly on our ability to predict the effects of anthropogenic activity on natural populations of wildlife. We hope that this volume will find a useful niche in the continuing development of our knowledge of mammalian relationships and distribution patterns.

Ethics in Research

Rasanayagam Rudran and Thomas H. Kunz

Introduction

Mammalian inventory and monitoring projects typically require the collection of whole animals for voucher specimens and of tissue samples for related laboratory research. In addition, they often involve live animal capture and marking, which can cause pain, discomfort, or disruption to the normal activities of target organisms. As a result, the ethical treatment of animals is an important issue when conducting mammalian inventory and monitoring projects. Views on this issue vary widely, but they can be broadly classified into two categories. In one category are the views of those who support animal rights and oppose any use of animals by humans, even if such use can benefit the animals and/or humans. The second category includes the views of animal welfare advocates who do not oppose the use of animals by humans but promote high standards for the humane treatment of animals. For these people, an important ethical issue with regard to research is the balance achieved between the benefits of a study

and the cost of that study to its animal subjects. Biologists using animals for research should make every effort to mitigate any adverse effects on their subjects while pursuing their scientific objectives.

Several scientists have discussed the ethical treatment of animals in order to promote a better understanding of this issue among fellow scientists and lay people alike (Midgley 1981; Still 1982; Huntingford 1984; Driscoll and Bateson 1988; Bateson 1991). In addition, professional societies such as the American Society of Mammalogists (1985, 1987) and the Animal Behavior Society (1986, 1991) have developed comprehensive guidelines to assist researchers in maintaining high ethical standards when conducting research on animals. These guidelines extend beyond the humane treatment of animals to include other ethical considerations such as the choice of target species, the number of individuals used, and legislation related to animal research. At the same time, these societies fully endorse research on animals as a means to advance scientific knowledge.

In each situation it is the investigator's responsibility to weigh the potential gain in knowledge against any adverse effect that a given treatment may have on an animal. Ultimately, the investigator is responsible for the efficacy and appropriateness of a procedure and the manner in which the research is conducted. Bearing this in mind, in the following sections we present guidelines for dealing with ethical issues that we consider most relevant to inventory and monitoring projects.

Legal and cultural considerations

Regulations and Permits

Many countries have well-defined laws and regulations concerning biological research and the manipulation of free-ranging animals. In several countries, prior approval must be obtained to conduct research on animals (see "Permits," Chapter 4). National regulations may also control the capture, collection, and export of some species. Scientists must become familiar with international and national regulations before launching a research project. They should also abide by those regulations both in letter and in spirit at every stage of their investigations. Ignorance of the law or even inadvertent violation of regulations may result in prosecution (Choate and Genoways 1975). Violations of laws and regulations can also taint the professional reputation of an investigator and the scientific value of the research.

Copies of specific regulations pertaining to mammalian inventory and monitoring projects may be obtained from international and national authorities that regularly deal with wildlife issues. Investigators planning research in a foreign country must also consider immigration and residency regulations in the host country. All official approvals stipulated by these regulations must be secured before a research project is initiated. Investigators planning to carry out projects on privately owned land must obtain prior approval from the owner. Investigators should remember that approvals and permits expire and must be renewed periodically.

Investigators should take special care to adhere to the restrictions related to work on endangered or locally rare species (see "Permits," Chapter 4). From an ethical perspective, these species should not be collected or manipulated in the field except as part of an authorized program for conserving the species (Animal Behavior Society 1991).

Religious and Cultural Concerns

In some countries local tribes have guidelines for research conducted on their lands (Colvin 1992). Religious and cultural taboos against the collection or capture of certain animals may also exist in some countries. Investigators must become acquainted with such customs before undertaking field studies, and they must respect local practices. Information about local customs may not be available in books and other printed materials; thus, it is important for investigators to consult local people and other investigators who may have worked in the same area. At times, local sensitivities to animal collection or capture can be more difficult to deal with than the laws and regulations established by national or international authorities. If problems arise, they should be resolved as diplomatically as possible.

Trapping animals

Sample Size

Collections of whole animals and tissue samples and capture and marking of animals are standard protocols used in most inventory and monitoring projects (see "Voucher Specimens," Chapter 4). Information obtained through such protocols enables accurate identification of species and enhances our knowledge about their ecology, distribution, habitat preferences, and other biologically important phenomena. This knowledge can lead to legislation and other measures to help protect wild species. The potential benefits to wild species, however, do not justify the haphazard collection or capture of animals. From an ethical perspective, an investigator must have a clear idea of the number of animals that must be collected or captured for a specific project (see "Sample Size" under "Voucher Specimens," Chapter 4). As a general rule, however, investigators should use the smallest number of animals necessary to achieve the research objectives. The number of animals required can often be substantially reduced by carefully designed experiments and by the use of powerful statistical tests (Hunt 1980; Still 1982). Ultimately, investigators must be able to justify the number of animals used in their research.

Investigators should always take steps to minimize specimen loss when collecting. If animals are lost, additional individuals will have to be obtained to ensure an adequate sample for research.

Kill-Trapping and Shooting

Humane methods of kill-trapping are those that kill animals swiftly and avoid damaging the body parts required for research. The most common kill-trapping devices used for collecting small mammals are snap-traps and pitfalls. For medium-size mammals, spring-powered conibear traps can be used for humane kill-trapping. Leg-hold traps and foot and neck snares should be used only in the absence of alternative methods, and only if the investigator is experienced in their operation. Shooting is a humane method of collecting mammals larger than rabbits but should be employed only by those experienced in dispatching animals swiftly with firearms. The proper use of traps and guns for capturing and collecting mammals is discussed in detail in Chapter 8.

Live-Trapping and Netting

A captive animal is helpless against the vagaries of climate and predators. Therefore, humane methods of live capture require that captive animals be maintained alive and uninjured in a nonstressful environment until they are processed and released. Traps should not injure the animal when their trigger mechanism is activated, and they should be large enough to contain the entire animal with room for movement. Traps should be sheltered from sun and rain, and they must be provided with sufficient food, water, and nesting materials to sustain the animal while it is held.

Traps should be set in areas where the risk of trapping nontarget species is minimal, and investigators should deactivate the traps when the target species is inactive. Some traps (e.g., harp traps, mist nets) have the potential for capturing several individuals simultaneously. They should be used only when a sufficient number of people is available to deal promptly with all individuals trapped.

During the reproductive season, pregnant females may abort their fetuses or suffer other reproductive problems if held in traps for prolonged periods (Ramsay and Stirling 1986). Some females abandon a den or home range, or even newly born young, when disturbed. If such problems are encountered, investigators must change their capture techniques or terminate live-trapping altogether.

Use of Capture Guns

Guns that fire darts to deliver immobilizing drugs can be used for the capture of medium-size and large mammals (see "Guns" under "Large Terrestrial Mammals," Chapter 8). These devices are considered lethal weapons in some countries and may be illegal; in other countries a special permit is required for their use. In addition, lethal or immobilizing drugs are controlled substances in most countries and can be used only by qualified veterinarians or by biologists with special permits. Biologists who use guns and drugs for animal capture must be experienced and aware of the potentially lethal nature of the equipment and the drugs used.

Processing and handling animals

General Considerations

A captive animal can be processed immediately in the field or transported to a laboratory for processing. Field processing cannot always be carried out under sterile conditions, but cleanliness is important to reduce the risk of infection and the contamination of samples (see "Frozen Tissues," Appendix 4). Under field as well as laboratory conditions, the primary concern of the investigator should be the comfort of the captive animal. Processing should be terminated at the first sign of stress or injury.

If an animal is physically restrained while being processed, the restraint should not cause discomfort or pain and should not hinder breathing. When medium-size or large mammals are physically restrained, their eyes should be covered with an opaque cloth so that they do not become excited by movements around them. Large mammals that cannot be restrained physically must be sedated before they are processed. If an animal is to be subjected to several procedures, the researcher should consider using an anesthetic, particularly if a procedure is likely to cause more than momentary discomfort. The heart rate and respiration of animals under anesthesia should be monitored constantly, and steps should be taken to ensure that saliva and regurgitated food do not block air passages. The eyes of anesthetized animals should be covered to prevent dehydration of the cornea.

Precautions should be taken to prevent dehydration and overheating of captive animals. Small animals are particularly susceptible to dehydration and should be provided with drinking water and held in containers that minimize evaporative water loss. Small nectarivorous bats are especially susceptible to dehydration and thermal stress and should be offered small amounts of sugar solution. Medium-size and large mammals can be doused periodically with water to prevent dehydration or given intramuscular doses of isotonic saline or Ringer's solution.

Marking Animals

The purpose of marking animals is to permit the identification of individuals in the field. If individuals can be identified by natural markings (see "Problems Related to Marking," Appendix 7), it is unreasonable to subject them to marking procedures. Therefore, an investigator must observe a target species carefully and evaluate the feasibility of relying on natural markings before initiating a marking program.

Several techniques are available for marking species. These procedures are discussed in Appendix 7, along with guidelines for the ethical application of marks. Marks must not cause discomfort or injury and are best applied under the supervision of someone experienced with the technique.

Collecting Tissue and Fluid

Samples of body tissues and fluids such as blood and lymph are usually obtained from kill-trapped specimens (see "Frozen Tissues," Appendix 4). They can be collected from live animals as well, but only trained personnel should carry out these procedures. A conscious effort must be made to minimize any pain involved, including use of an anesthetic. Procedures must be carried out under sanitary conditions,

and target animals should be observed until they recover completely.

Holding, Transporting, and Releasing Animals

Animals that have been processed may have to be held in cages until their release. Different species must be held in separate cages and kept singly or in groups depending on their natural propensities. Holding cages must be provided with adequate food, water, and nesting material and should be kept away from climatic extremes. If animals are kept in a laboratory for extended periods, their maintenance should incorporate aspects of natural conditions important for their welfare and survival (Animal Behavior Society 1991). In the United States, conditions in captivity should comply with standards in the *Guide for the Care and Use of Laboratory Animals* (National Research Council 1985). Other countries require compliance with similar guidelines. Investigators transporting animals by vehicle should comply with guidelines for transportation of household pets. Investigators transporting animals by air or other commercial transport should follow national and international regulations. In the United States these guidelines can be obtained from the Department of Agriculture.

Wild animals should be held in captivity only until the required research is completed. Release, however, should coincide with the time of day or night when the captive is most active. An animal should be released only at the site where it was captured, with the exception of animals released to establish populations in areas from which their species has been previously extirpated. Investigations must always precede such releases to ensure that the release area is suitable.

Human Health Concerns

Thomas H. Kunz, Rasanayagam Rudran, and Gregory Gurri-Glass

Introduction

Field biologists preoccupied with their scientific objectives often fail to recognize the health risks to which they may be exposed (Constantine 1988). Investigators conducting mammalian biodiversity projects are particularly vulnerable because of the wide variety of potentially harmful species, techniques, and environments that can be involved in such studies. Those investigators, therefore, must actively seek to avoid potential health hazards or run the risk of undermining their field research through injury or illness.

In this appendix we discuss general precautions that one should take during a mammalian biodiversity project. We also consider disease risks encountered in different parts of the world and discuss preventive measures that can be taken against certain of them. Our treatment of disease risks is not intended to be comprehensive, however; rather, we highlight certain illnesses that are either common or particularly relevant to mammalian field studies.

Medical knowledge is increasing, and recommended treatments for injuries and illnesses of various types are constantly changing. In addition, none of us is a trained physician. Readers should, therefore, consult a physician and relevant medical textbooks for more detailed and up-to-date information.

General precautions

Mammals are known to transmit a wide range of pathogens, some of which can be life-threatening. The environment in which mammals live may also harbor certain diseases. Therefore, investigators should seek medical advice on disease risks related to their research and take appropriate precautions before commencing fieldwork. They should also consult a physician for an evaluation of their existing health conditions and assurance that they are physically and mentally capable of undertaking the field investigation. Medical advice is especially important when research involves an area or region that is new to the investigator, such as a foreign country.

Investigators must learn to protect themselves from hazards that they may encounter in the field. Training and experience in the use and handling of equipment and chemicals are important to avert potential problems. Investigators should also protect themselves against injury from the animals that they study by wearing special protective clothing, including gloves, chaps, headgear, and footwear. In addition, investigators should take precautions to avoid being bitten by venomous animals or exposed to poisonous plants. Learning the appropriate antidotes to toxic and poisonous substances can be invaluable.

Physical environments that pose special hazards to researchers include caves and mines, which may contain life-threatening levels of noxious gases. Physically challenging environments, such as those at high elevations, with steep terrain, or in humid forests, as well as prolonged exposure to severe cold or heat, also pose certain health risks. Investigators who expect to work in such environments should know how to deal with potential health problems and must obtain special training to minimize injury.

Prior knowledge of the potential hazards and health risks in a given region and steps needed to respond to life-threatening situations should be part of the overall training of a field biologist. In particular, investigators should be trained in first aid procedures and rescue operations for emergency situations. They should carry appropriate first aid materials and reference works (e.g., Tilton 1994; Werner 1994) that will help them to diagnose (and if necessary, treat) serious injury or illness if medical assistance is unavailable.

Immunizations

Immunization against various diseases is an important first step in preventing illness during field investigations. Children are routinely immunized in many countries, but periodic booster shots are often necessary to maintain immunity as adults. For example, although children in the United States are immunized against tetanus and diphtheria, the immune status of a significant number of adults is deficient (Jong 1993). Researchers must review their medical records and ensure that their immunizations remain effective throughout a field study. Most immunizations and booster vaccines can be received at a physician's office or clinic, but vaccines cannot always be administered at one time. Thus, medical advice must be sought well in advance (at least 4–6 weeks) of a proposed project.

Some countries require specific immunizations. Relevant information can be obtained from a local health office, or from the World Health Organization (WHO) or the U.S. Centers for Disease Control, which regularly publish health warnings pertaining to different regions and countries and recommend immunizations for travelers. Member nations of WHO normally require foreign visitors to carry a valid certificate of immunization against yellow fever. They no longer require smallpox and cholera immunizations.

Investigators planning to carry out research in countries where it may be difficult to avoid contaminated food and water should obtain typhoid and immune globulin vaccinations. These vaccines are effective against the serious and sometimes life-threatening *Salmonella typhimurium* bacterium and hepatitis A virus, respectively. Investigators planning to work in Saudi Arabia, Nepal, India, Kenya, or Tanzania and other sub-Saharan countries should consider immunization against meningococcal meningitis, a bacterial infection that enters through the respiratory system and infects the brain. Those intending to work for long periods in some rural areas of Asia should consider immunization against the Japanese encephalitis virus, which is spread by *Culex* mosquitoes. Vaccination against plague, which is caused by the bacterium *Yersinia pestis,* is important for biologists working with rodents and rabbits in areas where this disease is suspected. Plague still occurs in pockets throughout Asia, Africa, South America, and the United States (Kusinitz 1990). Immunization against rabies (see "Rabies," below) is an important precaution for investigators working with bats and carnivores in most parts of the world, but especially in Central and South America, Southeast Asia, the Philippines, India, and Africa. Immunizations against numerous other diseases are also available.

Certain immunizations may not be compatible with particular health conditions (e.g., allergies, pregnancy). Other immunizations may be an important precaution for people at high risk for a particular disease because of their existing health status (e.g, people lacking a spleen or suffering from chronic illnesses). Furthermore, each vaccine has a specific period of effectiveness, and some vaccines (e.g., for rabies and hepatitis B) are relatively expensive. To learn the details of relevant immunizations, investigators should consult a physician before commencing field investigations.

Disease risks

Most, if not all, mammals (including humans) and their arthropod parasites carry pathogens that may be transmitted through direct contact, bites, scratches, or exposure to body fluids (urine, blood, saliva). In rare instances pathogens may be inhaled. Other pathogens may be transmitted to humans when contaminated food or water is ingested. Field researchers should take precautions to avoid exposure to these pathogens, which include viruses, bacteria, fungi, protozoans, platyhelminths (flatworms), and nematodes (roundworms). In most instances, risks are minimal, and attention to personal hygiene and the type of food and water consumed, as well as the use of protective clothing and insect repellents, is sufficient to prevent infections. Nevertheless, it is important to know the symptoms and treatment of some of the potentially serious illnesses, so that one can deal with them effectively if necessary.

Common diseases

Malaria

Malaria is the single most important disease hazard for investigators conducting research in the tropics. It infects over 250 million people each year and is a leading cause of death in developing countries (Keystone 1993). Malaria is caused by the protozoan blood parasite *Plasmodium,* which is transmitted by the bite of *Anopheles* mosquitoes. The four species of *Plasmodium* that cause malaria are *P. falciparum, P. malariae, P. vivax,* and *P. ovale.* All species have an initial phase of development in the liver, after which they enter the bloodstream to undergo further development within red blood cells. *Plasmodium vivax* and *P. ovale* leave behind dormant forms in the liver that can cause relapses of malaria for up to 3 years after exposure. *Plasmodium falciparum* has the greatest potential to kill, because it can parasitize up to 80% of the red blood cells (Hall 1988). This condition can lead to severe anemia, and liver and kidney failure during the final stages before death. Principal

symptoms of the early stages of malaria include recurring bouts of fever and chills with sweating, fatigue, and abdominal pains.

The battle against malaria continues, but there is still no vaccine to prevent the infection. As the first line of defense, investigators must avoid being bitten by the *Anopheles* mosquito, which is active between dusk and dawn. Preventive measures include sleeping under mosquito netting, wearing tightly woven clothing, and applying an insect repellent containing *N, N*-diethyl meta-toluamide (DEET) to exposed areas of the skin every 3 to 4 hours. For additional protection clothing can be impregnated with DEET-containing repellents or with permethrin (Duranon or Permanone). Permethrin kills insects and other arthropods that alight on the treated fabric (Rose 1992).

Protection against mosquito bites alone may not be sufficient to prevent malaria, especially in areas with a high incidence of the disease. Before visiting such areas, an investigator must begin a course of antimalarial drugs. Medical advice is essential in choosing the appropriate drug(s), because *P. falciparum* strains that are resistant to certain drugs occur in some tropical regions; certain drugs may have adverse side effects on some people; and some drugs may be incompatible with a person's existing health condition (e.g., allergies, pregnancy) or medications taken for other ailments.

Antimalarial drugs do not prevent the disease but usually act on the parasite after it has been released from the liver into the bloodstream. Consequently, these drugs must be taken regularly during and also after a period of exposure. Postexposure prophylaxis should be continued for 4 weeks, to help kill the parasites released from the liver after the exposure. Ideally, investigators must also take the drugs for up to 2 weeks before entering a malaria area. This precaution ensures adequate drug levels in the blood before exposure and also enables one to switch drugs in case of adverse side effects.

Chloroquine (Aralen) is the prophylactic drug of choice for all four species of the malaria parasite, except in areas where chloroquine-resistant *P. falciparum* strains occur. Adult dosage for chloroquine-sensitive areas is 300 mg of chloroquine (found in a 500-mg tablet of chloroquine phosphate) taken each week during and after exposure (Rose 1992). In areas with chloroquine-resistant *P. falciparum* strains, adults may take weekly dosages of 250 mg of mefloquine (Lariam). However, the efficacy of mefloquine is low against the chloroquine-resistant strains of *P. falciparum* found along Thailand's borders with Myanmar and Cambodia (Keystone 1993). In such areas, where multidrug-resistant *P. falciparum* strains occur, daily doses of 100 mg of doxycycline (Vibramycin) are effective as a prophylactic. Those who cannot tolerate or take the above-mentioned drugs can take daily doses of 200 mg of proguanil (Paludrine), but this drug is effective only in areas where chloroquine-sensitive strains occur. Proguanil is one of the safest anti-malarial drugs available and can be used even during pregnancy (Keystone 1993). Primaquine is used as a postexposure prophylactic to eradicate *P. vivax* and *P. ovale* parasites that lie dormant in the liver. Its use must be preceded by a blood test because of the risk of primaquine toxicity in people whose red blood cells are deficient in the enzyme glucose-6-phosphate dehydrogenase.

Pyrimethamine/sulfadoxine (Fansidar) is no longer used as a prophylactic, but it is sometimes taken with other drugs for the actual treatment of malaria when immediate medical assistance is unavailable. Halofantrine (Halofan) is a new drug that is also effective in the treatment of malaria. Other drugs that can be used in the self-treatment of malaria include mefloquine and combinations of quinine with either doxycycline or tetracycline (Rose 1992). It is wise to consult a physician and be prepared for the self-treatment of malaria, especially when working in remote areas. Nevertheless, investigators must seek expert medical attention as soon as possible if afflicted with the disease.

Diseases from Contaminated Water or Food

As indicated earlier, typhoid and hepatitis A infections are transmitted via contaminated water or food, and both diseases can be avoided by taking appropriate immunizations. Cholera is another health hazard that can be transmitted by contaminated water or food. Immunization for cholera is not recommended by WHO or the Centers for Disease Control, because the vaccine that is currently available is not very effective in preventing the disease (Jong 1993). Instead, biologists working in countries with recent cholera outbreaks should pay particular attention to the water and food that they consume. Undercooked or raw pork, sausage, beef, and fish can cause diseases such as trichinosis, tapeworm, and fluke infections. Meat and fish must be cooked for at least 1 hour at a temperature of 55° C or more before being eaten.

One of the most common ailments in the field is diarrhea. Noninfectious types of diarrhea are caused by food poisoning, overindulgence in food, and even exhaustion at high elevations. Infectious diarrhea is caused by a variety of microorganisms, the most common of which is the *Escherichia coli* bacterium (Kammerer 1993). Other organisms causing infectious diarrhea include certain viruses, bacteria such as *Shigella* and *Campylobacter,* and protozoans including *Giardia lamblia* and *Entamoeba histolytica.* Diarrhea is often accompanied by stomach cramps, nausea, vomiting, and fever. High fever and blood, pus, or mucus in the stool are symptoms of dysentery caused by bacteria such as *Shigella* or *Campylobacter.*

Microorganisms causing diarrhea may be found in untreated tap water and ice or fruit drinks made from it, raw vegetables, fruits, seafood, meat, dairy products, and foods left in the open for several hours, such as those sold by street vendors. Such foods and drinks must be avoided if there is even the slightest suspicion of contamination. As a precaution against diarrhea, researchers should eat only hot, cooked food and fruits with unbroken skins that can be peeled, and they should drink only carbonated beverages or beverages such as beer, tea, and coffee. Water used for making tea or coffee and brushing teeth must be boiled for at least 5 minutes to kill dangerous microorganisms including hepatitis viruses (Kammerer 1993). Water can also be chemically treated with iodine tablets (Potable-Aqua), which are superior to chlorine tablets (Halazone) in killing parasitic cysts (Wolfe 1993).

Even with precautions, it may be difficult to avoid diarrhea in the field. The first step in treating diarrhea is the replacement of lost fluids and electrolytes. Commercially available rehydration salts are ideal for this purpose, but fruit drinks and caffeine-free beverages can also be used. Dairy products, alcoholic drinks, and fatty or spicy foods must be avoided during treatment. This may be the only treatment necessary, but if bowel movements are frequent and accompanied by abdominal cramps, an antimotility agent such as loperamide (Imodium) or diphenoxylate (Lomotil) may be taken. Commonly used treatments such as Kaopectate, activated charcoal, and yogurt are not recommended, because they have not been shown to be effective in reducing cramps or frequency of stools. If diarrhea continues for more than 3 days without improvement, medical advice must be sought. A physician should also be consulted if blood, mucus, or pus occurs in the stool, or if diarrhea is accompanied by severe cramps or high fever with chills. If medical advice is unavailable, antibiotic drugs such as trimethoprim (Trimpex), trimethoprim-sulfamethoxazole (Bactrim or Septra), or ciprofloxacin (Cipro) may be taken. It is crucial, however, that these drugs be taken only on the basis of prior consultation with a physician, because they may be incompatible with certain health conditions. Furthermore, medical attention must be sought as soon as possible after a bout of severe diarrhea.

Recurrent diarrhea may be a symptom of a parasitic infection. Intestinal parasites that cause diarrhea include *Giardia lamblia, Entamoeba histolytica, Cryptosporidum,* and *Dientamoeba fragilis.* Stool examination is required for identification of the parasite causing the infection and determination of the appropriate treatment.

Skin Diseases

Investigators working in the humid tropics are often susceptible to skin diseases. Most of these diseases can be prevented by maintaining high standards of personal hygiene. Perhaps the most important preventive measure is keeping one's skin and clothes clean and dry. One of the most common skin problems is "athlete's foot," a fungal infection that can be prevented by regular use of foot powder and frequent changes of socks. Foot infections can also be caused by sand fleas and hookworms. The best way to prevent these problems is to wear shoes that adequately protect the feet.

In tropical Africa, Tumbu fly larvae may pose health problems by penetrating the skin and causing painful swellings. The larvae hatch from eggs deposited on clothes left out in the sun to dry. To prevent the larvae from penetrating the skin, clothes that are sun-dried must be pressed with a hot iron before they are worn. A similar fly-larva infection also occurs in the tropical areas of Latin America.

Diseases involving wild mammals

Rabies

One of the most feared and respected diseases of animals that can be transmitted to humans is rabies (*Lyssavirus*), an acute viral infection of the central

nervous system. Rabies occurs mostly in warm-blooded animals, although susceptibility to the virus differs considerably among species. The rabies virus typically enters a host's body via the saliva of an infected animal, which contaminates a bite wound or contacts broken skin. Liquid saliva remains infectious for 24 hours at room temperature. Rabies virus in dry saliva can survive for up to 14 hours. The virus may also enter the body via the digestive tract, the respiratory system, or through contact with mucous membranes (Constantine 1988). The rabies virus progresses along the nerves to the spinal cord and brain and then reverses direction, moving centrifugally and eventually invading the nerves in all organs including the skin. Infected animals may engage in savage attacks, thereby transmitting the virus and continuing the cycle (Constantine 1988). Infected bats are only rarely involved in unprovoked attacks and may appear normal before developing paralysis, followed by death.

The incubation period of the disease is usually from 2 to 12 weeks in humans, although this can vary from 10 days to 15 months. Because of this relatively long incubation period, vaccine treatment (active immunization) can be administered after infection. If the disease appears, it is generally fatal to humans and most mammals within 3 to 7 days following the appearance of the first symptoms. Only two humans are known to have survived an infection with the rabies virus.

The virus cannot survive for long outside the body and is typically inactivated at temperatures in excess of 56° C, and by ultraviolet and solar rays. It is not killed by low temperatures (e.g., by freezing infected animals). Treatment should be initiated as soon as possible following exposure. National and international health authorities recommend immediate prophylactic treatment with rabies vaccine following bites or other potential exposures from bats and wild carnivores. The decision to treat an individual, however, may be influenced by knowledge of the presence or absence of the disease in the geographic area or in the species of interest, and the circumstances of the bite (Constantine 1988).

WHO and the Centers for Disease Control recommend that all individuals who have regular contact with bats or with other high-risk mammals (carnivores) be immunized against rabies. Investigators should follow the immunization recommendations of the Advisory Committee on Immunization Practices of the U.S. Public Health Service. Preexposure immunization resulting in a serum rabies neutralizing antibody titer of at least 1:5 is an appropriate pre-caution for most bat researchers (Constantine 1988). Prophylactic vaccination with human diploid cell vaccine consists of one intramuscular injection on days 0, 28, and 56, or in an accelerated form on days 0, 7, and 28. Reinforcing (booster) doses are advisable every 1 to 3 years. Individuals who do not have regular contact with bats should wear nonpenetrable gloves when handling these animals.

If a person is bitten by a suspect mammal, the wound should be washed thoroughly with detergent and water. The detergent inactivates the virus particles, whose outer membranes are high in lipids. Alcohol (70%) or other skin disinfectants should then be applied. Postexposure vaccination against rabies is advisable. Treatment consists of six injections on days 0, 3, 7, 14, 30, and 90. If bites have occurred on or near the head (neck or face), or if there has been mucosa contact with the animal's saliva, rabies hyperimmunoglobulin should be administered as passive immunization.

Rabies-related viruses, including Mokola, Dubenhage, and Lagos bat virus, have been reported from wildlife, humans, and domesticated animals. Diseases produced by these viruses resemble paralytic rabies, but whether rabies vaccines provide protection against these viruses is questionable. All such viruses appear to be restricted to Africa (Constantine 1988).

Arboviral Fevers

There are several types of arboviral fever, of which the best known is yellow fever. These fevers are caused by arboviruses, which are transmitted by arthropod vectors such as mosquitoes, ticks, and sand flies (Kusinitz 1990). In Latin America and Africa, the yellow fever virus is spread among humans by the mosquito vector *Aedes aegypti,* which is usually active in the morning and at twilight. The virus can also be transmitted from infected nonhuman primates to humans by *A. simpsoni* in Africa, and by *Haemagosgus* and *Sabethes* mosquitoes in Latin America. Symptoms of the disease include the characteristic yellow-green color of the skin, mucous membranes, and eyes. This is accompanied by nausea, muscle aches, and fever, which can be very high and followed by vomiting of blood in the most serious cases. Fortunately, the vaccine available for yellow fever is highly effective against the disease and provides 10 years of protection.

Other arboviral fevers include dengue and Japanese encephalitis. The latter illness occurs regularly in Asia and the eastern part of the former Soviet Union, whereas dengue is present in Asia, Pacific islands, several Latin American countries, and parts of tropical Africa. Dengue, which is spread by the *Aedes aegypti* mosquito, causes as much illness as all other arboviral fevers put together (Welsby 1988). The symptoms of dengue include sudden high fever, severe headache, fatigue, and muscle and joint pains. The *Culex* mosquitoes that spread Japanese encephalitis are most abundant during the summer months in temperate areas and during the rainy season in the tropics. Symptoms of the illness include nausea, vomiting, headache, and fever. The most serious cases may result in permanent brain damage and sometimes death. Because the viruses that cause both illnesses are biologically similar, the vaccination against Japanese encephalitis may provide some protection against the hemorrhagic form of the dengue fever, which can sometimes be fatal (Rose 1992). Nevertheless, avoiding mosquito bites is still the best protection against disease.

African hemorrhagic fevers, such as Lassa, Crimean-Congo, Ebola, and Marburg, are also caused by arboviruses. They produce high fevers, muscle aches, headaches, and internal bleeding that can be fatal. The vector for Crimean-Congo fever is a tick that spreads the virus to humans from wild and domestic animals. The vectors for the other fevers are not known, but the rodent *Mastomys natalensis* serves as an intermediate host for Lassa fever virus (Kusinitz 1990). There is no reliable cure for these illnesses, although Lassa fever has responded to the antiviral drug ribavirin in experimental trials. As there is no cure, it is essential to prevent arthropod bites by taking the previously mentioned precautions (see "Malaria," above). Investigators intending to work in western and central Africa should be especially careful, because Lassa fever is a major health problem in those areas.

Hantavirus

The hantavirus receives its name from the Hantaan river in South Korea, where it was first identified among U.S. military personnel nearly 40 years ago (Anonymous 1994). Since that time about 200,000 cases of hantavirus infection have been reported annually, mainly from the People's Republic of China and to a lesser extent from Korea, Russia, Scandinavia, Europe, and the Balkan region (Morris et al. 1994). The best-known hantavirus infection is hemorrhagic fever with renal syndrome (HFRS), an illness of varying severity characterized by high fever, hemorrhage, and kidney malfunction.

In May 1993, another type of hantavirus infection, called the hantavirus pulmonary syndrome (HPS), was discovered in the southwestern United States. This infection is caused by a recently identified strain of hantavirus and is 10 times more likely to cause death than HFRS. The onset of HPS occurs within 45 days after exposure, and early symptoms include fever, headache, muscle ache, fatigue, vomiting, shortness of breath, and a dry cough. Later symptoms include rapid heartbeat and breathing difficulty, which may lead to abrupt respiratory failure.

Rodents are the primary reservoirs of hantaviruses. The hantavirus strains causing HFRS are carried by murine rodents, particularly *Apodemus* and *Rattus*. The principal carrier of the HPS-causing hantavirus strain is the deer mouse (*Peromyscus maniculatus*), although other species of *Peromyscus* have also been implicated. The 1993 outbreak of HPS in the United States was attributed to a sharp increase in deer mouse populations resulting from high precipitation and heavy seed set in previous years (Anonymous 1995). Hantaviruses are transmitted from rodent to rodent and from rodent to human through inhalation of viral particles released into the air when infected rodents, their nests, or materials contaminated with their urine, feces, or saliva are disturbed. Disease transmission may also occur through contact of contaminated materials with broken skin or the conjunctiva of the eye.

An experimental drug, ribavirin, shows promise in the treatment of hantavirus infections, but its effectiveness has still to be proven. Therefore, disease prevention is essential, and adequate precautions must be taken to avoid contracting the infection. We recommend that investigators wear disposable aprons or coveralls, shoe covers, gloves, and protective goggles when working with rodents. In addition, investigators must wear air-purifying respirators with high-efficiency, particulate air filters, and they must be trained in the proper use and care of this equipment. Similar precautions must be taken when investigators expect to come in contact with the nests or excreta of rodents. Those who fall ill after working with rodents should consult a physician immediately, because early diagnosis and prompt hospitalization can greatly reduce mortality risk.

Leishmaniasis

Leishmaniasis is one of the most common parasitic diseases in the world and is an important health concern in Latin America, Africa, the Middle East, Mediterranean countries, central Asia, northern China, and India and neighboring countries. The disease has several forms, each of which is caused by a different species of protozoan parasite belonging to the genus *Leishmania*. The parasites may be found in domestic dogs, some rodents, foxes, jackals, and rock and tree hyraxes. They are transmitted to humans by blood-sucking sand flies (Manson-Bahr 1988; Kusinitz 1990). They then invade macrophage cells located in the liver, spleen, bone marrow, skin, and mucous membranes. The form of the disease depends on the organs that are affected and the species responsible for the infection. The three main forms of the disease are visceral, cutaneous, and mucocutaneous leishmaniasis.

Visceral leishmaniasis (kala azar) results in anemia and enlargement of the liver and spleen. It develops relatively slowly after the infection; symptoms include muscle aches, chills and fever, weight loss, cough, and diarrhea. Warty skin nodules or skin ulcers may also occur. This form of leishmaniasis can be confused with several other diseases (lymphoma, leukemia, malaria, typhoid) and can be fatal if untreated. Cutaneous leishmaniasis is characterized by ulcerative skin lesions or nodules, which may heal by themselves. However, ulcers caused by *L. mexicana mexicana,* which usually infects the ear, can persist and ultimately destroy the pinna. This form of leishmaniasis is common among forest dwellers of Guatemala and Mexico. Mucocutaneous leishmaniasis is caused by *L. braziliensis,* which can affect the mucous membranes of the mouth, nose, and throat and cause severe disfigurement of these areas. The disease is becoming increasingly common in Paraguay.

Leishmaniasis can be prevented by avoiding the bite of sand flies. Application of DEET-containing insect repellents is effective for this purpose. Because sand flies are active from dusk to dawn, sleeping under a mosquito net is also effective, but the net must either have a very small mesh or be treated with permethrin. It is also advisable to sleep at least 60 cm above the ground, because sand flies can rarely jump to this height. Camping in areas with sand flies should be avoided as well. Treatment of the disease involves a course of antimony drugs such as sodium stibogluconate (Pentostam) or sodium antimony gluconate. Other drugs such as amphotericin

B and allopurinol are also available for treatment. These drugs can have serious side effects and should be taken only under the supervision of a physician.

Histoplasmosis

Histoplasmosis is a disease of humans and other mammals caused by a fungus, *Histoplasma capsulatum* (Constantine 1988). The fungus occurs naturally as a soil saprophyte in warm, humid regions on all continents. Its development is enhanced by suitable organic matter such as the feces of bats and birds. Infection occurs upon inhalation of the fungal spores, which become airborne when dry deposits of guano are disturbed.

Researchers entering potentially contaminated bat roosts should wear respirators with filter cartridges capable of filtering out particles as small as 2 microns in diameter, or they should use a self-contained air supply unit (Constantine 1988). There is no vaccine against histoplasmosis. Most people who have visited contaminated bat roosts have positive histoplasmin skin tests, probably reflecting resistance to subsequent infection. Reinfection can occur, however, after exposure to large doses of spores. Care should be taken to avoid creating airborne dust from guano or soil.

Lyme Disease

Lyme disease is caused by the bacterium *Borrelia burgdorferi,* a spirochete similar to the organism that causes syphilis. In the United States, the primary vectors that transfer the pathogen to humans are ticks in the *Ixodes ricinus* complex. The disease is usually passed to humans by tiny juvenile (nymphal) ticks, which, unlike the adults, may feed unnoticed for the 24 to 48 hours it takes to transmit the disease.

The incidence of lyme disease has increased steadily in the United States following its original discovery in 1975 in the town of Lyme, Connecticut. Currently Lyme disease is most commonly reported in the northeastern United States, although it has been reported from nearly every part of the country. In the Northeast, the disease is maintained by the white-footed mouse (*Peromyscus leucopus*) and the deer tick (*Ixodes dammini*). The deer mouse provides a meal for the deer tick, and the feeding process keeps the disease thriving by infecting each

generation of mice with the Lyme disease–causing spirochete. In southern California, Lyme disease appears to be maintained by the dusky-footed wood rat (*Neotoma*) and by two species of ticks: *Ixodes pacificus,* which feeds on *Neotoma* and humans, and *I. neotomae,* which feeds only on the wood rat. Lyme disease is much more common in the Northeast because only one tick is involved in the cycle. In the East, 25% to 50% of *I. dammini* are infected with the Lyme spirochete, whereas in the West only 1% to 5% of *I. pacificus* are infected.

The occurrence of the nymphal (and biting) stage of the deer tick in the Northeast is highly seasonal. The spring and early summer nymphal stage is about the size of a sesame seed or the period at the end of this sentence. By early fall, when it has become an adult, the tick is larger, but still only about half the size of a dog tick (ca. 2 mm in diameter, although adults engorged with blood may reach 6 mm in diameter). Deer ticks can be found wherever white-tailed deer, on which the adults feed, are found. Because the deer population in the United States has exploded in recent years, the risk of Lyme disease has increased.

The clinical symptoms of Lyme disease usually include a red, ring-shaped expanding rash that develops 2 to 5 days after the bite. Other symptoms include weakness, dizziness, muscle aches, sore throat, and swollen lymph glands. Common sense is the best preventive measure. Investigators must wear light-colored clothes so that ticks will be visible, long-sleeved shirts, and long pants tucked tightly into socks and apply DEET-containing insect repellents to the skin. The body should be searched daily for ticks. Ticks are found most commonly on the legs or thighs, in the groin and armpits, along the hairline, and in or behind the ears. A tick can be removed by grasping its body firmly with forceps and pulling straight out. Antiseptic should be applied to the bite, and the tick drowned in alcohol. Investigators who suspect that they have been exposed to a tick-borne disease should consult a physician.

Other health hazards

Hazards in Caves and Mines

It is sometimes necessary to census or monitor bat populations in caves and mines. Gases such as ammonia, sulfur dioxide, carbon monoxide, carbon dioxide, and methane can accumulate in bat caves and mine tunnels and may threaten the health of an investigator (Constantine 1988). These gases may irritate the skin, lungs, and eyes or block access to atmospheric oxygen. Some of the gases are odorless and, if present in high concentrations, may overcome an investigator without warning. Oxygen may be deficient in some caves and mines, and strenuous exercise in such places, coupled with excessive heat and humidity, may adversely affect the depth and frequency of breathing.

Methods for detecting and measuring atmospheric gases are summarized by Constantine (1988), along with recommended limits for human exposure. Researchers are advised to adhere to these standards or risk brain damage, pulmonary impairment, or death. Devices that sound an alarm when they detect a noxious gas or oxygen deficit can be worn. Some species of bats can live in cave environments with concentrations of ammonia or carbon dioxide that would soon kill a person. Thus, one cannot assume that the presence of bats signifies a safe atmosphere (Constantine 1988).

Investigators working in caves with noxious gases or low concentrations of oxygen should wear a respiratory apparatus appropriate for the contaminant gas (Constantine 1988). Protective wear ranges from a simple respirator mask with a filter for each type of gas to a self-contained breathing apparatus with a supply of atmospheric oxygen. Various kinds of respiratory protection systems for use with particular concentrations of gas are described in Mackison et al. (1981). Respirator masks and facepieces are usually available in different sizes and must be individually fitted to ensure effectiveness. If cartridge filters are used, care should be taken not to exceed their useful lives.

Appropriate clothing and at least two sources of lighting should be carried by individuals working in caves. In addition, knowledge of first aid as practiced by cavers and familiarity with cave rescue techniques is recommended (Halliday 1982). No one should ever enter a cave alone, and someone should always be notified if caves and mines are to be entered by research personnel. The National Speleological Society recommends that individuals in the United States notify the National Cave Rescue Commission (by contacting the United States Air Force Rescue Coordination Center) prior to entering caves, to ensure a successful rescue in the event of an emergency.

Climatic Hazards

In cold climates, proper clothing is essential to prevent health problems such as hypothermia and frostbite. Clothing should fit well and be worn in layers, which can be shed when physical activity leads to sweating. During physical activity, the neck and wrist areas should be opened to permit ventilation of the inner layers, so that they do not become wet with sweat and cause chills. Appropriate protection for the head, feet, and hands is also essential in cold climates. In addition, investigators working at elevations over 2,000 m must take precautions against acute mountain sickness (AMS), which is triggered by low blood oxygen levels brought about by rapid ascent. Symptoms include headache, insomnia, fatigue, and gastrointestinal discomfort. Severe cases of AMS can lead to pulmonary and cerebral edema. They can be prevented by eating high-carbohydrate diets, maintaining high fluid intakes, reducing strenuous activity, and ascending gradually. Acetazolamide (Diamox) may be taken as a prophylactic by those who are not allergic to sulpha drugs.

In hot climates, investigators must take precautions against heat exhaustion and heat stroke. Symptoms of water-deficiency heat exhaustion include lack of appetite, restlessness, giddiness, and the passage of small quantities of deeply colored urine. Salt-deficiency heat exhaustion leads to severe muscle cramps, headache, and fatigue. Symptoms of both types of heat exhaustion can be prevented by adding salt to food and regularly drinking fluids in excess of those require to quench thirst. Heat exhaustion due to a sweating disorder (anhidrotic heat exhaustion) can result from working in hot climates for extended periods. Symptoms include a skin rash mainly in the trunk and upper arms, fatigue, rapid breathing, and a frequent urge to urinate (Adam 1988). Individuals with these symptoms must be moved to a cool environment and rest for at least several weeks.

Heat stroke is a life-threatening condition that can be brought about by continuous heat stress (during day and night), a sweating disorder, overindulgence in alcohol and overly strenuous exercise in hot climates. Symptoms include a pounding headache, confusion, a staggering walk, and a high fever without sweating. In the absence of sweating, the person suffering from heat stroke must be fanned to promote cooling by evaporation. The patient must also be wrapped in wet sheets and transferred to a hospital immediately. Heat stroke can occur without direct exposure to the sun, although it is often referred to as sunstroke.

Investigators should take precautions against the harmful ultraviolet rays of the sun, regardless of whether they work in hot or cold climates. Protection against ultraviolet A and ultraviolet B rays can be obtained by applying a broad-spectrum sunscreen cream, oil, or lotion. Sunscreens with a sun protective factor of at least 15 should be used and reapplied every few hours after sweating. A hat or a cap should also be worn to protect the head from the sun.

Venomous Animals

Venomous animals such as snakes, spiders, scorpions, bees, wasps, hornets, and centipedes can cause serious health problems. The probability of encountering these animals, however, is low, unless one goes looking for them.

Victims of venomous snake bites should be reassured; the speed with which snake venom kills has been greatly exaggerated (Warrell 1988). They should then be quickly transported to a hospital, along with the dead snake. Antivenom should never be administered unless epinephrine (0.5 ml of a 1:1,000 solution) is available to combat potential allergic reactions to the antivenom. Additional information on treatment of snake bite can be found in Hardy (1992, 1994a, 1994b).

Bee or wasp stingers should be scraped rather than pulled from the affected site, because the latter action may inject more venom into the skin. Those who are allergic to bee or wasp stings should always carry supplies for the self-administration of epinephrine.

Conclusions

Health problems during field investigations will be minimal for anyone who takes appropriate precautions against potential diseases and health hazards. Seeking timely medical advice, taking the required immunizations, preparing a kit containing all the necessary medications, training in administering first aid and the use of equipment, and mental and physical conditioning are important preliminaries to fieldwork. When fieldwork commences, special attention must be paid to personal hygiene and precautions must be taken against arthropod bites and the consumption of contaminated food and water. A schedule must be developed for medicines that are to be taken at regular intervals. Even ailments that may seem minor must be attended to promptly. In the case of serious illnesses, investigators must seek expert medical attention immediately or as soon as possible. If several investigators work as a team, at least the team leader must be aware of the health status (including allergies and ailments) of every member. The team leader must also develop a plan for emergency evacuations and discuss it with team members.

After extended periods of fieldwork, investigators should consult a physician and have a complete medical examination, including a complete blood count, a test for liver function, and stool examination for intestinal parasites. Serologic tests that screen for malaria, filariasis, hepatitis, schistosomiasis, and other infections should be carried out if investigators suspect that they may have been exposed to these diseases.

Acknowledgments. We thank P. M. Wijeyaratne and D. L. Hardy, Sr., for commenting on an early draft of the manuscript.

Preservation of Voucher Specimens

Terry L. Yates, Clyde Jones, and Joseph A. Cook

Introduction

The degree to which voucher specimens contribute to research and education depends on the quality of the preparation and data associated with each specimen. These data are of increasing importance as analyses, such as those involving molecular data, become more and more removed from the actual voucher specimens. Research specimens and their associated label and field data often represent the only means of verifying the identity and other attributes of individual pieces of DNA, chromosomes, or other materials being used to address questions in biological diversity research. In addition, these specimens provide documentation of historical biotic diversity for the assessment of change due to natural or human-induced perturbations. Renewed efforts should be made for the proper preservation, storage, and curation of these valuable scientific materials.

Types of specimen preparations

The nature of scientific specimens has changed over the past few decades. There has been a rapid transition from the standard skin and skeletal preparations of the past to preservation of a primary voucher specimen and a host of accessory materials, which must be cross-referenced with the primary voucher through a manual or a computerized system. In this appendix we provide an overview of procedures most commonly employed by mammalogists using this comprehensive approach to the preparation of scientific specimens. Most of these procedures can be varied to accommodate local conditions and resources.

Historically, the most common type of mammalian voucher specimen was the museum study skin and accompanying skull (Ingles 1954; Hall 1962). Hafner et al. (1984) proposed that the body skeleton also be retained. These authors focused on a procedure originally described by Miller (1932) that maximizes the utility of the skin and skeletal preparation for the widest group of users. We will describe that technique as the standard method for preparing mammalian dried specimens. Because it incorporates all of the procedures required to prepare a voucher consisting of a skin plus skull only, however, it can be modified easily for such preparations. We describe methods for preparing skeletons and fluid-preserved specimens as well. Additional details may be found in Nagorsen and Peterson (1980), DeBlase and Martin (1981), Hall (1981), and Williams and Hanks (1987).

Sometimes, when a large series of animals is collected or when field personnel are minimal, only the skeleton is retained as a voucher. Skeletal preparations require less time and fewer supplies than skin preparations while retaining a large amount of information. Nevertheless, such preparations should be considered only after a series of skin-plus-skeleton or fluid-preserved vouchers has been prepared.

Methods of specimen preparation

Record Keeping

The initial and most important step common to all methods of specimen preparation is accurate recording of all associated data. Record keeping for voucher specimens has increased in complexity as the number and diversity of materials preserved from each specimen have increased. We describe a method of recording data appropriate for specimens accompanied by accessory materials of all kinds. It can be used by single investigators, but it works especially well when several investigators (the optimal number is six) prepare large numbers of specimens in assembly-line fashion.

The investigator should assemble and record all information associated with a specimen before it is prepared as a voucher (see "Data Standards," Chapter 4). This information always includes the collector's field number. Besides this traditional number, many investigators also include a second number associated with an additional, separate field-cataloguing system used to facilitate cross-referencing and tracking of individual data sets. One such method, originally developed at Texas Tech University and since modified at the Museum of Southwestern Biology at the University of New Mexico, is the New Mexico Kryovoucher, or NK, system (Fig. 49). In this system, each specimen is assigned a unique NK number

in the field at the beginning of the processing procedure. A wide variety of information is recorded on the NK data sheet, including the type of specimen prepared and the kind and amount of other materials (e.g., chromosomes, tissues, parasites) that were taken from that specimen.

Use of a single unique numbering system for tracking associated materials with traditional vouchers in the field has the advantage of allowing specimens to be given a number as soon as they are removed from the field. As a result, each worker in the assembly-line process can label associated materials (e.g., parasites, tissues) with the same number and, thus, reduce confusion and errors that might occur using individual collector or preparator numbers. Such a system also has advantages for storage and retrieval of these materials once they are in the collection. Because ancillary materials are stored numerically by NK number, the problem of wasted storage space common to taxonomically arranged collections is eliminated. Such a system is particularly valuable when ancillary materials and voucher specimens are deposited at different institutions. It also saves time once the collection is returned to the repository, because ancillary materials can be installed immediately without the need for "recataloguing." However, the traditional approach of using the field-catalogue number for the primary voucher specimen and all ancillary materials works as well.

Processing Procedures

Once the NK book is completed and processing is to begin, specimens are euthanized and labeled individually with a skull tag (Fig. 50) that contains the NK number. The weight of the specimen is also recorded on the skull tag, which is attached to the front leg with a loose knot so that it can be removed easily prior to voucher preparation. One person is assigned to manage the NK book, sacrifice the animals, and weigh, sex, and tag the specimens. In assembling animals for preparation, care should be taken to avoid ectoparasite exchange between animals in traps or bags (see "Procedures for Collecting Ectoparasites and Ectosymbionts," Appendix 6).

We recommend the use of appropriate euthanizing chemicals with properties similar to chloroform for killing most mammals, because these materials also will kill any ectoparasites that may spread disease to humans. Investigators should not breathe chemical fumes or use any chemicals near an open flame. Best

MUSEUM OF SOUTHWESTERN BIOLOGY
BIOLOGICAL MATERIALS DATA SHEET

MSB 75,000 NK 33503

Collector/Project: T. L. Yates / Japan Date 7/4/94

Preparator: T L Yates # TLY 4100 Date 7/4/94

DIV: (Mammals) Fish Birds Herp Plant Invert OT Family: Muridae

Genus, species Mogera tokudae

Country/State: Japan County: Honshu

Specific locality: Niigata, 37° 34' 48" N, 139° 01' 12" E

Nature of voucher: (skin) skull (skeleton) fluid OT Sex: (M) F ?

Material: heart ✓ kidney ✓ liver ✓ spleen ✓ muscle ✓ cell susp. ✓

karyotype ✓ DNA ___ OT ___ (indicate multiple samples by number; e.g., "x2")

Parasites: endo ✓ ecto ✓ feces ✓ other ___

Remarks: Karyotype produced using velban
3 cysts removed from liver

2N 36 FN ___ Published? N (Y) where? Tsuchiya 1988
J. Mamm. Soc. Japan

USE

Loan 1# Moore Date 8/15/94 Amt Remain: 50%

Name/Instit: Emporia State University

Type used H+K Tissue ground? (N) Y

Loan 2# ___ Date ___ Amt Remain: ___

Name/Instit: ___

Type used ___ Tissue ground? N Y

Figure 49. Page from a New Mexico Kryovoucher system field catalogue, used to facilitate cross-referencing of ancillary materials with the primary voucher specimen.

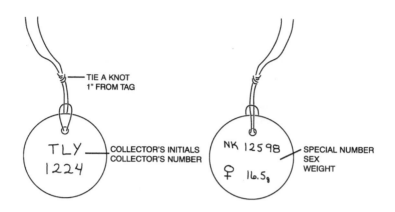

Figure 50. Front and rear view of a mammal skull tag used at the University of New Mexico.

results are obtained when chemicals are used at moderate temperatures in a well-ventilated area. Small mammals can be sacrificed by placing a piece of cotton moistened with a small amount of chemical in a small container with a tightly fitting lid and placing each specimen inside the container after the drug has had a few moments to vaporize. For larger mammals, enclosing the trap containing the specimen within a large plastic bag that includes chemical-soaked cotton works well under most conditions.

After the first member of the team has weighed, sexed, and tagged a specimen, it is passed to the next station in the line. Normally, the second person to receive a specimen is the parasitologist who removes the ectoparasites (see "Procedures for Collecting Ectoparasites and Ectosymbionts," Appendix 6). The specimen then goes to a cytogeneticist for karyotyping and to the next person for tissue removal (see Appendix 4). This individual passes the specimen to another person for removal of endoparasites (see "Procedures for Collecting Endoparasites and Endosymbionts," Appendix 6). At each station, items that are removed from the specimen are labeled with the same NK number.

What now remains of the specimen is passed to the person who will prepare the voucher. At this point the original skull tag with the NK number, sex, and weight is still attached to the specimen. The preparator of this specimen now enters specific information into his or her individual field catalogue (Fig. 51) and prepares a skin tag (Fig. 52). The sex of the specimen is recorded on the skin tag and in the individual catalogue, as are all obvious reproductive characteristics (e.g., placental scars, number and size of embryos, lactation status, size of testes; see Appendix 5). Individual embryos should be pre-

served in alcohol or frozen. In the latter case each embryo receives a separate NK number with the number of the mother noted for each in the NK book. Next the name and individual field number of the preparator (and the collector if different from those present in the field party) is recorded in the preparator's catalogue and on the tag. The exact locality and date of collection should be recorded in the same order each time, with the day listed first, the month (spelled out completely) listed next, and the year (four-digit number) listed last. The exact locality should list the country first followed by a colon; then the state, department, or province followed by a semicolon; any other normal geographic subdivisions, such as county, followed by a comma; and the exact locality. We strongly recommend the inclusion of longitude, latitude, and elevation in meters. The use of a global positioning system is recommended for this purpose (see "Global Positioning Systems," Chapter 11). Notes on habitat use should also be included (see "Habitat" under "Data Standards," and "Microhabitat Description," Chapter 4).

Standard measurements (in millimeters) for each specimen should be recorded on the tag and in the preparator's catalogue, in the following order: total length (from the end of the snout to the end of the last caudal vertebra), tail length (from the base of the tail to the end of the last caudal vertebra), hindfoot length (from the heel to the end of the longest claw), ear length (from the notch to the tip), and weight (in grams). On the back side of the tag and in the field catalogue the NK number is listed along with notations of other items collected with the specimen. Some of the more common notes are "+ tissue," which indicates that heart, liver, kidney, muscle,

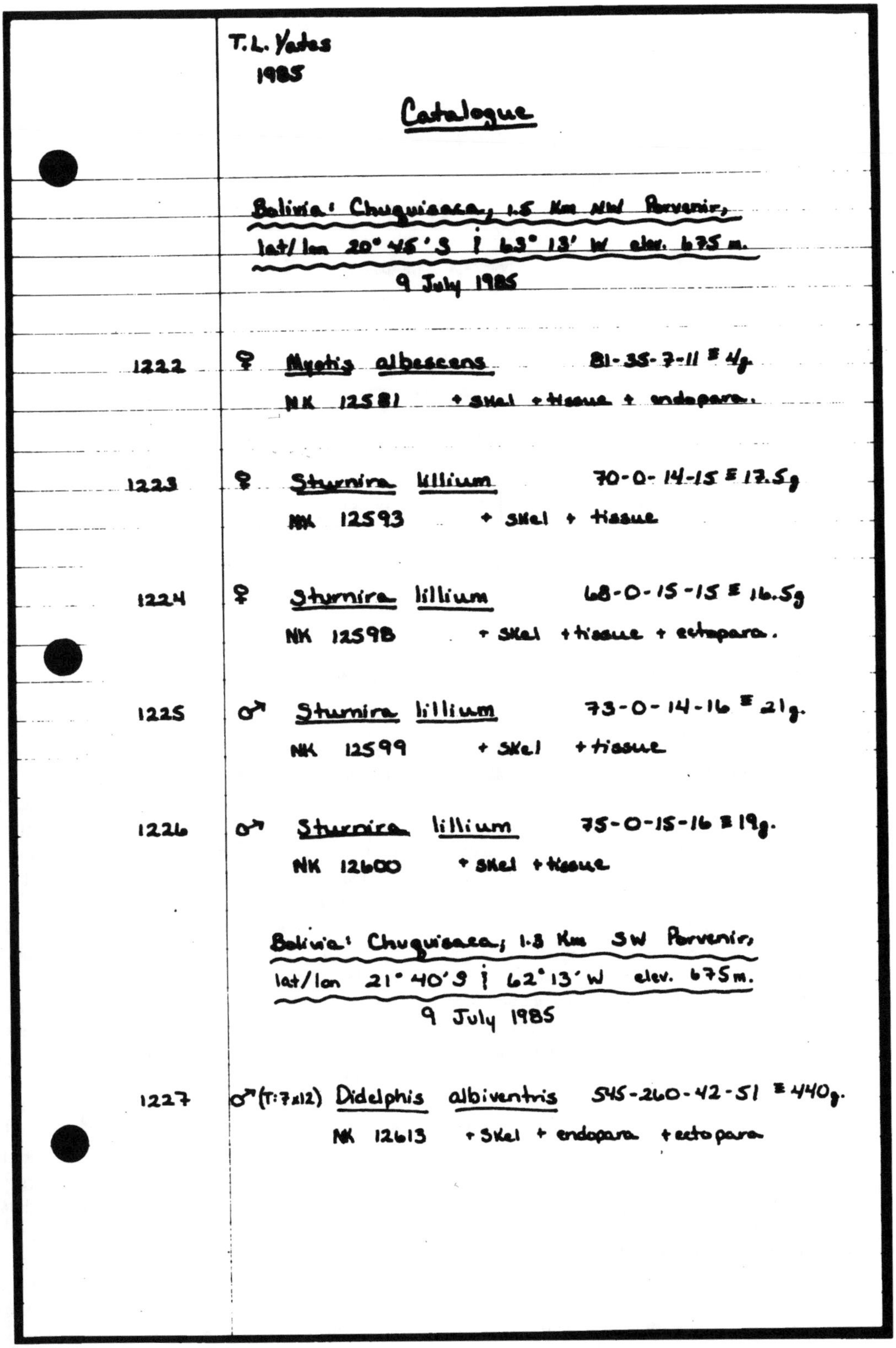

T.L. Yates
1985

<u>Catalogue</u>

<u>Bolivia: Chuquisaca, 1.5 Km NW Porvenir,
lat/lon 20° 45' S | 63° 13' W elev. 675 m.</u>
9 July 1985

1222   ♀ <u>Myotis albescens</u>   81-35-7-11 ☰ 4g
NK 12581   + skel + tissue + endopara.

1223   ♀ <u>Sturnira lillium</u>   70-0-14-15 ☰ 17.5g
NK 12593   + skel + tissue

1224   ♀ <u>Sturnira lillium</u>   68-0-15-15 ☰ 16.5g
NK 12598   + skel + tissue + ectopara.

1225   ♂ <u>Sturnira lillium</u>   73-0-14-16 ☰ 21g.
NK 12599   + skel + tissue

1226   ♂ <u>Sturnira lillium</u>   75-0-15-16 ☰ 19g.
NK 12600   + skel + tissue

<u>Bolivia: Chuquisaca; 1.8 Km SW Porvenir,
lat/lon 21° 40' S | 62° 13' W elev. 675m.</u>
9 July 1985

1227   ♂ (T:7×12) <u>Didelphis albiventris</u>   545-260-42-51 ☰ 440g.
NK 12613   + skel + endopara + ectopara

Figure 51. Sample page from a collector's field catalogue.

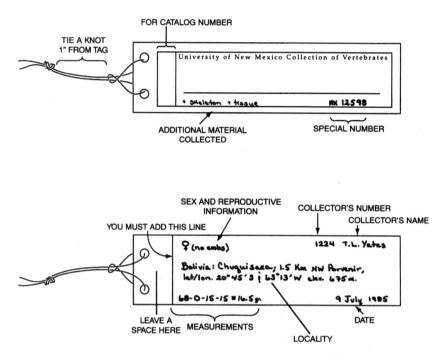

Figure 52. Front and rear view of a mammal skin tag used at the University of New Mexico.

spleen, or blood tissues were saved; "+ chromosomes" (= cell suspension of chromosomes and/ or prepared slides); "+ endoparasites"; "+ ectoparasites"; "+ embryo"; and "+ skeleton" (= body skeleton in addition to skull). More detailed information can be obtained by referring to the NK book (Fig. 49), which lists specifics regarding what was preserved from each animal; what type of voucher was maintained; any unusual procedures that were carried out; the repository and collection catalogue number; the collector's number; the family, genus, and species names; and the specific locality. Information on use of the material is recorded when it is utilized.

Additional information associated with a particular specimen should be included in the field catalogue. It is tied to the voucher through its individual specimen number, regardless of the system used.

Skin and Skeleton Preparation

SKINNING THE SPECIMEN

The skinning procedure for each specimen is similar to that used in the preparation of the traditional skin and skull (see Hall 1962; Davis 1974; Nagorsen and

Peterson 1980; DeBlase and Martin 1981), except that only the right fore- and hindlimbs remain in the skin. The limbs on the left are skinned completely and remain with the skeleton. In most cases, the abdomen will have been opened when tissues and internal parasites were removed. The skin should be separated from the carcass by gently pushing the flesh toward the midline of the body. Fine hardwood sawdust or cornmeal should be applied liberally during the entire process to facilitate absorption of lipids and other body fluids and drying of the skin. For male specimens, the phallus should be everted (extracted from the protective covering of skin) and should remain with the skin.

In animals with a tail, the vertebrae should be removed by applying pressure with the thumb and forefinger at the base of the skin of the tail and then gently pulling the vertebrae anteriorly. For larger specimens, it may be necessary to hold the skin at the base of the tail with two sticks in order to remove the vertebrae because of the large amount of connective tissue holding them. This technique works best with two people. One person holds one stick tightly on each side of the vertebrae at the base of the tail, providing the same type of pressure applied by thumb and forefinger on smaller specimens.

The second individual pulls the tail vertebrae away from the person holding the sticks.

After the tail vertebrae are removed, the skin is separated from the body of the animal until the cranium is reached. Special care must be taken at this point not to damage the skin when it is removed from the ears, eyes, and mouth. Fine scissors or a scalpel are often helpful for separating the skin.

STUFFING THE SKIN

Once the skin has been removed, the mouth on most species is sewn shut with a single triangular stitch. This step may be omitted in species with very small mouths such as small mice.

The skin of small animals is stuffed with cotton. Skins of large mammals are stuffed with a combination of cardboard and cotton. The cotton can be rolled or folded to form the body, but folding usually produces a more uniform and realistic body shape and promotes even drying of the skin. In both cases, a rectangular piece of nonabsorbent cotton roughly the length of the head and body portion of the skin is selected. Beginning at one corner, the piece of cotton is folded or rolled with medium pressure until the other side is reached. The result should be a firm but not tight body that is pointed at the front and wide at the back. It may be necessary to remove some cotton from the broad end to obtain the appropriate length for the skin.

The pointed end of the cotton body is grasped tightly with forceps and placed on the nose of the skin, which is inverted back over the cotton body. A gentle tug on the skin below the neck will help secure the point in the nose region of the skin. When the skin is in place over the cotton, the nose should be pointed and the posterior end should be square. Next, straight pieces of wire are placed securely in the legs with the distal tips extending into the fleshy portion of the hands and feet and the proximal ends extending into the body cavity. In larger specimens, portions of these wires should be wrapped with cotton to help fill the empty skin of the limbs. Wire size should increase in diameter as the size of the specimen increases but must be small enough to extend into the hands and feet. For specimens with a tail, a wire covered with cotton is placed in the tail. This wire can be prepared easily by wetting a straight piece of wire and covering it with a layer of cotton to a diameter slightly smaller than that of the animal's tail vertebrae. The cotton will adhere best if it is very thin when applied and the wire is rotated between the thumb and forefinger. The cotton should

also taper from narrowest at the distal end of the tail to widest at the proximal end, and the wire should be long enough to extend anteriorly well into the abdominal cavity. After the wires have been installed, the edges of the abdominal cavity are sewn together loosely with cotton thread. For bats, the limb bones from which the flesh has been removed are left attached and used in place of wire.

The skin tag should now be securely attached to the right hindfoot (the only one remaining). For bats, we recommend that the skin tag be attached through the patagium above the knee to prevent interference with the calcar. The specimen is now pinned to a flat surface for drying. Initially, two pins should be crossed at the base of the tail and two more at the distal end. The right forefoot, with bones and support wire intact, should then be pinned under the chin and the right hindfoot pinned parallel to the tail. An additional pin should be used to bring the ankle close to the tail. Keeping the appendages close to the body helps prevent breakage once they are dry. The entire skin should be perfectly straight (Fig. 53). Foot pads in large skins should be injected with 10% buffered formalin using a syringe.

Additional pins are used to secure the patagia of bats and to ensure that the bones of the forearm and fingers are available for study (Fig. 54). Normally, the tail (if present) is secured first, followed by the forearm. The latter should be pinned parallel to the body and far enough away to allow measurement at a later date. Individual fingers should be separated with pins to facilitate measurement, and the thumb should be pinned against the wing to prevent it from snagging other objects once the specimen is dry. The wing should have a slightly rounded appearance (Fig. 54). The hindlimbs are pinned at varying distances from the body, depending on the species, to help extend the uropatagium. Placing a pin behind the calcar also helps extend the uropatagium in a symmetrical fashion.

PREPARING THE SKELETON

Once the skin has been removed from the carcass, the eyes and any organ tissues remaining in the body cavity should be removed. Removal of large muscle masses speeds drying and facilitates cleaning of the skeleton at the repository.

When climatic conditions make drying difficult, as is usually the case in the wet tropics, insects can pose a serious problem. A portable oven is fine for drying, if one is available. If not, skulls and skeletons can be soaked for several days in either ethyl or

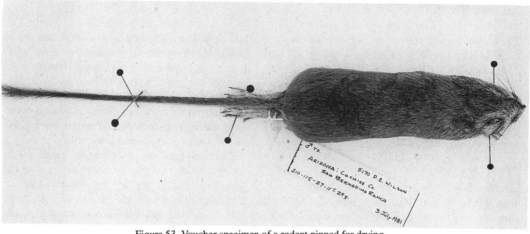

Figure 53. Voucher specimen of a rodent pinned for drying.

denatured alcohol and then air-dried. Skeletons should never be sprayed with insecticide of any kind, as it will hamper future cleaning by dermestid beetle colonies. Dermestids may also fail to clean bones properly if they are dried over an open fire.

Fluid-Preserved Specimens

A diversity of specimen preparations is important for all inventory and long-term monitoring projects because different preparations provide material for different types of investigation. Fluid preservation (preservation in formalin and storage in alcohol) offers the option of retaining the entire specimen intact

and may provide material for anatomical studies of soft body parts, analyses of stomach contents, or a host of other research projects.

As with other preserved materials, associated locality, reproductive, ecological, mensural, and other data should be carefully documented. Labels (made only of 100% rag paper) and ink (permanent black, such as Pelican fine drawing ink) should be tested for durability in formalin and ethanol prior to use (Jones and Owen 1987). Tags are prepared in the same manner as discussed previously and permanently tied to the hindfoot of each specimen using 100% cotton string. Tissues, parasites, chromosomes, and other materials may also be recovered

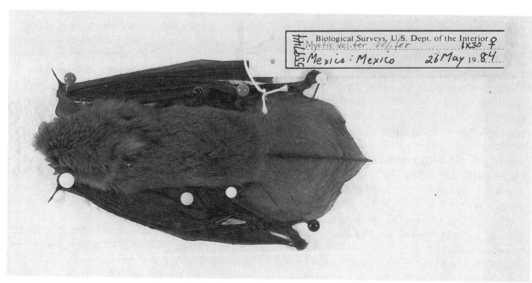

Figure 54. Voucher specimen of a bat pinned for drying.

from a specimen before fixation, although we recommend preservation of a few entire specimens when a large series from a single locality is available.

An opening several centimeters long should be made in the abdominal wall of small and medium-size mammal specimens to ensure that the fixative penetrates quickly to the internal organs. Very small specimens can be injected using a syringe and hypodermic needle. Fleshy footpads, large skeletal muscles, and brain tissue should be injected with buffered formalin using a syringe.

If powdered paraformaldehyde is used, it should be prepared as liquid formalin according to package instructions. Specimens should be fixed in 10% formalin (1 part formalin to 9 parts water) buffered to a pH of 7 with a mixture of 4 parts monobasic sodium phosphate to 6.5 parts dibasic sodium phosphate or hexamine (Jones and Owen 1987). A piece of cotton should be placed in the mouth prior to fixation to maintain the mouth in an open position. If the mouth is open, dental and palatal characters can still be examined after fixation. Fixation time varies among species but generally is related to the size of the animal. Adding a small piece of soap or a few drops of liquid soap to the formalin solution helps a great deal in increasing the rate of fixation, because it breaks down the surface tension of oily hairs (i.e., it breaks the bubble that often surrounds a specimen). This is an especially good thing to do in hot climates, where the specimens can decompose significantly in 24 hours. Fixation can take from 2 to 14 days, but most specimens can safely remain in formalin for up to a month if necessary. Specimens left in formalin for more than a month often become excessively brittle, which renders them worthless for histological work and difficult to manipulate during

routine examination. After fixation, as much formalin as possible should be removed from the specimen by rinsing with water for at least 24 hours. Rinsing is best accomplished in the museum but can be conducted in the field if necessary. Specimens are then transferred to alcohol for permanent storage.

In species for which cleaned cranial material may be essential to identification, the skull may be removed before the specimen is fixed. The skin is carefully peeled away from the mouth starting with the chin and upper lips and skinning backwards over the rest of the skull. Care should be taken not to tear the skin around the eyes and ears. Once the skin has been removed from the cranium, the skull can be removed by separating it from the cervical vertebra. When skull removal is completed, the space left in the skin should be filled with cotton and the type of preparation (skull plus alcoholic body) should be noted in the special field book and in the collector's personal field catalogue.

Skeletons as Vouchers

The procedures for such preparations are similar to those for preparation of skin and skeleton vouchers. Measurements are taken, tissues are taken and prepared, the skull tag with appropriate data is attached, ancillary materials such as parasites are removed, and a "skin" tag is prepared and attached to the voucher specimen. The primary difference is that the skin is removed from the carcass as quickly as possible and discarded. Organs are removed from the body cavity, and the skeleton is loosely wrapped with thread to help prevent the loss of small bones during cleaning.

Tissues, Cell Suspensions, and Chromosomes

Terry L. Yates

Introduction

Fueled by the knowledge that the diversity of life on earth is being lost at an ever-increasing pace, global efforts to inventory what remains have increased rapidly. In order to maximize the amount of data collected from each specimen, collection of tissues for genetic and molecular analyses is becoming routine practice in mammalian fieldwork. Summaries of the procedures for collecting and preserving such materials and for maintaining records about them have not, however, been widely available. In this section, I describe such procedures for field use with a wide range of research projects that require collection of mammals or their tissues, including chromosomes. I attempt to balance practicality (in terms of time, effort, and resource use) with the need for maximum data preservation. Voucher specimens should always include the widest possible range of materials.

Chromosomes and cell suspensions

Routine preservation of chromosomal preparations is highly desirable, especially in regions where the mammalian fauna is poorly known. The method described in this section is a relatively easy and reliable means of obtaining standard karyotypes from large samples of individuals in the field. In addition, it produces a cell suspension consisting of late-prophase and early- to middle-metaphase cells that can be frozen in liquid nitrogen and used for more complex cytogenetic studies such as chromosome banding analyses and in situ hybridization. Standard chromosomal preparations can also be processed from the suspensions. The procedure can be carried out by one individual or as part of assembly-line specimen preparation (see "Processing Procedures," Appendix 3). A mammal is karyotyped immediately after it is weighed, sexed, and tagged and its ectoparasites have been removed. It should be noted directly on the specimen tag that the specimen has been karyotyped.

The initial step in karyotyping is the removal of one or both femurs, depending on the size of the animal, or one or both humeri, in the case of bats or moles. The ends of the bone(s) are carefully removed only to the extent necessary for the insertion of a small needle into the medullary cavity. The needle is attached to a syringe containing 0.075 molar potassium chloride that has been prewarmed to 37° C. The potassium chloride solution is used to flush marrow from the bone into a 15-ml centrifuge tube. Tubes should be clean, dry, and labeled with the field number of the animal being processed.

The researcher shakes the tube or gently aspirates its contents with a pipet so that the marrow cells are suspended uniformly in the solution. The mixture is then incubated at 37° C for approximately 30 minutes. If no incubator is available, the researcher can cap the centrifuge tubes and place them in the waistband of his or her clothing for incubation against the body. Approximately 1 minute prior to the completion of incubation, 1 ml of Cornoy's fixative (3:1 absolute methanol to glacial acetic acid) is added to the suspension. This and all subsequent operations should be carried out gently, because the mitotically active cells are extremely fragile.

After incubation, the cell suspension is centrifuged for approximately 1 minute at 1,500 rpm. Numerous suitable electric centrifuges that can be powered by a small generator or via a power converter attached to a 12-volt battery are available. Hand-powered models that hold two or four tubes are also available. Field crews working in remote areas should carry a hand centrifuge as an emergency backup, even if power is available.

After centrifugation, the liquid is decanted from the tube, and the remaining button of cells is gently resuspended in Cornoy's fixative. The amount of fixative added depends on the volume of cells, but 2–3 ml is usually sufficient. The suspension is centrifuged and the liquid decanted, and then the process of resuspension and centrifugation is repeated two more times. Lipids will float on the surface of the suspension and should be removed with a pipet.

After the final centrifugation, the investigator resuspends the cells in 1.5 ml of fixative and prepares one or more test slides. Preparing such slides in the field reduces the likelihood that the karyotype of a specimen will be lost, provides a rapid means of obtaining a standard karyotype (for reference in the field, as well as in the laboratory), and provides a

means for quickly choosing the best cell suspension for chromosome banding or in situ hybridization.

Test slides are prepared by labeling a precleaned frosted-end microscope slide in pencil with the appropriate field number. All slides should be cleaned thoroughly with alcohol, even if they are new and of the "precleaned" variety. The researcher places labeled slides on a flat surface in an area free of wind, adds six to eight drops of the reserved cell suspension to each slide, and ignites the mixture with a match. Once the flame is extinguished, she or he shakes the remaining moisture from each slide and places the slide in a clean slide box. Nonwooden boxes should be used in the field because changes in humidity and temperature can cause wood to swell and contract.

Occasionally it is advisable to stain a test slide and examine the preparation to assess the quality of the results. The preferred stain for such preparations is a 2% solution of Giemsa blood stain in phosphate buffer. The buffer is made by dissolving 0.469 g of sodium dihydrogen phosphate (NaH_2PO_4) and 0.937 g of disodium hydrogen phosphate (Na_2HPO_4) in 1,000 ml of distilled water. If buffer is not available, distilled water can be used alone, but the staining quality is not as good. Staining normally takes about 10 minutes if the solution is buffered and 15 minutes if it is not. Before the slides are immersed in the stain, the oily film should be skimmed from its surface with a clean paper towel. After staining is completed, excess stain should be rinsed from the slide with distilled water. The preparation is allowed to dry before examination.

One common problem affecting the quality of preparations, especially in humid climates, is contamination of methanol with water. This condition prevents proper fixation and inhibits spreading of the chromosomes during slide preparation. Making fixative with new methanol often will correct the problem. In fact, fixative should be made fresh at the beginning of each processing cycle. It should be maintained in a covered container to prevent evaporation of the methanol or contamination by condensation. Parafilm stretched over the top of a graduated cylinder works well for this purpose. One can also examine stained preparations to obtain information on chromosome variation. If unexpected karyotypes are recorded, the investigator may wish to focus collecting efforts on certain organisms or geographic areas.

Once test slides have been made, the researcher places the remaining cell suspensions in individual plastic cryotubes suitable for freezing at the ultralow temperatures associated with liquid nitrogen. An air space should be left at the top of each tube to accommodate expansion of the liquid during freezing; otherwise, the tubes may rupture. The field number should be written on the side of the tube, as well as on the white label area of the tube and on the colored top of each tube. A permanent marker, such as a Sharpie fine-point pen (which can withstand freezing at ultralow temperatures) should be used to guard against accidental erasure of the number. A unique color of top (i.e., different from colors used with other frozen materials) should be chosen for closing tubes containing cell suspensions to facilitate rapid sorting of those tubes after return from the field. The methanol used in the fixative thaws at temperatures only slightly above that of liquid nitrogen, so cell suspensions must be sorted and placed back in liquid nitrogen quickly. Green tops are used for tubes containing cell suspensions in my collection. Protective eyeware should be worn when thawing cell suspensions for protection against release of pressurized fixative.

Frozen tissues

Tissue samples should be salvaged from mammalian specimens whenever possible, in order to maximize the information available from each specimen. Tissues should be removed and frozen rapidly to minimize damage due to decomposition, although tissues from animals that have been dead up to 48 hours are still of value (Moore and Yates 1983). For short collecting trips, tissues can be frozen and maintained on dry ice and then transferred to liquid nitrogen or ultracold freezers in the laboratory. On longer trips investigators most commonly freeze tissues in liquid nitrogen contained in nonpressurized Dewar flasks, generally referred to as "nitrogen tanks." These flasks come in a variety of sizes, and many are small enough to be transported easily in the field. Some flasks have static holding times of more than a year. It may be difficult to find liquid nitrogen in some countries, so investigators should try to locate a source well in advance of their trips.

Types of tissues to be collected depend on the purpose of the collection and the species involved. Many scientists working on mammals in the field preserve only heart, liver, and kidney samples if other tissues are not required for a specific purpose.

Blood, skeletal muscle, spleen, and brain are also routinely saved. For extremely rare or endangered species, samples from as many different types of tissue as possible should be preserved, because collection of such species may be difficult or impossible in the future. Investigators should always collect multiple sets of tissue from rare or endangered forms and, when time and space permit, from other mammals as well, especially when working in remote or poorly known areas.

Tissues should be collected under conditions that are as sanitary as possible, with special care being given to the appropriate labeling of vials. Samples should be placed in plastic cryotubes or aluminum foil. Cryotubes should be labeled as described earlier for cell suspensions. Tubes should not be overfilled so the tissues have room to expand during freezing, but excess space should also be avoided as this promotes tissue dehydration during storage. Most commonly, liver tissue is stored in one container and heart and kidney tissues are stored in another. Muscle is also normally placed in its own container, as are other tissue types. The gallbladder should be removed from the liver samples before freezing because bile salts adversely affect tissue stability (Dessauer et al. 1990). If bile is to be saved, it should be stored separately. Tissue type should be clearly noted on each tube. Tops of tubes containing different tissues should be colorcoded to facilitate sorting. Red tops are frequently used for heart and kidney samples, yellow for liver, and white for muscle. If plastic cryotubes are not available, samples can be tightly wrapped in aluminum foil (Dessauer et al. 1990).

Most tissue can be removed from a specimen with a good pair of forceps or a pair of scissors. Equipment should be cleaned thoroughly with alcohol between specimens to prevent transfer of tissue between individuals. Even minute amounts of tissue can contaminate samples when DNA is amplified using the polymerase chain reaction. Surgical gloves should also be worn any time tissues are handled to prevent contamination from the investigator and to protect the investigator from disease.

The substantial increase in the amount and types of materials taken from each individual mammal sometimes makes it more difficult to keep track of these items, especially when several investigators, working as a group, maintain personal field catalogues. Although the traditional field catalogue number can accommodate ancillary materials as well, some investigators use a unified field cataloguing system that is separate from the cataloguing systems maintained by individual collectors. One such system is the New Mexico Kryovoucher (NK) system used by the Museum of Southwestern Biology at the University of New Mexico. The NK system is a sequential numbering system that allows materials—such as frozen tissue samples, chromosomes, parasites, and purified DNA and RNA—that are associated with primary voucher specimens to be tracked. It is discussed in detail in Appendix 3 (see "Record Keeping").

Sex, Age, and Reproductive Condition of Mammals

Thomas H. Kunz, Christen Wemmer, and Virginia Hayssen

Introduction

Knowledge of the sex, age, and reproductive condition of individual mammals is vital for evaluating the viability of a species, because these variables reflect the structure and dynamics of populations. Necropsy is the most accurate method for assessing these variables, but it is not always compatible with efforts to determine the dynamics of natural populations. Noninvasive field methods that rely on external genitalia, secondary sexual characteristics, body size, and behavior can be used, but the differentiating criteria for each of these characters vary across mammal species. Reliable assessment is influenced by the detection limits of the methods, the biology of the animals, and the expertise of the individual making the determination. In this appendix, we provide information on common methods of determining the sex, age, and reproductive condition of a mammal in the field. We include techniques requiring necropsy, be-

cause some survey and census protocols entail collection of specimens.

Sex determination

Potentially, any sexually dimorphic character can be used to distinguish males from females, including differences in genitalia, body size, pelage, ornamentation, scent glands, and behavior. In practice, many sex differences either are statistical in nature (e.g., males on average are 20% larger than females) or vary seasonally or with the age and condition of the animal. Thus, accurate sexing of an animal often requires some knowledge of the natural history and morphology of the species (or related taxa) and may require capture of the individual.

Genitalia

Female mammals are distinguished by the presence of a single vulva or vaginal opening, whereas the presence of testes and a glans penis distinguishes males (Rowlands 1966; Brown and Williams 1972; Racey 1988; Hayssen et al. 1993). Because the visibility of sexually dimorphic genitalia depends on age, reproductive condition, and physiological state, generally the sex of an animal cannot be reliably determined from a distance, based solely on a superficial examination of the genitalia.

When the penis, testes, or epididymides are conspicuous and readily identifiable, they provide a reliable means for sex determination (e.g., the pendulous, blue scrota of the patas monkey, *Erythrocebus*). In fact, males with permanently scrotal testes are often easy to identify. Testes, however, are not externally visible in many mammals. Nonscrotal testes are common, for example, in monotremes, edentates, insectivores, bats, lorisids, marine mammals (pinnipeds, cetaceans, sirenians), elephants, tapirs, rhinoceroses, aardvarks, hyraxes, pangolins, beavers, hystricomorph rodents, and elephant shrews. In addition, the testes of other mammals (e.g., soricids, pteropodids, many rodents, lagomorphs) may be scrotal for only part of the year, during the breeding season, and are retracted into the body cavity during the nonbreeding season. Scrota, when present, are usually located posterior or lateral to the penis, but they are prepenial in marsupials, pandas (*Ailuropoda*), and lagomorphs and postanal in vespertilionids. Some female primates (e.g., lorisids, aye-aye [*Daubentonia*]) and female spotted hyenas (*Crocuta*) have scrotal-like appendages lateral to the clitoris.

The glans penis in males of many species protrudes from the prepuce or cloaca only during erections and therefore may not be visible externally. In handheld animals, the penis can be physically expressed by applying pressure to the sides of the urogenital sinus, or it can be detected by palpation. Males of other species (e.g., bats, primates, canids, and ungulates) have a pendulous glans. Females of some species (e.g., hedgehogs [Erinacidae], moles [*Talpa*], spider monkeys [*Ateles*], spotted hyena, viverrids) have a large, pendulous, and occasionally erectile clitoris, which mimics and may even be larger than a penis. The clitoris of many rodents superficially resembles the male urinary papilla, but in these and most other mammals, the anal-genital distance is diagnostic, typically being shorter in females than in males (Fig. 55). In addition, the perineal area between the anus and the clitoris in female rodents is sparsely haired and unpigmented, whereas in males the area between the anus and genital papilla is usually hairy and darkly pigmented. Species with distinctive genitalia as adults may be difficult to sex earlier in life (e.g., Callitrichidae; Stein 1978). In rodents, however, the anal-genital distance can be used to distinguish the sexes of juveniles before the testes descend into the scrotum (Watts and Aslin 1981).

Echidnas and some marsupials have a pouch in which to carry their young. Pouches, when present, are usually characteristic of females but may be rudimentary or completely functional in males of some species. Water opossums (*Chironectes*) use the pouch to protect the scrotum during submersion. Pouch size and shape vary with season and age (Woolley 1974).

It is extremely difficult to determine the sex of many insectivores. Shrews and some moles can be sexed only by counting the number of openings in the perineal region: three in females (urinary papilla, vagina, and anus) and two in males (urinary papilla and anus). The common mole (*Talpa europaea*), however, has no external vaginal opening during anestrus. At the onset of the breeding season, the vagina, which lies between the base of the peniform clitoris and the anal papillae, opens spontaneously. It closes again following parturition (Harrison Matthews 1935).

For marsupials, but not eutherians, the presence of mammae (teats, nipples) usually distinguishes females from males. Mammary glands are located in axillary, pectoral, thoracic, or inguinal regions; the number and position of mammae vary across species. Cetaceans have two mammae, each located in a small pouch on either side of the genital vent. Thus, the sex of cetaceans is externally determined by the number of ventral slits on the inguinal surface of the animal: three in females, one in males.

Mammary glands are enumerated as the number of pairs located in each anatomical region, moving from anterior to posterior. P2, A2, I2, for example, indicates 2 pairs of pectoral, 2 pairs of abdominal, and 2 pairs of inguinal mammae for a total of 12 (Smithers 1983). When mammae are not paired, the total numbers on each side and in the center are noted. The nipples of parous females are usually more prominent than those of juveniles, nonparous females, or males. As some male mammals may produce small quantities of milk (e.g., *Dyacopterus;* Francis et al. 1994), notation of the presence of discharge from mammary glands should be supple-

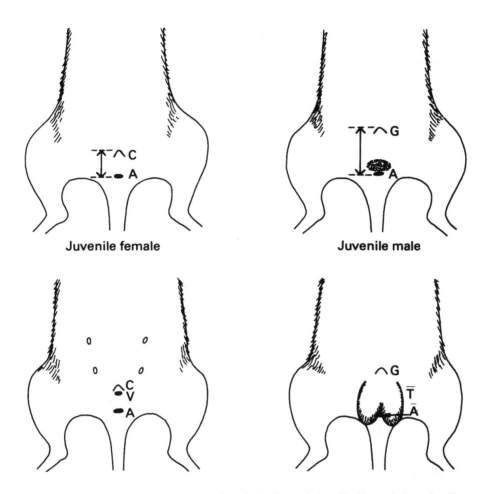

Figure 55. External sex differences in rodents. A = anus; C = clitoris; G = genital papilla; V = vaginal opening; T = testes. The A-to-C distance in female rodents is less than the A-to-G distance in male rodents and can be used to sex both adults and juveniles. Reprinted with permission of HarperCollins from Watts and Aslin (1981).

mented by inspection of the genitalia to confirm that the animal is, in fact, a female. In addition, some female rhinolophid bats have inguinal papillae that can be mistaken for nipples or even penes.

Body Size, Pelage, and Ornamentation

The standard measurements for size are head-body or total length, tail length, hindfoot length, ear length, and mass. Although males of many mammalian species are larger than females (Clutton-Brock et al. 1977), in several groups (including honey possums, bats, many phocid seals, and baleen whales) females can be larger than males (Ralls 1976; Myers 1978a; Williams and Findley 1979; Renfree et al.

1984). Even when dimorphism does occur, the ranges of male and female body mass or measurements usually overlap greatly. Thus, body size, except at the extremes, is not a reliable criterion for sex determination.

Pelage dimorphisms can be used to distinguish adult males from adult females in some marsupials, flying lemurs (*Cynocephalus*), bats (Pteropodidae, Emballonuridae, Phyllostomidae), primates (*Alouatta*, perhaps *Brachyteles, Erythrocebus, Gorilla, Hylobates, Homo, Pithecia, Rhinopithecus, Theropithecus*), pinnipeds (*Callorhinus, Neophoca, Otaria, Phocarctos*), and artiodactyls (*Antilocapra, Antilope, Tragelaphus*) (Wemmer 1987; Racey 1988; Skinner and Smithers 1990; Nowak 1991). The fur color may be completely different in males and females, as in cuscuses (*Spilocuscus*) or red kangaroos (*Macropus rufus*), or different only in particular body regions, as in the dark crown of female oribi (*Ourebia*) or the visually striking epaulets of male epomophorine and some phyllostomid bats (Bradbury 1977; Racey 1988). Integumentary pigmentation may also be sexually dichromic; male mandrills (*Mandrillus*), for example, have brilliant scarlet and purple facial stripes and male beaked whales (*Mesoplodon*) have white patches on the lower jaw. Finally, the texture or length of fur may differ between the sexes, as in the manes of male baboons (*Papio*), lions (*Panthera*), otariids, and gorals (*Nemorhaedus*) or the hair tufts on the chins of some goats (*Capra*) or on the throats of male nilgai (*Boselaphus*) and oryx (*Oryx*).

Males often have sex-specific ornaments, such as antlers (cervids except for *Rangifer*), tusks (narwhal, *Monodon;* elephant, *Elaphas;* some deer, *Elaphodus, Hydropotes, Moschus, Muntiacus*), headshields (African buffalo, *Syncerus*), or the postanal keel of some odontocetes (*Phocoenoides, Stenella*). Alternatively, morphological features present in both sexes may be elaborate or enlarged in males. Examples include teeth (pongids, tragulids, beaked whales), thickened necks (otariids, suids, *Bison*), noses (proboscis monkeys, *Nasalis;* hooded seals, *Cystophora;* gray seals, *Halichoerus;* elephant seals, *Mirounga*), fins (*Stenella, Orcinus*), or warts (warthogs, *Phacochoerus*). Horns are sex-specific in some bovids (*Aepyceros, Ammoodorcas, Antilope, Boselaphus, Dorcatragus, Kobus, Litocranius, Madoqua, Neotragus, Ourebia, Pantholops, Pelea, Procapra, Raphicerus, Redunca, Saiga, Tetracerus,* some *Tragelaphus*). When both male and female bovids have horns, they are often larger in males (Geist 1966; Chapman 1975; Janis 1982; Packer 1983; Georgiadis 1985; Estes 1991a).

Scent Glands

Many mammals have specialized, sex-specific skin glands; examples of such structures are the civet gland of female toothed palm-civets (*Arctogalidia*), the scent pouch of female broad-striped mongooses (*Galidictis*), and the musth gland of male elephants (Poole 1989; Nowak 1991). Alternatively, glands may differ in size and appearance between the sexes; for example, the sternal glands of brush-tail possums (*Trichosurus*) and opossums (*Didelphis*) are more prominent in males. Some mammals have conspicuous accessory sex glands. Examples include the preputial glands of male rodents, which lie between the prepuce and the penis or between the pubic skin and the body wall (Brown and Williams 1972; Dixson 1983), and the lateral and ventral glands of some adult male insectivores, which typically enlarge during the breeding season (Hamilton 1929; Eadie 1938; Poduschka and Wemmer 1986).

Behavior

Behaviors that are sexually dimorphic (sexual diethisms) can be used for sex determination. Postures used for urination, such as hindleg lifting in male canids, are often sex-specific in adults. Mating behaviors also have sex-specific components (e.g., lordosis in female rodents). Behavior patterns and frequencies of inter- and intrasexual encounters vary with the mating system. In polygynous species, males use visual or olfactory displays to advertise social status, which they fight to determine (Gorman and Trowbridge 1989; Estes 1991b). Thus, both display and aggressive behaviors tend to be exhibited by males rather than females (Clutton-Brock et al. 1982; Estes 1991b).

Knowledge of the natural history and behavior of the species in a particular area is necessary to make full use of behavioral cues for sex determination, because sexual diethisms can be idiosyncratic. Female pronghorns (*Antilocapra*), for example, hold their heads more horizontally than do males when running (Nowak 1991); female baboons initiate allogrooming more commonly than do males (Hausfater 1975). Reviews of sex- and age-specific differences in behavior are available for African primates and hoofed mammals (Estes 1991b), kangaroos (Jarman 1991), bats (Bradbury 1977; Racey 1988), carnivores (Gittleman 1989; Estes 1991b), and cervids (Wemmer 1987).

Age estimation

Methods of varying accuracy are available for estimating the age of a mammal in the field (Madsen 1967; Morris 1972; Anthony 1988). Assigning animals to broad age classes will suffice for most studies, but more accurate determinations of age are needed to construct life tables for demographic analyses.

Table 18. Commonly Used Age Categories for Mammals

Neonate: A newborn mammal with a detectable umbilical cord. Sometimes used to refer to any young animal early in lactation.

Nestling: A young animal with limited locomotory and sensory development that has not left the nest; usually young of an altricial species prior to weaning.

Suckling: A mammal before weaning.

Pouch young: A young marsupial that has not left the pouch or, if a pouch is not present, has not detached from a teat.

Juvenile: A weaned young mammal that still associates with its mother or siblings and may nurse infrequently; usually smaller than a subadult.

Immature: A young mammal that is neither fully grown nor sexually mature.

Subadult: A young mammal that is not fully grown but that may or may not be sexually mature or have adult pelage.

Adult: A fully grown mammal that is sexually mature.

Old adult: An animal that shows extreme tooth wear and/or poor body condition.

In general, broad age classes are defined relative to developmental or reproductive milestones and include the following: neonate, nestling, suckling, pouch young, juvenile, immature, subadult, adult, and old adult (Table 18). Subdivisions of these categories have been described for particular taxa. Suckling marsupials, for example, are assigned to one of three stages: continuously attached to a teat; intermittently attached to a teat but continuously resident in the pouch; and intermittently attached to a teat and intermittently resident in the pouch or outside the pouch (young-at-heel). These age classes may correlate with pelage or sensory development as well as behavior. Similar stages are defined for eutherian young based on whether they have opened their eyes, whether they are furred, and whether they have eaten solid food.

Accurate assessment of absolute age relies on detection of age-related differences in characteristics such as body size, ossification of long bones, incre-

mental lines on horns or teeth, successional replacement of teeth, tooth wear, or changes in color and condition of pelage. These age-related differences must be verified by measurement on a statistically appropriate number of known-age individuals from a wild population over an average lifetime. Criteria for aging are often highly specific to particular taxa (e.g., forearm length for bats, Anthony 1988; girth and fin measurements for cetaceans, Jonsgard 1969). Establishing the physical changes that mark the absolute age of an individual is challenging, because it requires marking a sample of individuals at birth and reexamining and measuring them (or a subset of them) periodically through their lives. Such data allow the ages of conspecifics from different populations to be estimated. The more distant and environmentally different the study population is from the referent population, the less reliable is the estimate of age.

A combination of body measurements and reproductive criteria offers the best means of determining the age of many mammals in the field (Spinage 1973; Pucek and Lowe 1975; Anthony 1988). Dental characteristics are especially valuable for estimating age (Pucek and Lowe 1975; Malcolm and Brooks 1985; Arnbom et al. 1992; Malcolm 1992).

Body Mass and Size

Body mass and size change rapidly during early stages of development and, therefore, are good indicators of age, especially for medium-size and large mammals. The dimensions of different body parts can be calibrated to age by monitoring the growth of known-age individuals marked as infants. For small mammals, however, large diurnal variations in gut contents often make body mass an unsuitable character for estimating age. In fact, for very young neonates, ingested milk can be a large proportion of body mass. Microtines, in which body mass is a good predictor of age (McCravy and Rose 1992), are an exception. Another serious problem with using body size to estimate the age of small animals in the field is the difficulty of taking accurate linear measurements of live animals without anesthesia. Bats are an exception. Before weaning, the forearms, fingers, hindlimbs, and epiphyseal gap lengths can be measured easily without anesthesia (Kunz 1987; Anthony 1988).

Sometimes only rough estimates of age can be obtained using body mass. In ground squirrels (*Sper-*

mophilus) and marmots (*Marmota*) body mass can distinguish adults from young of the year, but it does not correlate with age after adult size is reached (Armitage et al. 1976; Boag and Murie 1981; Michener 1984). In addition, maternal nutrition influences growth rates, as does nourishment during early development. Poor nutrition will result in significantly reduced body size and subsequent underestimation of age. When physical growth slows or undergoes pronounced seasonal fluctuations, other criteria are required for age determination.

Tooth Characteristics

ERUPTION SEQUENCE

Teeth are "high-growth-priority" tissues, meaning that they receive nutrients even during times of food shortage; thus, tooth development is far less affected by nutritional factors than is development of skeletal parts. Consequently, the appearance of deciduous, or milk, teeth and their replacement with permanent teeth follow a predictable sequence in many mammals. The sequence of eruption of deciduous and permanent teeth and relative crown heights (as a measure of tooth wear) provide useful indices of age in bats (Twente 1955; Fleming 1988), carnivores (Mills 1982b), ungulates (Mason 1985; Anderson 1986), and rodents (Willner et al. 1983; Hench et al. 1984; Taylor et al. 1985). Unfortunately, the sequence of tooth eruption has not been documented in known-age animals for most mammalian species. This subject is a fertile area for investigation.

TOOTH WEAR

The relative wear of incisors, canines, or molars can often be correlated with age (Pucek and Lowe 1975; Anthony 1988). Tooth wear can be influenced, however, by nutrition and diet (Morris 1972; Pucek and Lowe 1975). Repeated tooth impressions taken from live animals in the field are needed to quantify wear and to reduce subjectivity in assignment to age classes.

Examination of skulls and mandibles, making reference to the appropriate literature on related species, promotes familiarity with tooth nomenclature (which can be confusing in some species) and with patterns of tooth wear. Some teeth are better indicators of wear than others. Museum collections provide excellent sources of material, although a field collection can often be assembled in a relatively

short time. If dental casts or impressions are taken in the field from live animals (see "Dental Casts," below), they must be compared with museum collections. If marking a large number of immature animals and recapturing them periodically during their lives is possible, an aging system can be established based on tooth eruption and tooth wear. Even if animals cannot be marked and recaptured, relative age categories can be established by examining a large series of skulls or mandibles. This is frequently done for large herbivores, because mandibles are easily acquired from carcasses cleaned by carnivores and scavengers. These techniques are not useful for short-lived mammals whose teeth are not functional long enough to show measurable wear (Morris 1972; Pucek and Lowe 1975).

Tooth wear can be measured with a millimeter ruler or calipers directly from live animals or indirectly from dental casts. Not all teeth are good indicators of tooth wear. In general, wear on those teeth most directly associated with the animal's diet will be the best indicator of age. Thus, canines and carnassial teeth are good tooth-wear indicators for carnivores, whereas the last premolars and first molars are preferred for bovids and cervids. The actual dimensions measured should be clearly defined. For example, the gum-to-tip height of a canine tooth can be measured on the anterior, posterior, lingual, or labial side. An anterior or posterior measurement is preferred, because the lateral gum line often erodes with age. In addition, labial surfaces are easier to measure in live, unanesthetized animals than are lingual surfaces.

A promising approach for estimating age in live small mammals is to make plasticine impressions over time to record the sequence of tooth eruptions. Using a temporal series of impressions of the right maxillary tooth row, Malcolm (1992) distinguished 15 age classes in lemmings (*Dicrostonyx*) based on the length of the tooth row, stage of tooth eruption, and wear of the occlusal surface.

DENTAL CASTS

Dental casts are an excellent means of recording the tooth eruption and wear patterns of mammals. Briefly, dental putty is placed in a holder (dental tray), and the investigator holding the tray pushes the putty against the upper or lower jaw to produce a negative impression of the teeth. After the mold sets, a permanent cast, which will match the dentition, can be made from plaster or dental stone. Measurements are taken from this cast.

At least two people are required to prepare the putty and trays and to make the impressions. Timing is critical because casts can be made only while an animal is anesthetized and immobile (except for reflex movements of the tongue), and long periods of anesthesia may result in death.

MAKING THE DENTAL IMPRESSION.   The first step is the preparation of dental trays to hold the dental putty. Trays of different sizes and shapes are made from dental wax or sheets of beeswax to fit the upper or lower jaws of the study animals. Skulls of different sizes can be used to model several crude forms.

As an anesthetized animal becomes immobile, its body and head are positioned to take the mold. Silicone dental putty is put into the dental tray, and the tray is positioned in the animal's mouth. Overfull trays will spill. The teeth are allowed to settle into the putty, and the body is held quietly for about 1 minute while the impression is made. The impression is complete when the putty forms a gel. The tray is gently extracted from the mouth, and any putty remaining in the mouth is removed.

Dental putty that has been thoroughly mixed with hardening catalyst can also be used. The putty is pressed onto the end of a thin wooden template (e.g., ice cream stick) and shaped into a thin wedge. The template is pressed against the tooth row as described previously. The putty imprint usually hardens about 5 minutes after removal from the animal's mouth. Finally, impressions of an entire tooth row can be taken using plasticine (modeling clay) attached to the end of an aluminum rod. The rod is used as a handle, and the clay is pressed tightly against the tooth row as noted above.

A label with the date, species, sex, locality, name of the investigator, and characteristics identifying the individual is placed with the mold in a plastic bag, which is sealed. A sealed bag minimizes evaporation and resulting distortion of the impression until the casts are made.

MAKING DENTAL CASTS.   Dental plaster or plaster of Paris is prepared according to package instructions (see also "Plaster of Paris Method" under "Casting Tracks," Chapter 9) and poured smoothly into the molds. Gentle and repeated tapping of the mold against a surface releases bubbles trapped when the liquid is poured. Data are etched into the cast before it completely solidifies. Casts should be stored in a dry, protected place. The dental trays can be cleaned, repaired, and reused or stored for future use.

Cementum and Dentine Deposits

Often, secondary dentine and cementum are deposited in bands (annuli) with a regular periodicity that is assumed to be annual (Morris 1972; Pucek and Lowe 1975; Grue and Jensen 1979; Lieberman and Meadow 1992). The number of annuli has been used to estimate chronological age in bats (Schowalter et al. 1978), carnivores (Stephenson 1977; Harris 1978), pinnipeds (Scheffer 1950; Bowen et al. 1983; Arnbom et al. 1992), cetaceans (Molina and Oporto 1993), dugongs (Mitchell 1978), and cervids (Ohtaishi et al. 1990). Annuli occur in the teeth of tropical mammals with a periodicity that presumably reflects seasonal changes in rainfall and, therefore, food supply. Annuli are used primarily to estimate the ages of dead animals, because the procedure requires the removal of a tooth (usually a lower incisor or postcanine tooth). Tooth extraction is not appropriate for most species, because the surgery is technically difficult to perform and is traumatic for the animal.

Teeth are generally prepared for examination of cementum and dentine growth annuli by decalcification, followed by sectioning, staining, and mounting on a glass slide for microscopic examination (Klevezal and Kleinenberg 1967; Thomas 1977; Fancy 1980; Molina and Oporto 1993), although decalcification is not advised in some situations (Lieberman and Meadow 1992). Equipment costs can be prohibitive unless a histology laboratory is locally available; some commercial concerns provide aging services. Teeth saved for analysis can be stored in small labeled envelopes or placed in formalin in labeled containers to prevent drying and cracking.

Although growth annuli are widely used, significant errors in estimating age can result, because secondary dentine and cementum layers may be deposited asymmetrically or deposition of acellular cementum and dentine may not be annual (Roberts 1978; Phillips et al. 1982; Lieberman and Meadow 1992). For example, the annuli of 78% of Antarctic fur seals (*Arctocephalus gazella*) sampled by Arnbom et al. (1992) corresponded to the actual age of the animal, but the annuli counts of the remaining 22% under- or overestimated the animals' true ages by 1 to 4 years. As with other methods of age estimation, assignment based on dentine or cementum layers should be verified by comparison with animals of known age (Klevezal and Kleinenberg 1967). Dyes such as tetracycline or alizarin red can be fed to animals to validate annular analysis, because they stain only those bands under deposition at the time of exposure to the dye.

Periosteal Growth Lines

Deposition of dense appositional material varies seasonally and is reduced when food supply is limited or when the physiological condition of the animal is poor. Consequently, the succession of growth layers in periosteal bone tissue can be used to estimate the number of winters an animal has survived (Klevezal and Kleinenberg 1967; Morris 1972; Pucek and Lowe 1975; Ohtaishi et al. 1976; Frylestam and von Schantz 1977; Iason 1988). Whether this technique applies to animals in highly seasonal tropical environments is not known.

The method requires dead animals, because bone layers are examined in histological sections of limb bones (including digits) and mandibles (Morris 1972; Pucek and Lowe 1975; Ohtaishi et al. 1976; Frylestam and von Schantz 1977). Alternatively, bones may be cut with a diamond-impregnated circular saw into several sections that can be mounted in 25×40-mm blocks of resin (Iason 1988). Blocks are then smoothed, first with coarse waterproof carbide paper, then with aloxite optical smoothing powders (nos. 600, 175, and 50), and finally with a succession of diamond polishing compounds. Each block is cleaned and then polished briefly with magnesium oxide powder on a soft cloth to add relief. Annual lines can be distinguished by microscopic examination of the polished surface under incident light (Iason 1988).

Epiphyseal Closure

The presence or absence of cartilaginous zones in long bones and tail vertebrae can be used to distinguish younger from older individuals (Thomsen and Mortensen 1946; Hale 1949; Watson and Tyndale-Biscoe 1953; Lechleitner 1959; Carson 1961; Walhovd 1966). The long bones of immature mammals typically have two terminal cartilaginous zones (growth plates) covered with bony caps (epiphyses). As mammals grow, the long bones lengthen in the re-

gion of the growth plates, and the cartilage is replaced by bone. When bone growth is complete, the epiphyses fuse with the shaft (diaphysis). Cartilaginous epiphyseal plates can be examined visually or by X-ray (Pucek and Lowe 1975; Myers 1978b). Portable X-ray machines are available for use in the field on living animals (Page 1993).

Immature bats can usually be distinguished from adults by transilluminating the wing bones; cartilaginous zones appear more translucent than dense bone (Anthony 1988). This technique is less effective for large pteropodids because of their densely pigmented wing membranes.

Secondary Sexual Characteristics

Often secondary sexual characteristics can be used to determine age as well as sex. All juvenal monotremes possess a hindfoot spur, but the weapon is lost in adult females (Griffiths 1978). In contrast, the thickened neck and shoulder manes of some suids and bovids and the antlers of cervids appear only after puberty; these characters can be used to distinguish mature males from either females or immature males (Geist 1966).

In addition, horns, like teeth and bone, exhibit annular patterns of growth, which can be used to estimate age. The conspicuous "annual" ridges on horns reflect annual and multiannual variation in food quality or the cessation of growth during the breeding season. The shape and size of horns or antlers can also indicate age. In musk-oxen, the horns of yearlings are small and straight, whereas in adults they are larger and curved. Among cervids, mass, size, and often branching complexity of the antlers correlate with age. Yearling deer usually have small, simple, and unbranched antlers, whereas two-year-olds have small, branched antlers. Antler size can be used to place animals in general age classes (e.g., yearling, young adult, and prime). Antler condition is also an indicator of a male's nutritional status when the antlers were developing.

Mass of the Lens in the Eye

Eye lenses increase in size and mass throughout life (Friend 1967) and may be reliable indicators of age in bats (Anthony 1988), carnivores (Grau et al. 1970), ungulates (Kauffman et al. 1967), and rodents (Williams 1976; Malcolm and Brooks 1985), at least through some portions of the life cycle. The technique is unsuitable for live animals, but it holds promise for field studies in which the animals are kill-trapped. The protocol requires removing the eyes, fixing them, excising the lenses, and drying each lens to constant mass. Use of a standardized fixative (e.g., 10% buffered formalin) is important (Pucek and Lowe 1975). Because much of the dry mass of the lens is proteinaceous, analyzing water-soluble protein fractions or specific amino acids (e.g., tyrosine) that are proportional to the total protein content has improved the accuracy of earlier techniques (Pucek and Lowe 1975). As with other characters, the growth of eye lenses can be influenced by genetic factors, habitat, or diet. Age-specific mass tables have to be constructed on a species-by-species basis.

Pelage or Color

Pelage condition and molt patterns have been used to assign relative ages to some live mammals (Anthony 1988). The natal and juvenile coats or colors of many mammals (e.g., primates, carnivores, otariids, phocids, cetaceans, tapirids, suids, cervids, rodents) are distinctive. Generally, the pelage of many young mammals is darker in color, longer, duller, and less dense than that of adults. Little information is available on the timing of molts for most species. The presence of molt can sometimes be determined by external examination of live animals (e.g., those, such as ermine [*Mustela*] and arctic hares [*Lepus*], whose pelage changes between winter and summer). Information on sequences and patterns of molt, however, which is necessary to determine relative age, is revealed only by examination of the internal surface of the skin (Pucek and Lowe 1975). The molts of immature mammals usually proceed in a uniform and relatively symmetrical pattern across the body of the animal, whereas those of adults are often patchy (Rowsemitt et al. 1975).

Tail Collagen

The denaturation of tail-tendon collagen fibers over time is a generalized measure of physiological age that can be correlated with absolute age. Tail biopsies can be taken in the field and age determined from the amount of collagen present (Sherman et al. 1985; Austad 1993).

Reproductive condition

Size by itself is not a clear or consistent determinant of the ability to reproduce, because sexual maturity can be achieved before or after adult body size is reached. The most reliable indicators of reproductive maturity are the condition of the reproductive organs and the presence of young.

Males

Reproductively active males usually have some or all of the following attributes: large descended testes, mature sperm, adult body size, sex-specific behaviors, hypertrophied accessory glands, breeding pelage, and well-developed body ornamentation (e.g., antlers). Of these, testis size (length × width; Kenagy 1979) and position and presence of sperm are most important. Testes measurements should indicate if the scrotal skin was included.

In live animals, large, turgid, scrotal testes with visually conspicuous, convoluted tubules in the cauda epididymides indicate ongoing spermatogenesis, as does the presence of mature sperm (Jameson 1947, 1950; Christian 1950; Taitt and Krebs 1985; Racey 1988). Small, flaccid, abdominal testes are associated with reproductive quiescence. In many mammals, the precise position of the testes is difficult to judge, especially when the organs shift position more than once a day, as in *Peromyscus* (Jameson 1950). The testes of hibernating bats typically involute shortly before the male enters hibernation, but large stores of sperm are often retained in the cauda epididymides well into the winter.

Ejaculation to obtain sperm for microscopic examination can be stimulated mechanically or electrically in some large mammals; such procedures must be carried out under the supervision of a veterinarian. In most mammals, however, a necropsy or testicular biopsy will be required. Testes and accessory reproductive glands removed from dead mammals are measured, and the presence or absence of sperm-filled convolutions of the cauda epididymides is noted. The cauda epididymides and testes are then sliced longitudinally, and their contents are collected (Christian 1950).

Sperm should be stained as soon as possible after being collected. The amount of stain solution should be varied according to stain density. Normally one volume of semen is mixed with two volumes of eosin-nigrosin stain for 2 minutes. The stain-semen

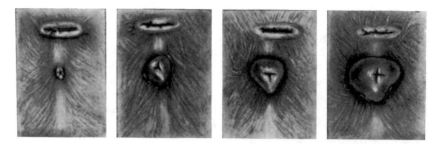

Figure 56. Cyclic changes in the vulva (lower opening) of the American marten (*Martes americana*). Conditions range from least development during anestrus (left) to maximum swelling during estrus (right). Reprinted with permission of John Wiley & Sons, Inc., from Enders and Leekley (1941).

mixture is placed on a glass slide and allowed to air-dry before a permanent slide with a suitable mounting medium and cover slip is prepared. The number of eosinophilic sperm heads should be counted and expressed as a percentage of the total number of sperm heads (Mann 1964). The higher the percentage, the more advanced the stage of reproductive activity.

Females

Characteristics associated with gestation or lactation—such as large body mass, presence of suckling young, enlarged nipples, milk-filled mammary glands, and an enlarged or swollen pouch or pouch area in marsupials—are used to identify reproductively active females. The external condition of the vagina or vulva often indicates the reproductive status of female mammals before mating. The vagina of most eutherian mammals is sealed by a vaginal membrane until puberty (Weir 1974). Thereafter, it may remain open, as in carnivores, primates, and ungulates; it may reseal during anestrus, as in seasonally breeding rodents; or it may become plugged by cornified detritus, as in many temperate-zone hibernating bats and rodents (Racey 1972; Michener 1980). The position and condition of this membrane and the duration of perforation during estrus vary among species (Weir 1974). For example, the vaginal membrane is perforate during estrus in all hystricomorphs except coypus (*Myocastor*). In chinchilla (*Chinchilla*), the membrane is very sturdy and distinct, whereas in degus (*Octodon*) and tuco-tucos (*Ctenomys*) it is more fragile (Weir 1974).

Characteristic vulvar changes in many mammals are correlated with the stage of the estrus cycle. During anestrus in most seasonal breeders, the vulva is

small, inconspicuous, and pale. As estrus approaches, the vulva swells, protrudes, and changes color. For example, the vulva of female martens (*Martes americana*) typically changes from a pale-colored structure with fine wrinkles during anestrus to a turgid, wrinkle-free organ during estrus (Fig. 56). The timing and duration of vulvar swellings vary across species, but the general pattern is diagnostic. As rodents and carnivores approach estrus, the clitoral part of the vulva often becomes more prominent and rises above the level of the surrounding perineal skin. For some mammals, the vulvar swelling is maintained throughout the breeding season.

Characteristic changes in the vulva and the contiguous perineal area (sexual skin) are diagnostic of the estrous (menstrual) cycle in primates (Gillman 1935; Hausfater 1975; Dixson 1983). The distribution and degree of color and edema in the perineal skin vary considerably among Old World primates (Fig. 57). Most New World primates do not have a well-developed sexual skin.

The time of estrus cannot readily be determined from the appearance of the vulva in some species. For example, in plains viscachas (*Lagostomus*) the vulva remains open for about 13 days (Weir 1974). By contrast, the vulva of the chinchilla can open and close within 12 hours, but this usually occurs over a period of 2 to 4 days, and cuis (*Galea*) have no obvious vaginal swelling associated with estrus (Weir 1974). Vaginal opening can also occur at times other than during estrus and parturition. Vaginal opening occurs late in pregnancy in guinea pigs and occurs sporadically throughout gestation in plains viscachas, degus, and tuco-tucos (Weir 1974).

In large mammals pregnancy can be determined visually once the abdomen increases in size. In most small mammals (e.g., microtines; Nadeau 1985), pal-

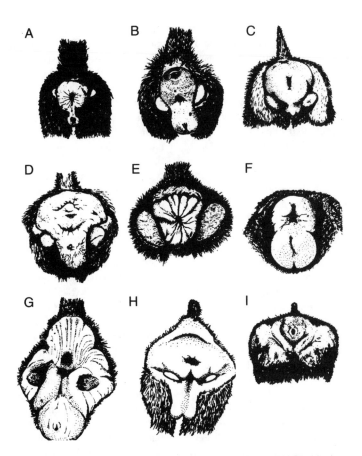

Figure 57. Appearances of the sexual skin during the mid–menstrual cycle in selected Old World primates. A. White-cheeked mangabey (*Lophocebus albigena*). B. White-collared mangabey (*Cercocebus torquatus*). C. Mandrill (*Mandrillus sphinx*). D. Hamadryas baboon (*Papio hamadryas*). E. Olive colobus (*Procolobus verus*). F. Chimpanzee (*Pan troglodytes*). G. Talapoin (*Miopithecus talapoin*). H. Pigtail macaque (*Macaca nemestrina*). I. Celebes ape (*Macaca nigra*). Drawings at different scales. Reprinted with permission from Dixson (1983).

pable embryos can be detected after about 1 week of gestation. Bats have relatively prolonged periods of gestation for their body size, and pregnancy usually is not detectable by palpation until the third trimester, except in large pteropodids (Racey 1969). Females of most mammals also become less aggressive and exhibit increased roost fidelity and nest-building behavior as the time for parturition approaches.

In most small mammals, lactational condition can be determined by observing the nipples (teats), or by the presence of suckling behavior (Doidge et al. 1986; Numan 1988). Nulliparous females (those that have not borne young) usually lack readily observable nipples, which remain small until the female's first parturition. The nipples of parous females are typically longer and more darkly pigmented than

those of nonparous females. In many long-lived mammals, the size of nipples can often be used to distinguish yearlings from older females before breeding begins (Racey 1974, 1988). Species with longer lactation periods often have larger teats than congeneric taxa with shorter periods of lactation (Sanderson 1950; Racey 1974; Bronson 1979).

In the field, the nipples and the underlying mammary glands should be examined for signs of lactation. Reddening of the area around the teats often indicates increased vascularization associated with mammary tissue development. Nipples of lactating or recently lactating females also are often enlarged and highly cornified, whereas teats of nonlactating and postlactating females are usually smaller and show less cornification. In actively lactating females, the mammary glands often appear "whitish" if they contain milk; those recently depleted of milk are usually flaccid but retain a "pinkish" color. Mammary glands should be palpated to express milk as evidence of lactation.

In many marsupials, the appearance on the abdomen of a pouch or of lateral ridges of skin, anterolateral folds of skin, or circular folds of skin is indicative of breeding condition (Woolley 1974). The appearance of the skin folds and the pouch changes during estrus, pregnancy, and lactation. Immediately prior to mating, the skin folds and pouches become increasingly vascularized. During lactation, changes that occur in the pouch are mostly related to the enlargement of the mammary area, nipples, and folds of skin (Woolley 1966, 1974). Once the young are weaned, the nipples and skin folds or pouch regress to conditions seen at the beginning of the breeding season, rather than the condition seen in immature females, and the nipples often remain slightly discolored, enlarged, and elongated. Nipples of marsupials are located on the abdominal skin beneath the skin folds or within the pouch. In some species, however, the nipples are located in two anteriorly directed pockets, which are extensions of the area covered by skin folds (Woolley 1974). Because pregnancy is relatively brief in marsupials and the young of most species are born at an early stage of development, reproductive condition is easily assessed by examining the pouch or the nipples in those species lacking pouches. In contrast to eutherian mammals, pregnancy in marsupials cannot be determined by palpating embryos.

Female mammals have two ovaries, located in the abdominal mesenteries. In platypuses and bats one side of the female reproductive tract may be reduced. The condition of the ovaries in small mammals can be determined by necropsy or laparoscopy; the ovaries are best located by tracing along the oviducts beginning distally from the uterus or paired uteri. In freshly killed specimens, ovaries are whitish pink or pale yellow and often have a "pebbled" appearance, reflecting the presence of active follicles and/or corpora lutea. For most small mammals, histological analysis is the most accurate way of assessing reproductive status. Ovaries and uteri can be palpated through the wall of the rectum in anesthetized large mammals such as bovids and perissodactyls. This method should only be attempted by experienced personnel.

Hormone Assays

Blood sera, urine, and feces can be assayed for reproductive steroids to ascertain the reproductive status of wild mammals (Seabloom 1985; Risler et al. 1987; Wasser et al. 1988; Monfort et al. 1993). Serum can be separated from blood collected from immobilized animals and then frozen in liquid nitrogen. Urine samples must also be frozen, preferably in liquid nitrogen, until analyses. Feces, however, can be preserved in formalin or dried for analysis, with less cost. Determination of reproductive status by analysis of reproductive steroids or their metabolites by radioimmunoassay is possible from all three materials, but with varying accuracy depending on the compounds to be measured and the species in question. The assays are available as kits but are relatively costly, and collaboration with a reproductive physiologist with a fully equipped laboratory is desirable.

Acknowledgments. We thank Frank Bonaccorso, Paul Racey, Bill Rainey, Greg Richards, and Chris Tidemann for making suggestions and comments on early drafts of the manuscript.

Appendix 6

Field Parasitology Techniques for Use with Mammals

Scott Lyell Gardner

Introduction

Obtaining parasites from mammals that are collected during surveys or bioinventories is time-consuming, and in the past such collections have rarely been made. Parasites and other symbionts are important components of the biology of the host, however, and must be sampled for a complete picture of its ecology and other aspects of its life history. For example, work on parasite diversity and biogeography of mammals in Bolivia (Gardner and Duszynski 1990; Gardner 1991; Gardner et al. 1991; Gardner and Campbell 1992a, 1992b) could not have been accomplished without proper collections of the parasite and symbiont fauna of the hosts.

Many researchers are reluctant to assign one or two persons from a field crew to process parasites from mammals when the specimens and associated data are considered "auxiliary" or "collateral." The most effi-

cient and cost-effective method of obtaining data on the parasite fauna of a host group, however, is to collect the parasites when mammals are being processed in the field. Examples of the types of data that may be obtained when a host is collected are prevalence, intensity of infection, and distribution of parasites in or on individuals and through populations of hosts.

Studies of the systematics and ecological characteristics of hosts and parasites require proper identification of both groups. If the parasites associated with a host are not collected and preserved properly, species-level diagnostic characters (i.e., morphological characters) will almost certainly be destroyed; in addition, improper preservation of parasite material will severely limit studies based on intact DNA molecules of those parasites.

Problems with identification of correct hosts may occur when parasites accidentally transfer from one host to another at the time of collection. Anthropogenic host-transfer can occur anywhere and is difficult to avoid in the field. For example, ectoparasites from one host may remain in a trap after the animal is removed. Those parasites may then transfer to another host of a different species at the same or a different collecting locality. Anthropogenic host-transfers occur even more frequently when hosts are placed in a killing jar that has not been cleaned completely after each use.

In this appendix, I outline methods for collecting parasites that maximize the amount of morphological and molecular information available from each parasite obtained. The procedures are designed to minimize the likelihood of mistakes in recording data due to anthropogenic host-transfers. My collaborators—Dr. Robert L. Rausch, Ms. Virginia Rausch, Dr. Terry L. Yates, and Dr. Sydney Anderson—and I have been collecting parasites in the field for more than 25 years. The protocols I present here have evolved considerably over that time and are based on our collective experience. Additional information on collecting techniques is presented in Anderson (1965) and Pritchard and Kruse (1982).

Data are collected in the field laboratory in assembly-line fashion, with different individual researchers in the field crew assigned to different tasks. Every part of every animal that is collected is processed in some way. Because of the possibility of contracting various viral diseases from handling wild-caught mammals, I recommend that surgical gloves and lab coats be worn during all laboratory procedures.

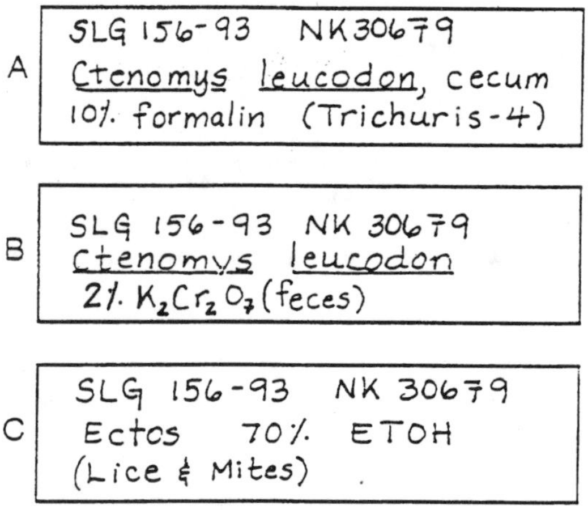

Figure 58. Examples of data tags to be placed in vials or plastic bags with specimens of parasites. The collector's number, the field identification number, the type of preservative, and the general kinds and approximate number of parasites in the vial or bag are noted on the tag. A. Tag for a vial with nematodes from the cecum of *Ctenomys leucodon* NK30679. B. Tag for a vial containing feces from the cecum of *Ctenomys leucodon* NK30679, in potassium dichromate. C. Tag for a vial or plastic bag containing ectoparasites.

Procedures for collecting ectoparasites and ectosymbionts

1. Each live host is placed in a *new* bag made of thin plastic (1.5–2 mil). Dead hosts that are removed from traps should be placed immediately in *new* plastic bags to prevent loss of ectoparasites and cross-contamination of hosts. A 200 × 350-mm bag is a good, all-purpose bag for most small mammals.

2. A bag with a host is placed into a glass or plastic jar containing cotton moistened with chloroform or ether. Such chemicals should be used in well-ventilated areas and away from any open flame. Carbon dioxide introduced into the jar kills ecto-

parasites too slowly to be effective. Dead hosts also must be treated with chloroform or ether to ensure that their ectoparasites are killed.

3. After the ectoparasites die, while the animal is still in the bag, the investigator shakes and brushes the dead ectoparasites from the fur into the bottom of the bag. She or he also examines the pinna, ears, and body of the mammal for ticks and trombiculid mites, which are removed and stored. The host is then passed to the next team member for other processing (see "Processing Procedures," Appendix 3). A labeled tag (Fig. 58) is placed in the plastic bag with the ectoparasites; the bag is tied shut and set aside. The label on the tag is written in permanent India ink and includes the host identification number or the field collection number of the collector. The investigator repeats the procedure with the next animal until all animals have been processed.

4. The specimen preparator should inspect the mammal carefully for small ectoparasites. Generally, the external pinnae and the fur from the skin to the tips of the hairs are examined using a dissecting microscope. The fur is parted with jeweler's

forceps and dissecting needles. Examination begins at low power (20×–40×). The magnification is increased (50×–70×) to see if a suspicious object has legs. Several types of mites live in the subcutaneous connective tissue of mammals. These are often found by pulling the skin off the host and examining the underside of the skin. Larvae of flies (botflies or warbles) may also be found in the subcutaneous layer. A reliable identification can be obtained for these larvae only if they are reared to adulthood in an artificial brood chamber. Larval botflies cannot be identified at the present time. Live botfly larvae are removed from the host and placed head-up in a small, highly porous box. Ammunition cartridge case boxes (about .45 caliber or 10–12 mm in diameter) made of styrofoam work especially well for this purpose; several larvae can be placed in the box, stored, and transported at one time. Parasites and other symbionts can be removed with forceps, a swab, or a needle dipped in ethanol. If a tick or mite is embedded in the skin of the host, it is carefully pried from the skin, rather than scraped, to ensure that mouthparts are included. Small mites that are attached to a hair or are embedded in a follicle are collected by removing the complete hair and preserving both. The locations of all ectoparasites collected are also recorded in the field notebook of the collector (Fig. 59). Recording the numbers of ectoparasites collected from each body region of the host and the type of vial or other container used to store them facilitates locating the organisms during later laboratory analysis. Ectoparasites and other ectosymbionts are preserved in 70% ethanol (EtOH). Parasites from each host individual should be stored in a separate vial.

5. A small amount of 70% EtOH (2–6 ml) is squirted into each plastic bag used to hold a host mammal. The contents are washed down the sides and into a corner of the bag. The bag is held over a Whirl-Pak (plastic bag with integral wire twist-ties), the corner is cut from the bag holding the ectoparasites and ethanol and the contents drain into the smaller Whirl-Pak bag. This method produces an uncontaminated sample of ectoparasites and ectosymbionts with minimal time expenditure. To avoid contamination, bags should never be reused. The contents of the Whirl-Pak bags should be transferred to vials as soon as possible after returning to the laboratory.

Procedures for collecting endoparasites and endosymbionts

The following procedures are appropriate for collecting protozoa (coccidia), acanthocephala (spiny-headed worms), cestoda (tapeworms), trematoda (flukes), and nemata (nematodes).

1. The researcher collects endoparasites from a host specimen after its femur has been removed to obtain tissue for karyotyping but before its organs have been removed and preserved for genetic analyses (see "Processing Procedures," Appendix 3). Scissors, forceps, and probes are always rinsed in distilled water or 70% EtOH and dried with a tissue between animals to avoid contamination with blood or other sources of foreign DNA.

2. If the animal is still warm and fresh (i.e., the blood has not coagulated), the investigator makes a blood smear. He or she places a drop of blood (approximately 250 µl) in the middle of a slide using a disposable plastic pipet. The edge of another slide held at an angle is placed on the slide with the blood so as to contact the blood drop. The inclined slide is pushed evenly and rapidly away from the blood drop, drawing the blood out into a thin (one cell layer–thick) smear (Fig. 60). The slide is labeled with the field identification number of the host using a diamond-point pencil and allowed to dry. Depending on the humidity, a slide should dry for from 10 to 30 minutes. New blood smears are fixed at the end of each day, or sooner if the temperature and humidity are high. Dry blood smears are fixed in 100% methanol (MeOH) for from 2 to 5 minutes. Blood smears that are well formed and of even thickness are important for documenting and detecting protozoa and microfilariae (juvenile nematodes).

3. To obtain endoparasites from the digestive system, the investigator cuts across the esophagus just above the stomach and across the colon just anterior to the rectum. Care must be taken not to perforate the organs during this procedure or parasites may transfer from one organ to another and invalidate information on the distribution of parasites within an individual host. The digestive tract is removed intact and placed in a clean petri dish with a tag bearing the field identification number of the host and a small amount of water. If the or-

Bolivia, La Paz, 11.5 Km w. San Andres de Machaca;
16°59'47"S, 69°03'31"W; 3800m.

— 4 August 1993—

SLG 156-93   NK 30679   _Ctenomys leucodon_
Feces, Blood smear.
Stomach ⊖; Saved contents in Snap Cap.
Small intestine, 5 cestodes in duodenum.
      (4 → 10% formalin - 15ml vial).
      (1 → 95% ETOH) dram vial.
      Many trichostrongylids in duodenum.
      approx. 20 pres. in liquid N_2.
      Rest in 10% formalin. Few in
      GAA → 10% formalin. (dram vial)

Cecum and Large Intestine: whipworms →
Trichuris, 10 in 10% formalin, 2 in
GAA → 70% Etoh (dram vial)

Ectos: 10 small mites from perianal
      skin (dram vial, 70% etoh)
      Estimated 500 small lice in nape
      region of neck, 10 collected
      (dram vial) 70% ETOH.
      — 4 August 1993 —
SLG 157-93   NK 30699   _Galea musteloides_
Feces, Blood smear.
stomach ⊖, small intestine ⊖,
cecum + Large Intestine ⊖, no ectos.

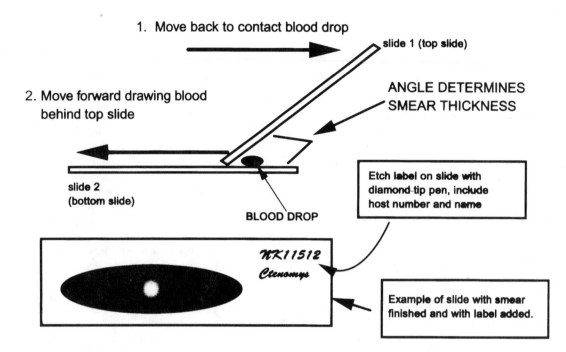

1. Move back to contact blood drop

slide 1 (top slide)

ANGLE DETERMINES
SMEAR THICKNESS

2. Move forward drawing blood
   behind top slide

slide 2
(bottom slide)

BLOOD DROP

Etch label on slide with
diamond-tip pen, include
host number and name

NK11512
Ctenomys

Example of slide with smear
finished and with label added.

Figure 60. Preparation of a blood smear. A smear must be made from fresh, uncoagulated blood. The angle of the upper slide determines the thickness of the smear: a steep angle creates a thin smear; a shallow angle provides a thicker smear. A 45° angle is good for general purposes.

gans are large, buckets and porcelain pans are used in place of petri dishes. If organs are very large (for example, those of a moose, whale, or elephant), they can be subsampled.

4. At this point, the researcher examines the body cavity, liver, kidneys, and lungs for helminth cysts or filariid worms. All organs to be removed from the host for future molecular or biochemical-genetic work must be carefully examined for parasites. Filariid nematodes may be encountered in

the heart, aorta, pleural cavity, mesenteries, or subcutaneous tissues. Larval or juvenile acanthocephalans, pentastomes, nematodes, cestodes, and trematodes may be found in the liver, mesenteries, or any other tissues.

If cestode cysts are encountered, some are fixed in situ in the host tissue. Part of the organ is removed with the cyst intact and preserved in 10% formalin (assume for fixation purposes that 37% formaldehyde = 100% formalin for the 10%

Figure 59. Sample field-notebook entries for parasites. The collecting locality is written at the top of every page or more frequently if the locality changes. The date is included with every record. The field collector number is next, followed by the field identification number and field identification of the host. *Above:* Parasite examination record for a specimen of *Ctenomys leucodon.* The collection of feces and blood or organ smears is recorded. The present entry indicates that the stomach had no parasites but that contents were saved for later analysis of diet. Five cestodes were recovered from the duodenum of the small intestine; 4 were fixed in 10% formalin in a 15-ml vial, and 1 cestode was preserved in 95% ethanol in a 1-dram vial. Small trichostrongylid nematodes were also found in the duodenum; approximately 20 were preserved in a cryotube in liquid nitrogen; the rest were preserved without counting in 10% formalin in a 1-dram screw-cap vial. The cecum and large intestine contained whipworms (genus *Trichuris*); 10 were preserved in 10% formalin and 2 in 70% ethanol in 1-dram vials. Ectoparasites were preserved in 70% ethanol in two 1- dram vials. *Below:* Parasite examination record for a specimen of *Galea musteloides.* Feces and a blood smear were taken. No parasites were found. Negative data are always recorded to ensure accurate estimations of prevalence, intensity of infection, and other ecological parameters.

dilution). If more than 10 cestode cysts are encountered in one host, a few cysts are removed and carefully cut open; the cestode strobila is relaxed in distilled water and fixed in 10% formalin. Additional cysts are stored in 95% EtOH, and some are frozen in liquid nitrogen. These various methods of preservation ensure that adequate material will be available for future investigations of both morphology and genetic attributes.

5. The collector next examines the intestinal tract after freeing it from attached mesentery and straightening it. The stomach, small intestine, cecum, and large intestine are cut apart and placed in separate petri dishes, each with a tag bearing the field identification number of the host. Petri dishes must always be cleaned and dried between uses. Each organ is opened (the intestine is cut lengthwise) and examined for parasites. It is important to use scissors with blunt ends because scissors with sharp points perforate the organ while cutting, making it difficult to open the organ quickly. For small specimens, iris scissors are appropriate. An enterotome is appropriate for opening the intestines of larger animals. If water is abundant, intestinal contents can be washed in a soil screen and then placed in a petri dish and searched. If the water supply is limited, one can take advantage of the fact that worms sink and other materials float. The intestinal contents are covered with water and gently stirred, and the lighter materials (plant parts and other food items), which float or do not settle quickly, can be decanted. This procedure is repeated several times. The remaining material is searched for helminths with either a dissecting microscope or a 10× jeweler's magnifying visor. If many helminth specimens are encountered, the intestinal contents should be preserved in 10% formalin so that accurate counts of the numbers of parasites can be made in the laboratory.

Organs such as skin, eyes, urinary bladder, lungs, and gallbladder must also be examined, preferably with a dissecting microscope. Small trematodes commonly occur in the bile ducts and gallbladder of the liver and in the mesenteric veins; small nematodes can be found in the urinary bladder.

6. Cestodes, trematodes, and acanthocephala are relaxed and killed by placing them in distilled water, although tap water or filtered river water can also be used. Osmotic imbalance causes water to move into the body cavity of the worm, leading to osmotic shock and death. The increasing pressure within the body cavity also causes the scolex or proboscis to evert and the strobila of a cestode to relax; it is especially important to leave a specimen in water long enough for eversion and relaxation to occur. The relaxation process can take from 10 minutes to more than an hour, depending on the size of the worm and the species. The relaxed, dead helminths are fixed in 10% formalin and placed in vials with both the field identification number and the location of the parasite in the host written on the tag. If abbreviations are used (e.g., SI for small intestine, C for cecum), they must be fully defined in the collector's field notebook. Formalin can be premixed, or specimens can be placed in a vial 90% full of water and enough 100% formalin added to make a 10% solution. Parasites from each organ must be preserved separately, in their own vials.

7. Saline is never used to kill cestodes or other platyhelminths because it prevents both osmotic imbalance and subsequent death. Nematodes, however, are never placed in distilled water, because the resulting osmotic imbalance causes them to burst, and in many cases the specimens are destroyed. During the dissection of the host, nematodes can be placed temporarily in saline and then transferred directly to a vial 90% full of very hot (not boiling) water. A 10% formalin solution is made by filling the vial with 100% formalin. Alternatively, nematodes can be placed in glacial acetic acid (GAA) for a few minutes before being transferred to either 10% formalin or 70% ethanol for storage. The GAA treatment causes the nematodes to uncoil. Such specimens are much easier to identify than those fixed without straightening, because the morphological characters are more readily seen. Specimens to be saved for molecular analyses should be washed in saline, placed in a cryotube, labeled, and stored in liquid nitrogen or placed in a vial containing 95% ethanol.

8. If sufficient numbers of helminths are available, representative individuals should be preserved in liquid nitrogen for future allozyme and DNA studies. Because GAA and formalin destroy DNA, individuals preserved for future genetic analysis *should not* be treated with these chemicals before freezing or preservation in EtOH. Investigators should preserve parasites using many different methods, thus ensuring availability of adequate material for future studies.

9. Fecal material should be preserved for later collection and study of coccidian parasites. A fecal pellet or some material from the cecum is placed in a vial half full of a 2.0% solution of potassium dichromate ($K_2Cr_2O_7$), along with a tag bearing the field identification number and the generic name of the host. Wheaton snap-cap vials (15 ml) are best for this purpose, because one vial half full of 2.0% $K_2Cr_2O_7$ contains a sufficient quantity of oxygen to keep the coccidia alive. The vials can be used many times, and they rarely leak.

10. If many conspecific hosts are available, from two to five entire gastrointestinal tracts should be preserved individually in 10% formalin. This approach will allow for future examination of the morphological characteristics of the intestines and any associated worms in situ.

Recording data

Each researcher should maintain a field notebook (Fig. 59) for recording data on the collections and the preparations that are made. The notebook should contain 100% cotton rag paper. Only permanent India ink should be used for recording data. For each record, the field collection number of the host should be recorded as well as the individual collection number. The host collection number is necessary to allow accurate cross-referencing with the voucher or symbiotype specimen (Frey et al. 1992).

Collectors should number their specimens sequentially, beginning with 1 and continuing indefinitely, rather than beginning the sequence anew with each collecting trip. The collector precedes the number with his or her initials so that he or she can be identified at a later date. The date and geographic locality of the collection must be provided for each specimen. The species name of the host should be entered, acknowledging the provisional nature of field identifications. The location of a parasite in a host must be noted, as should negative searches. The collector should also record the general kinds of parasites encountered in each organ.

Materials

In this section I list the equipment, disposable supplies, and reagents needed to collect endo- and ecto-

Table 19. Estimated Quantities of Disposable Supplies Required for Collecting Parasites from 100 Mammals in the Field[a]

144 (1 gross) 15-ml Wheaton snap-cap vials

144 20-ml Wheaton snap-cap vials

144 1-dram vials with Teflon-lined screw caps

2 boxes of standard precleaned microscope slides (not frosted)

10 boxes of tissues or Kim-Wipes for cleaning equipment

100 cryotubes with brown-colored caps for parasites

2 rolls of paper towels

200 disposable plastic pipets

500 ml of 100% formalin

500 ml of 95% ethanol

1,000 ml of 70% ethanol

500 ml of 100% methanol

200 rectangular plastic bags, ca. 200 mm × 350 mm, and 1.5–2 mil thick

400 Whirl-Pak or other plastic bags with twist tie closures

200 sheets of 100% cotton-rag field-notebook paper

[a]For general collecting, investigators should plan on using approximately 1 large plastic bag, 2 Whirl-Pak bags, 2 1-dram vials, and 2 15-ml snap-cap vials per host (actual use may be less, because not all hosts are infected with parasites).

parasites from mammals in the field. An estimate of the quantities of disposable materials required for collection of parasites from 100 small mammals is provided in Table 19.

Equipment

Durable equipment needed for collecting parasites includes the following items: dissecting microscope with 0.5×–30× magnification; bright light source (headlamps [e.g., Justrite] that use four size D batteries work well); two pairs of jeweler's forceps (100 mm); two pairs of gross dissection forceps with blunt tips (120 mm and 140 mm); scissors of different types and sizes, including 100-mm iris scissors for fine work, 120-mm blunt scissors for coarse work, and 120-mm sharp-tipped scissors for cutting tissue; scalpel and disposable scalpel blades (preferably size 21) for cutting through host tissue; Copelin staining jar for fixing blood smears; diamond-point pen for scribing field collection number on blood smear slides; small (300 × 200-mm) and large (400 × 300-mm) porcelain dissection trays; soil sampling sieve (no. 325, USA standard sieve series, 45-μm mesh size, 20-cm in diameter) for catching the nematodes and allowing the small colloidal particles in the water to escape; Rapidograph pens with indelible India ink, or disposable black Uniball Deluxe (Faber-Castell) permanent-ink pens; fluorescent workshop light for detailed work with the microscope or jeweler's visor; and liquid nitrogen tank (non-pressurized Dewar flask) with static holding capacity sufficient for the length of the field trip.

Expendable Supplies and Reagents

The following materials are expendable and will need to be replenished before each field trip: microscope slides; small, medium-size, and large plastic petri dishes; small insect pin probes; dissection probes and needles; many disposable plastic pipets; 15- or 20-ml snap-cap Wheaton vials; 1-dram screw-cap vials with Teflon cap inserts; 15-ml screw-cap vials with Teflon cap inserts; cryotubes with brown lid inserts to indicate parasite samples; paper for labels placed inside vials (100% cotton-rag notebook paper or museum-quality label stock); 100% cotton-rag field-notebook paper; 100% MeOH; 95% EtOH for making 70% EtOH; and 100% formalin (37% formaldehyde).

Acknowledgments. I thank the many people who have contributed to our studies of mammalian-parasite biodiversity. Special thanks go to Mariel Campbell and Bill Gannon for input on various methods of collecting and maintaining data and specimens.

Methods for Marking Mammals

General marking techniques

RASANAYAGAM RUDRAN

Mammalian inventory and monitoring projects frequently require the marking of at least some individuals in a population. The objective of marking is to facilitate the identification of animals, either upon recapture or from a distance. In achieving this objective, an investigator must address the ethical issues related to marking, particularly when selecting a marking technique (Appendix 1). Selection of a marking technique and also the type of mark most appropriate for a study depend on factors such as (1) the distance at which the mark should be visible; (2) the need for individual identification; (3) the size, shape, and habits of the target species; (4) the number of animals that must be marked; (5) the period for which the mark should be functional; (6) the effect of the mark on the survival, behavior, and reproduction of the marked animal; and (7) the objectives of the study. An investigator can select among

techniques that leave permanent, semipermanent, or temporary markings on animals.

Permanent Markers

Animals can be permanently marked with brands, tattoos, or surgical alterations of the shape or length of an extremity. Surgical procedures include toe clipping and ear punching (or ear notching), which cause only brief and minor discomfort to small mammals (American Society of Mammalogists 1987).

FREEZE BRANDING

Freeze branding is carried out with a copper branding iron that is supercooled in liquid nitrogen or a mixture of dry ice and alcohol and applied to an area of the body. Branding kills pigment-producing melanocytes of the skin, but not the hair follicles, so the hair and skin that grow back in the branded area remain permanently white (Day et al. 1980). Some investigators have used commercial refrigerants such as Arcton 12, Quick-Freeze, or Freon for branding (Thorington et al. 1979; Rood and Nellis 1980; Rood 1983). These materials, which can be obtained in aerosol cans from electronic parts suppliers, are sprayed on the area selected for branding, through a cardboard template of a distinctive pattern or number. Animals can be individually marked by varying the branding site and branding pattern.

A branding site should be naturally devoid of hair or it should be shaved before branding. Sites are selected according to the manner in which an animal will be viewed after marking. The sides of the body are marked for lateral viewing; the back is marked for viewing from above (Newsom and Sullivan 1968; Rood 1983). It takes 1 to 2 months for the white hair to grow and for the mark to become fully visible. Therefore, freeze branding is not appropriate if the study period is limited. Freeze branding is also inappropriate for animals with white or very light-colored fur, because the brand is not very visible, even from a short distance.

Freeze branding can cause serious injury and scarring if the brand is applied for too long. If application time is too short, however, the brand may not be permanent (Thorington et al. 1979; see also "Other Marking Methods," under "Methods of Marking Bats," below). For best results, a minimum branding time should be determined during preliminary trials. Hooded rats and house mice were branded successfully in 20 to 35 seconds, fox and Abert's squirrels (Hadow 1972) in 25 to 40 seconds, mantled howler monkeys in 20 to 30 seconds (Thorington et al. 1979), and white-tailed deer in 20 to 25 seconds (Newsom and Sullivan 1968). An exposure of 5 seconds produced the best results in dwarf mongoose, whereas an application of more than 10 seconds caused extensive damage to the skin and surrounding tissue (Rood and Nellis 1980). If properly executed, this technique leaves virtually no chance for infection. It should be used only in the dry season in tropical areas, however, so that the wound will heal and insect attack can be avoided.

HOT BRANDING

Branding can also be carried out using hot irons (Bradt 1938; Aldous and Craighead 1958; David et al. 1990). The technique is used extensively to mark domestic livestock, but it is seldom applied to wild animals because of the elaborate preparations required, the availability of easier alternative methods, and concerns for the ethical treatment of animals.

TATTOOING

A tattoo is applied with a special set of pliers or with a battery-powered, field tattoo kit, which is used to make tiny perforations in the skin. A dark-colored dye is rubbed into the perforations to produce a visible pattern. Any body part that is relatively free of hair and remains fairly clean can be tattooed. Such markings have been applied successfully, for example, to the ears of agoutis (Smythe 1978), cottontail rabbits (Thompson and Armour 1954), snowshoe hares (Keith et al. 1968), and white-tailed deer (Downing and McGinnes 1969); to the upper lips and groin areas of polar bears (Lentfer 1968); and to the chests and upper arms of monkeys (Kaufmann 1965; Thorington et al. 1979). Tattoos on bats, in contrast, are only temporary (see "Other Marking Methods," under "Methods of Marking Bats," below). Tattoos are most visible on animals with light skin and virtually useless for marking those with dark skin. Different numbers or tattoo patterns can be used to distinguish individuals. Generally, however, animal tattoos are relatively small, and animals must be recaptured or observed close-up for positive individual identifications.

TOE CLIPPING AND EAR PUNCHING

Toe clipping and ear punching have been used to mark a variety of mammals of various sizes. Recent recommendations suggest, however, that these techniques are suitable only for small, mouse- or rat-size mammals, and that they should be used only when no other marking methods are feasible (American Society of Mammalogists 1987). The site selected for marking should be treated with a local anesthetic such as ethyl chloride (Twigg 1975). Marking is then carried out with a clean, sharp instrument, and an antiseptic solution is applied immediately to the site to prevent infection (American Society of Mammalogists 1987). In toe clipping, the nail and the first joint of a toe are removed with scissors. In ear punching a hole is made in the pinna with a poultry punch (Blair 1941).

Animals can be individually marked with ear punches if the site of the hole or notch is varied. This technique has been used to identify individual hyenas (Kruuk 1972) and bush babies (Bearder and Martin 1980). Toe clipping has greater scope than ear punching for individual identification. For this purpose, each toe is assigned a number so that the clipped toes provide a unique numerical series. Some numbering systems are based on toe clipping alone (Melchior and Iwen 1965; Twigg 1975); others combine toe clipping with ear punching to increase the number of animals that can be individually identified. Blair (1941) combined ear punching and toe clipping only of the forefeet to produce a numbering system with 899 unique combinations. In a system developed for voles, toe clipping restricted to one toe per foot was combined with ear notching to produce a numerical series for identifying 624 individuals (Meunier et al. 1982).

A disadvantage of these techniques is that the marks they produce can be confused with an animal's natural scars and injuries. On the other hand, the procedures facilitate identification of numerous individuals and can be applied fairly easily and cheaply with a minimum of equipment.

Semipermanent Markers

Semipermanent markers are devices that are attached to the animal. They are broadly classified as collars, tags, rings, or bands, depending on their shape, placement, and type of attachment. Collars are sometimes referred to as neck bands, but true bands are used almost exclusively to mark bats (see "Methods of Marking Bats," below). Semipermanent markers come in a variety of types, sizes, and shapes and are usually made of either plastic or metal, which can be colored and/or numbered for individual identification. These markers are perhaps the most widely used marks in long-term field studies, because they can function for more than a year if properly applied.

COLLARS

Collars are marking devices placed around the necks of animals whose heads are wider than their necks. A collar must be flexible and have a smooth inner surface to prevent injury, and it must "fit" the neck. An animal's front foot can become inextricably caught in a loose collar (Mech 1983), and loose collars may come off. A collar should also be tight enough to prevent it from moving up and down and chafing the necks of animals with long necks. On the other hand, collars should be able to accommodate growth or seasonal changes in neck size. Expandable collars have been used on immature animals and adult male ungulates for this purpose (Hawkins et al. 1967b; Strathearn et al. 1984; Burger 1988).

In general, animals must be captured before they can be collared. Some collars, however, can be fitted with the help of automatic devices that make animal capture unnecessary (e.g., Beale 1966). One such device consists of a modified snare that locks securely around an animal's neck without choking it, and an anchor section that breaks under stress, permitting the animal to escape (Verme 1962; Siglin 1966). Automatic collaring devices are particularly useful for marking relatively large mammals that are difficult or dangerous to capture.

Some of the earliest collars used in animal marking were made of leather and sheet aluminum (Progulske 1957; Fashingbauer 1962). Recently, metal ball-chain collars (or necklaces) have been used on bats and primates (Barclay and Bell 1988; Fedo 1991; see also "Methods of Marking Bats," below). The collars most commonly used today, however, are made from a wide array of plastics such as nylon, neoprene, and polyethylene. Some plastic materials are affected by extremes of temperature and tend to lose their flexibility (Twigg 1975). This is an important consideration when marking animals living in very hot or extremely cold areas.

Several animals can be individually identified by varying collar color or by combining two or more colors. Some colors fade rapidly when exposed to insolation and other climatic factors. Preliminary trials must be conducted to identify the colors best suited for individual identification given the study period and the body color of the target species. Plastic collars of different colors or with attachments bearing numbers or letters have been used for individual identification of white-tailed deer, mule deer, elk, and bighorn sheep (Harper and Lightfoot 1966; Hawkins et al. 1967b; Haas 1990). Transmitters with unique frequencies can also be attached to collars for individual identification (Danner and Smith 1980; Person and Hirth 1991).

For individual identification at night, collars can be fitted with a combination of light-emitting diodes (LEDs) and lenses, which produce a variety of colors. Brooks and Dodge (1978) used LED collars for individual identification of beavers at night. Batchelor and McMillen (1980) developed LED collars for wallabies; the collars shut down during the day to conserve power. They varied the sizes and colors of the LEDs and programmed their flash patterns to provide individual identification. Flashes were visible at 100 m with binoculars and at 800 m with an image intensifier. The service life of LED collars depends primarily on battery capacity. Some batteries currently available can support flasher units with 5-year service lives yet weigh only 150 g, a characteristic that makes LED collars suitable for long-term marking of numerous nocturnal species. The materials for each collar cost about U.S. $10 (Brooks and Dodge 1978).

TAGS

Tags vary in size, shape, and materials used for their construction. They can be hung from a collar, but usually they are attached directly to an animal's body with specially designed pliers. The most common tagging site for mammals is the ear. Ear tagging is probably no more painful than human ear piercing. For a secure attachment, the tag should be crimped close to the base of the ear where the cartilage is relatively thick and the tag cannot be easily pulled away. Some rectangular tags can be crimped to grip the upper, middle, or lower margin of the ear and thereby increase the number of positional marks for individual identification from a distance (Thorington et al. 1979). Experience in the use of tag pliers will ensure that tags are securely attached.

Ear tags are usually circular or rectangular and numbered and/or color-coded for individual identification. Tags made of plastic or aluminum have been used to mark small terrestrial mammals such as white-footed mice (Korytko and Vessey 1991), pikas (Smith and Gao 1991), rabbits (Thompson and Armour 1954; Davey et al. 1980), and prairie dogs (Slobodchikoff et al. 1991), and medium-size or large terrestrial mammals such as white-tailed deer (Hölzenbein and Schwede 1989), dingoes (Whitehouse 1980), pronghorns (Miller and Byers 1991), elk (Steigers and Flinders 1980), wild goats (Edge and Olson-Edge 1990), bighorn sheep (Festa-Bianchet 1991), bison (Rothstein and Griswold 1991), polar bears (Lentfer 1968; Ramsay and Stirling 1986), and guanacos and vicunas (Franklin 1983). Ear tags have also been used to mark diurnal arboreal species such as gray squirrels (Nixon et al. 1980; Fischer and Holler 1991) and red howler monkeys (Thorington et al. 1979) and the nocturnal tree-dwelling marsupials *Marmosa* and *Caluromys* (Charles-Dominique 1983). Richter (1955) covered the ear tags of marked rabbits with colored reflective tape for nighttime identification.

Ear tags can be pulled out during grooming bouts in both group-living species, which exhibit frequent grooming behavior (Rood 1983), and solitary species (Linduska 1942; Cooley 1948). To overcome the problem of lost tags in red howler monkeys, Thorington et al. (1979) used a redundant coding system based on the color and position of two or more ear tags, and on the age, sex and group membership of individuals. The system permitted individual identification even when a tag was lost or its color or position was altered. Careful planning is needed to develop such systems, but they do preclude problems related to tag loss.

Although ear tagging is convenient, the ear is not always the best site for attaching tags. Tags were applied to the basal joint of the outer toe of the hindfoot in fox squirrels because ear tags caused irritation and injury (Linduska 1942). Similarly, in species in which the pinna is small or nonexistent (e.g., sea otters, monk seals, and elephant seals), tags have been attached to rear limbs or flippers (Loughlin 1980; Henderson and Johanos 1988; Huber et al. 1991). Keith et al. (1968) attached tags to the interdigital webbing of the hindfeet of snowshoe hares, because they were interested in predation mortality in these animals and the hindfeet were always present in the remains.

BETALIGHTS

A Betalight is a phosphor-coated glass capsule containing a small quantity of mildly radioactive tritium gas. When low-level beta radiation from tritium strikes the phosphor, the phosphor produces visible light of a characteristic color. For a given color and capsule size, the brightness of the light can be altered within certain limits by varying the pressure of the tritium gas. Thus, investigators can individually identify numerous animals with Betalight attachments by varying the color of the phosphor, capsule size, and gas pressure. Davey et al. (1980) identified individual rabbits at night with Betalights glued onto ear tags with a cold-curing epoxy resin.

Betalights pose no radiation hazard, because the glass capsule is impervious to tritium and absorbs any residual radiation not taken up by the phosphor. Their use is controlled in many countries, however, and may require permits. Betalights are visible from as far away as 300 m and are supposed to function for 15 years. They are potentially useful for marking nocturnal animals for long-term monitoring studies. Betalights have also been used in conjunction with radiotelemetry in studies of rats (Cheeseman and Mallinson 1980; Hardy and Taylor 1980).

RINGS

Rings are most often used to mark bats (see "Methods of Marking Bats," below), but they have sometimes been used with other mammals as well. Chitty (1937), for example, marked mice and voles with numbered zinc rings placed just above the ankle.

However, the numbers were barely visible after a few weeks and lost their function as individual markers. Godfrey (1954) placed monel rings above the ankles of voles and at the bases of the tails of moles (Godfrey 1955). He soldered the rings to brass tubes containing radioactive cobalt, which enabled him to locate the ringed animals with the help of a Gieger-Muller counter. Cook (1943) placed monel and aluminum rings around the Achilles tendons of opossums, skunks, and muskrats. Takos (1943) fitted rings around the entire legs of muskrats, an approach that yielded better tagging results than Cook's technique. Cooley (1948) successfully fitted monel rings to the inner toes of the hindfeet of fox squirrels.

Temporary Markers

Temporary markers usually persist for less than a year. They include chemical products that gradually fade after application and attachments that become dislodged from animals after a short periods.

DYES, PAINTS, AND POWDERS

Nyanzol dyes have been used as temporary markers on ground squirrels (Melchior and Iwen 1965), chipmunks (Elliott 1978), prairie dogs (Slobodchikoff et al. 1991), polar bears (Ramsay and Stirling 1986), and Hawaiian monk seal pups (Henderson and Johanos 1988). The retention time (visibility) of Nyanzol A and Nyanzol D dyes on ground squirrels varied from 2 weeks to 9 months (Melchior and Iwen 1965). Best results were obtained when fresh dye mixed with an optimal proportion of fresh hydrogen peroxide (3.7 ml 30% H_2O_2 added to 36 ml of fresh dye concentrate in 55 ml water) was applied to body areas with old guard hairs or with a high proportion of underfur. Too much or too little hydrogen peroxide caused the mark to fade in a few days.

Black hair dye and clothing dye have been used to mark rock squirrels (Shriner and Stacey 1991) and mountain goats (Singer 1978), respectively. Colored paints have been used to mark bighorn sheep (Haas 1990), and yellow picric acid and pink Rhodamine B have been used to mark snowshoe hares and cottontail rabbits (Keith et al. 1968; Brady and Pelton 1976). The acid was visible on the rabbits for 7 months after application. Aniline red dye applied to ground squirrels was visible for from 30 to 60 days (Evans and Holdenried 1943). Bison marked with hair bleach were identifiable for about 6 months (Rothstein and Griswold 1991).

Paints and dyes can be rubbed directly onto the body of an animal or can be delivered from a distance with solenoid-triggered spray devices (Simmons and Phillips 1966), marking darts (available from the Palmer Chemical & Equipment Company; see Singer 1978), or paint pistols (available from the Nelson Paint Company). Bison were marked with hair bleach delivered with a bow and a set of blunt arrows (Rothstein and Griswold 1991). The position and color of the paint or dye mark may be varied to facilitate individual identification.

Marking small mammals with fluorescent powders and paints has become increasingly common in recent years because of the ease with which they can be applied (McCracken 1984; Lemen and Freeman 1985; Jike et al. 1988; Mullican 1988; Kaufman 1989). Animals marked with such materials are viewed with an ultraviolet light for the detection of fluorescence and individual identification. The fluorescent trails left by powdered animals can also be used to study their movements (Jike et al. 1988).

BODY ATTACHMENTS

Streamers and colored discs can be attached to the bodies of some animals for temporary identification. Colored plastic streamers, for example, have been attached through a slit or with a metal tag to the ears of bighorn sheep (Aldous and Craighead 1958), white-tailed deer (Downing and McGinnes 1969), elk (Harper and Lightfoot 1966), and mule deer (Harper and Lightfoot 1966). In young pronghorn antelope, white-tailed deer, and mule deer the streamers were passed through a slit in front of the Achilles tendon and fastened with a jesse knot (Queal and Hlavachick 1968). The streamers were visible at distances up to 300 m and persisted for nearly 9 months.

Problems Related to Marking

ANIMAL CAPTURE

Many of the techniques available for marking mammals require that the animals be captured. Animal capture involves considerable human effort (see Chapter 8) and poses certain risks to the animals and their captors. To avoid the problems of animal capture, investigators can use automatic tagging devices or remote marking techniques (i.e., those that allow marking from a distance).

Natural markings can also be used for individual identification. In many species, for example, individuals can be identified by pelage marks (e.g., tigers,

Schaller 1967; lions, Pennycuick and Rudnai 1970; hyenas and wild dogs, van Lawick-Goodall and van Lawick-Goodall 1970, Frame et al. 1979; giraffes, Foster and Dagg 1972; and zebras, Klingle 1965), by facial characteristics and head-hair patterns (e.g., various primate species, van Lawick-Goodall 1967; Oppenheimer 1969; Dittus 1977; Harcourt and Groom 1972), or by horns, scars, and other features (e.g., African antelopes, Jongejan et al. 1991; Williamson 1994). The feasibility of using natural markings can be assessed only through careful observations of target animals. Such observations should be conducted before launching a capture and marking program.

If there is no alternative to animal capture, steps should be taken to minimize the need for recapture. Investigators should employ techniques that will ensure that marks are functional for the entire study period. If the clarity of a mark or its period of usefulness cannot be ascertained until sometime after the animal is released, investigators should use several marking techniques to help ensure that an individual remains recognizable by one type of mark or another for as long as is necessary. Redundant marking systems have been used with polar bears (Lentfer 1968), snowshoe hares (Keith et al. 1968), ground squirrels (Evans and Holdenried 1943; Melchior and Iwen 1965), and red howler monkeys (Thorington et al. 1979).

ADVERSE EFFECTS OF MARKS

Marks themselves may also adversely affect individual animals. Physical injury and irritation can generally be avoided with the proper use of equipment and techniques. Some marked individuals, however, may also encounter social problems. Beale and Smith (1973) reported, for example, that the number of pronghorn fawns abandoned by their mothers increased among fawns whose appearance was altered by collars. In such cases, alternative marking techniques must be used, or marking should be terminated. Preliminary trials conducted before a marking program is initiated can identify the technique and type of mark most appropriate for the target species and the investigation and so prevent many problems.

Methods of marking bats

THOMAS H. KUNZ

A wide range of methods and devices are available for marking bats. Each method has advantages and

disadvantages, and some methods are better than others. Research during the past four decades has shown, for example, that some marking methods and devices, long considered appropriate for temperate species, are unsuitable for most tropical species (see "Wing Bands," and "Ball-Chain Necklaces," below).

Devices considered suitable for marking bats include numbered lip-end (flanged) and butt-end metal bands, plastic split-rings, ball-chain necklaces (with numbered butt-end bands or split-rings), ear tags, light tags, tattoos, and punch marks (Greenhall and Paradiso 1968; Stebbings 1987; Barclay and Bell 1988; Handley et al. 1991).

Wing Bands

Butt-end (bird) bands, used in early studies, caused several types of wing injuries to bats, including local edema, wing punctures, ingrown bands, formation of bone and cartilage spurs, and irreparable nerve damage (Hitchcock 1957; Davis 1960; Herreid et al. 1960; Dwyer 1965; Perry and Beckett 1966; Greenhall and Paradiso 1968). Bats often chewed on these soft aluminum bands, causing the identifying inscriptions to become obscured and in some situations causing serious injury to the wearer.

Lip-end (flanged) bands, which were introduced in Europe and the United States in the early 1960s, were designed to remedy many of the problems caused by butt-end bands (Davis 1966; Greenhall and Paradiso 1968). Lip-end bands have been used successfully on several species of microchiropterans, but they should not be used on the forearms of megachiropterans.

A lip-end band is attached to a bat by placing an opened band over the shaft of the forearm and closing it with pressure from a thumb and forefinger (Fig. 61A). Ideally, when a lip-end band is closed around a bat's forearm, it should form a nearly round closed ring. The edges of the band should be smooth, and its inside diameter should be large enough to prevent irritation of the underlying tissue. When the band is closed, a small space should remain between the lip-ends and the wing membrane. This gap minimizes the possibility that the wing membrane will become irritated, leading to an overgrowth of tissue.

Lip-end wing bands for bats are currently available in aluminum and anodized (hardened) aluminum and in at least two alloys, magnesium-aluminum and nickel-chromium (incoloy). Alloys tend to be harder than aluminum and thus sustain much less damage if chewed by bats.

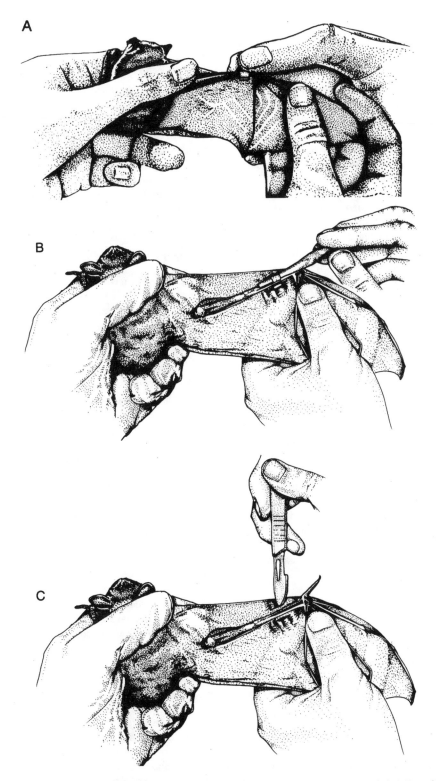

Figure 61. Methods for attaching bands to the forearms of bats. Bats are held with one wing extended. A. Bat with a small propatagium. The bander slides an open lip-end band over the forearm near the wrist and closes it. B–C. Bats with a large propatagium. B. The lip-end band is inserted through a slit (~8 mm) in the propatagium and then closed. C. A butt-end band is inserted through two slits (each ~8 mm) made parallel to the forearm, and the band is closed around the shaft of the forearm. Illustrations prepared by Susan Thomson.

Use of lip-end bands to mark small microchiropterans generally does not require physical modification of the wing membrane. For species with large propatagia, however, the membrane must be slit to accommodate the bands and prevent distortion of the propatagium or antebrachial membrane (Bateman and Vaughan 1974; Bonaccorso et al. 1976). For most species the lip-end band is inserted through a small (~8 mm long) slit made in the propatagium so that the band can be closed over the forearm (Fig. 61B). This technique is often more effective for temperate microchiropterans than for neotropical species. If butt-end or plastic split-rings are used, slits are made in both the propatagium and the adjacent chiropatagium (Fig. 61C). Each slit should be located near the distal third of the radius, to avoid the muscular part of the forearm, and at least 1 cm from the wrist (proximal to the carpus), so that the band will not irritate the adjacent metacarpals.

The membranes generally do not bleed when cut, and the band can usually be attached to the bat without complication. If bleeding does occur, however, it can be stopped with a small amount of pressure applied with a thumb and forefinger, or small vessels can be cauterized by applying alum from a styptic pencil or small amounts of silver nitrate. Application of a topical antibiotic to the cut areas will reduce the risk of infection.

A special aluminum applicator is required to open plastic split-rings for attachment or removal (Fig. 62). Special tools are also needed to open and close stainless steel and monel butt-end bands (Fig. 63), especially if the bat has been injured by a band that was improperly applied or chewed. Two pairs of high-quality fine Spencer-Wells hemostat forceps (H. Spencer, pers. comm.), small needle-nose pliers, or tissue forceps with claws can be used for this purpose. Other methods for opening and closing metal bands are discussed in Greenhall and Paradiso (1968). If bands are not supplied in a preopened position, they should be opened in advance of fieldwork and strung on metal or plastic rods, ready for use (Greenhall and Paradiso 1968).

Plastic split-rings are available in several sizes and colors (single or bicolored) and are suitable temporary markers for bats. However, they are easily damaged and broken if bats chew on them. If plastic split-rings are improperly applied, the rings can injure the wing bones and underlying tissue. Wing injuries can be reduced by rounding the sharp edges of the split-rings with a fine file, by slightly widening the gap before using a band (Stebbings 1987), and by ensuring that the band diameter is sufficient to prevent irritation of underlying tissue. Bands embedded in or overgrown with tissue should be cut with scissors and removed immediately.

For some taxa, bands placed on the forearms of adults can also be used on neonates and young of the year. In fact, newborn bats appear to adjust better to banding than adults (T. H. Kunz, pers. obs.).

Thumb Bands

Stainless steel or monel butt-end bands and plastic split-rings can be used to mark the thumbs of large ($\geq$100 g) megachiropterans (Fig. 63). Metal bands are preferred over plastic ones because they are harder and more resistant to chewing, and therefore, less likely to be damaged or cause injury to the underlying tissue (Parry-Jones and Martin 1987; H. Spencer, pers. comm.).

Thumb banding is less effective for most small (<100 g) megachiropterans because their thumb pads and claws are too small to retain the bands. Small-diameter aluminum butt-end bands can be used for some small megachiropterans after the ends of the bands are cut to reduce their size and the edges are smoothed with a fine file.

Before attaching a thumb band, different band sizes should be tested to ensure a proper fit. Thumb bands should be loose enough so that they can be moved freely over the shaft of the thumb without causing irritation, but they should not be so loose as to slip over the thumb claw. Investigators must ensure that the ends of the bands do not overlap when the band is closed. Thumb bands should never be used on animals that are still growing. Other temporary marking devices should be employed until the young have reached adult size (Fleming 1988).

Ball-Chain Necklaces

Necklaces fashioned from stainless steel ball-chains onto which butt-end bands or split-rings are strung for identification purposes have proven highly effective for marking bats for long-term studies. Many neotropical species and some megachiropterans are more effectively marked with ball-chain necklaces than with forearm bands. The necklaces are durable, lightweight, flexible, and rarely lost, and they have been used successfully on both megachiropterans

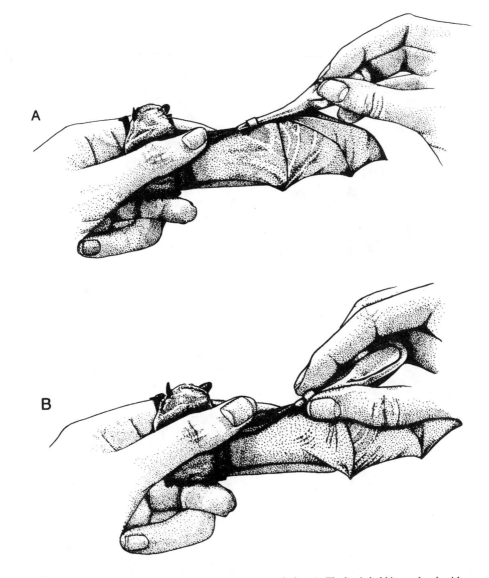

Figure 62. Method for attaching a plastic split-ring to the forearm of a bat. A. The bat is held in one hand, with one wing extended. The bander slides the open split-ring over the forearm near the wrist using the applicator supplied by the manufacturer. B. With the split-ring positioned and firmly held, the bander pulls the applicator away, allowing the band to close around the forearm. Illustrations prepared by Susan Thomson.

and microchiropterans with little injury or mortality (Fleming 1988; Heideman and Heaney 1989; Spencer and Fleming 1989; Handley et al. 1991; C. Tidemann, pers. comm.; R. C. B. Utzurrum and T. H. Kunz, unpubl. data). They should not, however, be used on young bats until the bats have achieved adult size (Fleming 1988).

To prepare a ball-chain necklace, the investigator cuts the chain into appropriate lengths with cutting pliers. A clasp or connector is attached to one end of the chain. Investigators may find it helpful to expand the clasp slightly to facilitate closure. A hemostat or pair of needle-nosed pliers is ideal for closing the clasp. Closed necklaces of a given size can be strung

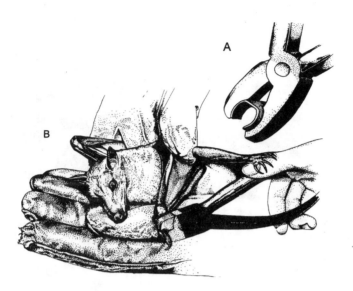

Figure 63. Method for attaching a butt-end band to the thumb of a megachiropteran. A. An open butt-end band is inserted into appropriate closing pliers and placed around the thumb. B. The pliers are used to close the band around the thumb. Illustrations prepared by Susan Thomson.

on a safety pin and removed in the field as needed. Unconnected chains may be stored in separate compartments such as those used for storing fishing lures (Handley et al. 1991; R. C. B. Utzurrum and T. H. Kunz, unpubl. data).

The most suitable size and length of a ball-chain necklace should be determined empirically for each sex of each species being studied. Handley et al. (1991) recommended sizes and lengths of ball-chain necklaces for 46 species of microchiropterans captured in Panama, which can serve as guidelines. Ball-chains are available in several sizes. Handley et al. (1991) recommended no. 2 or no. 3 ball-chains for most microchiropterans and for megachiropterans weighing less than 60 g. I recommend a no. 6 ball-chain for species weighing more than 60 g (Utzurrum and Kunz, unpubl. data).

Ball-chain necklaces are best fitted to bats by two people working together (Fig. 64). Although a single person can attach a necklace, considerable practice and, sometimes, a restraining device are required (Handley et al. 1991). A necklace should be small enough to prevent removal by the bat and large enough so that it does not abrade the underlying skin or choke the bat. If skin irritation occurs, necklaces should be enlarged or replaced with another type of marking device. Necklaces should not be used on species with large neck and chest glands or enlarged

facial ornaments, and they should not be used on growing animals. Although ball-chain necklaces are judged suitable for many free-ranging bats, they may not be appropriate for marking bats maintained in captivity. Skin irritation has been observed in some captive bats, especially if such animals become obese or if food accumulates under the necklace when bats feed from dishes (K. Atkinson, pers. obs.).

Monel, stainless steel, plastic split-ring, or aluminum butt-end bands that are strung on ball-chain necklaces should fit the chain snugly so that they do not move and irritate the underlying skin. Aluminum butt-end bands with an inside diameter of 2.28 mm (National Band and Tag Company; Appendix 9) fit firmly around a no. 3 ball chain. A no. 5 monel band (National Band and Tag Company) can be used on no. 3 and no. 6 ball-chains. Sharp edges of the butt-end should be rounded with a file before stringing. Ball-chain necklaces usually become buried in the pelage and cannot be seen without parting the fur. When bats with necklaces are recaptured, band numbers are most easily read if the bands are rotated to the back of the neck.

Color Bands

Investigators can use color bands to identify individual bats in roosts (e.g. Gaisler and Nevrly 1961;

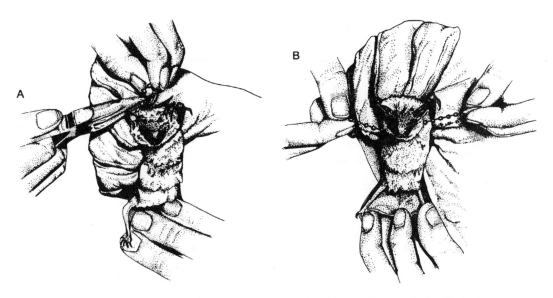

Figure 64. Method for attaching a ball-chain necklace to a bat. A numbered, butt-end band or plastic split-ring is strung onto a chain of appropriate length. A. One person holds the bat. A second person grasps the second ball from the end of the chain firmly with needle-nose pliers and the clasp-end of the chain with the forefinger and thumb of the other hand. B. The chain is placed under the chin and around the neck of the bat so that it can be closed at the back of the neck. Illustrations prepared by Susan Thomson.

Bradbury and Emmons 1974; Bradbury and Vehrencamp 1976; McCracken and Bradbury 1981; Fleming 1988) and during flight (Barclay and Bell 1988). Plastic split-ring bands can be used, but color-anodized butt-end and lip-end bands are more durable. On the other hand, colored plastic split-ring or color-anodized aluminum, alloy, or incoloy bands have relatively poor reflective properties, at least compared to reflective tape (e.g., Scotchlight) or paint. Colored reflective tape can be attached to metal or plastic bands to provide different color combinations and to enhance visibility from a distance. Underlying numbers and inscriptions may, however, become obscured (Gaisler and Nevrly 1961). The number of individuals that can be uniquely marked in a study will depend upon the number of color combinations available and upon whether bands are placed on one or both forearms.

Most reflective tape comes with an adhesive backing, but the bond between the band and the tape can be greatly enhanced by using a cyanoacrylate adhesive (e.g., Super Glue). One disadvantage of applying paint and reflective tape to bands is that they are often chewed off by the bats, making regular replacement necessary.

Other Marking Methods

Most other methods of marking bats produce temporary marks or pose risks of injury to the animals. Freeze brands, punch marks, and tattoos on wings generally are not effective for long-term studies (Griffin 1934; Bonaccorso and Smythe 1972; Kleiman and Davis 1974; Marshall and McWilliam 1982; Parry-Jones and Martin 1987; Heideman and Heaney 1989). Scars and tattoos usually disappear within a few months, making it necessary to remark on a regular basis (Kleiman and Davis 1974; Bonaccorso et al. 1976).

Ear notching has been used successfully to mark some small megachiropterans (H. Spencer, pers. comm.), but this technique does not permit individual recognition and thus is of limited use. Toe clipping should not be used because toes of bats are important for normal roosting. For short-term studies, wings and ears of bats can be marked with waterproof felt-tipped pens or typewriter correction fluid.

Several types of collars other than ball-chain necklaces have been used on bats, but most are made of materials that deteriorate over time and thus cannot be used for long-term monitoring studies (e.g., Spencer et al. 1991; Balasingh et al. 1992). Radioactive tags (including Betalights) have also been used to mark, locate, and follow the movements of hibernating bats (Punt and van Nieuwenhoven 1957; Gifford and Griffin 1960; Cope et al. 1961; Harvey 1965).

Ear tags (fingerling tags) have sometimes been used to mark both microchiropterans (Mohr 1934; van Heerdt and Sluiter 1958) and megachiropterans (Sowler 1984; D. Makin, pers. comm.). They have been applied to the hindlimbs, uropatagium, wings, and pinnae of some species, but I generally do not recommend their use. The attachment of such tags to the pinna in particular can adversely affect a bat's ability to echolocate. Ear tags placed on other body parts (e.g., trailing edge of uropatagium or hindlimb) may be suitable as temporary markers for some species (e.g., megachiropterans). Bats marked with fingerling tags should be carefully evaluated (both in the wild and in captivity) to ensure that the tags are not lost and the bats are not injured.

Small pieces of plastic or reflective tape attached with adhesive to the backs or heads of bats may enhance their visibility in some situations. Such marks were used to establish the relative numbers and fidelity of hibernating bats (Daan 1969) and to quantify movements of mothers and pups in maternity roosts (McCracken and Gustin 1991). In these instances, the reflective tape was marked with unique numbers or identifying marks so that individual bats could be distinguished at a distance. Such visually conspicuous bands may obviate the need to handle and disturb bats in roosts. Night-vision devices or low-light-level video cameras may be useful for observing bats marked in this manner (Frantz 1989; McCracken and Gustin 1991).

Conventions and Management of Banding Data

Numbered bands or multiple color bands applied to a bat must be read correctly at the time of attachment and when the bat is recaptured. Because accuracy is critical, each number should be read by the observer and then repeated by the recorder. To avoid ambiguity in reading band numbers, bands should be positioned on the forearms or thumbs so that the numbers can be read in a distal to proximal direction. Otherwise, number combinations that include 6s, 0s, and 9s may be read incorrectly. Similarly, colors of split-ring bands on the forearm or thumb should be read in a distal to proximal direction. Such conventions are extremely important if individuals other than the bander have the opportunity to record data on marked bats. By convention, I also recommend that males be banded on the right wing (forearm or thumb) and females be banded on the left (Fleming 1988).

Regional programs for issuing bands and managing records of banded bats are necessary for a successful banding program (Stebbings 1987; Lowe 1989). Well-established programs can set standards for band management and prevent different investigators from using bands with duplicate number series or colors in the same or adjacent research areas. When researchers use bands without unique marks, for example, those who recapture a banded individual cannot identify the original bander or locate the original banding records. Important information is thus lost. Although bird-banding schemes are highly developed in many countries, similar programs for bats are not well established or maintained. In most parts of the world they do not even exist.

Precautions

Marking methods should be chosen to minimize disturbance and injury to an animal. If new marking devices are employed, a proven device should be used concurrently to allow for validation of the new method. In doing so the investigator can establish rates of band or mark loss to quantify permanency on a particular species.

Because many species of bats have never been marked with bands or other devices, investigators should be alert to evidence of injuries or possible behavioral modifications that may be caused by a particular marking method. A band should be removed from a bat if it impedes the animal's normal movement or if injury or abnormal behavior (e.g., intensive chewing on the band or grooming) is observed.

Recording Mammal Calls

William L. Gannon and Mercedes S. Foster

Introduction

Mammals as Noisemaker

Through the processes of evolution, mammals have developed a great diversity of methods and mechanisms for producing sounds that carry information. Depending on the ecological niche occupied and the sound-transmitting medium used, many parts of the mammalian body can be applied to the production of acoustic signals. For example, quills can serve as a stridulating organ (tenrecs), and tails can serve as rattles (Bornean rattle porcupine, *Hystrix*); flippers and tails can be used for slapping water (Cetacea, Castoridae) and arms or legs for branch shaking, ground thumping, or chest beating (some heteromyid rodents and primates). Many mammalian sounds are produced by the soft cartilaginous or bony structures of the face: hedgehogs smack their lips and tongue; the Tasmanian devil (*Sarcophilus harrisii*) claps its jaws; rodents chatter their teeth; and many artiodactyls produce sounds with their nasal cartilages

(McComb 1991). In almost all mammals, however, the larynx with its vocal folds is the basic generator of sounds for communication.

The sounds mammals make are as diverse as the structures that produce them, and they cover a wide range of frequencies. Whales (*Megaptera, Balaenoptera*), for example, produce intense, low-frequency calls (10–20 Hz) that can travel hundreds of kilometers underwater (Payne 1983). Calls of bats in the suborder Microchiroptera range from 4 kHz to more than 220 kHz. The highest-frequency mammalian vocalization known is that of the rough-toothed dolphin (*Steno bredanensis*), which reaches about 350 kHz (Sebeok 1977).

A Need for Better Documentation

Most of the more than 4,600 known species of mammals are probably capable of acoustic communication (Sebeok 1977). Yet the existing inventory of mammalian sounds is woefully inadequate. Consider, for example, the holdings of the two largest ar-

Table 20. Numbers of Species of Mammals with Calls Archived in the Borror Library of Bioacoustics at Ohio State University and the Library of Natural Sounds at Cornell University[a]

Mammalian order	Borror Library of Bioacoustics	Library of Natural Sounds
Didelphimorphia	3 (5)	1 (1)
Xenarthra	1 (1)	3 (7)
Insectivora	1 (3)	2 (2)
Chiroptera	3 (4)	2 (11)
Primates	10 (36)	54 (309)
Carnivora	16 (90)	34 (108)
Cetacea	1 (1)	8 (14)
Proboscidea	1 (2)	1 (1)
Perissodactyla	6 (13)	4 (5)
Hyracoidea	0	2 (7)
Artiodactyla	9 (15)	16 (38)
Rodentia	17 (85)	39 (95)[b]
Lagomorpha	1 (1)	3 (3)
Totals	69 (256)	169 (601)

[a]Numbers as of 1995. Numbers in parentheses indicate numbers of available cuts or usable tape segments.

[b]Species (cuts) represented by Sciuridae = 30 (81).

chives of mammalian sounds. Recordings of only 69 species of mammals are deposited in the Borror Library of Bioacoustics at the Department of Zoology, Ohio State University, and those of only 169 species are archived in the Library of Natural Sounds (LNS) at Cornell University (Table 20). These recordings combined represent only about 5% of the species of mammals, and most of the species represented are primates and sciurids.

This poor documentation of mammalian vocal diversity may in part reflect the difficulty encountered in locating mammals to record, a lack of knowledge of recording techniques and archiving procedures, and lack of sound-recording gear. In this chapter we discuss common types of recording equipment and where they can be purchased, equipment for field re-

cording, field methods for recording, methods for handling and vouchering tapes, and analysis of calls.

Equipment

A cassette tape recorder (weighing ca. 1.3 kg) equipped with a high-quality microphone will suffice for recording most mammalian vocalizations. A number of such inexpensive, lightweight, audio-recording systems that are simple to operate are now on the market (Appendix 9). Specifications of some available equipment are provided in Table 21. Several major companies supply audio equipment, including Saul Mineroff Electronics in New York and Full Compass Systems in Wisconsin. Both companies are accustomed to selling to biologists, and both will provide customized equipment. Other companies and sources are listed in Appendix 9.

Audio Tape Recorders

Ideally, a recorder used in the field must be portable, easy to use, and of high quality. It must have a VU meter, a peak meter, a recording-speed control, and a control for adjusting gain (signal intensity). The VU meter indicates the strength of the incoming signal in decibels (dB). The peak meter, or light-emitting diode, alerts the researcher to acoustical overload, which can lead to poor-quality recordings and playback distortion. Because some calls may not register on the peak meter, recorders equipped with such meters should also have VU meters so that the investigator can determine if the strength of the incoming call is excessive. Recorders with automatic gain control should be avoided, because the investigator may have to adjust signal intensity in some recording situations. Recorders to be used with the audible calls of mammals should have a flat recording response of 20 Hz to 17 kHz.

Tape speed may fluctuate with battery use, and some recorders cease to function when batteries begin to lose their charge. Therefore, batteries must be replaced or recharged frequently to avoid loss of recording quality. Rechargeable batteries lose their charge more quickly than alkaline batteries.

OPEN-REEL RECORDERS

Open-reel recorders have exposed, detachable tape reels, which can be replaced with spools of audio

Table 21. Sources and Specifications of Audio Equipment for Recording Mammal Vocalizations

Name	Source[a]	Accuracy range[b]	Frequency[b]	Weight (kg)	Power	Other features
Mini-2, bat	1	± 1 kHz	15–160 kHz	0.2	2 AA batteries	3 output jacks
S200 detector	1	± 2 kHz	10–180 kHz	1.7	6 AA batteries	<0.5% harmonic distortion
Anabat II	5	± 0.5 kHz (calibration tone allows for accurate compensation for variable tape recorder speeds)	10–200 kHz	0.7	9-V battery	4, 8, 16, 32 divide-by ratios, output jacks for headphones and tape recorder
Recorder, cassette, Sony TCD-D5 Pro II	2,3	4 tracks, 2 channels, stereo	40 Hz–14 kHz, S/N 55 dB	1.7	2 D batteries, AC/DC	Speaker, counter, 2 Cannon input jacks, 2 output jacks, VU meter
Recorder, cassette, Sony TCD-D5M	2,3	4 tracks, 2 channels, stereo	40 Hz–14 kHz, S/N 55 dB	1.7	2 D batteries, AC/DC	Same as TCD-D5 Pro II, except input/output jacks are minijacks
Recorder, cassette, Uher CR-160AV	2,3	2 or 4 tracks, stereo, 2 speeds	30 Hz–16 kHz, S/N 55 dB	2.7	6 D batteries, Uher 7217 or 7131 rechargeable battery	Speaker, counter, clock timer, DIN microphone, VU meter, 3 input jacks, 3 output jacks
Recorder, cassette, Marantz, PMD 201 or 221	2	2 tracks, 1 channel, 2 speeds	40 Hz–13 kHz, S/N 55 dB	1.3	3 D rechargeable batteries, AC/DC	Auto shut-off, memory rewind, 2 input jacks, 2 output jacks, speaker
Recorder, cassette, Marantz PMD 420 or 430	2	2 or 4 tracks, stereo, 2 speeds	35 Hz–17 kHz, S/N 51 dB	1.4; 1.3	3 D rechargeable batteries, AC/DC	Same as 200 series except 3 input jacks, 3 output jacks, built-in microphone attenuator, with both DIN and minijack inputs
Recorder, DAT, Sony TCD-D10 Pro II	2	Portable, digital audio, absolute time recording,	Defines frequencies in real time, records at 48/44/33 kHz (switchable)	2	4 D batteries, AC/DC, or its own rechargeable battery pack	Tape time-clock, speaker, digital input and output

(Continued)

Table 21. (*Continued*)

Name	Source[a]	Accuracy range[b]	Frequency[b]	Weight (kg)	Power	Other features
Recorder, DAT, Marantz PMD 700	2	Portable, digital audio, absolute time recording; 1-bit, 256K sampling, analog to digital	Defines frequencies in real time, records at 48/44/33 kHz (switchable)	1.5	4 D rechargeable batteries, AC/DC, or its own rechargeable battery pack	RCA line in/out, digital play/record
Microphone, SMI hydrophone	1	–100 dB sensitivity	10–180 kHz	0.25	PSM2 power supply	2–10 m lead, spherical hydrophone (0.5-in. diameter)
Microphone, Audio-Technica 835, electret, condenser	2	Unidirectional	40 Hz–20 kHz	0.212	1 AA battery	Maximum input = 115 dB, good long-distance pickup
Microphone, Audio-Technica 815A (long shotgun) or 835A (short microphone), electret, condenser	2	Unidirectional (shotgun)	40 Hz–20 kHz	0.26	1 AA battery	Distortion-free, maximum input = 140 dB
Microphone, Sennheiser	3			Variable	K3U power module with 5.6-V battery	
ME 20		OC	50 Hz–16 kHz S/N 64 dB			
ME 40		OC	80 Hz–16 kHz S/N 64 dB			
ME 80		Shotgun	40 Hz–15 kHz S/N 70 dB			
ME 88		Shotgun	40 Hz–16 kHz S/N 70 dB			
MKE 10-3		Spot head lavalier	60 Hz–18 kHz S/N 64 dB			
Amplifier, SME-BA high-gain microphone amplifier	3	High-gain booster	—	0.15	None: power from recorder	Amplifies input ×28 dB
Reflector, Dan Gibson	3	Diameter is variable	Variable	Variable	—	Focal length at plane of dish
Reflector, Atherstone, MK3	4	Diameter = 51 cm, focal length = 12.7 cm	Variable	0.5	—	Focal length at plane of dish

[a]Sources: 1 = Ultra Sound Advice; 2 = Full Compass Systems, Ltd.; 3 = Saul Mineroff Electronics, Inc.; 4 = Richard Margoschis Natural History Sounds; 5 = Titley Electronics, Ltd.

[b]OC = omnidirectional cardioid; S/N = signal-to-noise ratio; all ratios with Dolby noise reduction systems inactivated.

tape of variable lengths. Such recorders are preferred in studios and laboratories, because they permit recording at several speeds and because tape on an open-reel spool is easily spliced for editing. Edited tapes can be used for playback and also are more easily archived and conserved than other types of tapes. Because blank tape before and after recordings can be discarded, edited tapes require less space.

Uher and Nagra IV recorders have been the standard open-reel recorders for monitoring mammalian calls for the last four decades (Struhsaker 1975; Slobodchikoff and Coast 1980; Randall 1989). Their quality is excellent, but they are costly (U.S. $2,500 to $6,000). Archivists at the LNS (G. Budney, pers. comm.) consider the open-reel Nagra recorder to be the best machine for recording animals and producing archival tapes, despite technical advances in other types of recorders (see the two sections that follow).

CASSETTE TAPE RECORDERS

The Sony TCD, WM, and Pro models and the Marantz PMD series are typical of high-quality cassette tape recorders. These recorders are considerably cheaper (U.S. $250 to $700) than either Uher or Nagra open-reel tape recorders and also more widely available. They are now commonly used for recording mammal vocalizations (e.g., Gannon and Stanley 1991; Tamura and Yong 1993). They are also frequently used in the field with ultrasonic bat detectors (e.g., MacDonald et al. 1994; C. Corben, pers. comm.). Any cassette recorder that has inputs for a microphone, remote operation, and auxiliary equipment such as a timer can be used.

DIGITAL-AUDIO TECHNOLOGY AND COMPACT DISKS

The new digital-audio technology (DAT) is also available for portable recording. Sony produces the TCD-D10 Pro II professional portable DAT recorder-player, which has a liquid-crystal multifunction display, peak-level meters, tape time-clock, high-speed spooling, built-in speaker, XLR microphone inputs, and digital input and output; it retails for about U.S. $900. Aside from real-time recording, DAT machines are able to play at 3× and 16× speed. Some researchers have begun using DAT in a system that periodically activates via a relay switch and samples the audio realm at regular intervals. Unattended taping periods of 2 to 4 hours (depending on tape speed) are possible.

Compact disks (CDs) can also be used for portable recording. They are made with a compact, noncontact recording system that produces near-perfect recordings. CDs are not used for permanent archival recordings, however, because CD surface materials may degrade and flake from their plastic base over time.

Microphones

The microphone is the most important part of the recording system and should be of the best possible quality (Table 21). A high-quality microphone produces no distortion in the 20 Hz to 10 kHz range, has its own power source, and can use a variety of heads. Sennheiser microphone systems, for example, offer a power module with several interchangeable heads, an arrangement that enables an investigator to switch within seconds from a head that is sensitive 360° around the microphone (omnidirectional) to one that focuses on a narrow 20° zone (unidirectional or shotgun; e.g., Sennheiser ME 88). An investigator can focus an omnidirectional microphone like a shotgun microphone by mounting it at the focal point of a parabolic dish (e.g., Dunford and Davis 1975). The parabola helps to concentrate animal sounds and insulate them from background noise while they are being recorded. Parabolas are bulky, however, and vary in diameter from 50 to 240 cm. With the advent of high-quality shotgun microphones, fewer researchers are using parabolas.

High-quality shotgun microphones equipped with preamplifiers are now commonly used for recording mammalian calls. We recommend the Sennheiser ME 80 and ME 88 with a K3U power module, and the Audio-Technica microphones (Table 21). Both brands of microphone perform well at a variety of temperatures and humidities, produce high-quality audio recordings, and record for as many as 500 hours on one set of batteries. An impedance-matching transformer is required when using Sennheiser microphones with Sony recorders. Audio-Technica microphones do not require a transformer and operate on one AA alkaline battery. They perform as well as Sennheiser microphones and retail for about half the price; models AT 815A and 835A are long and short shotgun microphones, respectively (Table 21).

The microphone should be placed as close as possible to the sound source to minimize attenuation of high frequencies (>20 kHz) in air (ambient temperature and humidity also affect signal attenuation). Furthermore, the microphone should be placed as high as possible from the ground to avoid absorption, attenuation, or scatter of the sound signal. Effective microphone placement for a chipmunk (*Tamias*), for example, would be 1 m above the ground and 2 m from the calling animal.

Tape

Except for tape thickness, tape selection is largely subjective, because most tapes available are high-quality and will make fine audio recordings. Thin tapes stretch and degrade faster than thicker tapes. We recommend that investigators use reel tapes of 1.5 mil (0.0038 cm) thickness. Cassette tapes usually allow 30, 60, 90, or 120 minutes of recording. Longer tape times often mean thinner tapes. We recommend using 60-minute tapes (30 minutes on each side) for field recording, because they are relatively easy to edit, are sufficiently thick, and provide enough time to record long call sequences.

Recordings should be made only on one side of a tape to prevent audio "bleeding." Bleeding causes recorded segments from one side of reel tapes to be heard on the other side. With cassette tapes, this problem results from the transfer of recordings from one portion of the tape to another portion on the same side. Stored tapes may bleed when part of the magnetized portion of one side is transferred by contact to the back of the tape above it. Bleeding can also occur when playback heads on tape recorders are improperly aligned.

The greatest concern regarding cassette tapes, open-reel tapes, DAT, and CDs is their longevity. After 30 years, tapes become so brittle that they disintegrate during one playing. For this reason it is best to archive research tapes at a sound-recording library. These institutions regularly curate their collection, produce new voucher copies of tapes, store them in light- and humidity-controlled environments, and make recordings available for use by others.

Other Considerations

No equipment is sturdy enough to withstand field conditions indefinitely. Thus, an investigator should always plan for equipment failure by carrying backup equipment. In addition, an investigator should select a recorder with a hard case or purchase a carrier that will effectively protect the machine. All recorders should

have sturdy connector ports that will withstand repeated use and movement (e.g., Cannon XLR instead of ¼-in. phone minijack connectors). Minijack connectors are universal, but loosening of internal solder joints may lead to degraded signals. High humidity can cause malfunction of microphone membranes as well as greater attenuation of the sound signal. Equipment should be stored in desiccant.

Field Gear Checklist

The following equipment is needed for recording mammal vocalizations in the field: two cassette tape recorders (Nagra, Uher, Sony, or Marantz; one for spare or playback); two microphones with interchangeable heads (Audio-Technica or Sennheiser); one set of headphones with head-mounted microphone; connector cables and adapters; shoulder straps and cases; windscreens to fit microphone heads; binoculars; audio tapes to match the recorders (the number of tapes will vary with the study); batteries appropriate for the recorders and microphones and the expected duration of the work; a field notebook and pen; and a sturdy field storage box.

Recording calls of free-ranging mammals

Before commencing fieldwork the recording gear should be assembled and tested. If possible, the researcher should listen to taped vocalizations of the target species and become familiar with its calls.

Documenting Calls

It is convenient to document calls directly on tape. Sound archivists recommend that the researcher stop the recorder after a call and then start it again to record his or her audio notes. This procedure separates the calls from the investigator's notes by "click" sounds audible during tape playback; this separation helps in editing the tape. Stereo systems have the advantage of allowing an investigator to record audio notes on one channel, via a head-mounted microphone, and the animal vocalizations on the other channel, with a second microphone. The researcher speaks between the calls of the subject without interrupting the call sequence or losing track of call duration. Written documentation must follow soon after the recording is completed. Many recording protocols require that each subject be recorded for a set amount of time or sampled at regular time intervals. Such protocols can be helpful in editing.

Recordings should be thoroughly documented during each recording bout or session. This information should be provided both on the tape and in a notebook. Documentation should include the following: name of the investigator; type of recorder and type of microphone used; notation of other accessories used (e.g., parabola, preamplifier, wind screen); species, sex, age, and general behavior of the individual recorded; calling location and distance from the microphone; date and time of recording; specific locality; and habitat description (e.g., dominant plant species, percent cover, slope, aspect, direction of travel while sampling). A map of the study area should be included in the notebook. The time is noted at several points during a recording. Temperature, weather, and comments on the subject's behavior are repeated for each recording. Because habitat features and climatic conditions can cause distortion, absorbance, and attenuation of sound and can affect call morphology, such information may be important during analysis to explain characteristics of individual recordings (Marten and Marler 1977; Marten et al. 1977; Wiley and Richards 1982).

Recording Techniques

Recording techniques vary with the research animal. For diurnal recording, an investigator walks through a site with the recorder on to record the noise of his or her own movements and other nontarget animals, and to document the first calls of the target species. In addition, diurnal mammals are usually recorded after an individual is located visually. Nocturnal mammals are most often recorded after capture in mist nets or with trip-lines (see "Miscellaneous and Ancillary Methods for Capturing Bats," Chapter 8) or after trapping. Whales and porpoises can be recorded in aquaria after acclimation to the recording arena (Nachtigall and Moore 1988; Au 1992) or by locating wild individuals or pods and placing hydrophones in the water.

Recordings of individual calls are the most useful for research. A recording of many individuals calling simultaneously is difficult to analyze, although it may be useful for estimating relative abundance. The proper placement of the microphone is the most important part of recording. Microphones have been developed for use with and without wires, for work underwater, and for detection of ultrasound, but they will not function efficiently if improperly placed. In general, a microphone should be located from 0.5 to 2 m from the subject. Good recordings can be made from as far away as 20 m, however, with a high-quality system.

For high-quality recordings, it is imperative that the caller be clearly observed while calling. When an investigator hears an animal, he or she should stop moving and scan the area for the caller, both visually (a good pair of binoculars is essential) and acoustically (by scanning the area ahead). Some calling animals are difficult to locate because of the ventriloquial characteristics of their calls. To obtain the best audio-signal input, the investigator should adjust the gain level down to the point at which the peak meter (VU meter) just hits the red zone or at which the peak light flickers on and off.

An animal generally stops calling when an investigator invades its "safe zone." The size of the safe zone varies from species to species and from individual to individual. For example, prairie dogs (*Cynomys*) and marmots (*Marmota*) tolerate invaders up to a distance of about 5 m, whereas chipmunks and squirrels (e.g., *Tamias, Tamiasciurus,* and *Spermophilus*) may tolerate them only to 10 m. An investigator must be careful not to invade the safe zone of a caller and unwittingly interrupt a recording session.

An investigator should record an entire call, as well as intervals between calls, to provide data on call rate. Most protocols call for a minimum of 20 calls per individual. Some calls contain long sequences that use a great deal of tape. An investigator recording bat (Microchiroptera) calls at high speed (e.g., 7.6 cm/sec) can fill a 270-m tape in 3553 seconds, or about 1 hour. A tape of the same length can last for about 1.5 hours at a tape speed of 4.76 cm/sec, which is appropriate for recording pikas (*Ochotona*). Information on tape speed will help an investigator determine the number of tapes needed for a recording session. Once a session has ended, the investigator again notes the locality, time, species identification, temperature, and so forth and finally says, "the end," to facilitate the editing process.

After a series of calls is recorded, the investigator should walk carefully in the direction of the calling animal, stopping occasionally to note its approximate distance and its behavior. For accurate distance measurements, a marker (e.g., a colored chip, such as a poker chip, with a number) should be placed at the starting point and at each stop. The investigator can approach the animal until it flees and then measure the distance between the point of flight and the recording point. This distance is helpful in assessing sound quality, attenuation, and the width of the animal's safe zone. The investigator keeps the recorder off when moving to another subject.

The specific questions to be answered and the variables to be measured must be clearly identified before recording. The design used to sample calls will depend on the habitat and size of the area to be surveyed, species that inhabit the area, season, and goals of the study. For purposes of statistical analyses, an investigator should carry out ten 100-m transects through an area and record calls of at least 20 individuals. If an area is to be sampled repeatedly over the long term, then sampling points must be established to ensure repeatability.

Some primates locate *sound windows,* or structural paths through the forest that allow for optimal sound transmission with the least amount of distortion or attenuation (Waser and Brown 1984). These individuals then position themselves at the center of a window when vocalizing. It may be possible for investigators to monitor the biodiversity of primates and perhaps of other mammals in tropical forests by monitoring vocalizations at acoustic stations along sound corridors. This technique merits study.

Playback Techniques

Playback recordings are prerecorded calls that are designed to elicit a response from a target individual (Kroodsma 1989; see "Call Playbacks," Chapter 6). For example, grasshopper mice (*Onychomys leucogaster*) are easily located at night by their responses to playbacks of their long calls (Hafner and Hafner 1979; W. L. Gannon, unpubl. data). When playbacks are used, both the playback recording and the elicited response should be recorded on a second recorder. Playbacks can be used to determine the distribution of individuals or the presence of a species in an area. Prerecorded calls of prey can also be used to attract ("phonotaxis"; Busnel 1963) predators.

Recording calls from captive mammals

Mammals in captivity often produce vocalizations that are rarely heard or recorded under natural conditions. Such vocalizations include calls used for con-

tacting conspecifics and for communicating distress or alarm. Investigators can increase the reference collection of vocalizations that can be used for species identification by recording calls from captive mammals. Investigators should be aware, however, that the context in which a call is given in captivity and the form it takes in such situations may differ substantially from the context and form in the wild.

Calls can be recorded from animals live-trapped during surveys. In such circumstances, the microphone should be placed no closer than 0.5 m from a trap or other container housing the subject. Vocalizations of bats captured in mist nets can also be recorded. Neotropical fruit bats (e.g., *Artibeus lituratus, A. jamaicensis*) call when entangled in mist nets and often draw other bats into the nets (Handley et al. 1991). Squeaks made with an Audubon bird-call device have also been used to attract neotropical fruit bats (Handley et al. 1991), and distress calls may draw some North American insectivorous bats (e.g., *Antrozous pallidus, Myotis lucifugus*) to nets.

Handling tapes

When a tape is full, it should be removed from the recorder and labeled with the date, time of recording, location, species, and name of the person making the recording. The investigator should review tapes soon after a recording session so that she or he can supplement taped comments with comments written in a field tape log (Fig. 65). If a tape is to be archived with a known institution, that institution's data form should also be completed (Fig. 66). All logs should be made with archival-quality materials such as 100% rag stock white paper and permanent black ink. When tapes have been fully reviewed and documented, the tape, tape log, and other pertinent data sheets should be placed in a sturdy envelope or tape carrier and stored in a cool, dry place away from magnetic fields. The tape and all accompanying materials should be duplicated prior to shipment from the field. Tapes can be mailed in any sturdy container and should be labeled "Magnetic Media— Avoid Excessive Heat or Magnetism."

Tapes should be deposited in a research, archival library of natural sounds such as the Library of Natural Sounds at Cornell University or the Borror Library of Bioacoustics at Ohio State University. Tape vouchers must be cross-referenced with specimen vouchers, because vocalizations are being used with

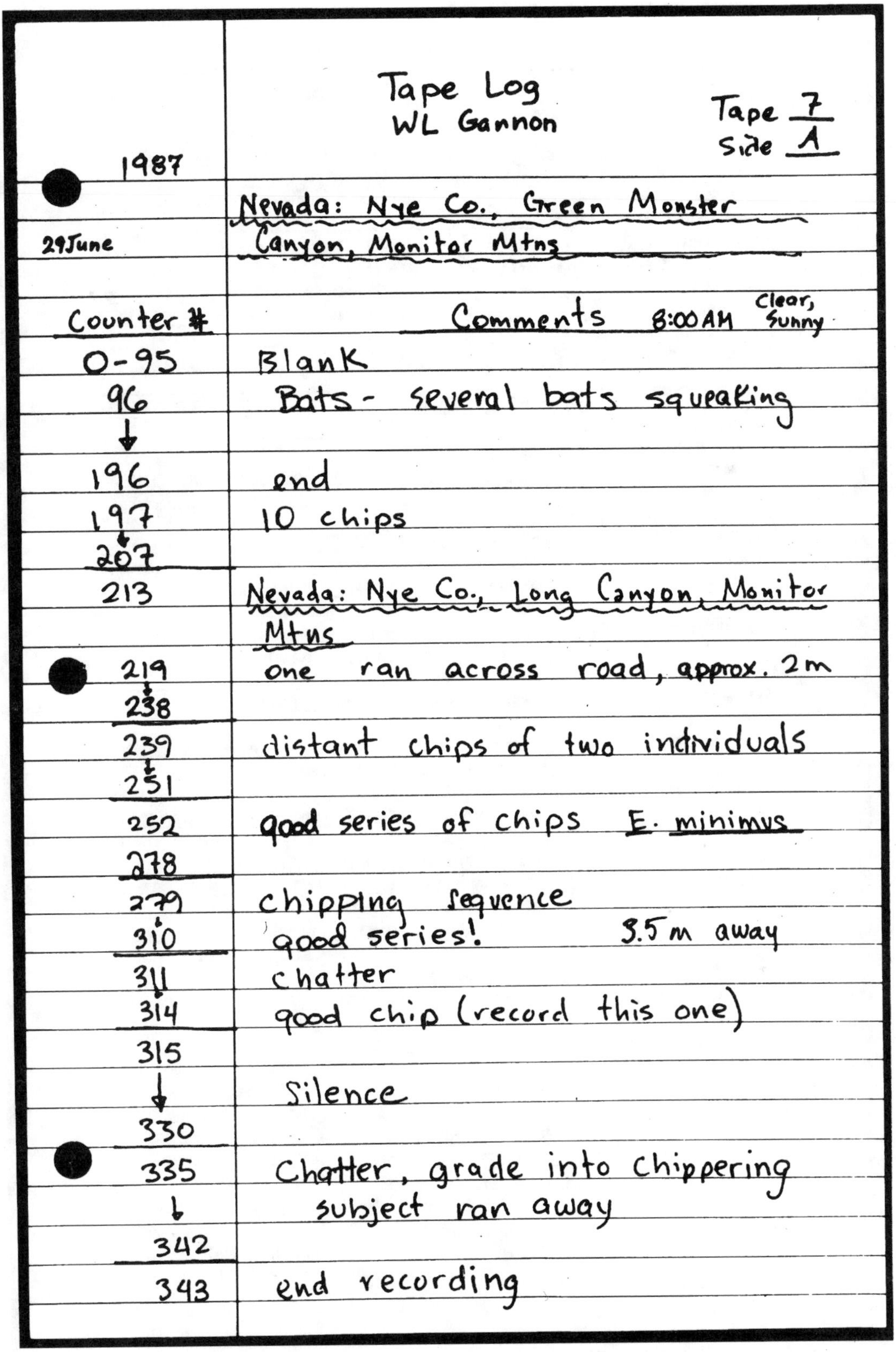

	Tape Log WL Gannon	Tape _7_ Side _A_

1987

29June

Nevada: Nye Co., Green Monster
Canyon, Monitor Mtns

Comments   8:00AM   Clear, Sunny

Counter #	Comments
0-95	Blank
96 ↓	Bats - several bats squeaking
196	end
197	10 chips
207	
213	Nevada: Nye Co., Long Canyon, Monitor Mtns
219 ↓ 238	one ran across road, approx. 2m
239 251	distant chips of two individuals
252 278	good series of chips E. minimus
279	chipping sequence
310	good series! 3.5m away
311	chatter
314	good chip (record this one)
315 ↓ 330	silence
335 ↓ 342	Chatter, grade into chippering subject ran away
343	end recording

Figure 65. A raw tape log prepared in the field shortly after a recording session. This log, the audio log made on the second tape channel, and the transcription notes (spectrograph log) made in the laboratory together produce a complete record.

DATA FORM
LIBRARY OF NATURAL SOUNDS
CORNELL UNIVERSITY
159 SAPSUCKER WOODS RD
ITHACA, NEW YORK 14850

TAXON CODE #66-
RECORDIST'S
REFERENCE
NUMBER (RRN) #69-

**** IDENTIFICATION and DATE (COMPLETE THIS SECTION FOR EACH DATA FORM) ***********

SPECIES, SOUND, or SUBJECT: *Eutamias umbrinus (WLG 428)*
SAME SOUND SOURCE AS RRN: _____. CROSS REFERENCE TO CATALOG NO #08-
RECORDIST(S) #47- *WLGannon*
TIME (24HR) #13- 8 AM . DAY #14- 22 . MONTH #15- July . YEAR #16- 1990
RELATIVE TIME #17- (DAWN)(+)(-)+ 1 HR; ___ NOON (+)(-) ___ HR; ___ SUNSET (+)(-) ___ HR
MOONLIGHT #18- ___ NONE; ___ SOME; ___ BRIGHT.   SPECIMEN COLLECTED #19- X YES; ___ NO
HOW IDENTIFIED #20- X SIGHT; ___ SOUND.   CONFIDENCE IN IDENTIFICATION #21- 100 %
DISTANCE TO SOUND SOURCE #57- ___ M.   RECORDIST'S: TAPE # ___ ; CUT #
BACKGROUND SOUNDS #22- *Some water noise, chainsaw.*

**** GEOGRAPHIC (EXCEPT AS NOTED, SAME AS RRN: _____) **************************
COUNTRY or ARCHIPELAGO #07- *USA*
STATE, DEPT, PROV. or ISLAND #08- *Nevada; Snake Range HNF*
LOCALITY #09- ___ KM ___ N ___ S ___ E (W) of *Mt. Moriah, White Pine Co.*
LATITUDE #10- _____. LONGITUDE #11- _____. ALTITUDE #12- 11,000 ft N
REFERENCE FOR ANIMAL NAMES: *Chipmunk* (3352 m)

**** BIOLOGY AND BEHAVIOR (EXCEPT AS NOTED, SAME AS RRN: _____) *******************

NUMBER OF ANIMALS #23- ___ PRENATAL; ___ NESTLING(S); ___ FLEDGLING(S); ___ JUVENILE(S); ___ IMMATURE(S); X ADULT(S); ___ UNKNOWN AGE

SEX #24- X MALE(S); ___ FEMALE(S); ___ UNKNOWN SEX

SPECIES SOUND OR SOCIAL CONTACT #25- ___ ISOLATED; ___ INFREQUENT; X FREQUENT; ___ CONSTANT; ___ FAMILY; ___ MIXED SPECIES; ___ COLONY; ___ FLOCK; ___ TROOP; ___ HERD; ___ OTHER SOCIAL UNIT: _____

RANGE STATUS #26- X NORMAL; ___ RANGE EXTENSION; ___ MIGRATION; ___ ACCIDENTAL; ___ INTRODUCED; ___ CAPTIVITY

BREEDING STATUS #27- ___ NOT TERRITORIAL; ___ TERRITORIAL; ___ TERRITORIAL SOLITARY; ___ TERRITORIAL PAIRED; ___ BREEDING; ___ NOT BREEDING

SOUND CATEGORY #28- ___ SONG; X CALL; ___ MECHANICAL; ___ DEVELOPMENTAL SUBSONG; ___ OTHER SUBSONG; ___ OTHER: *Chips + chippering*

SPECIAL SONG TYPE #29- ___ DUET; ___ COUNTER SINGING; ___ FLIGHT; ___ WHISPER; ___ DAWN; ___ MIMICRY; ___ OTHER: _____

STIMULUS FOR SOUND #30- X NATURAL(NO PLAYBACK); ___ SQUEAK-SPISH; ___ HUMAN IMITATION; ___ PLAYBACK OWN SONG; ___ PLAYBACK SAME SPECIES; ___ PLAYBACK ARTIFICIAL SOUND; ___ OTHER: _____

RESPONSE TO PLAYBACK #31- ___ NONE; ___ ORIENTATION; ___ APPROACH; ___ NORMAL SONG(SOUND); ___ DIFFERENT SONG(SOUND); ___ ATTACK

BEHAVIORAL CONTEXT OF SOUND #32- ___ EXPERIMENT; ___ ADVERTISING; ___ COURTSHIP; ___ COPULATION; ___ MATING INVITATION; ___ LEK; ___ MATE CONTACT; ___ NEST INVITATION; ___ INCUBATION; ___ NEST RELIEF; ___ CARE OF YOUNG; ___ PARENT-YOUNG CONTACT; ___ BEGGING; ___ ANNOYANCE; ___ ALARM; ___ THREAT; ___ SCOLDING; ___ DISTRESS; ___ AGGRESSION; ___ MOBBING; ___ FIGHTING; ___ FLYING; ___ FORAGING; ___ FLOCK CONTACT; ___ ROOSTING; ___ CONTENTMENT; ___ ARRIVAL; ___ DEPARTURE; ___ FLUSHED; X OTHER: *unknown; prob. alarm*

SOUND DELIVERY RATE #33- ___ SPORADIC; X LOW; ___ NORMAL; ___ HIGH; ___ AGITATED

SOUND SOURCE #34- ___ SYRINX; X LARYNX; ___ AIR SAC; ___ BILL; ___ BILL DRUMMING; ___ WINGS; ___ TAIL; ___ FEET; ___ HORNS; ___ OTHER: _____

VISUAL DISPLAY WITH SOUND #35- ___ NO; X YES(DESCRIBE IN NOTES (#67) OR VERBALLY ON TAPE) *see tape*

Figure 66. Front and back of an audio spectrograph comment log used by the Library of Natural Sounds, Cornell University, to track the number and quality of spectrographs made from a tape.

**** HABITAT AND ENVIRONMENT (EXCEPT AS NOTED, SAME AS RRM: _____) **************

GENERAL CLIMATE   #36- ___WET; ___HUMID; _X_ARID; ___CYCLIC WET-DRY

ENVIRONMENTAL ZONE   #37- ___TROPICAL; ___SUBTROPICAL; _X_TEMPERATE; ___BOREAL;
                                    ___ARCTIC; _X_MONTANE; ___ALPINE; ___OTHER:_____

SEASON   #68- ___SPRING; _X_SUMMER; ___FALL; ___WINTER; ___WET; ___DRY

GENERAL HABITAT   #38- _X_WOODS; ___FOREST; ___RAINFOREST; ___CLOUDFOREST;
                                   _X_BRUSH; ___CHAPARRAL; ___GRASSLAND; ___SAVANNAH;
                                   ___RIPARIAN; ___SWAMP; ___MARSH; ___TUNDRA; ___MUSKEG;
                                   ___TAIGA; ___DESERT; ___DUNES; ___BEACH; ___SHORE;
                                   ___MARINE; ___OPEN WATER; ___RURAL; ___URBAN; ___SUBURBAN;
                                   ___ISLAND; ___OTHER:_____

HABITAT TYPES   #39- _X_CONIFEROUS; ___DECIDUOUS, _X_EVERGREEN; _X_SECOND GROWTH;
                                   _X_UNDERGROWTH; ___SCRUB; ___THICKET; ___GROVE; ___SANDY;
                                   _X_ROCKY; ___CANYON-RAVINE; ___CLIFF; ___BANK; ___CAVE;
                                   ___BURROW; ___SALT; ___FRESH; ___BRACKISH; ___BOG;
                                   ___MUDFLAT; ___SANDSPIT; ___PASTURE; ___MEADOW; ___ORCHARD;
                                   ___HEDGEROW; ___EDGE; ___TUSSOCK; ___YARD; ___PARK-CAMPUS;
                                   ___CULTIVATED; ___FALLOW; ___BARREN; ___FIELD; ___ROADSIDE;
                                   ___BURN; ___CLEARING; ___EXOTIC; ___OTHER:_____

DOMINANT PLANT(S)   #40- Cercocarpus, Pinus

COVER DENSITY   #41- ___NONE; ___OPEN; _X_SPARSE; ___MEDIUM; ___THICK

STRATA IN HABITAT   #42- ___SURFACE; _X_LOW; ___MEDIUM; ___HIGH; ___CANOPY;
                                   ___TRUNKS-LIMBS; ___LOW FLIGHT; ___HIGH FLIGHT; ___SONG PERCH

WATER ASSOCIATION   #43- _X_NONE; ___MOUNTAIN STREAM; ___CREEK; ___RIVER; ___POND;
                                   ___LAKE; ___LAGOON; ___ESTUARY; ___SEA-OCEAN; ___BAY

WEATHER   #44- _X_CLEAR; ___CLOUDS; ___OVERCAST; ___FOG; ___RAIN; ___SNOW;
                                   ___WIND; ___OTHER:_____

TEMPERATURE (AIR)   #45- 25 DEGREES (C)(F)   WATER #46- _____ DEGREES (C)(F)

**** TECHNICAL (EXCEPT AS NOTED, SAME AS RRM: _____) ******************************

TAPE SPEED #48- Cassette CMS. FORMAT #49- C TRK MONO; ___TRK STEREO; _X_CASSETTE _X_
FIELD RECORDER #50- Marantz. MICROPHONE #51- AudioTech. TAPE: Sony
NOISE REDUCTION SYSTEM: none. BIAS: ——. EQUALIZATION: ——
COPY RECORDER: _____ FILTERING: _X_NO; ___YES(DESCRIBE IN NOTES)
PARABOLA (D/FL) #56- directional microphone   AT B15A

**** EDITING AND CATALOGING (FOR USE BY LIBRARY OF NATURAL SOUNDS ONLY) ****************

TAPE CATEGORY   #58- ___FIELD; ___RESEARCH; ___HABITAT; ___SOUND EFFECT;
                                   ___INTERVIEW; ___PROGRAM; ___COMPILED; ___FROM DISC

CUT LENGTH #59- _____. QUALITY #60- _____. SAFETY ARCHIVE REEL NO #61- _____
SPECTROGRAM FILED #62- ___YES; ___NO.   NOTES ON TAPE #63- ___YES; ___NO
EDITED: _____. CATALOGED: _____|__. MICROFILMED: _____

**** NOTES **

#67- good series of calls on this entire
tape. All E umbrinus

02-82

<u>SPECTROGRAPH LOG</u>

System Used *Canary 1.1*   Recorder Used **Marantz**   (Cass)/Reel

Date 20 NOV 1993 Tape Number 17A   Rec'd By Gennon

Locality: Country/State/Co: U.S.A. Nevada, White Pine Co.

Locality: Specific: Mt. Moriah, elev 11000 feet

Counter#: 102   X-ref No. WLG 292, NK 7901

Comments: HiFreq 12 KH₃   LoFreq 1 KH₃   DomFreq 5 KH₃

Narrow/Wide   Smooth/(Boxy)   Filter Band _____

MaxF _____ kHz MinF _____ kHz   Peak/Peak _____ s   Harn ____

Amplitude _____   Duration Total 5.65

Other: This print with 5 syllables of the
Chip call and one multi-syllab Chipper cd
(w/ 5 syllables to it) about 3 S duration

_____

Tamais umbrinus

- -

stored as: NV umbrinus

Quality? excellent

Use? Submitted in manuscript

Archived? _____

Figure 67. Spectrograph log used when analyzing tapes and producing prints of calls. This form has been completed for the spectrograph in Figure 70B.

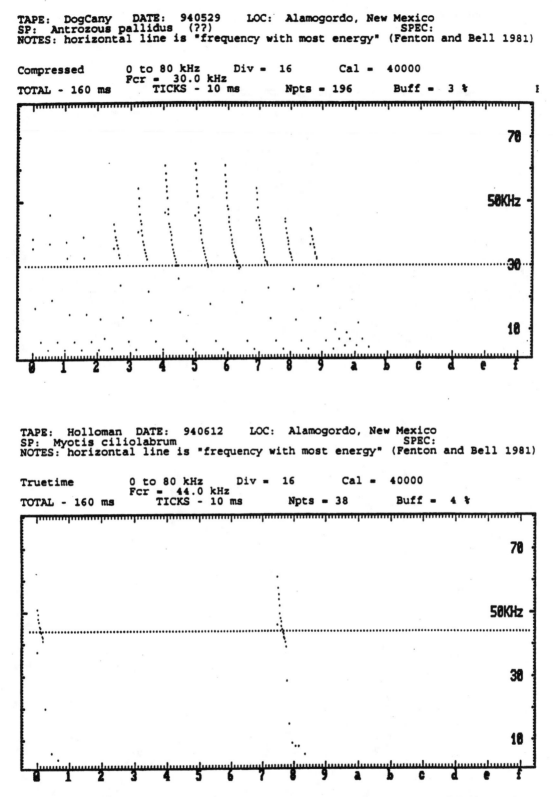

TAPE: DogCany   DATE: 940529   LOC: Alamogordo, New Mexico
SP: Antrozous pallidus   (??)   SPEC:
NOTES: horizontal line is "frequency with most energy" (Fenton and Bell 1981)

Compressed   0 to 80 kHz   Div = 16   Cal = 40000
                  Fcr = 30.0 kHz
TOTAL - 160 ms   TICKS - 10 ms   Npts = 196   Buff = 3 %

70

50KHz -

30

10

0 1 2 3 4 5 6 7 8 9 a b c d e f

TAPE: Holloman DATE: 940612   LOC: Alamogordo, New Mexico
SP: Myotis ciliolabrum   SPEC:
NOTES: horizontal line is "frequency with most energy" (Fenton and Bell 1981)

Truetime   0 to 80 kHz   Div = 16   Cal = 40000
                  Fcr = 44.0 kHz
TOTAL - 160 ms   TICKS - 10 ms   Npts = 38   Buff = 4 %

70

50KHz -

30

10

0 1 2 3 4 5 6 7 8 9 a b c d e f

Figure 68. Spectrographs of bat calls. *Above:* Anabat spectrograph of a call of the pallid bat (*Antrozous pallidus*) from southern New Mexico, displayed in compressed call mode (i.e., after removing the nonsignal portions in a call sequence). The line at 30 kHz represents a frequency containing the most energy deemed important by Fenton and Bell (1981). *Below:* A truetime (actual temporal scale) spectrograph of a call of a western small-footed myotis (*Myotis ciliolabrum*) from New Mexico.

A: MULTI-HARMONIC CALL OF *Onychomys leucogaster*

B: COMMON CALLS OF *Tamias palmeri*

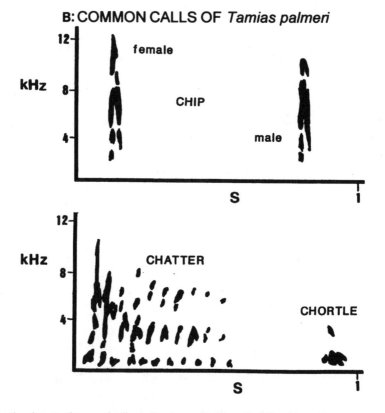

Figure 69. Acoustic printouts of mammal calls. A. Spectrograph of the call of *Onychomys leucogaster* produced using the Kay Real Time Sonic analyzer (W. L. Gannon, unpubl. data). B. Spectrograph of a *Tamias palmeri* call made with a Multigon Uniscan II system, which emphasizes the most intense parts of the call signal stream. Reprinted with permission of *The Southwestern Naturalist* from Gannon and Stanley (1991).

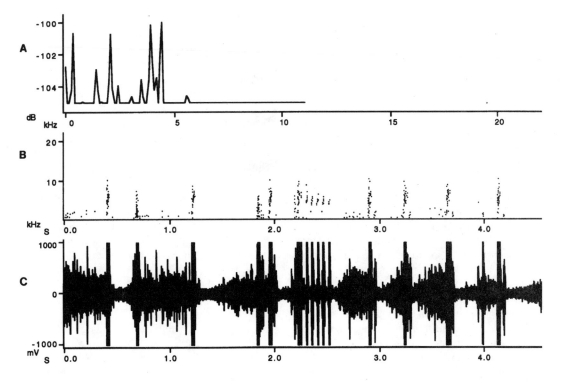

Figure 70. Vocalizations of *Tamias umbrinus* from White Pine County, Nevada, analyzed three ways using CANARY 1.2 (W. L. Gannon, unpubl. data). A. Call spectrum with frequency (kHz) plotted against intensity in decibels (dB). B. Call spectrograph with time in seconds (S) plotted against frequency (kHz). C. Power spectrum with time (S) plotted against power in millivolts (mV).

increasing frequency to discriminate between species (e.g., between *Myotis ciliolabrum* and *M. californicus;* W. L. Gannon, unpubl. data). At least one voucher specimen per species per calling site must be collected. Furthermore, most systematics collections require that the collector's field notes and field catalogue be deposited with the tapes and specimen voucher. Voucher specimens with complete locality information, carefully recorded calls, and field notes are a valuable resource that can remain useful for generations. If tapes and voucher specimens are deposited at separate institutions, both should carry the same tracking number (e.g., collector's number or cross-reference number [NK number in Fig. 67]) to assure that they can be associated in the future.

Call analysis

Kay Elemetrics Corporation Digital Sona-Graph machines have been used routinely to analyze calls. The Model 7800 has a DC 32-kHz spectrum analyzer with printer, a time-frequency audio spectrum, high-speed dual-channel usage, a waveform printer, and 128K of digital data storage. Kay Elemetrics machines are expensive (U.S. $8,000–$10,000), however, and their technology is at least 12 years old. Although they can manipulate and enhance calls, they cannot store, cut, or paste them or analyze them in ways that other systems allow. For less than half the cost, a researcher can purchase a personal computer system that can be used for audio analysis and other functions.

Signal Technology, Inc., provides an interactive laboratory system with modern technology for processing digital signals. Although this system is designed to operate with digital recorders, it also interfaces with IBM-compatible personal computers. Macintosh computers can be used for audio analysis with programs produced by Apple Corporation. Programs that are now available can store, analyze, and synthesize sound down to 0.0001 msec over a wide range of frequencies (up to 300 kHz) with the aid of a "sound board." One recent addition to the growing list of sound software is CANARY (version 1.2; Cornell Laboratory of Ornithology, Bioacoustics Research Program). This software runs on Macintosh computers with a math coprocessor, a minimum of 4-bit color, and MOS 6.05 or later. The program provides a long sampling time, stores call files, and includes a measuring device that stores call measurements. It also allows researchers to display, edit, analyze (e.g., waveform, amplitude, and spectrum analyses), and print natural sounds easily and rapidly.

When printing sound spectrographs, a researcher should use a spectrograph production log to track particularly good spectrographs for analysis (Fig. 67). Data entered in the log include call characteristics (e.g., maximum and minimum frequency, number of harmonics, amplitude, and duration) and characteristics of the spectrogram (e.g., high and low frequency range, dominant frequency or most intense portion of the call, narrow or wide bandwidth setting), as well as possible uses of the spectrogram.

Analysis of mammalian calls may totally bypass the tape recorder in the near future. Instead, signals may be fed directly into CANARY or other systems such as Anabat II, where they can be stored digitally. The Anabat II ultrasonic acoustics system contains IBM-compatible software that allows the investigator to analyze bat calls in the field on a laptop computer. Spectrographs from the Anabat can be displayed in compressed or truetime mode on a variety of time and frequency scales (Fig. 68). Acoustic printouts from laser printers have already replaced the Kay Elemetrics heat-sensitive paper methods of a decade ago (Figs. 69 and 70). The price of digital tapeless systems is also decreasing. Developments such as these will allow researchers to identify, monitor, and verify mammalian calls more readily in the future.

Additional information on call analysis techniques can be found in Sebeok (1977), Koeppl et al. (1978), and Sparling and Williams (1978). Techniques for monitoring chiropterans and cetaceans acoustically have also been described (Vincent 1963; Sebeok 1977; Ford 1983; Payne 1983; Fenton and Taylor 1989; also see "Ultrasonic Bat Detection," Chapter 7).

Acknowledgments. We thank L. A. Valle, M. L. Campbell, R. D. Forbes, and C. A. Parmenter for critically reading early versions of the manuscript and C. Corben and G. Budney for providing important technical advice. WLG also thanks the Museum of Southwestern Biology, University of New Mexico, and especially T. L. Yates, for continued support of his efforts to document and voucher the diversity of mammal vocalizations.

Vendors of Supplies and Equipment for Mammal Biodiversity Studies

Rasanayagam Rudran and F. Russell Cole

In this appendix we provide information on selected vendors in the United States and other countries that sell equipment and supplies commonly used in mammalian biodiversity studies. Some vendors carry only certain types of specialized equipment. Others deal in general field supplies and sell a wide range of items needed to study mammals and their habitats. Vendors are grouped by type of equipment or supply. Additional information on products offered is indicated in brackets. Inclusion of vendors in this list does not constitute an endorsement of their products.

General field and laboratory supplies

Ben Meadows Co.
3589 Broad Street
Atlanta (Chamblee), GA 30341
USA

Telephone: 800-241-6401, 770-455-0907
Fax: 800-628-2068, 770-457-1841
Telex: 804468 ATL
Cable address: BENCO

Carolina Biological Supply Co.
2700 York Road
Burlington, NC 27215
USA
Telephone: 800-334-5551, 919-584-0381
Fax: 919-584-3399
Cable address: SQUID, Burlington, NC

Fisher Scientific
711 Forbes Avenue
Pittsburgh, PA 15219-4785
USA
Telephone: 800-766-7000, 412-562-8300
Fax: 800-926-1166

Forestry Suppliers, Inc.
205 West Rankin Street
P.O. Box 8397
Jackson, MS 39284-8397
USA
Telephone: 800-674-5368, 601-354-3565
Fax: 800-543-4203, 601-355-5126
Telex: 585330
Cable address: JIM-GEM, Jackson, Mississippi

Lab Safety Supply
P.O. Box 1368
Janesville, WI 53547-1368
USA
Telephone: 800-356-0783
Fax: 800-543-9910
[general safety supplies]

Nasco
901 Janesville Avenue
P.O. Box 901
Fort Atkinson, WI 53538-0901
USA
Telephone: 800-558-9595, 414-563-2446
Fax: 414-563-8296

Nasco West
4825 Stoddard Road
Modesto, CA 95356-9318
USA
Telephone: 209-545-1600
Fax: 209-545-1669

Recreation Equipment, Inc. (REI)
P.O. Box C-88125
Seattle, WA 98188-0125
USA

VWR Scientific Products
405 Heron Drive
Bridgeport, NJ 08014
USA
Telephone: 800-932-5000
Fax: 609-467-3336
Cable address: VANROG

Ward's Natural Science Establishment, Inc.
5100 West Henrietta Road
P.O. Box 92912
Rochester, NY 14692-9012
USA

Telephone: 800-926-2660, 716-359-2502
Fax: 716-334-6174

Whatman Labsales
P.O. Box 1359
Hillsboro, OR 97123
USA
Telephone: 800-528-5114
Fax: 800-858-2243

Maps

Map Link
25 East Mason Street
Santa Barbara, CA 93101
USA
Telephone: 805-965-4402
Fax: 800-627-7768

Map Store
1636 I Street NW
Washington, DC 20006
USA
Telephone: 202-628-2608

Omni Resources
104 South Mebane Street
P.O. Box 2096
Burlington, NC 27216
USA
Telephone: 800-742-2677, 910-227-8300

Spot Image Corp.
1897 Preston White Drive
Reston, VA 22091
USA
Telephone: 703-620-2200

U.S. Department of Agriculture
517 Gold Avenue
Albuquerque, NM 87102
USA
(office in every state)

U.S. Geological Survey Map Distribution
P.O. Box 25286, Building 810
Denver Federal Center
Denver, CO 80225
USA
Telephone: 800-872-6277, 303-236-7477

Global positioning systems

IImorrow
2345 Turner Road SE
Salem, OR 97302
USA
Telephone: 503-391-3411
Fax: 503-364-2138

J.P. Instruments
P.O. Box 7033
Huntington Beach, CA 92615
USA
Telephone: 800-345-4574
Fax: 714-557-9840

Magellan Systems Corp.
960 Overland Court
San Dimas, CA 91773
USA
Telephone: 909-394-5000
Fax: 909-394-7050

Capture equipment

General

Allcock Manufacturing Co.
Northwater Street
Ossining, NY 10562
USA
Telephone: 914-979-1366
[Havahart humane traps]

Deer Farmer's Supply
P.O. Box 334
Warwick, Queensland 4370
Australia
Telephone: 61-76-612-099

Fuhrman Diversified, Inc.
905 South 8th Street
La Porte, TX 77571
USA
Telephone: 713-474-1388
Fax: 713-474-1390
[professional capture, handling, and safety
equipment for mammals, birds, fish, and reptiles;
Flexinets; Q-nets]

A.C. Hughes
1 High Street
Hampton Hill
Middlesex TW12 1NA
United Kingdom
Telephone: 44-81-797-1366
[bird rings and cages]

Longworth Scientific Instrument Co., Ltd.
Radely Road
Abingdon, Oxon
United Kingdom
Telephone: 44-235-24042
[Longworth small mammal trap]

H. B. Sherman Traps, Inc.
P.O. Box 20267
Tallahassee, FL 32316
USA
Telephone: 904-575-8727
Fax: 904-575-4864
[folding and nonfolding aluminum traps]

Tomahawk Live Trap Co.
P.O. Box 323
Tomahawk, WI 54487
USA
Telephone: 715-453-3550
Fax: 715-453-4326
[collapsible or folding live traps for small mammals;
carrying cages; collapsible fish, turtle, and bird
traps; beaver traps]

Woodstream Corp.
69 North Locust Street
Lititz, PA 17543
USA
Telephone: 717-626-2125
[museum specials and rat traps]

Nets

AFO Mist Nets
c/o Manomet Bird Observatory
P.O. Box 1770
Manomet, MA 02345
USA
Telephone: 508-224-9220

AviNet, Inc.
P.O. Box 1103

Dryden, NY 13052-1103
USA
Telephone: 607-844-3277

Dart Guns

Palmer Chemical & Equipment Co., Inc.
P.O. Box 867
Palmer Village
Douglasville, GA 30133
USA
Telephone: 770-942-4395
Fax: 770-949-3562
[cartridge-fired capture pistols, rifles, and
accessories]

Pneu Dart, Inc.
P.O. Box 1415
Williamsport, PA 17703
USA
Telephone: 717-323-2710
Fax: 717-323-2712
[air pump rifles, cartridge-fired rifles, CO_2-powered
guns, pistols, and accessories]

Scientific Marking Materials, Inc.
P.O. Box 24122
Seattle, WA 98124
USA
Telephone: 206-524-2695
[airguns]

Simmons Dart
Zoolu Arms of Omaha
10315 Wright Street
Omaha, NE 68124
USA
Telephone: 402-397-4983
[darts, blowguns, pole syringes]

Light aircraft

Cessna Aircraft Co.
P.O. Box 7704
Wichita, KS 67277-7704
USA
Telephone: 316-941-6000
Fax: 316-941-7850

Pilatus Aircraft Ltd.
3740 20th Street, Suite A
Vero Beach, FL 32960
USA
Telephone: 407-567-4555
Fax: 407-567-4555

The New Piper Aircraft Inc.
2926 Piper Drive
Vero Beach, FL 32960
USA
Telephone: 407-567-4361
Fax: 407-562-0299

Trade-A-Plane
410 West Fourth Street
Crossville, TN 38555
USA
Telephone: 800-337-5263
Fax: 615-484-2532

Marking supplies and equipment

Split-Rings, Bands, Tags, and Ball-Chain Necklaces

Ball-Chain Manufacturing Co., Inc.
741 South Fulton Avenue
Mount Vernon, NY 10550-5013
USA
Telephone: 914-664-7500
Fax: 914-664-7460
[ball-chains and connectors (stainless steel)]

Frank Ferris Co.
6726 North Figueroa Street
Los Angeles, CA 90042
USA
Telephone: 213-254-6161
Fax: 213-254-4545
[ball-chain connectors for necklaces (stainless steel)]

Gey Band and Tag Co.
P.O. Box 363
Norristown, PA 19404
USA
Telephone: 610-277-3280
Fax: 610-277-3282
[lipped bat bands (aluminum), butt-end bird bands

(monel, aluminum, and stainless steel), closing and opening pliers, ball-chains and connectors (stainless steel), ear tags (monel) for marking bats]

A.C. Hughes
1 High Street
Hampton Hill
Middlesex TW12 1NA
United Kingdom
Telephone: 44-81-797-1366
[plastic split-rings (colored) and anodized aluminum rings (colored) for marking bats]

L & M Bird Leg Bands
P.O. Box 2943
San Bernardino, CA 92406
USA
Telephone: 909-882-4649, 909-882-5231

Lambournes, Ltd.
Shallowford Court, off High Street
Henley-in-Arden, Solihull
West Midlands B95 5BY
United Kingdom
Telephone: 44-1564-79-4971
Fax: 44-1564-79-3075
[lipped bat bands (incoloy and color-anodized incoloy) for marking bats]

I.O. Mekaniska HB
P.O. Box 98
F-564 Bankeryd
Sweden

National Band and Tag Co.
721 York Street
Newport, KY 41071
USA
Telephone: 606-261-2035
Fax: 606-261-8247
[single-fold spiral leg bands (aluminum and colored), butt-end bird bands (monel, stainless steel, aluminum, or anodized aluminum), ear tags (monel), ear notchers]

Nelson Paint Co.
P.O. Box 2040
Kingsford, MI 49802
Telephone: 906-774-5566
Fax: 906-774-4264
[paint balls and pistols]

Paxar Graphics Group
815 South Brownschool Road
Vandalia, OH 45377
USA
Telephone: 513-898-1334
Fax: 513-898-9248
[numbered field tags]

Betalights

Saunders-Roe Developments
North Hyde Road
Hayes, Middlesex UBW 4N3
United Kingdom

Self-Powered Lighting, Ltd.
8 Westchester Plaza
Elmsford, NY 10523
USA
Telephone: 914-592-8230

Cyalume

American Cyanamid Corp.
Organic Chemicals Division
1100 West Main Street
Bound Brook, NJ 08805
USA
Telephone: 908-560-2000

Punch-Marking and Tattooing Equipment

Joe Kaplan Tattoo Supply Corp.
Department AM-EC1
P.O. Box 1374
Mt. Vernon, NY 10550
USA
Telephone: 914-668-5200
Fax: 914-668-2300

Panther Products Canada
P.O. Box 298
West Lorne, Ontario NOL 2PO
Canada
Telephone: 519-768-2841

S & W Tattoo Supply Co.
P.O. Box 263
East Northport, NY 11731

USA
Telephone: 516-842-9777

Spaulding & Rogers Manufacturing, Inc.
Route 85
New Scotland Road
Voorheesville, NY 12186
USA
Telephone: 518-768-2070
Fax: 518-768-2240

Weston Manufacturing and Supply Co.
1942 Speer Boulevard
Denver, CO 80204
USA
Telephone: 303-860-7704

Fluorescent Powder and Hand-Held Ultraviolet Lights

Forestry Suppliers, Inc.
205 West Rankin Street
P.O. Box 8397
Jackson, MS 39284-8397
USA
Telephone: 800-674-5368, 601-354-3565
Fax: 800-543-4203, 601-355-5126
Telex: 585330
Cable address: JIM-GEM, Jackson, Mississippi

Radiant Color Co.
2800 Radiant Avenue
Richmond, CA 94804
USA
Telephone: 510-233-9119

Scientific Marking Materials, Inc.
P.O. Box 24122
Seattle, WA 98124
USA
Telephone: 206-524-2695
[fluorescent marking materials, pigments, and ultra-violet lights]

Tapes

3M Manufacturing Co.
3M Twin Cities Sales Center
P.O. Box 33211
St. Paul, MN 55133-3211

USA
Telephone: 800-480-1704, 612-733-3300
Fax: 612-736-1379
[reflective tape]

Chemicals and drugs

Aldrich Chemical Co.
940 West Saint Paul Avenue
Milwaukee, WI 53233
USA
Telephone: 800-558-9160

American Cyanamid Corp.
Organic Chemicals Division
Bound Brook, NJ 08805
USA
Telephone: 908-560-2000

Bayer Corp.
12707 Shawnee Mission Parkway
Shawnee, KS 66216
USA
Telephone: 913-631-4800
Fax: 913-962-2803
[xylazine (Rompun)]

Bristol-Myers Squibb Co.
P.O. Box 4755
Syracuse, NY 13221
USA
Telephone: 315-432-2000
[ketamine hydrochloride veterinary injectable
(Ketamine, Vetalar, Ketaset)]

Fisher Scientific
711 Forbes Avenue
Pittsburgh, PA 15219-4785
USA
Telephone: 800-766-7000, 412-562-8300
Fax: 800-926-1166

Laboratoires Reading Z.A.C.
17, rue des Marronniers
94240 l'Hay-les-Roses
Fabrique par virbac
06516 Carros
France
[tiletamine, zolazepam (Cl 744, Zoletil 100)]

Lemmon Co.
P.O. Box 904
Sellersville, PA 18960
USA
Telephone: 800-523-6542
Fax: 215-721-9669
[etorphine (M99), Diprenorphine (M5050)]

Sigma Chemical Co.
P.O. Box 14508
St. Louis, MO 63178
USA
Telephone: 800-325-3010, 314-771-5750
Fax: 800-325-5052, 314-771-5757
Telex: 434475
Cable: SIGMACHEM
[tolazoline, naloxone, yohimbine hydrochloride]

Sigma Aldrich Co., Ltd.
Fancy Road, Poole
Dorset BH17 7NH
United Kingdom
Fax: 44-0120-271-5460

Sigma Aldrich Chemie GmbH
Grünwalder Weg 30
D-82041 Deisenhofen
Germany
Fax: 49-89-613-5135

Sigma Chemie Sarl
L'Isle D'Abeau Chesnes
B.P. 701
38297 St. La Varpilliere
Cedex
France
Fax: 33-74-956808

Tech America
15th and Oak
P.O. Box 338
Elwood, KS 66024
USA
[acepromazine maleate injectable]

Teva Pharmaceutical Inc., Ltd.
(Distributor for Lemmon Co.)
P.O. Box 1423
Tel Aviv 61013
Israel

VWR Scientific Products
405 Heron Drive
Bridgeport, NJ 08014
USA
Telephone: 800-932-5000
Fax: 609-467-3336
Cable address: VANROG

Warner-Lambert Co.
201 Tabor Road
Morris Plains, NJ 07950
USA
Telephone: 201-540-2000
Fax: 201-540-3761
[ketamine hydrochloride veterinary injectable (Vetalar)]

Wildlife Pharmaceuticals
1401 Duff Drive, Suite 606
Fort Collins, CO 80524
USA
Telephone: 303-484-6267
Fax: 303-482-6184
[carfentanil, yohimbine injectable solution, naloxone injectable solution]

Electronic equipment

Radio Transmitters and Radiotelemetry Equipment

Advanced Telemetry Systems, Inc.
470 1st Avenue North
Box 398
Isanti, MN 55040
USA
Telephone: 612-444-9267
Fax: 612-444-9384

AVM Instrument Co.
6575 Trinity Court
Dublin, CA 94566
USA
Telephone: 415-829-5030

Biotrack
Stoborough Croft
Grange Road, Wareham
Dorset BH20 5AJ
United Kingdom

Cubic Communications Inc.
OAR D.F. Products Division
4285 Ponderosa Avenue
P.O. Box 85587
San Diego, CA 92186
USA
Telephone: 619-277-6780
Fax: 619-277-9353

Custom Electronics
2009 Silver Court West
Urbana, IL 61801
USA
Telephone/fax: 217-344-3460

Custom Telemetry and Consulting
1050 Industrial Drive
Watkinsville, GA 30677
USA
Telephone: 404-769-4024
Fax: 404-769-4026

Holohil Systems, Ltd.
3387 Stonecrest Road
Woodlawn, Ontario KOA 3MO
Canada
Telephone: 613-832-3649
Fax: 613-832-2728

Karl Wagener
Herwarthstrasse 22
5000 Cologne
Germany

L.L. Electronics
P.O. Box 420
Mahomet, IL 61853
USA
Telephone: 800-553-5328, 217-586-2132
Fax: 217-586-5327

Mariner Radar (Lowestoft), Ltd.
Bridle Way, Camps Heath
Lowestoft, Suffolk
United Kingdom

Mini-Mitter Co., Inc.
P.O. Box 3386
Sunriver, OR 97707
USA
Telephone: 503-593-8639
[custom service in mammal tracking devices]

Smith-Root, Inc.
14014 Northeast Salmon Creek Avenue
Vancouver, WA 98686
USA
Telephone: 360-573-0202
Fax: 503-286-1931

Telemetry Systems, Inc.
Box 187
Mequon, WI 53092
USA
Telephone: 414-241-8335
Fax: 414-241-9593

Televilt HB
PL 5226
71050 Stora
Sweden

Telonics
932 East Impala Avenue
Mesa, AZ 85204-6699
USA
Telephone: 602-892-4444
Fax: 602-892-9139

Tracktech AB
Lundagatan 5
S-171 63 Solna
Sweden

Wildlife Materials, Inc.
1031 Autumn Ridge Road
Carbondale, IL 62901
USA
Telephone: 618-549-6330

Ultrasonic Bat Detectors

Dr. Lars Petersson
Petersson Electronik AB
Tallbacksvagen 51
S-75645 Uppsala
Sweden
Telephone: 46-18-303880
Fax: 46-18-303840

Skye Instruments, Ltd.
Units 5/6, Ddole Industrial Estate
Llandrindon Wells

Powys, Wales LD1 6DF
United Kingdom

Stag Electronics
1 Rosemundy
St. Agnes
Cornwall TRF OUF
United Kingdom

Titley Electronics, Ltd.
P.O. Box 19
Ballina, New South Wales 2478
Australia
Telephone/fax: 11-61-066-866617

Ultra Sound Advice
23 Aberdeen Road
London, N5 2UG
United Kingdom
Telephone: 44-71-3591718
Fax: 44-71-3593650

Motion Detectors

EME Systems
2229 Fifth Street
Berkeley, CA 94710
USA
Telephone: 510-848-5725

Audio Equipment

Audio-Technica U.S., Inc.
1221 Commerce Drive
Stow, OH 44224
USA
Telephone: 216-686-2600
[high-quality Japanese audio recorder and compatible microphones]

Boyton Studio
Melody Pine Farm
Morris, NY 13808
USA
Telephone: 607-263-5695
[Sony recorders and Audio-Technica microphones]

Cornell Laboratory of Ornithology
159 Sapsucker Woods Road

Ithaca, NY 14850
USA
Telephone: 607-254-2408
[CANARY program; professional assistance and archival capabilities for recorded sound; publishes a newsletter, *The Library of Natural Sounds Bulletin,* that provides telephone numbers of technicians who can assist the recordist, as well as advertisements for field equipment and technical recording equipment]

Florida State Museum
Bioacoustics Laboratory and Archive
University of Florida
Gainesville, FL 32611
USA
Telephone: 904-392-1721

Forty-Seventh Street Photo
36 East 19th Street
New York, NY 10003
USA
Telephone: 800-221-1774

Full Compass Systems, Ltd.
8001 Terrace Avenue
Middleton, WI 53562
USA
Telephone: 800-356-5844, 608-831-7330
Fax: 608-831-6330
[large inventory of common audio systems; catalogue available]

GW Instruments
35 Medford Street
Somerville, MA 02143
USA
Telephone: 617-625-4096
Fax: 617-625-1322
[Soundscope (for the Macintosh computer)]

Himelblau, Byfield, and Co.
1530 North Mannheim Road
Stone Park, IL 60165
USA
Telephone: 708-343-3384
Fax: 708-343-5718
[acoustical foam and other noise-control substances]

Kay Elemetrics Corp.
2 Bridgewater Lane
Lincoln Park, NJ 07035
USA

Telephone: 201-628-6200
Fax: 201-628-6363
[6000 series sonagraphs, 7800 series digital analyzers, and DSP Sona-Graph work stations (true real-time signal analysis system)]

Bernie Krause
Recordist
California Academy of Sciences
Golden Gate Park
San Francisco, CA 94118-4599
USA
Telephone: 415-750-7037, 415-386-6677
[professional consultant and recordist of vocalizations and natural sound]

Richard Margoschis Natural History Sounds
80 Mancetter Road
Mancetter, Atherstone CV9 1NH
Warwickshire
United Kingdom
Telephone: 08-27-2925
[inexpensive parabolic dishes]

Marice Stith Recording Services
59 Autumn Ridge Circle
Ithaca, NY 14850
USA
Telephone: 607-277-5920
Fax: 607-277-5942
[parabolic dishes, books, microphones, and tape recorders]

Mineroff (Saul) Electronics, Inc.
574 Meacham Avenue
Elmont, NY 11003
USA
Telephone: 516-775-1370
Fax: 516-775-1371
[authorized Uher and Sennheiser dealer; common brands of audio equipment; will customize and build systems for any research use]

Mini-Mitter Co., Inc.
P.O. Box 3386
Sunriver, OR 97707
USA
Telephone: 503-593-8639
[custom service in acoustic systems]

Multigon, Inc.
1 Odelle Plaza

Yonkers, NY 10701
USA
Telephone: 800-289-6858
[maker of Uniscan II, FFT real-time sonogram spectral display; this machine will interface with either IBM or Apple personal computers]

Newark Electronics
4801 North Ravenswood Avenue
Chicago, IL 60640-4496
USA
Telephone: 312-784-5100
Fax: 312-784-3850

QMC Instruments, Ltd.
229 Mile End Road
London E1 4AA
United Kingdom

Security Products International
30 Old Budd Lake Road
Budd Lake, NJ 07828
USA
Telephone: 201-426-0800
[electronic surveillance, detection, and countermeasures equipment; minitransmitters, microrecorders, and minimicrophones]

Signal Technology, Inc.
120 Cremona Drive
P.O. Box 1950
Goleta, CA 93116
USA
Telephone: 800-235-5785, 805-968-3000
Fax: 805-683-1394

Skye Instruments, Ltd.
Units 5/6, Ddole Industrial Estate
Llandrindon Wells
Powys, Wales LD1 6DF
United Kingdom
[standard receivers and receivers with switchable bandwidths]

Titley Electronics, Ltd.
P.O. Box 19
Ballina, New South Wales 2478
Australia
Telephone/Fax: 11-61-066-866617

Uher Corp.
Ubostrasse 7

8000 Munich 60
Germany

Ultra Sound Advice
23 Aberdeen Road
London N5 2UG
United Kingdom
Telephone: 44-71-3591718
Fax: 44-71-3593650
[ultrasound speakers and high-frequency cassette recorders]

Remote cameras, triggering devices, and film

Camtrak South, Inc.
1050 Industrial Drive
Watkinsville, GA 30677
USA
Telephone: 800-654-8498
[35 mm camera traps; Camtrak cameras]

Eastman Kodak Co.
343 State Street
Rochester, NY 14650
USA
Telephone: 716-724-4000

Edmund Scientific Co.
101 East Gloucester Pike
Barrington, NJ 08007
USA
Telephone: 609-547-6279
Fax: 609-547-9238
[photic cells and light sensors]

Forty-Seventh Street Photo
36 East 19th Street
New York, NY 10003
USA
Telephone: 800-221-1774

Lanson Industries, Inc.
582 W18717 Gemini Drive
P.O. Box 906
Muskego, WI 53150
Telephone: 414-679-0045
Fax: 414-679-2505
[pressure mats, motion and presence sensors]

Penn Camera Exchange, Inc.
915 E Street NW
Washington, DC 20004
Telephone: 800-347-5777
[Mazof Trigger System, a three-mode system responding to optical sound and vibration signals]

Telonics
932 East Impala Avenue
Mesa, AZ 85204-6699
USA
Telephone: 602-892-4444
Fax: 602-892-9139
[intervalometer camera and pressure mats]

Trailmaster
Goodson & Associates, Inc.
10614 Widmer
Lenexa, KS 66215
USA
Telephone: 913-345-8555
Fax: 913-364-8272
[active and passive infrared trail monitors, 35 mm camera trap, video camera trap, intervalometer]

Environmental monitoring equipment

Monitoring Equipment and Supplies

Belfort Instrument Co.
727 South Wolfe Street
Baltimore, MD 21231
USA
Telephone: 410-342-2626

Campbell Scientific, Inc.
815 West 1800 North
P.O. Box 551
Logan, UT 84321-1784
USA
Telephone: 801-753-2342 (USA), 05-09-672516 (UK)
Fax: 801-750-9540 (USA), 05-09-674928 (UK)

Cole-Parmer
625 East Bunker Court
Vernon Hills, IL 60061-1844
USA
Telephone: 800-323-4340

Hach Co.
P.O. Box 389
Loveland, CO 80539
USA
Telephone: 800-227-4224
Fax: 303-669-2932

LI-COR, Inc.
4421 Superior Street
P.O. Box 4425
Lincoln, NE 68504
USA
Telephone: 402-467-3576

Omega Engineering, Inc.
P.O. Box 4047
Stamford, CT 06907
USA
Telephone: 800-826-6342 (USA), 203-359-1660
(other countries)
Fax: 203-359-7700 (USA), 203-359-7807 (other
countries)

Physical Chemical Scientific Corp.
36 West 20th Street
New York, NY 10011
USA
Telephone: 212-924-2070
Fax: 212-243-7352

Qualimetrics, Inc.
1165 National Drive
Sacramento, CA 95834
USA
Telephone: 800-247-7234

Solar Light Co.
721 Oak Lane
Philadelphia, PA 19126
USA
Telephone: 215-927-4206
Fax: 215-927-6347

Solomat Neotronics
26 Pearl Street
Norwalk, CT 06850
USA
Telephone: 203-843-3111

Texas Electronics
P.O. Box 7225
Dallas, TX 75209

USA
Telephone: 214-631-2490

Data Loggers

LI-COR, Inc.
4421 Superior Street
P.O. Box 4425
Lincoln, NE 68504
USA
Telephone: 402-467-3576

Onset Computer Corp.
536 MacArthur Boulevard
P.O. Box 3450
Tocasset, MA 02559
USA
Telephone: 508-563-9000
Fax: 508-563-9477

Museum and collecting supplies

General Museum Supplies

Archivart
7 Caesar Place
Moonachie, NJ 07074
USA
Telephone: 201-804-8986
Fax: 201-935-5964
[archival-quality paper and board, including
pH-neutral materials; glassine envelopes]

Art Preservation Services
253 East 78th Street
New York, NY 10021
USA
Telephone: 212-794-9234
Fax: 212-794-3045

Conservation Materials, Ltd.
1165 Marietta Way
Sparks, NV 89431
USA
Telephone: 702-331-0582
Fax: 702-331-0588

Conservation Resources International, Inc.
8000-H Forbes Place

Springfield, VA 22151
USA
Telephone: 800-634-6932, 703-321-7730
Fax: 703-321-0629
[see next entry for products]

Conservation Resources, Ltd.
Units 1 and 4, Pony Road
Horspath Industrial Estate
Cowley, Oxfordshire OX4 2RD
United Kingdom
Telephone: 08-65-747755
Fax: 08-65-747035
[pH-neutral paper for specimen labels, archival-quality paper supplies; excellent catalogue with information on paper chemistry available]

Dow Chemical
Customer Information Center
Dow North America
P.O. Box 1206
Midland, MI 48641
USA
Telephone: 800-441-4369
[see next entry for products]

Dow Europe S.A.
Bachtobelstrasse 3
8810 Horgen
Switzerland
Telephone: 41-1-728-2111
Fax: 41-1-728-2935
[Ethafoam for specimen storage]

Kapak Corp.
5305 Parkdale Drive
Minneapolis, MN 55416
USA
Telephone: 612-541-0730
[fluid-safe plastic pouches for shipping wet specimens]

Light Impressions
439 Monroe Avenue
P.O. Box 940
Rochester, NY 14603-0940
USA
Telephone: 800-828-6216
Fax: 800-828-5539
[slide and negative storage containers; Tyvek tape]

Matra Plast, Inc.
420 Notre-Dame
C.P. 600
Berthierville, Quebec J0K 1A0
Canada
Telephone: 800-661-7662, 514-836-7071
Fax: 514-836-2315
[polycarbonate board for making strong, rigid boxes and trays]

Museum Services Corp.
1107 East Cliff Road
Burnsville, MN 55337
USA
Telephone: 612-895-5199
Fax: 612-895-5298
[buffered and unbuffered acid-free paper and board, mylar, sealers for mylar film, pest-monitoring traps, miscellaneous conservation and museum supplies]

National Bag Co., Inc.
2233 Old Mill Road
Hudson, OH 44236
USA
Telephone: 800-247-6000, 216-425-2600
Fax: 216-425-9800
[polyethylene heat sealers, polyethylene bags and sheeting]

Paper Technologies, Inc.
929 Calle Negocio, Unit D
San Clemente, CA 92673
USA
Telephone: 714-366-8799
Fax: 714-366-8798

Rubbermaid, Inc.
1147 Akron Road
Wooster, OH 44691
USA
Telephone: 216-264-6464
[storage containers]

Texas Tech University
PrinTech
Box 43151
Lubbock, TX 79409-3151
USA
Telephone: 806-742-2768
[tags, fieldnote paper, catalogue paper]

University Products, Inc.
517 Main Street
P.O. Box 101
Holyoke, MA 01041
USA
Telephone: 800-628-1912, 413-532-3372
Fax: 413-532-9281
[specimen storage boxes, specimen trays, blotting paper, polyester felt, glassine envelopes]

Jars, Bottles, and Vials

ABICO Scientific Co.
P.O. Box 12
Kashiwa 277
Japan
Telephone: 0471-46-2497, 0471-52-7866
Fax: 0471-44-5443
[specimen jars with polyethylene inserts and plastic caps (excellent seals)]

Acme Vial and Glass Co.
1601 Commerce Way
Paso Robles, CA 93446
USA
Telephone: 805-239-2666
Fax: 805-239-9406
[vials]

Chelsea Bottle Co.
10 Wesley Street
P.O. Box 6330
Chelsea, MA 02150-6330
USA
Telephone: 800-345-5277 (northeastern U.S. only), 617-884-2323
Fax: 617-889-1626
[polypropylene screw-top lids]

Cole-Parmer
625 East Bunker Court
Vernon Hills, IL 60061-1844
USA
Telephone: 800-323-4340
[polyvials and polypropylene bottles]

Kew Scientific
3824 April Lane
Columbus, OH 43227
USA

Telephone: 614-231-3685
[polyethylene stoppers]

Kol's Container Corp.
1408 DeSoto Road
Baltimore, MD 21230
USA
Telephone: 410-646-2300
Fax: 410-646-5671
[polypropylene screw lids with polyethylene liners for wide-mouth jars]

Nalge Nunc International
P.O. Box 20365
Rochester, NY 14602-0365
USA
Telephone: 716-586-8800
Fax: 716-586-8431
[plastic products]

Twinpack, Inc.
7544 Côte de Liesse
St. Laurent, Quebec
Ontario H4T 1E7
Canada
Telephone: 514-733-3133
[glass jars with polypropylene lids and polyethylene liners]

Wheaton Science Products
1501 North Tenth Street
Millville, NJ 08332
USA
Telephone: 800-225-1437
Fax: 609-825-1368

Wilkens Anderson Co.
4525 West Division Street
Chicago, IL 60651
USA
Telephone: 312-384-4433
[vials with caps]

Cryogenic Supplies

Cardinal Health Pharmaceuticals
3530 Pan American Freeway NE
Albuquerque, NM 83107
USA
Telephone: 505-884-0101
[Velban]

Linox Co.
12000 Roosevelt Road
Hillside, IL 60162
USA
Telephone: 708-449-9300
[Nunc tubes]

Southland Cryogenic
2424 Lacy Lane
Carrollton, TX 75006
USA
Telephone: 214-243-1312
[Nunc tubes]

Vanguard International
1111-A Green Grove Road
P.O. Box 308
Neptune, NJ 07754-0308
USA
Telephone: 908-922-4900

Specimen Storage Cases, Drums, and Tanks

Delta Designs, Ltd.
P.O. Box 1733
Topeka, KS 66601
USA
Telephone: 913-234-2244
Fax: 913-233-1021
[museum storage cases; specializes in custom designs]

Interior Steel Equipment Co.
2352 East 69th Street
Cleveland, OH 44104
USA
Telephone: 216-881-0100
Fax: 216-991-0090
[museum storage cases; will provide custom designs]

Lane Science Equipment Co.
225 West 34th Street
New York, NY 10122-1496
USA
Telephone: 212-563-0663
[museum cases]

Steel Fixture Manufacturing Co.
612 Southeast Seventh Street
Topeka, KS 66607
USA
Telephone: 913-233-8911
Fax: 913-233-8477
[museum storage cases, specialty case designs, stainless steel tanks for fluid-preserved specimens]

Computer equipment

Computer Hardware and Supplies

Apple Computer, Inc.
20525 Mariani Avenue
Cupertino, CA 95014
USA
Telephone: 408-996-1010

IBM Corp.
1 Culver Road
Dayton, NJ 08810
USA
Telephone: 908-329-7000

Computer Software

Colorado Cooperative Fish and Wildlife Research Unit
Room 201, Wagar Building
Colorado State University
Fort Collins, CO 80523
USA
Telephone: 303-491-6942
[CAPTURE program]

Dr. James E. Hines
National Biological Service
Patuxent Wildlife Research Center
11510 American Holly Drive
Laurel, MD 20708-4017
USA
Telephone: 301-497-5661
[JOLLY, JOLLYAGE, and RBSURVIV programs]

Table of Random Numbers

Lee-Ann C. Hayek

Instructions for use

This table of random numbers consists of a sequence of 32,768 random digits (integers 0 through 9), generated using SAS-PC Release 6.04. It is presented as an *octal* table with eight pages of eight columns and eight rows, divided into blocks that themselves consist of eight columns and eight rows of digits. Random number tables are commonly organized in groups of 10 with instructions for use that include a nonrandom component. Use of this table of random numbers to select a sample of size k includes the selection of a random starting point and a rule for proceeding from that point to selecting the k sample elements.

The first step is to list or number all K elements in the available population from which the sample of size k is to be taken. The lowest integer power of 10

that produces a number at least as large as k is then identified. This integer will be the length (in digits) of the random number to be selected, as shown in

k	Length of integer (number of digits)
$1 \leq k \leq 10$, or 10^1	1
$11 \leq k \leq 100$, or 10^2	2
$101 \leq k \leq 1{,}000$, or 10^3	3
$1{,}001 \leq k \leq 10{,}000$, or 10^4	4
.	.
.	.
.	.
$10^i \leq k \leq 10^{i+1}$	$i + 1$ ($i = 0, 1, 2, \ldots$)

the following table:

The k elements corresponding to the set of k random numbers identified are the random sample.

This appendix is reprinted from Heyer et al. (1994), with permission.

For simplicity, the value 0 is taken to correspond to the largest value of k for which the indicated length is appropriate. Thus, 0 corresponds to the 10th element of the group for which selection is being made; 00 corresponds to the 100th such element; 000 corresponds to the 1,000th such element; 0000 corresponds to the 10,000th such element; and so on. In general, the number consisting of i zeros corresponds to selection of the element of length 10^{i+1}.

Numbers larger than K may be ignored, and selection is continued until k numbers less than or equal to K are selected. Similarly, duplicates should be ignored and additional numbers selected.

The method of random table entry may be based on any dichotomous experiments—assumed here to be flips of a coin. Each decision to be made in our octal layout will require three coin flips—referred to as a *tripflip*. Alternatively, three coins may be flipped once, but their order must be determined prior to flipping. The result of a tripflip is an ordered set of results of three individual flips, which we denote with H for heads and T for tails. The set of possible tripflips is HHH, HHT, HTH, THH, TTH, THT, HTT, and TTT.

One tripflip is used to identify the page of entry, as designated in the upper left-hand corner of each page. Two additional tripflips identify the row and column of the block, and a final two tripflips identify the row and column of the digit within the selected block (the arrangements of heads and tails shown at the top of the page and along the left side identify, in connection with these two pairs of tripflips, both the row and column of the block to be selected and the row and column within the block). The process identifies a specific digit that serves as the first digit of the first number to be selected and that may be referred to as the *entry point*. For convenience, the investigator may want to omit the final two tripflips and begin with the first digit of the selected block; however, this procedure reduces the number of entry points from 32,768 to 512, and I do not recommend it. Aisles between blocks of numbers and between pages may be ignored for the purpose of selecting the number to be used.

The entry process could be repeated $k-1$ times. This would, however, be a tedious procedure and is not generally considered to be worth the effort. Other approaches are limited only by one's imagination; many researchers select numbers simply by continuing down the page (and onto the next page or back to the first if necessary). A slightly more complex but recommended procedure is to follow one of the eight possible straight paths away from the initial digit, selecting numbers contiguously. The path is chosen by use of an additional tripflip. For convenience, the following diagram could be used when selecting a path. Here O_1 denotes the number of the first observation.

HHH		HHT		HTH
•		•		•
	•		•	•
		•	•	•
THH	•	• • O_1 • •	•	TTH
		•	•	
	•		•	•
•		•		•
THT		HTT		TTT

Example

Assume that a random sample of size 30 is to be taken from a collection of 100 elements. Thus, 30 unique two-digit numbers are needed. A list of the 100 elements is compiled, and each is assigned a unique integer from 1 to 100. A coin is flipped 18 times, yielding, for example,

$$\text{T H H H H T T H T H T H T H T T H H}$$

These flips can be considered as six tripflips. The tripflips and their corresponding uses are as follows:

THH	HHT	THT	HTH	THT	THH
Page	Block row	Block column	Row	Column	Path

This sample set of tripflips thus identifies the starting digit as 4—corresponding to the fifth page (THH), second row of blocks (HHT), sixth column of blocks (THT), third row (HTH), and sixth column (THT)—and the first two-digit number obtained is 43, which is ignored because it is larger than 30. Because the path selected (THH) is to the left (see diagram, above), the next two-digit number is 31, which is also ignored (because it is larger than 30), and the third is 09, which selects element 9 as the first observation in our sample. The second element would be 7 (ignoring 79 and 50), the third 25 (ignoring 77, 84, and 79), and so forth.

An alternative, *modulo system* could be used to reduce the number of steps required to select the sample. A modulo system is one in which all multiples of a given number (in our example, the sample size) are subtracted from any number identified in the table. For example, if an investigator wished to select one item from 25 items, the numbers 4, 29, 54, and 79 in the table would all identify the fourth item for selection (by subtraction of multiples of 25).

Note that if 30 items were to be sampled, as in our first example, the numbers 91–99 would be ignored to avoid giving the numbers 1 to 9 a greater chance of being selected than those from 10 to 30.

If a modulo system had been used in our example above, it would not have been necessary to ignore as many numbers. The numbers from 31 through 60 and those from 61 through 90 would have been used to identify sample elements by subtracting 30 in the first case and 60 in the latter. For example, the first value of 43 would have designated element 13 as part of the sample. The numbers from 91 through 100 (00) would still have been ignored and replaced if obtained.

Table of random numbers

This table was designed and prepared by Charles R. Mann Associates, Inc., a Washington, D.C., statistical consulting firm.

HHH	HHH	HHT	HTH	HTT	THH	THT	TTH	TTT
HHH	18690397	11918532	66601968	77040187	00201468	70870951	12737739	86205251
	64523500	33711091	76092869	96283931	92186107	35644871	51983555	60242068
	57243033	95196310	39966207	14969833	71395689	17927888	64891645	40958903
	51391391	55243302	74000551	78307991	73248209	67038085	70604928	25323583
	83790183	14817791	57159452	00531460	10411732	30234717	60416789	55547259
	67240122	17766145	10400875	35438031	56967025	51938632	58123846	92861914
	24352731	04526465	98603628	79320580	25689996	70729599	21209421	89471768
	56603342	89568542	99818849	90561616	66899505	28322136	20554683	10280424
HHT	03893607	22146827	58172988	62780377	26692825	63196686	56447502	42567921
	02964028	48830405	12102002	52645552	57926882	16958796	03690513	98974571
	89657939	70063123	93979399	77604203	57704768	71607186	49620181	42404821
	23033568	95818740	49596983	39507849	68619050	48098746	76378230	46290172
	32802477	69098720	87360118	31977607	86751205	04846084	15397880	67685301
	53465825	82867460	49476748	75107052	63936299	52863835	64186622	27236440
	25553512	00462346	76776282	45111008	76801591	64239665	65594816	55082358
	67309539	68952203	57473647	84289966	87469779	44387800	52772492	09949150
HTH	91871904	29477762	47523801	28864948	66388733	79473688	00752299	24187281
	33543542	37755356	03718422	68190433	31608699	40287823	03293426	24102150
	72468782	70901965	63798056	22880533	22887385	06049915	19624923	94043056
	08267875	80444780	53634671	52338408	40408108	29215395	49215374	69027430
	57169577	49205816	64311776	10209865	35463887	99596507	80017106	42438347
	17684277	82052858	80215206	85959428	12936885	18665638	85013718	95750486
	38369168	41144361	20690462	80750652	60931407	72666292	74966858	52364045
	02659718	39526540	96246568	68967550	24376044	89118294	17072857	23363067
HTT	85411990	74348133	19375142	53094239	24183746	82676023	94314849	37410897
	73073380	46440119	55925855	48390049	52234227	86015837	64208763	16528798
	21752062	06673699	07927457	51521828	94661191	12177070	24616401	31564081
	33765125	36754911	38706141	64945825	39416528	40554986	53063534	91728617
	52005521	59096943	91111113	87415865	15046314	75292149	31754987	68589941
	67332967	09824980	08793541	05059043	37078129	38721263	83949456	09341968
	15123639	86853897	37361460	73878164	01872046	88610435	83684736	77273035
	80757533	45908014	39422294	99124467	15469754	96427171	18513656	04934016
THH	07833863	21154462	96604886	85089435	05543667	15675984	13033379	55466515
	34475768	08844848	54987067	99804055	40695910	30456498	50012261	81230259
	49343172	44058395	99603373	31154048	36444164	47288397	03800061	04302745
	17348702	89269966	75088544	38177324	39673979	07510058	83338795	95730819
	36714017	29283194	41034932	08662498	36856531	28438189	21078478	18296628
	76859836	34502954	86077485	59428569	42354316	82109604	92958752	84884159
	62806656	41350678	43793484	21302209	97600827	07163827	40641782	07355352
	49669489	43960710	47045149	91766235	12757256	14875488	77906943	69764367
THT	81215857	85931312	98384939	93218670	13007947	02159537	94403555	04104237
	87067798	57417777	12155227	37383701	04108061	55179947	16977319	89304125
	96879127	80958269	51559316	24840974	01239435	20360555	45449981	41456475
	06724434	31338454	60260577	12475807	87448267	76519786	40446530	46340186
	60170362	92470154	17219186	54386302	29223022	84680177	79021236	19637138
	06159657	63492321	31571839	02659646	99802016	11557743	76077121	99165063
	85840530	57577791	41676650	45379159	91129889	48156230	32238799	90962283
	03933858	27734649	68629669	16044711	82441066	99783803	33571778	02124116
TTH	91116129	74744189	31629207	02840329	92383931	26436908	44450351	01392724
	19847546	51453538	45375387	64840532	50497355	75482461	80355682	77931340
	02707513	57241737	40905668	15972963	64119250	79627096	84190852	99822201
	32642938	68465890	23694936	46064239	13475645	65329387	31937239	22024670
	75201619	66074592	36335851	24073983	89478010	90829348	10855075	84536560
	85042075	02417343	72761972	82158255	56839852	22691341	86710967	15202731
	61442981	89466398	12759382	35473039	05973735	27563540	69490782	31141590
	21836172	81825353	64245169	56703800	30621510	30765640	27481773	70501702
TTT	70214420	12615149	62028975	73347647	66679176	99989425	07558056	00223205
	90096558	27064177	44655377	26748456	71488190	15128909	52767803	40329869
	65602885	38522179	88503369	04668675	09681815	76477163	49964402	53116795
	81592129	01000062	03824803	26213884	56823398	62767963	86987793	23259217
	40007925	23178796	82244569	37879390	44793771	80544493	73673242	19109496
	18895326	20548512	30816371	97207184	30495239	29504865	08937725	72470706
	40300358	18229182	57194404	43052249	93337532	22790339	15219833	07759809
	49690630	00119404	65762324	35209983	89076943	48391372	60368891	87119103

HHT	HHH	HHT	HTH	HTT	THH	THT	TTH	TTT
HHH	44939377	49659364	39797583	02118489	91843795	59890564	94345466	74346755
	82198413	14424939	24378277	55166792	13402303	01033751	20337519	13587196
	25773059	58144069	13810459	10175872	34333189	96433820	65514459	74784643
	58942388	07870574	89256553	68716160	93838324	20308673	66748226	19166720
	21731767	27504373	40406619	88842244	45680153	30545169	19098953	07366957
	53901969	48772479	48717756	57888747	33318189	00286779	27497662	33835372
	58307766	37710516	81027461	51387781	69913270	55250209	20339895	37129657
	27373696	41806445	87879577	50047370	96588419	13281477	97176767	08968148
HHT	61518690	25422093	19495932	30471839	49393764	68923027	61195132	02731001
	35729875	88802954	13813985	65250822	43096970	06061959	57997114	41664501
	23962446	00467579	12150914	14062009	42423701	82456948	66723842	06227931
	50065723	07856559	22822562	32570235	61838508	91759860	56680135	03349353
	43174149	76369298	49764616	82085472	59206606	80364100	27809757	99411466
	41240483	64278275	33612787	93842983	59842201	22295086	21454560	49539571
	08334374	78724479	69682100	02351928	89900964	77507789	98074449	64146953
	46815979	90097854	07369791	80613135	17669317	59497324	80224914	30126308
HTH	31454165	35851697	50152934	89813061	41773593	08052428	70220285	71670829
	20429216	80430134	70437287	60938419	46561388	85802774	06852986	97762475
	70342755	89940459	77867248	72492398	62667921	42923587	12739548	65789010
	59183805	51140248	39830157	22298940	34689894	09732557	74326708	10033886
	68296003	47843006	76942114	51867498	83476541	55614692	63442369	37759427
	28007549	86733783	55710102	77305840	60868872	49765947	42102538	66773680
	58195764	41017972	23899523	29761545	07350970	14894949	15128690	20359987
	98633513	48179023	21158852	35291772	44314291	09470435	81367438	46772085
HTT	66837543	01863957	81854972	08498702	49462789	55610323	81604813	59446931
	56332680	60522590	64645080	11065936	38862219	24360799	84507459	52699871
	75230314	92559654	49153960	17498960	00923673	00562083	25846025	38548540
	86047310	99941528	23491409	67250708	01339709	16235873	21265374	47951421
	88463062	18609904	18490541	44471792	06980070	02617757	21908532	40963916
	80974241	75222928	82743449	75026147	64197646	53069491	69914295	20873873
	91585300	70462498	11912440	04759812	73470911	38611638	43732000	99974355
	96843530	65278277	95762339	75621419	79312544	48577661	13870066	32948405
THH	58326880	66840204	33476691	59802817	24530988	31142064	69221022	23164501
	05336059	02344387	98526551	45890892	43173791	92183049	75471308	97334644
	74129395	30699659	78105918	14043976	44952257	32836900	90681364	82407351
	84314774	79354592	02767343	32573985	28420771	22169255	61852211	88760841
	24413188	73011381	52450154	89190992	40540854	27490831	36394909	23946752
	54946922	88238297	55333325	37397493	48057197	87627029	13538614	69284918
	39626491	83587200	94240896	72749856	26016625	88636041	93895509	52755561
	59331816	03357133	57339396	57515066	94499159	77434184	96994238	06263202
THT	17961813	13517094	35098133	71685904	43175865	41484008	62575331	54864244
	31936444	79940442	51710773	26089514	87209006	95512020	51475574	99593278
	94069188	56257407	15998561	34026224	11459511	26033042	32456515	08752092
	98297784	03311720	37605207	65787338	28520713	02739776	70395602	40560684
	03037933	73324297	22074275	68440687	57535942	40489310	42592309	80565946
	51294889	30168571	04817418	08622091	24296589	29288381	46618895	41821778
	79889907	40249716	91392037	14056841	50379961	82757578	16838403	10963575
	22252701	13912986	25490190	20292801	73044612	36320366	30206003	02496308
TTH	93020835	61159151	63457106	77680385	04653861	06425607	26219384	79209365
	59650620	91744028	62238810	33176903	36450862	89344058	21001817	96893132
	44682037	18833340	20547932	56476830	84614253	09515268	63153539	39164157
	71326505	62463110	11327515	83961333	61189023	30807449	65798276	73871923
	55466375	23749737	18123694	27906073	68932068	53999457	84118753	04685308
	35683346	51382478	94213053	20849915	41352982	94077735	11293866	48214649
	89480743	06677835	82465130	73511984	14112738	42825034	83033521	46840048
	83376873	28975338	19205726	22370454	90824118	63300777	72011394	38576634
TTT	29313558	95325320	83543961	80383699	35907255	60373378	86441909	19803391
	56672841	05144431	64558805	48386006	07203032	75786554	89583295	17489639
	68171030	40723152	62746173	03984310	82925653	81376072	00786642	87696687
	64358298	40259168	57574397	26949072	23863852	71499263	80528049	53452601
	82275966	43209532	24304602	94663678	12299920	98357978	93358598	22531436
	94520715	79753817	73730443	73216384	40654936	18449083	52895504	20590165
	84716899	56751204	94815486	30814877	22245765	55984146	92955237	17482002
	93003490	67839703	84020892	21839322	53610199	14406548	68134495	62457482

HTH	HHH	HHT	HTH	HTT	THH	THT	TTH	TTT
HHH	72204903	88752845	24608058	64719294	69202447	84750555	63477608	50756914
	12502346	11220081	66619917	07264035	45817213	18608609	41042693	94851153
	35391572	00323474	58902323	46859340	53170803	25402663	60329994	26669351
	40503296	45406860	19069554	54468838	60963538	00961557	78400014	69378794
	64829884	70360832	14945869	57702365	72265619	06803063	49073247	96324102
	65715663	93833588	95484812	73483440	86026499	59688694	52806703	98976436
	27757562	81351563	10360425	17299207	91708606	33537473	72612809	61416368
	38149726	81544591	56608435	36172228	42155154	08891898	65523175	48677801
HHT	88495331	99284663	35281117	49287880	36487778	41577932	12908444	48811074
	94474666	86712883	91666319	67534500	86433247	16004700	06561730	73051169
	49754802	40437701	48497644	63165975	23538853	48602170	30311254	15310398
	71046850	69802441	49797248	37624581	93053445	08420048	73223026	13818253
	17729442	27113812	15950254	65002938	84168837	72378778	24504073	94606348
	41456387	78272536	63633297	73343740	67710883	49682855	61594827	97468226
	28788779	83955227	57988488	39532346	09567110	35074288	33824366	71428849
	58013689	73359776	49742818	06748246	51293904	04859598	88121715	00337698
HTH	14582566	94714390	46530025	60785146	79711938	91028467	56538815	81757068
	97517689	56414106	74429276	35588035	48982748	49150858	41370035	78873390
	51087624	34250933	45694024	69560057	09796255	33726469	24509902	15002979
	70123110	38690191	79034224	84657074	38563368	23143182	22949493	88799934
	02074890	60867723	02594967	24267641	20157849	99557214	23939150	07460889
	63035137	83378292	87970502	15415856	07241941	67218205	80432441	96963269
	44542213	55962922	57970344	31946826	60885966	40741545	28765374	17225914
	23068890	50048850	26131570	52228687	78803949	03803019	06671406	85767340
HTT	61606118	99936132	99788456	24080670	37544287	15635282	08189198	08106470
	31335846	77388288	30623803	71422014	14406355	05209192	08542899	05311139
	52976504	45775408	48414062	35568499	47824690	44183303	52414793	94045129
	28696097	69247171	51761630	28190062	56200126	07277262	15131724	03532321
	50545007	83673750	95814563	38804506	10826046	84400165	27588229	15017609
	94667981	06896417	79205925	89718461	03232954	00423547	62526081	07163269
	95934475	02189333	79922392	21601788	65240216	57561787	29580748	39390790
	89395780	09572066	02682447	70078270	41413141	93435618	42557549	50142638
THH	59875579	65314905	92556946	95870333	99475504	82864873	82401386	32291610
	39915973	93388769	61866577	91558567	76246667	52942084	70960696	67916866
	62142029	63492905	60198027	44758256	99780532	56481348	98985382	25190157
	58659065	95122277	98735671	06671729	23775824	91210266	37934280	26207805
	84910740	03735955	60373117	14921523	85078268	55149278	12203697	06313911
	68431470	67376482	62575914	80031605	06937467	40112221	11295869	35234940
	47029015	71053464	04159166	01878870	82138551	87652915	82795023	00987964
	72801998	54283680	81544460	00622572	38252002	31507574	43812412	87359052
THT	78670256	65645593	47079723	48139155	50743535	89901302	86348452	94446525
	76663452	66473064	04928354	15014771	66301172	13183395	90042741	80914756
	27681119	51904543	58763652	99281857	06324140	85713981	15166575	83160487
	12254997	74284673	64871019	71898690	47802598	57464519	18522515	29403485
	08462822	38946770	11096019	38767920	71640265	54188888	31507401	82247667
	23330215	24081099	91529920	34439984	86181843	99597371	41072897	48032494
	11719364	44359625	29477921	86226423	02196829	66318985	67870542	96673671
	20113161	78122842	23155388	28295982	27294425	56627980	51619113	78808307
TTH	94508956	78925698	81518308	69768738	09821347	78793845	48970506	27701465
	18008952	77919009	82881082	25998066	53385838	84204206	74895815	33839252
	69898351	83986908	56952534	46239568	76354884	40333427	20087666	77425084
	65973490	47537590	51460062	15489824	16720189	69078375	72684592	55990810
	62818161	78029584	98770920	40274634	93390926	71497479	48471720	82467611
	35587739	49780855	66745447	33076223	53214660	96168617	67711961	19535736
	19962688	50754392	10712961	01762729	38125429	82020269	22736264	03265557
	28734667	11987880	65883838	95849970	75319643	21276221	25601516	73890109
TTT	06224351	13317702	56837248	93504639	53604174	57700648	78278171	95690714
	16891834	41638870	32690556	74389582	22677575	19763060	85493065	15499631
	61373268	83589676	92254585	58380000	71992657	03714592	17519959	25002313
	45212114	82893985	72341523	38616687	55884066	77900417	11829595	72910365
	48193623	56069310	93891169	75589847	20537434	00754102	65723945	90130388
	17310410	28732100	42814250	45856554	44327513	92166734	71386251	78066477
	09793957	61940654	33707036	22034859	63748041	22128681	92240933	36735499
	93080427	94603248	23673998	33585988	53438493	45283094	53426936	07232736

HTT	HHH	HHT	HTH	HTT	THH	THT	TTH	TTT
HHH	97664164	87639349	90203481	70541358	71081865	04826301	39566587	06725065
	77575647	94576743	54199726	83891127	53935887	74388263	34694811	40484850
	23128498	94776467	14231720	02424238	58044556	94744287	35392825	60613838
	79600654	14137304	91166053	89298668	97967237	66432627	10252714	21213775
	23033292	13014835	86988602	43694338	45222341	37770071	94909827	59640494
	39610845	41241179	13395019	76065254	31531974	42019748	74943840	52537740
	17445083	24076543	21205607	53543640	44977801	28589517	95213892	04647605
	00766361	57404455	99498931	72308767	90552720	93394506	96666896	36677334
HHT	51416577	59290956	89284128	15709570	34321907	56224797	02271352	51180461
	15479768	79498680	82325990	46671607	23822956	00270952	12155834	43612179
	29704555	26250782	13412245	37463084	65265854	82820883	45874150	06457123
	37456171	92458018	48739455	14243517	48227999	33424916	03995864	25636961
	65757367	01135512	79743195	18565204	91488180	88479010	72189941	07850775
	47254031	69606518	24185280	23366045	54579690	18604878	35970708	73972641
	43226878	38978072	57445884	27455777	61056357	79627710	60797989	28891374
	40797253	70918329	70456688	04454972	28792533	85726820	40134223	63246163
HTH	47987354	65614723	75468452	80841523	78160509	67984441	49371927	15677534
	23954517	01738817	69713725	65356254	24846978	45607235	38277903	60017558
	79469689	34683952	87341627	42315000	51888186	41580998	98949437	97959278
	37618111	97093365	83840277	74243943	32798850	40153605	75327297	80510185
	52882774	65886313	23337457	40011888	47129267	22259118	42636364	56965248
	79283420	08558307	43362585	81302381	92608555	45453770	07902089	30568201
	41787381	83278547	24010871	09532259	80860033	03318422	00413830	85574118
	59951521	42104170	82535990	83250580	04999983	27429942	52720058	87594671
HTT	08471846	32647240	08174677	94276736	42951184	15981226	06424486	06877453
	26519977	02844542	78226910	98109505	34824556	62516497	03491303	82130889
	58040291	52041085	71908278	49671380	28986702	01248435	58413819	23137214
	05662139	37802406	73803188	55948849	59636522	71897749	90478932	32099216
	53740463	77222262	12370893	54001041	15803746	64442724	98078078	69934862
	33221406	20827361	30712532	90303733	80694610	67683264	19666708	99869560
	17874066	95131498	35062002	18374695	44772853	73486689	51379891	71648920
	87007858	97385364	56783075	32493048	21753572	77280146	99650056	61711337
THH	76347876	53657605	84917117	21618439	16360573	28740709	27346293	78933924
	56291757	68983945	46082999	11503608	11470053	93651244	70890141	31217648
	24230463	64072026	07526244	71288268	97433623	89699099	70557294	93680003
	07773469	67127701	13708359	31553053	31690342	98781990	59506903	60286154
	55007251	17430423	65987092	15179635	24897460	23365277	06790824	33339384
	45637259	03426008	97056751	91411536	47683112	27312365	65645347	31354086
	50270677	23562628	49596950	03706565	25222651	20766277	73735898	98744308
	47905560	20300336	52386903	41828317	87302835	31977145	44333282	37955488
THT	09457838	81447326	21791933	19797061	20009909	28117255	32994535	05381164
	65278750	28710820	67809750	40702115	53240550	15050725	00302163	50954304
	84645622	54083734	31359174	43789838	97914190	29877435	23005271	15789703
	29728342	35671316	28353694	22332922	59726403	36748325	13081148	00240372
	24255507	65513426	27804777	02171074	97559668	43061108	54198453	36676574
	92502628	74507013	76016591	77120316	78914830	95590001	60260772	30288820
	51628996	42886468	54418807	89980569	84307643	55041621	49220219	23062686
	27264798	61093106	33156784	77192817	63215792	52191045	85562214	54260444
TTH	49310339	73711581	05378143	39103286	75176905	47606426	17930099	48676903
	25171943	77214273	54071685	50880000	10236879	84064405	86168860	87661961
	76653929	26431431	99666674	88007952	86183629	63985265	77457107	93930098
	72441945	49130381	52813450	88717963	96998943	16038740	70415724	58165266
	06529933	81414724	96939639	37432849	01096077	13540271	73627060	30232420
	58556113	04921823	73920614	30188367	37044520	87199439	95003796	69839136
	29171964	26719522	07775926	03631111	89807621	64521703	22702807	42623841
	54515135	35595360	73213536	50630307	97739959	67685433	75316939	73107063
TTT	99491830	74021662	59869702	12269840	41761383	93531582	91953968	38646277
	72909611	86274366	67483029	78251763	38882305	06765158	25363529	53932203
	47077004	25678992	90768334	75707769	61417556	25635398	53929189	28272714
	69635500	87613353	10832996	39777217	19616092	52891409	63811038	72409518
	62084564	28090213	83181000	46388286	28368793	45238006	49145087	00242230
	69420326	28462259	06838782	24946533	15364580	68903414	78205567	27143107
	55728990	55077935	11317626	63339821	66196841	41896044	98345467	45047590
	90883836	55134984	53190383	87027312	41672307	59039088	12563681	67081737

350   APPENDIX 10

THH	HHH	HHT	HTH	HTT	THH	THT	TTH	TTT
HHH	83855133	05635177	94032166	55113759	88731692	22430035	29677768	50361879
	12186359	83974184	73228403	69093273	17173423	98288379	21807739	70770844
	14365099	24380721	34673330	91123986	18710998	59466288	58479960	44148317
	81611434	82678572	08895854	82102472	91201927	76203962	53558897	51096736
	00313468	38690757	71564044	48428847	45130406	40617585	85761084	66924413
	13174432	59972604	92378004	89314526	28357388	97466695	27347975	23284781
	55617358	14605941	56771860	41931509	81196803	28551735	50193725	22508078
	76078328	77634812	46818112	22126466	80785060	15741761	48803652	13066786
HHT	36691863	72016423	64405573	47265159	98182473	70116736	45019495	92729878
	49485045	93664913	95698385	42297027	07069534	18449627	00154599	66263275
	19400995	90008191	79587980	28325798	47707507	90931433	82254263	07154203
	48685068	62931092	60612555	33735766	31690199	15273054	42271198	14224185
	19623120	85439018	97840718	20573478	95538554	36599231	70859483	02538217
	17387168	11372577	71391674	85087251	09136118	35493121	57248652	64541325
	12462483	34941691	31089989	27593964	46960163	53449628	62733964	99839944
	05952097	67771729	09282567	34829426	60019294	49403784	04478000	12231742
HTH	62044015	31420274	86035019	05392360	16886155	04603633	03794319	76994392
	30257344	84165728	51285104	30185156	27644312	40460667	06501157	86479943
	18596290	74754307	94887474	23356409	35099903	54865139	11499385	93443052
	74044647	64346063	15203406	49650508	03148450	57915365	93208158	90512929
	64359355	60781435	49985540	48071187	36608450	85831792	30895191	79069402
	81994282	90587232	50346702	93614461	55544811	59461908	22503298	53795361
	13977525	90298535	20971541	22483525	80674291	21889190	50701216	05992241
	73020895	41210446	68749332	90920996	27154383	90823544	26407271	16731455
HTT	34996394	02062183	65269905	27704536	38638214	98043086	24091895	89763647
	83188153	52964496	94032322	06597108	86398206	23677992	72584782	34263917
	38532116	56209685	70940514	20215637	77028279	78418380	04110118	45828393
	80887579	53158705	16334942	52506249	31179581	89636279	02594162	90450731
	09848333	63173861	57140075	86046079	54847031	61069172	71504298	66359166
	73584120	02796389	79784291	46254684	42875052	23034591	77080896	31514140
	93490758	93382806	01122575	75643370	25831499	12686684	71553702	40730640
	56809024	82673925	97486460	26631559	44802849	18324459	38229218	93868861
THH	25186264	23605296	77388462	47677319	26477465	88942803	57144459	97663412
	58051656	67016420	54111330	01668227	38265626	25924087	38339602	27135016
	27018443	15785442	03718214	26914236	95970546	98787862	03978672	52070359
	99348913	67807402	03936301	74909623	80952130	34512581	54275931	97464967
	90198963	82185202	67371489	49391053	79849009	10674868	45656088	64936764
	34274104	87228259	31908911	27548371	95282571	94220384	05082881	45012349
	68110930	50107893	29758249	19725624	82098644	58426649	85845980	77991245
	57013503	29040060	18815241	76887830	51268031	73502742	77026909	21996790
THT	70206260	02526615	15411392	12137459	68577056	05055934	78991941	50384604
	37022559	56466518	59031745	15845073	74018176	74583243	37117861	37058706
	57452700	98575291	37622828	67925663	82529823	94464628	41076514	74267386
	54880285	59534189	90034935	79371801	42505173	37131845	55511748	06628060
	22399307	08642807	54140264	65981552	30523892	09358643	13771426	81708134
	05867936	83729909	37223876	28938645	42087713	38852165	74959626	54180909
	45843555	88175958	53704819	35811489	68640250	42473862	72433265	87280239
	60269043	83424055	43467364	42070672	00676901	32250685	64015819	85498440
TTH	82954316	04391414	35607573	52998572	40455533	49657640	49416504	75568947
	82508624	02947277	31358598	52609683	13128571	47298649	17181618	32599594
	49008024	98944543	59846705	02225489	83618797	30017235	23310830	24780691
	83341722	48489637	50274141	45082104	09641394	96543870	51144252	43945599
	97218054	63792540	34323337	87855324	07530813	71993419	83982564	85649893
	97584644	95456085	32551116	13157982	98246042	50373427	35947891	08051359
	96159974	47979694	23822435	01140349	70918507	00143388	66379837	21607181
	79768812	60952999	14904708	83223167	84529529	16704074	35890856	85581898
TTT	96071851	87356397	07059700	89877666	48710708	16461864	94447122	39730366
	62082327	19934732	00920040	70562387	69757488	29171876	65334999	93911081
	97030523	78125135	67871650	63078619	22283026	79918627	69750978	18912076
	98170040	09953828	37426940	03528890	42871732	02919245	50520560	04246862
	29269495	82682107	59763373	25542980	57509451	71321015	38727477	35458704
	55216846	10558679	58016989	55122049	92399665	41746056	01974771	65046871
	81277527	59354470	88468809	45142905	50427979	16039977	31001171	13794035
	39886907	44216844	73386451	38073790	53779962	86756079	26560669	30993472

THT	HHH	HHT	HTH	HTT	THH	THT	TTH	TTT
HHH	68679569	05026845	89210240	72424338	92410325	34368222	51915557	22938883
	16646538	30688780	64053478	04460063	09530967	35743749	19097004	28034709
	50573740	44156912	09747668	98138831	26043067	06659645	78622629	15959275
	37349386	97901504	68572800	84160274	62148633	36191808	85283547	93225340
	57910481	94798322	18009951	67142507	32149539	29187336	85660577	45252955
	84027828	61271906	89136282	77234916	22235752	38777887	21462279	15111914
	10849521	06492493	57868403	18041601	43539514	83090245	07395215	30007767
	27348694	86996329	75585452	03352113	42340457	04151476	48497085	34097160
HHT	29044586	99505140	48177344	23575928	50516208	77602992	07053175	64875521
	86564903	98407926	19049054	99287505	69363456	15675140	64315688	21811342
	92639870	63837935	24878810	35451389	59711254	89915644	78010201	35861965
	45343712	66931067	87845849	66603928	61940423	58738717	31111740	10734097
	93859355	06804631	47519637	64091801	54031280	70651622	76486290	08814200
	76987915	37903633	44148764	75152263	46907698	20106714	86381086	74798483
	60953869	93650542	47334849	37434446	89176880	90914798	30838822	01044842
	84840513	98581088	81165772	11320189	75369038	70381326	95087593	17211244
HTH	39713071	30284051	60928990	46402714	93594730	44092456	05269266	67269221
	25295500	87144326	16518817	92978093	83266635	33504766	64114931	55250823
	78450775	69189959	95545581	18319019	42481071	11510368	57607389	80509737
	00394348	43532985	52372568	72575171	73289102	80520921	42373876	16358723
	72450806	64879011	76965831	65473676	41942084	07172434	55248686	00255292
	61985120	01980631	78536741	86648666	28475006	83202633	15501996	45814053
	03061138	80133614	06881629	93402286	03313428	02867460	96063273	84338990
	87873049	26247348	95963550	36176286	21026014	91871367	34702131	89074967
HTT	11756276	12833053	12816768	79268805	91838128	35942030	35613891	32664633
	19105129	65930507	40537650	18544142	52951385	72878982	58807951	33060260
	54817416	52004969	30481648	59912094	75753704	20132627	38955315	62673651
	03886421	87288048	64068702	74176653	57098052	31413771	83747454	16478292
	70451618	02520950	23794422	18805105	26104312	09256555	77709109	28292051
	43806188	88628352	54950093	53864987	80424808	61044947	71253159	56060491
	13796281	77368645	88074686	46049859	65860712	58692494	17558463	06516934
	73834286	15309165	42191906	98812424	95175278	42492707	52145194	86576024
THH	29989187	68782849	23312822	05631932	04920507	36765671	13318249	13871761
	59710043	87234107	47349508	92666133	08865733	00228615	66736494	20753740
	63461093	94270576	58077578	82450593	30503536	81596413	67619967	02062022
	49724489	80000049	59445876	20171921	57439695	22872898	58096326	34459228
	60319852	67913599	20240929	12090511	12813796	81681888	12574999	97378693
	20000318	35570373	18709397	84166211	05440066	88254259	65309172	44374838
	45656466	02513515	22382185	64424249	71127873	06123661	23912449	27692319
	64115188	17596851	73404071	92242059	31718099	58621486	89722664	27418112
THT	45100041	56853398	31817521	43520230	24126315	75555900	45738578	66285549
	34003295	43543930	75621152	43730092	63814565	34402122	05043520	17331825
	46340676	40793815	25799019	32389192	41761238	55134937	64452116	93111389
	59329545	89475228	91504606	54708087	14448231	70617034	89732532	65175323
	47215349	73442628	63549528	98311911	31288670	15980787	75743602	31165365
	52304751	34282765	54034436	88002279	79676772	58206845	17848173	69932799
	49037094	27817027	09036557	54304715	81274664	69023863	21929656	05537470
	30337965	05307906	53794341	71964587	07975773	45811483	35690692	02387734
TTH	16782293	09120350	88442577	04185889	33835819	33352338	04930770	52255890
	29726530	61349568	43072380	06133622	80198597	59515271	85635913	25298474
	80070293	67163424	50785947	14294384	93490111	32403268	71825563	99822767
	13594154	69423966	69893016	08275797	11652649	78936301	83293045	37682169
	69552112	70911236	65104387	22111559	85014888	47740851	13187397	39871535
	26328190	77785220	97923044	19157574	71897891	15040269	91210719	40048404
	49448290	54118689	65159125	50239060	49433217	15132487	12673031	16976574
	15658722	28013929	21910402	47185494	90158709	29829446	42364772	17216020
TTT	15829166	60151769	59012243	38909696	49633322	52591513	99669086	97578457
	12828289	93697544	47778667	09046279	95318745	78524386	00798549	76547149
	84959442	87484763	95974507	79429227	96540771	17723371	88587299	66373182
	95506882	29968928	31731399	52623090	29423917	29635163	18532265	42156732
	99070970	20109544	33391693	82985596	73138988	40797547	71322344	14383034
	53697530	53698578	63991907	72575507	57316916	26449420	55768844	39022582
	79268416	26137243	47758003	02529139	23517369	75029518	56210644	01353062
	17925521	15186670	09928533	44619233	72530208	46494681	72261852	67131917

TTH	HHH	HHT	HTH	HTT	THH	THT	TTH	TTT
HHH	46319271	17775686	00975244	68177364	57086643	50756914	34143029	01119658
	96120977	02081174	53303617	94679605	53541858	94851153	43941647	67858029
	96933781	45297001	79070929	03734528	90547953	26669351	43351703	35519438
	98385890	31297476	43138692	23445280	08617821	69378794	60632786	57995597
	67145708	01847010	13036465	57905497	03421002	96324102	67629571	36568345
	32475922	16330711	30566902	81180147	05850596	98976436	46710304	31270365
	88789964	31499916	97359174	84412426	87991755	61416368	08366203	88267013
	49645258	13181192	13628656	94192139	18906580	48677801	64807511	51212445
HHT	17558053	99004150	34649182	50148208	34165735	48811074	71046942	14947411
	72229873	03100727	64887261	03326598	36222252	73051169	95152976	11639962
	96277644	61102057	24726306	64360861	26273413	15310398	40070343	46489088
	34805775	07639204	76438277	88164985	69694560	13818253	68621283	76553668
	97929465	23521621	16224619	12344227	16951409	94606348	05607225	56830999
	21832599	75004650	55583207	44697760	20454723	97468226	43943535	17093630
	18640260	72928471	36818473	33546743	22490257	71428849	40611046	30378101
	99126011	40373742	28555040	96018065	15202898	00337698	63279830	89755546
HTH	55827271	92658157	10421161	81822273	95793393	81757068	52590757	71884578
	16672799	57821382	23154879	16237111	44257246	78873390	21380800	29908637
	98943797	68306107	17111888	05062872	99327206	15002979	39394123	85004254
	31518853	85223559	12123095	43223185	08355529	88799934	20972029	40700553
	88781367	37498431	89579539	27897711	29599460	07460889	79212431	49751501
	88750288	65663366	28519552	54060696	45209741	96963269	70335280	33777161
	19785873	12766191	43732609	94023169	78865858	17225914	37310763	47472752
	00230550	56740306	60305169	53575880	14063064	85767340	51344668	28882550
HTT	88520522	26943043	39487338	44862242	95307388	08106470	25059305	21651352
	55758167	69251645	09124995	34587781	48097894	05311139	57020375	14164209
	03036462	84084944	26777175	23545597	87515875	94045129	10805768	37868380
	71296641	05942767	87797635	71598350	55384008	03532321	76948508	24943669
	95560174	23302160	58723118	25474041	99875115	15017609	89402757	59244481
	32935662	57603699	89136411	41340556	15872089	07163269	56466100	44147787
	79169547	91808877	08659830	45842679	29070894	39390790	86239027	17607891
	84505670	52521587	06349296	07645937	35798966	50142638	49634985	57387174
THH	53570264	09982239	10729439	50265939	03399716	32291610	51321828	47765403
	83177523	64413111	02849182	97302561	86001280	67916866	03100113	14295096
	86221939	68138214	03013880	79750193	65538289	25190157	28027935	96611711
	96512748	25760250	40706029	59580017	62847717	26207805	11606526	28876709
	81052160	40076172	63426892	96374910	28250123	06313911	85715484	74341961
	29534609	45115054	13111948	97992370	99183690	35234940	58345780	53639114
	06232052	68808649	92990071	67931453	10581818	00987964	96305166	22362508
	75096305	20306090	90350965	19114233	36200473	87359052	63868744	88949579
THT	65994197	55573723	41511993	02645158	85382114	94446525	82512040	25352005
	96194358	43684016	13470783	08130389	29674869	80914756	45824972	97879610
	05382560	87122965	02538397	01360516	02803121	83160487	57701522	83096588
	12032560	87465752	14098951	37524054	69420367	29403485	41743347	49292205
	27783883	83693102	40817430	06587434	29461493	82247667	42397374	64624459
	43060304	79329141	84491259	15873303	70866041	48032494	25920499	16776553
	23492209	80915718	89824170	34253179	75657047	96673671	82371816	91212713
	90223723	30444038	21568962	47376866	50560833	78808307	57824838	07829383
TTH	99729031	40169194	15247286	16515010	91037172	27701465	50118735	13271417
	95292129	91417322	37086950	58666431	62525522	33839252	07422465	73127442
	45908753	45546803	50854104	56766503	80498331	77425084	69593928	35696230
	43734371	84868660	79610240	90592175	31484190	55990810	30360686	61421090
	47630180	99068064	35177597	82797159	39333168	82467611	48551406	83732157
	70044665	44177965	33995637	82053262	42763242	19535736	60610265	17291180
	14448728	59212721	01312288	17868045	92111185	03265557	10132098	68408405
	67834368	01392127	42344263	06693458	50667689	73890109	79511758	71136874
TTT	18161953	89635169	51827706	98952619	28702616	95690714	56594443	28844270
	92498242	08001971	62545818	71700150	18111862	15499631	87747217	78223602
	97998517	12624056	70575761	39274007	86058761	25002313	45186701	46586057
	32826808	66830750	89850476	91777512	51310637	72910365	23127509	26759450
	30844035	75510258	25021099	71670188	74940627	90130388	32227247	76522651
	37093729	66314046	58914897	83613531	43803888	78066477	31650418	21880752
	21417325	83509000	56506709	67455546	90839691	36735499	16026502	02651917
	83549395	64099724	92156803	30148426	87054401	07232736	44850303	92009664

TTT	HHH	HHT	HTH	HTT	THH	THT	TTH	TTT
HHH	34779902	09816035	92263628	49982202	91859473	31721278	65153036	39224726
	08272033	92734465	78767849	43585049	84972810	26472902	29743575	75766564
	79393459	55086425	30394365	23794700	22963032	72058684	64384859	12760534
	32870779	05456255	92896805	21739444	75919230	13999458	12299725	69567424
	47124639	16270170	24909250	83151887	68026752	84857413	89638869	19896283
	14913798	52586093	16259039	87158938	97448605	67901840	89939011	40816274
	29489136	69269247	27797956	14959549	47026196	96678220	05021384	25635221
	27100255	52684968	96153215	79286213	92097016	71372601	47701138	54405149
HHT	11792716	70313082	68073743	90710785	88094911	10307511	99136303	27603179
	38321746	92883960	66243416	72715738	40480845	52546743	46777509	25287199
	44374678	15073621	24897744	43014153	45951729	26078252	08510261	41799796
	40878774	80196744	45479893	26565365	04525100	25341424	73224034	20789404
	65781084	66771634	62137765	50263502	75775580	13692812	30745632	88714284
	44727293	39114124	38538173	01347875	08882731	14929803	27088420	47469197
	88453813	84529383	01464636	34016398	60784925	93017474	19826432	60116408
	85821463	91171398	71178829	79236609	09835490	61964481	83442262	91526010
HTH	73666327	91413951	74547751	89444961	94391310	28368196	34201036	77464852
	65376407	75996140	86918982	74842810	51434059	05590959	62838231	79826346
	52326768	68042779	15608008	21000920	32968683	41851179	17149534	22516395
	18552892	44149571	48819451	37572078	55787291	90774742	82623604	46251374
	87876354	90567945	85667241	89100908	32610897	26969059	65330671	55245982
	59604849	50831077	02200031	02482769	24799515	28760466	11739989	88005928
	01862669	64409302	61057981	95537563	03318024	75912889	56731107	50152245
	82202238	04492173	48044277	12104857	84176103	83807053	79849707	73025859
HTT	75738436	41378329	58182033	83255316	44861274	50606013	86035972	90239098
	61566937	04800073	26280186	26913604	13341901	51214421	02764266	03371428
	49285336	22961354	40942073	52689139	05891111	63912300	18406658	05039239
	41991082	71626878	76246615	33074376	73166488	27313473	24531580	18568308
	51978614	99857540	01014333	59220855	50038085	72419965	23353494	80390359
	81558392	51237849	13562182	94868753	69703025	91366995	78480692	22420803
	91242275	50547481	75370846	28010001	25641439	54851234	92171984	50375967
	40299957	68915392	77907816	34131442	36442373	36733632	33626781	99968817
THH	47978178	82236267	25184602	03171143	12815532	14623470	88351304	42303947
	87849604	30781946	69307947	82013299	18028598	45280976	58696816	73665239
	96895763	83201531	20076769	27877336	58823208	39179195	26541367	95481833
	97398087	07574586	63479655	61251282	52407633	39045343	23666347	50593756
	19769373	12074601	26164484	39596337	95385858	05441165	69844179	14855540
	29385821	45256846	27802435	55482299	81231228	26977173	19613564	54107331
	58156025	46286898	34959086	63374463	82593726	03861423	67667261	03510917
	46023136	55858847	71585166	80069775	48334441	51018812	89894415	04978213
THT	12253330	75352084	01109300	81199139	33734588	61351119	60911173	86601098
	95050028	86703698	83021495	03511823	72206468	66446274	89790233	14194490
	33989422	83175020	01577860	58055786	81551915	73921568	03456503	18911605
	27026922	62226818	55123762	47738956	34024478	75807706	81384837	51238403
	67591388	57271818	00961090	82897566	34503455	20840044	04797894	53483362
	35653384	15729958	58035142	65417792	09952709	72371658	80345180	65064199
	67974406	39525949	30775408	50316416	27472172	04718395	66119483	34984352
	10457925	32198497	90920663	26887016	53462826	71887862	54218927	26435282
TTH	44178990	69544441	93476098	75744696	56211933	66768788	41437214	88971341
	69739748	87593219	50667912	87058454	75935582	67120046	62594172	99987904
	60256358	51593219	83684468	68306616	88719418	26984721	07381609	01161421
	90210442	20508858	36265518	58430739	12096727	67180809	13893885	27342402
	54256636	39850702	37345803	21904265	17959094	76019501	45211648	57316677
	77643615	94275459	09118419	59643067	28335313	02333050	62809148	72001907
	40652878	20882983	54790070	77122111	94584064	57677552	82785052	71465391
	51169247	76347112	82110471	93263761	90072701	35664594	41628369	67793470
TTT	15727448	24821232	17151716	57118581	54274433	20847905	90871686	67251957
	89448229	91180121	90036736	81058185	18737156	17389674	42930845	94841566
	59537763	40765805	16246680	14089548	69855380	70976353	16431937	78957394
	36277102	13332605	50951778	36689365	10836813	44428362	09898128	72570299
	87965970	78182605	00870096	17053774	38755502	52835736	34302161	36947749
	87293038	41463173	41862442	14344536	73237490	69132122	89485970	89569004
	64191855	21558670	52628715	25838485	58151668	36483643	84935454	07166691
	48867584	90142910	08782981	90366931	38264437	82949644	74928389	35523446

Literature Cited

Abbott, H. G., and A. W. Coombs. 1964. A photoelectric 35-mm camera device for recording animal behavior. Journal of Mammalogy 45:327–330.

Adam, J. M. 1988. Effects of climatic extremes. Pp. 232–253. *In* R. Dawood (ed.), How to Stay Healthy Abroad. Penguin Books, New York.

Ahlen, I. 1980. Field identification of bats and survey methods based on sounds. Myotis 18–19:128–136.

———. 1981. Identification of Scandinavian Bats by Their Sounds. Swedish University Agricultural Science Report, Department of Wildlife Ecology Report 6, Uppsala, Sweden.

———. 1990. Identification of Bats in Flight. Swedish Society for Conservation of Nature, Stockholm, Sweden.

Aldous, M. C., and F. C. Craighead, Jr. 1958. A marking technique for bighorn sheep. Journal of Wildlife Management 22:445–446.

Aldridge, H. D. J. N., and I. L. Rautenbach. 1987. Morphology, echolocation and resource partitioning in insectivorous bats. Journal of Animal Ecology 56:763–778.

Alford, B. T., R. L. Burkhart, and W. P. Johnson. 1974. Etorphine and diprenorphine as immobilizing and reversing agents in captive and free-ranging mammals. Journal of the American Veterinary Medical Association 164:702–705.

Alldredge, J. R., and C. E. Gates. 1985. Line transect estimators for left-truncated distributions. Biometrics 41:273–280.

Allen, S. H., and A. B. Sargeant. 1975. A rural mail-carrier index of North Dakota red foxes. Wildlife Society Bulletin 3:74–77.

Allen, T. J. 1970. Immobilization of white-tailed deer with succinylcholine chloride and hyaluronidase. Journal of Wildlife Management 34:207–209.

Altenbach, J. S., K. Geluso, and D. E. Wilson. 1979. Population size of *Tadarida brasiliensis* at Carlsbad Caverns in 1973. Pp. 341–348. *In* H. H. Genoways and R. J. Baker (eds.), Biological Investigations in the Guadalupe Mountains National Park, Texas. Proceedings of the Transactions Series 4. National Park Service, Washington, D.C.

Altmann, S. A., and J. Altmann. 1970. Baboon Ecology. University of Chicago Press, Chicago, Illinois.

American Society of Mammalogists. 1985. Guidelines for the use of animals in research. Journal of Mammalogy 66:834.

———. 1987. Acceptable field methods in mammalogy: preliminary guidelines approved by the American Society of Mammalogists. Journal of Mammalogy 68(suppl.).

Anderson, B. W., and R. D. Ohmart. 1977. Rodent bait additive which repels insects. Journal of Mammalogy 58:242.

Anderson, D. R., K. P. Burnham, G. C. White, and D. L. Otis. 1983. Density estimation of small-mammal populations using a trapping web and distance sampling methods. Ecology 64:674–680.

Anderson, J. L. 1986. Age determination of the nyala *Tragelaphus angasi.* South African Journal of Wildlife Research 16:82–90.

Anderson, R. M. 1965. Methods of Collecting and Preserving Vertebrate Animals. National Museum of Canada Bulletin 69, Biological Series 18, Ottawa, Ontario, Canada.

Anderson, S. 1982. Monodelphis kunsi. Mammalian Species 190:1–3.

Anderson, S., and B. D. Patterson. 1994. Biogeography. Pp. 215–233. *In* E. C. Birney and J. R. Choate (eds.), Seventy-Five Years of Mammalogy (1919–1994). American Society of Mammalogists Special Publication 11.

Animal Behavior Society. 1986. Guidelines for the use of animals in research. Animal Behaviour 34:315–318.

———. 1991. Guidelines for the use of animals in research. Animal Behaviour 41:183–186.

Anonymous. 1994. The hantavirus—an old enemy with a new look. P. 2. *In* Medical Update November '94. Medical Education and Research Foundation.

———. 1995. Hantavirus update. Peromyscus Newsletter 19(March):6. *Peromyscus* Genetic Stock Center, University of South Carolina, Columbia, South Carolina.

Antenucci, J. C., K. Brown, P. L. Croswell, M. J. Krevany, and H. Archer. 1991. Geographic Information Systems: A Guide to the Technology. Van Nostrand Reinhold, New York.

Anthony, E. L. P. 1988. Age determination in bats. Pp. 47–58. *In* T. H. Kunz (ed.), Ecological and Behavioral Methods for the Study of Bats. Smithsonian Institution Press, Washington, D.C.

Anthony, E. L. P., and T. H. Kunz. 1977. Feeding strategies of the little brown bat, *Myotis lucifugus,* in southern New Hampshire. Ecology 58:775–780.

Aranda Sánchez, J. M. 1981. Rastros de los Mamíferos Silvestres de México. Instituto Nacional de Investigaciones sobre Recursos Bióticos, Xalapa, Veracruz, Mexico.

Armitage, K. B., J. F. Downhower, and G. E. Svendsen. 1976. Seasonal changes in weights of marmots. American Midland Naturalist 96:36–51.

Arnason, A. N. 1973. The estimation of population size, migration rates and survival in a stratified population. Researches on Population Ecology 15:1–8.

Arnason, A. N., and C. J. Schwarz. 1987. POPAN-3. Extended Analysis and Testing Features for POPAN-2. Charles Babbage Research Center, St. Norbert, Manitoba, Canada.

Arnason, A. N., C. J. Schwarz, and J. M. Gerrard. 1991. Estimating closed population size and number of marked animals from sighting data. Journal of Wildlife Management 55:716–730.

Arnbom, T. A., N. J. Lunn, I. L. Boyd, and T. Barton. 1992. Aging live Antarctic fur seals and southern elephant seals. Marine Mammal Science 8:37–43.

Arnold, G. W., and R. A. Maller. 1987. Monitoring population densities of western grey kangaroos in remnants of native vegetation. Pp. 219–225. *In* D. A. Saunders, G. W. Arnold, A. A. Burbidge, and A. J. M. Hopkins (eds.), Nature Conservation: The Role of Remnants of Native Vegetation. Surrey Beatty Sons, Chipping Norton, New South Wales, Australia.

Aronoff, S. 1989. Geographic Information Systems: A Management Perspective. WDL Publications, Ottawa, Ontario, Canada.

Au, W. W. L. 1992. Application of the reverberation-limited form of the sonar equation to dolphin echolocation. Journal of the Acoustical Society of America 92:1822–1826.

August, P. V. 1991. Use vs. abuse of GIS data. Journal of the Urban and Regional Information Systems Association 3:99–101.

———. 1993. GIS in mammalogy: building a database. Pp. 11–26. *In* S. McLaren and J. Braun (eds.), GIS Applications in Mammalogy. University of Oklahoma Press, Norman, Oklahoma.

August, P. V., J. Michaud, C. LaBash, and C. Smith. 1994. GPS for environmental applications: accuracy and precision of positional data. Photogrammetric Engineering and Remote Sensing 60:41–45.

Austad, S. N. 1993. Retarded senescence in an insular population of Virginia opossums (*Didelphis virginiana*). Journal of Zoology (London) 229:695–708.

Austin, D. H., and J. H. Peoples. 1967. Capturing hogs with alphachloralose. Proceedings of the Southeastern Association of Game and Fish Commissioners 21:201–205.

Balasingh, J., S. Isaac, and R. Subbaraj. 1992. A convenient device for tagging bats in the field. Bat Research News 33:6.

Banfield, A. W. F. 1974. The Mammals of Canada. University of Toronto Press, Toronto, Ontario, Canada.

Baranga, J., and B. Kiregyera. 1982. Estimation of the fruit bat population in the Kampala Bat Valley, Uganda. African Journal of Ecology 20:223–229.

Barclay, R. M. R. 1982. Interindividual use of echolocation calls: eavesdropping by bats. Behavioral Ecology and Sociobiology 10:271–275.

———. 1989. The effect of reproductive condition on the foraging behavior of female hoary bats, *Lasiurus cinereus*. Behavioral Ecology and Sociobiology 24:31–37.

Barclay, R. M. R., and G. P. Bell. 1988. Marking and observational techniques. Pp. 59–76. *In* T.H. Kunz (ed.), Ecological and Behavioral Methods for the Study of Bats. Smithsonian Institution Press, Washington, D.C.

Barclay, R. M. R., D. W. Thomas, and M. B. Fenton. 1980. Comparison of methods used for controlling bats in buildings. Journal of Wildlife Management 44:502–506.

Barrett, R. H. 1982. Wild pigs. Pp. 243–244. *In* D. E. Davis (ed.), CRC Handbook of Census Methods for Terrestrial Vertebrates. CRC Press, Boca Raton, Florida.

Bartmann, R. M., G. C. White, L. H. Carpenter, and R. A. Garist. 1987. Aerial mark-recapture estimates of confined mule deer in pinyon-juniper woodland. Journal of Wildlife Management 51:41–46.

Batchelor, T. A., and J. R. McMillen. 1980. A visual marking system for nocturnal animals. Journal of Wildlife Management 44:497–499.

Bateman, G. C., and T. A. Vaughan. 1974. Nightly activities of mormoopid bats. Journal of Mammalogy 55:45–65.

Bateson, P. 1991. Assessment of pain in animals. Animal Behavior 42:827–839.

Bayless, S. R. 1969. Winter food habits, range use and home range of antelope in Montana. Journal of Wildlife Management 33:538–551.

Bayliss, P., and J. Giles. 1985. Factors affecting the visibility of kangaroos counted during aerial surveys. Journal of Wildlife Management 49:686–692.

Bayliss, P., and K. M. Yeomans. 1989. Correcting bias in aerial survey population estimates of feral livestock in northern Australia using the double-count technique. Journal of Applied Ecology 26:925–933.

Beale, D. M. 1966. A self-collaring device for pronghorn antelope. Journal of Wildlife Management 30:209–211.

Beale, D. M., and A. D. Smith. 1967. Immobilization of pronghorn antelopes with succinylcholine chloride. Journal of Wildlife Management 31:840–842.

———. 1973. Mortality of pronghorn antelope fawns in western Utah. Journal of Wildlife Management 37:343–352.

Bear, G. D., G. C. White, L. H. Carpenter, R. B. Gill, and D. J. Essex. 1989. Evaluation of aerial mark-resighting estimates of elk populations. Journal of Wildlife Management 53:908–915.

Bearder, S. K., and R. D. Martin. 1980. Social organization of a nocturnal primate revealed by radio tracking. Pp. 633–648. *In* C. J. Amlaner and D. W. Macdonald (eds.), A Handbook on Biotelemetry and Radio Tracking. Pergamon Press, Oxford, England.

Beardsley, K., and D. Stoms. 1993. Compiling a digital map of areas managed for biodiversity in California. Natural Areas Journal 13:177–190.

Becker, E. F. 1991. A terrestrial furbearer estimator based on probability sampling. Journal of Wildlife Management 55:730–737.

Beeman, L. E., M. R. Pelton, and L. C. Marcum. 1974. Use of M99 etorphine for immobilizing black bears. Journal of Wildlife Management 38:568–569.

Bell, G. P. 1985. The sensory basis of prey location by the California leaf-nosed bat *Macrotus californicus* (Chiroptera: Phyllostomidae). Behavioral Ecology and Sociobiology 16:343–347.

Bennett, L. J., P. F. English, and R. McCain. 1940. A study of deer populations by use of pellet-group counts. Journal of Wildlife Management 4:398–403.

Berry, J. K. 1993a. The application of GIS to mammalogy: basic concepts. Pp. 4–10. *In* S. B. McLaren and J. K. Braun (eds.), GIS Applications in Mammalogy. Special Publication of the Oklahoma Museum of Natural History, Norman, Oklahoma.

Berry, J. K. 1993b. Cartographic modeling: the analytical capabilities of GIS. Pp. 58–74. *In* M. F.

Goodchild, B. O. Parks, and L. T. Steyaert (eds.), Environmental Modeling with GIS. Oxford University Press, New York.

Bertram, B. C. R. 1979. Studying Predators. Rev. 2d ed. African Wildlife Leadership Foundation, Nairobi, Kenya.

Black, H. C. 1958. Black bear research in New York. Transactions of the North American Wildlife Conference 23:443-461.

Blair, W. F. 1941. Techniques for the study of mammal populations. Journal of Mammalogy 22:148-157.

Boag, D. A., and J. O. Murie. 1981. Weight in relation to sex, age, and season in Columbian ground squirrels (Sciuridae: Rodentia). Canadian Journal of Zoology 59:999-1004.

Bonaccorso, F. J. 1979. Foraging and reproductive ecology in a Panamanian bat community. Bulletin of the Florida State Museum, Biology Sciences 24:359-408.

Bonaccorso, F. J., and N. Smythe. 1972. Punch-marking bats: an alternative to banding. Journal of Mammalogy 53:389-390.

Bonaccorso, F. J., N. Smythe, and S. R. Humphrey. 1976. Improved techniques for marking bats. Journal of Mammalogy 57:181-182.

Boonstra, R., M. Kanter, and C. J. Krebs. 1992. A tracking technique to locate small mammals at low densities. Journal of Mammalogy 73:683-685.

Borell, A. E. 1937. A new method of collecting bats. Journal of Mammalogy 18:478-480.

Bossert, W. H., and E. O. Wilson. 1963. The analysis of olfactory communication among animals. Journal of Theoretical Biology 5:443-469.

Bowen, W. D., D. E. Sergeant, and T. Oritsland. 1983. Validation of age estimation in the harp seal *Phoca groenlandica* using dental annuli. Canadian Journal of Fisheries and Aquatic Sciences 40:1430-1441.

Bradbury, J. W. 1977. Social organization and communication. Pp. 1-72. *In* W. A. Wimsatt (ed.), Biology of Bats. Vol. 3. Academic Press, New York.

Bradbury, J. W., and L. H. Emmons. 1974. Social organization of some Trinidad bats. I. Emballonuridae. Zeitschrift für Tierpsychologie 36:137-183.

Bradbury, J. W., and S. L. Vehrencamp. 1976. Social organization and foraging in emballonurid bats. I. Field studies. Behavioral Ecology and Sociobiology 1:337-381.

Bradt, G. W. 1938. A study of beaver colonies in Michigan. Journal of Mammalogy 19:139-161.

Brady, J. R., and M. R. Pelton. 1976. An evaluation of some cottontail rabbit marking techniques. Journal of the Tennessee Academy of Sciences 51:89-90.

Braham, H. W. 1982. Coastal migrating whales. Pp. 341-343. *In* D. E. Davis (ed.), CRC Handbook of Census Methods for Terrestrial Vertebrates. CRC Press, Boca Raton, Florida.

Braham, H. W., R. D. Everitt, and D. J. Rugh. 1980. Northern sea lion population decline in the eastern Aleutian Islands. Journal of Wildlife Management 44:25-33.

Breininger, D. R., M. J. Provancha, and R. B. Smith. 1991. Mapping Florida scrub jay habitat for purposes of land-use management. Photogrammetric Engineering and Remote Sensing 57:1467-1474.

Bronson, M. T. 1979. Altitudinal variation in the life history of the golden-mantled ground squirrel (*Spermophilus lateralis*). Ecology 60:272-279.

Brooks, R. P., and W. E. Dodge. 1978. A night identification collar for beavers. Journal of Wildlife Management 42:448-452.

Brosset, A. 1976. Social organization in the African bat, *Myotis bocagei*. Zeitschrift für Tierpsychologie 42:50-56.

Brown, J. C., and J. D. Williams. 1972. The rodent preputial gland. Mammal Review 2:105-147.

Brown, T., Jr. 1983. Tom Brown's Field Guide to Nature Observation and Tracking. Berkeley Books, New York.

Brownie, C., J. E. Hines, and J. D. Nichols. 1986. Constant-parameter capture-recapture models. Biometrics 42:561-574.

Buchler, E. R., and S. B. Childs. 1981. Orientation to distant sounds by foraging big brown bats (*Eptesicus fuscus*). Animal Behaviour 29:428-432.

Buckland, S. T., K. P. Burnham, D. R. Anderson, and J. L. Laake. 1993. Density Estimation Using Distance Sampling. Chapman Hall, London, England.

Bueler, L. E. 1973. Wild Dogs of the World. Stein and Day, New York.

Burger, B. 1988. Expandable and breakaway collars. Telonics Quarterly 1:1.

Burnham, K. P., and W. S. Overton. 1978. Estimation of the size of a closed population when capture probabilities vary among animals. Biometrika 65:625-633.

———. 1979. Robust estimation of population size when capture probabilities vary among animals. Ecology 60:927-936.

Burnham, K. P., D. R. Anderson, and J. L. Laake. 1980. Estimation of Density from Line Transect Sampling of Biological Populations. Wildlife Monographs 72.

Burrough, P. A. 1986. Principles of Geographical Information Systems for Land Resource Assessment. Oxford University Press, Oxford, England.

Burt, W. H. 1943. Territoriality and home range concepts as applied to mammals. Journal of Mammalogy 24:346–352.

Busnel, R. G. 1963. On certain aspects of animal acoustic signals. Pp. 69–111. *In* R.-G. Busnel (ed.), Acoustic Behaviour of Animals. Elsevier Scientific Publishing, Amsterdam, Netherlands.

Buss, I. O., and J. M. Savidge. 1966. Change in population number and reproductive rate of elephants in Uganda. Journal of Wildlife Management 30:791–809.

Butterworth, D. S. 1982. On the functional form used for g(y) for minke whale sightings, and bias in its estimation due to measurement inaccuracies. Reports of the International Whaling Commission 32:883–888.

Butterworth, D. S., and P. B. Best. 1982. Report of the southern hemisphere minke whale assessment cruise, 1980/81. Reports of the International Whaling Commission 32:835–874.

Butterworth, D. S., and D. L. Borchers. 1988. Estimates of g(0) for minke schools from the results of the independent observer experiment on the 1985/86 and 1986/87 IWC/IDCR Antarctic assessment cruises. Reports of the International Whaling Commission 38:301–313.

Cabrera, A. 1957–1961. Catálogo de los mamíferos de América del Sur. Revista del Museo Argentino de ciencias naturales "Bernardino Rivadavia." Ciencias Zoológicas 4.

Calkins, H. W. 1990. Creating large digital files from mapped data. Pp. 209–214. *In* D. J. Peuquet and D. F. Marble (eds.), Introductory Readings in Geographic Information Systems. Taylor and Francis, Bristol, Pennsylvania.

Campbell, W. G., and D. C. Mortenson. 1989. Ensuring the quality of geographic information system data: a practical application of quality control. Photogrammetric Engineering and Remote Sensing 55:1613–1618.

Cant, J. G. H. 1977. A census of agouti (*Dasyprocta punctata*) in seasonally dry forest at Tikal, Guatemala, with some comments on strip censusing. Journal of Mammalogy 58:688–690.

Capen, D. E. 1978. Time-lapse photography and computer analysis of behavior of nesting white-faced ibises. Pp. 41–43. *In* A. Sprunt IV, J. C. Ogden, and S. Winkler (eds.), Wading Birds. National Audubon Society, New York.

Carey, A. B., and J. W. Witt. 1991. Track counts as indices to abundances of arboreal rodents. Journal of Mammalogy 72:192–194.

Carson, J. D. 1961. Epiphyseal cartilage as an age indicator in fox and gray squirrels. Journal of Wildlife Management 25:90–93.

Carthew, S. M., and E. Slater. 1991. Monitoring animal activity with automated photography. Journal of Wildlife Management 55:689–692.

Caughley, G. J. 1964. Density and dispersion of two species of kangaroo in relation to habitat. Australian Journal of Zoology 12:238–249.

———. 1974. Bias in aerial survey. Journal of Wildlife Management 38:921–933.

———. 1977. Analysis of Vertebrate Populations. Wiley, New York.

Caughley, G., and D. Grice. 1982. A correction factor for counting emus from the air, and its application to counts in western Australia. Australian Wildlife Research 9:253–259.

Caughley, G., R. G. Sinclair, and D. Scott-Kemmis. 1976. Experiments in aerial survey. Journal of Wildlife Management 40:290–300.

Caughley, G., R. G. Sinclair, and G. R. Wilson. 1977. Numbers, distribution and harvesting rate of kangaroos on the inland plains of New South Wales. Australian Wildlife Research 4:99–108.

Champion, F. W. 1928. With a camera in tiger-land. Doubleday, Doran & Co., Garden City, New York.

Chao, A. 1987. Estimating the population size for capture-recapture data with unequal catchability. Biometrics 43:783–791.

Chao, A., S-M Lee, and S-L Jeng. 1992. Estimating population size for capture-recapture data when capture probabilities vary by time and individual animal. Biometrics 48:201–216.

Chapman, D. G. 1951. Some properties of the hypergeometric distribution with applications to zoological censuses. University of California Publications in Statistics 1:131–160.

———. 1954. The estimation of biological populations. Annals of Mathematical Statistics 25:1–15.

Chapman, D. I. 1975. Antlers—bones of contention. Mammal Review 5:121–172.

Chapman, J. A., and G. A. Feldhammer. 1982. Wild Mammals of North America: Biology, Management, and Economics. Johns Hopkins University Press, Baltimore, Maryland.

Chapman, J. A., and J. E. C. Flux. 1990. Rabbits, Hares and Pikas: Status Survey and Conservation Action Plan. IUCNR, Gland, Switzerland.

Chapman, J. A., and G. R. Wilner. 1986. Lagomorphs. Pp. 453–473. *In* A. Y. Cooperrider, R. J. Boyd, and H. R. Stuart (eds.), Inventory and Monitoring of

Wildlife Habitat. U.S. Department of Interior, Bureau of Land Management Service Center, Denver, Colorado.

Charles-Dominique, P. 1983. Ecology and social adaptation in didelphid marsupials: comparison with eutherians of similar ecology. Pp. 395–422. *In* J. F. Eisenberg and D. G. Kleiman (eds.), Advances in the Study of Mammalian Behavior. American Society of Mammalogists, Special Publication 7.

———. 1993. Tent-use by the bat *Ryinophylla pumilio* (Phyllostomidae: Carolliinae) in French Guiana. Biotropica 25:111–116.

Cheeseman, C. L., and P. J. Mallinson. 1980. Radio tracking in the study of bovine tuberculosis in badgers. Pp. 649–656. *In* C. J. Amlaner and D. W. Macdonald (eds.), A Handbook on Biotelemetry and Radiotracking. Pergamon Press, Oxford, England.

Chitty, D. 1937. A ringing technique for small mammals. Journal of Animal Ecology 6:36–53.

Choate, J. R., and H. H. Genoways. 1975. Federal and state regulations pertaining to systematic collections. A case of inadvertent violation of federal regulations. SWANEWS, 1975 (3&4):10–13.

Christensen, P. E. S. 1980. The Biology of *Bettongia penicillata* Gray, 1837 and *Macropus eugenii* (Desmarest, 1817) in Relation to Fire. Forests Department of Western Australia Bulletin 91.

Christian, J. J. 1950. A field method of determining the reproductive status of small male mammals. Journal of Mammalogy 31:95–96.

Churcher, C. S., and M. L. Richardson. 1978. Equidae. Pp. 379–422. *In* V. J. Maglio and H. B. S. Cooke (eds.), Evolution of African Mammals. Harvard University Press, Cambridge, Massachusetts.

Clover, M. R. 1954. A portable deer trap and catch-net. California Fish and Game 40:367–373.

Clutton-Brock, T. H., F. E. Guinness, and S. D. Albon. 1982. Red Deer: Behavior and Ecology of Two Sexes. University of Chicago Press, Chicago, Illinois.

Clutton-Brock, T. H., P. H. Harvey, and B. Rudder. 1977. Sexual dimorphism, socionomic sex ratio and body weight in primates. Nature 269:797–800.

Cochran, W. G. 1977. Sampling Techniques. 3d ed. Wiley, New York.

Coggins, V. L. 1975. Immobilization of Rocky Mountain elk with M99. Journal of Wildlife Management 39:814–816.

Cole, F. R., D. M. Reeder, and D. E. Wilson. 1994. A synopsis of distribution patterns and the conservation of mammal species. Journal of Mammalogy 75:266–276.

Colvin, J. G. 1992. A code of ethics for research in the third world. Conservation Biology 6:309–311.

Conner, M. C., R. F. Labisky, and D. R. Progulske, Jr. 1983. Scent-station indices as measures of population abundance for bobcats, raccoons, gray foxes, and opossums. Wildlife Society Bulletin 11:146–152.

Conner, M. C., R. A. Lancia, and K. H. Pollock. 1986. Precision of the change-in-ratio technique for deer population management. Journal of Wildlife Management 50:125–129.

Conner, M. C., E. C. Soutiere, and R. A. Lancia. 1987. Drop-netting deer: costs and incidence of capture myopathy. Wildlife Society Bulletin 15:434–438.

Constantine, D. G. 1958. An automatic bat-collecting device. Journal of Wildlife Management 22:17–22.

———. 1966. Ecological observations in lasiurine bats in Iowa. Journal of Mammalogy 47:34–41.

———. 1988. Health precautions for bat researchers. Pp. 491–528. *In* T. H. Kunz (ed.), Ecological and Behavioral Methods for the Study of Bats. Smithsonian Institution Press, Washington, D.C.

Cook, A. H. 1943. A technique for marking mammals. Journal of Mammalogy 24:45–47.

Cook, R. D., and J. O. Jacobson. 1979. A design for estimating visibility bias in aerial surveys. Biometrics 35:735–742.

Cooley, M. E. 1948. Improved toe-tag for marking fox squirrels. Journal of Wildlife Management 12:213.

Cope, J. B., E. Churchwell, and K. Koontz. 1961. A method of tagging bats with radioactive gold-198 in homing experiments. Proceedings of the Indiana Academy of Science 70:267–269.

Corbet, G., and D. Ovenden. 1980. The Mammals of Britain and Europe. Collins, London, England.

Corbet, G. B. 1978. The Mammals of the Palaearctic Region: A Taxonomic Review. British Museum (Natural History), Cornell University Press, London, England.

———. 1984. The Mammals of the Palaearctic Region: A Taxonomic Review. Supplement. British Museum (Natural History), London, England.

Corbet, G. B., and J. E. Hill. 1992. Mammals of the Indomalayan Region: A Systematic Review. Oxford University Press, Oxford, England.

Cormack, R. M. 1979. Models for capture-recapture. Pp. 217–255. *In* R. M. Cormack, G. P. Patil, and D. S. Robson (eds.), Sampling Biological Populations. International Co-operative Publishing House, Fairland, Maryland.

Cotterill, F. P. D., and R. A. Fergusson. 1993. Capturing free-tailed bats (Chiroptera: Molossidae): the description of a new trapping device. Journal of Zoology (London). 231:645–651.

Coulson, G. M. 1979. Some methods for surveying kangaroo populations. Victorian Naturalist 96:184–189.

Coulson, G. M., and J. A. Raines. 1985. Methods for small-scale surveys of grey kangaroo populations. Australian Wildlife Research 12:119–125.

Craig, W. J., P. Tessar, and N. A. Khan. 1991. Sharing graphic data files in an open system environment. Journal of the Urban and Regional Information Systems Association 3:20–32.

Crockett, C. M., and J. F. Eisenberg. 1986. Howlers: variations in group size and demography. Pp. 54–68. *In* B. B. Smuts, D. L. Cheney, R. M. Seyfarth, R. W. Wrangham, and T. T. Struhsaker (eds.), Primate Societies. University of Chicago Press, Chicago, Illinois.

Crockford, J. A., F. A. Hayes, J. H. Jenkins, and S. D. Feurt. 1958. An automatic projectile type syringe. Veterinary Medicine 53:115–119.

Crome, F. H. J., and G. G. Richards. 1988. Bats and gaps: microchiropteran community structure in a Queensland rain forest. Ecology 69:1960–1969.

Croswell, P. L. 1991. Obstacles to GIS implementation and guidelines to increase the opportunities for success. Journal of the Urban and Regional Information Systems Association 3:43–56.

Daan, S. 1969. Frequency of displacements as a measure of activity of hibernating bats. Lynx 10:13–18.

———. 1973. Activity during natural hibernation in three species of vespertilionid bats. Netherlands Journal of Zoology 23:1–71.

Danner, D. A., and N. S. Smith. 1980. Coyote home range, movement, and relative abundance near a cattle feedyard. Journal of Wildlife Management 44:484–488.

Darlington, P. J., Jr. 1957. Zoogeography: The Geographical Distribution of Animals. Wiley, New York.

Daunt-Mergens, D. O. (ed.). 1981. Cave Research Foundation Personnel Manual. 3d ed. Cave Research Foundation, Mammoth Cave, Kentucky.

Davey, C. C., P. J. Fullagar, and C. Kogon. 1980. Marking rabbits for individual identification and a use for betalights. Journal of Wildlife Management 44:494–497.

David, J. H. M., M. A. Meyer, and P. B. Best. 1990. The capture, handling and marking of free ranging adult South American (Cape) fur seals. South African Journal of Wildlife Research 20:5–8.

Davis, D. E. 1956. Manual for Analysis of Rodent Populations. Edwards Brothers, Ann Arbor, Michigan.

———. 1982. CRC Handbook of Census Methods for Terrestrial Vertebrates. CRC Press, Boca Raton, Florida.

Davis, F. W., D. M. Stoms, J. E. Estes, J. Scepan, and J. M. Scott. 1990. An information systems approach to the preservation of biological diversity. International Journal of Geographical Information Systems 4:55–78.

Davis, R. B., C. F. Herreid II, and H. L. Short. 1962. Mexican free-tailed bats in Texas. Ecological Monographs 32:311–346.

Davis, W. B. 1974. Mammals of Texas. Rev. ed. Texas Parks and Wildlife Department Bulletin 41. Austin, Texas.

Davis, W. H. 1960. Band injuries. Bat Banding News 1:1–2.

———. 1966. The new bat bands. Bat Banding News 7:39.

Day, G. I. 1974. Remote Injection of Drugs. Arizona Game and Fish Department P-R Report, Project W-78-15. Phoenix, Arizona.

Day, G. I., S. D. Schemnitz, and R. D. Taber. 1980. Capturing and marking mammals. Pp. 61–88. *In* S. D. Schemnitz (ed.), Wildlife Management Techniques Manual. The Wildlife Society, Washington, D.C.

DeBlase, A. F., and R. E. Martin. 1981. A Manual of Mammalogy with Keys to the Families of the World. 2d ed. W. C. Brown Company Publishers, Dubuque, Iowa.

Dennis, B., G. P. Patil, O. Rossi, S. Stehman, and C. Taillie. 1979. A bibliography of literature on ecological diversity and related methodology. Pp. 319–353. *In* J. F. Grassle, G. P. Patil, W. Smith, and C. Taillie (eds.), Ecological Diversity in Theory and Practice. International Co-operative Publishing House, Fairland, Maryland.

Derleth, E. L., D. G. McAuley, and T. J. Dwyer. 1989. Avian community response to small-scale habitat disturbance in Maine. Canadian Journal of Zoology 67:385–390.

Dessauer, H. C., C. J. Cole, and M. S. Hafner. 1990. Collection and storage of tissues. Pp. 25–41. *In* D. M. Hillis and C. Moritz (eds.), Molecular Systematics. Sinauer, Sunderland, Massachusetts.

di Castri, F., J. Robertson Vernhes, and T. Younés. (eds.). 1992. Inventorying and Monitoring Biodiversity. Biology International Special Issue 27. IUBS, Paris, France.

Diefenbach, D. R., M. J. Conroy, R. J. Warren, W. E. James, L. A. Baker, and T. Hon. 1994. A test of the scent-station survey technique for bobcats. Journal of Wildlife Management 58:10–17.

Dieterlen, F. 1993a. Family Ctenodactylidae. P. 761. *In* D. E. Wilson and D. M. Reeder (eds.), Mammal Species of the World. 2d ed. Smithsonian Institution Press, Washington, D.C.

———. 1993b. Family Anomaluridae. Pp. 757–758. *In* D. E. Wilson and D. M. Reeder (eds.), Mammal Species of the World. 2d ed. Smithsonian Institution Press, Washington, D.C.

———. 1993c. Family Pedetidae. P. 759. *In* D. E. Wilson and D. M. Reeder (eds.), Mammal Species of the World. 2d ed. Smithsonian Institution Press, Washington, D.C.

Dinerstein, E. 1992. Effects of *Rhinoceros unicornis* on riverine forest structure in lowland Nepal. Ecology 73:701–704.

Dinerstein, E., and L. Price. 1991. Demography and habitat use by greater one-horned rhinoceros in Nepal. Journal of Wildlife Management 55:401–411.

Dittus, W. P. J. 1975. Population dynamics of the toque monkey, *Macaca sinica.* Pp. 125–151. *In* R. H. Tuttle (ed.), Socioecology and Psychology of Primates. Mouton Publishers, The Hague, Netherlands.

———. 1977. The social regulation of population density and age-sex distribution in the toque monkey. Behaviour 63:281–322.

Dixson, A. F. 1983. Observations on the evolution and behavioral significance of "sexual skin" in female primates. Pp. 63–106. *In* J. S. Rosenblatt, R. A. Hinde, C. Beer, and M-C Busnel (eds.), Advances in the Study of Behavior. Vol. 13. Academic Press, New York.

Dobie, J. F. 1961. The Voice of the Coyote. University of Nebraska Press, Lincoln, Nebraska.

Doidge, D. W., T. S. McCann, and J. P. Croxall. 1986. Attendance behavior of Antarctic fur seals. Pp. 102–114. *In* R. L. Gentry and G. L. Kooyman (eds.), Fur Seals: Maternal Strategies on Land and at Sea. Princeton University Press, Princeton, New Jersey.

Domning, D. P. 1982. Evolution of manatees: a speculative history. Journal of Paleontology 56:599–619.

Downes, C. M. 1982. A comparison of sensitivities of three bat detectors. Journal of Mammalogy 63:343–345.

Downing, R. L., and B. S. McGinnes. 1969. Capturing and marking white-tailed deer fawns. Journal of Wildlife Management 33:711–714.

Downing, R. L., W. H. Moore, and J. Kight. 1965. Comparison of deer census techniques applied to a known population in Georgia enclosure. Proceedings of the Southeastern Association of Game and Fish Commissioners 19:26–30.

Drew, G. S., D. B. Fagre, and D. J. Martin. 1988. Scent-station surveys of cottontail rabbit populations. Wildlife Society Bulletin 16:396–398.

Drickamer, L. C. 1984. Captures of two species of *Peromyscus* at live traps baited with male and female odors. Journal of Mammalogy 65:699–702.

———. 1987. Influence of time of day on captures of two species of *Peromyscus* in a New England deciduous forest. Journal of Mammalogy 68:702–703.

Driscoll, J. W., and P. Bateson. 1988. Animals in behavioural research. Animal Behaviour 36:1569–1574.

Drummer, T. D., and L. L. McDonald. 1987. Size bias in line transect sampling. Biometrics 43:13–21.

Drummer, T. D., A. R. Degange, L. L. Pank, and L. L. McDonald. 1990. Adjusting for group size influence in line transect sampling. Journal of Wildlife Management 54:511–514.

Dublin, H. T., and I. Douglas-Hamilton. 1987. Status and trends of elephants in the Serengeti-Mara ecosystem. African Journal of Ecology 25:19–33.

Dunford, C., and R. Davis. 1975. Cliff chipmunk vocalizations and their relevance to the taxonomy of coastal Sonoran chipmunks. Journal of Mammalogy 56:207–212.

Dupont, W. D. 1983. A stochastic catch-effort method for estimating animal abundance. Biometrics 39:1021–1033.

Dwyer, P. D. 1965. Injuries due to bat-banding. Pp. 19–24. *In* K. G. Simpson and E. Hamilton-Smith (eds.), 3rd and 4th Annual Report of Bat-Banding in Australia. Division of Wildlife Research Technical Paper 9. CSIRO, Melbourne, Victoria, Australia.

Eadie, W. R. 1938. The dermal glands of shrews. Journal of Mammalogy 19:171–174.

Easter-Pilcher, A. 1990. Cache size as an index to beaver colony size in northwestern Montana. Wildlife Society Bulletin 18:110–113.

Eberhardt, L., and R. C. Van Etten. 1956. Evaluation of the pellet group count as a deer census method. Journal of Wildlife Management 20:70–74.

Eberhardt, L. L. 1990. Using radio-telemetry for mark-recapture studies with edge effects. Journal of Applied Ecology 27:259–271.

Eberhardt, L. L., and M. A. Simmons. 1987. Calibrating population indices by double sampling. Journal of Wildlife Management 51:665–675.

Eberhardt, L. L., D. G. Chapman, and J. R. Gilbert. 1979. A Review of Marine Mammal Census Methods. Wildlife Monographs 63.

Edge, W. D., and S. L. Olson-Edge. 1990. Population characteristics and group composition of *Capra aegagrus* in Kirthar National Park, Pakistan. Journal of Mammalogy 71:156–160.

Edwards, D. K., G. L. Dorsey, and J. A. Crawford. 1981. A comparison of three avian census methods. *In* Estimating Numbers of Terrestrial Birds. Studies in Avian Biology 6:170–176.

Ehlers, M., G. Edwards, and Y. Bedard. 1989. Integration of remote sensing with geographic information systems: a necessary evolution. Photogrammetric Engineering and Remote Sensing 55:1619–1627.

Ehlers, M., D. Greenlee, T. Smith, and J. Star. 1991. Integration of remote sensing and GIS data: data and data access. Photogrammetric Engineering and Remote Sensing 57:669–675.

Eisenberg, J. F. 1979. Habitat, economy, and society: some correlations and hypothesis for neotropical primates. Pp. 215–262. *In* I. S. Bernstein and E. O. Smith (eds.), Primate Ecology and Human Origin. Garland STPM Press, New York.

———. 1989. Mammals of the Neotropics. Vol. 1. University of Chicago Press, Chicago, Illinois.

Eisenberg, J. F., and D. E. Wilson. 1978. Relative brain size and feeding strategies in the Chiroptera. Evolution 32:740–751.

Eisenberg, J. F., C. P. Groves, and K. MacKinnon. 1987. Tapirs. Grzimeks Enzyklopadie Saugetiere 4:598–608.

Ellerman, J. R., and T. C. S. Morrison-Scott. 1966. Checklist of Palearctic and Indian Mammals. 1758 to 1946. 2d ed. British Museum (Natural History), London, England.

Elliott, L. 1978. Social behavior and foraging ecology of the eastern chipmunk (*Tamias striatus*) in the Adirondack Mountains. Smithsonian Contributions to Zoology 265:1–107.

Emmons, L. H., and F. Feer. 1990. Neotropical Rainforest Mammals: A Field Guide. University of Chicago Press, Chicago, Illinois.

Enders, R. K., and J. R. Leekley. 1941. Cyclic changes in the vulva of the marten (*Martes americana*). Anatomical Record 79:1–5.

Engen, S. 1978. Stochastic Abundance Models. Wiley, New York.

Erickson, A. W. 1957. Techniques for live-trapping and handling black bears. Transactions of the North American Wildlife Conference 22:520–543.

Estes, J. A., and R. J. Jameson. 1988. A double-survey estimate for sighting probability of sea otters in California. Journal of Wildlife Management 52:70–76.

Estes, R. D. 1991a. The significance of horns and other male secondary sexual characters in female bovids. Pp. 403–451. *In* E. C. Mungall (ed.), Ungulate Behavior and Management. Applied Animal Behavioral Science (special issue) 29:1–531.

———. 1991b. The Behavior Guide to African Mammals. University of California Press, Berkeley, California.

Evans, C. D., W. A. Troyer, and C. J. Lensink. 1966. Aerial census of moose by quadrat sampling. Journal of Wildlife Management 30:767–776.

Evans, F. C., and R. Holdenried. 1943. A population study of the beechey ground squirrel in central California. Journal of Mammalogy 24:231–260.

Fancy, S. G. 1980. Preparation of mammalian teeth for age determination by cementum layers: a review. Wildlife Society Bulletin 8:242–248.

Fancy, S. G., L. F. Pank, D. C. Douglas, C. H. Curby, G. W. Garner, S. C. Amstrup, and W. L. Regelin. 1988. Satellite Telemetry: A New Tool for Wildlife Research and Management. U.S. Fish and Wildlife Service Resource Publication 172. Washington, D.C.

Fashingbauer, B. A. 1962. Expanding plastic collar and aluminum collar for deer. Journal of Wildlife Management 26:211–213.

Federal Geographic Data Committee. 1992. Draft Content Standards for Spatial Metadata. Federal Geographic Data Committee, Washington, D.C.

Federal Register. 1973. Code of Federal Regulations. Wildlife and Fisheries (Title 50), Chapter 1 (Bureau of Sport Fisheries and Wildlife Service, Fish and Wildlife Service, Department of Interior). Office of the Federal Register, National Archives and Records Administration, Washington, D.C.

———. 1975. Code of Federal Regulations. Animals and Animal Products (Title 9), Subchapter A: Animal Welfare, Parts 1,2,3. Office of the Federal Register, National Archives and Records Administration, Washington, D.C.

Fedo, B. 1991. Ball chain collars. Telonics Quarterly 4:2.

Fegeas, R. G., J. L. Cascio, and R. A. Lazar. 1992. An overview of FIPS 173, the spatial data transfer standard. Cartography and Geographic Information Systems 19:1–24.

Fenton, M. B. 1983. Just Bats. University of Toronto Press, Toronto, Ontario, Canada.

———. 1988. Bats. Facts on File, New York.

Fenton, M. B., and G. P. Bell. 1979. Echolocation and feeding behaviour in four species of *Myotis* (Chiroptera). Canadian Journal of Zoology 57:1271–1277.

———. 1981. Recognition of insectivorous bats by their echolocation calls. Journal of Mammalogy 62:233–243.

Fenton, M. B., and G. K. Morris. 1976. Opportunistic feeding by desert bats (*Myotis* spp.). Canadian Journal of Zoology 54:526–530.

Fenton, M. B., and I. L. Rautenbach. 1986. A comparison of the roosting and foraging behavior of three species of African insectivorous bats (Rhinolophidae, Vespertilionidae, and Molossidae). Canadian Journal of Zoology 64:2860–2867.

Fenton, M. B., and J. S. Taylor. 1989. Laboratory Exercises in Echolocation. Field Laboratory Manual. Department of Biology, York University, North York, Ontario, Canada.

Fenton, M. B., L. Acharya, D. Audet, M. B. C. Hickey, C. Merriman, M. K. Obrist, D. M. Syme, and B. Adkins. 1992. Phyllostomid bats (Chiroptera: Phyllostomidae) as indicators of habitat disruption in the neotropics. Biotropica 24:440–446.

Fenton, M. B., R. M. Brigham, A. M. Mills, and I. L. Rautenbach. 1985. The roosting and foraging areas of *Epomophorus wahlbergi* (Pteropodidae) and *Scotophilus viridis* (Vespertilionidae) in Kruger National Park, South Africa. Journal of Mammalogy 66:461–468.

Fenton, M. B., H. G. Merriam, and G. L. Holroyd. 1983. Bats of Kootenay, Glacier, and Mount Revelstoke national parks in Canada: identification by echolocation calls, distribution, and biology. Canadian Journal of Zoology 61:2503–2506.

Festa-Bianchet, M. 1991. The social system of bighorn sheep: grouping patterns, kinships and female dominance rank. Animal Behaviour 42:71–82.

Field, C. R., and R. M. Laws. 1970. The distribution of the larger herbivores in the Queen Elizabeth National Park, Uganda. Journal of Applied Ecology 7:273–294.

Findley, J. S., and D. E. Wilson. 1974. Observations on the neotropical disc-winged bat, *Thyroptera tricolor*. Journal of Mammalogy 55:562–571.

Finnemore, M., and P. W. Richardson. 1987. Catching bats. Pp. 18–24. *In* A. J. Mitchell-Jones (ed.), The Bat Worker's Manual. Nature Conservancy Council, Peterborough, United Kingdom.

Firchow, K. M., M. R. Vaughan, and W. R. Mytton. 1990. Comparison of aerial survey techniques for pronghorns. Wildlife Society Bulletin 18:18–23.

Fischer, R. A., and N. R. Holler. 1991. Habitat use and relative abundance of gray squirrels in southern Alabama. Journal of Wildlife Management 55:52–59.

Fisher, P. 1991. Spatial data sources and data problems. Pp. 175–189. *In* D. J. Maguire, M. F. Goodchild, and D. W. Rhind (eds.), Geographical Information Systems: Principles and Applications. Wiley, New York.

Flannery, T. 1990. Mammals of New Guinea. Robert Brown & Associates, Carina, Queensland, Australia.

Fleming, T. H. 1988. The Short-Tailed Fruit Bat: A Study in Plant-Animal Interactions. University of Chicago Press, Chicago, Illinois.

Floyd, R. B. 1980. Density of *Wallabia bicolor* (Demarest) (Marsupialia: Macropodidae) in eucalypt plantations of different ages. Australian Wildlife Research 7:333–337.

Floyd, T. J., L. D. Mech, and M. E. Nelson. 1982. Deer in forested areas. Pp. 254–256. *In* D. E. Davis (ed.), CRC Handbook of Census Methods for Terrestrial Vertebrates. CRC Press, Boca Raton, Florida.

Forbes, B., and E. M. Newhook. 1990. A comparison of the performance of three models of bat detectors. Journal of Mammalogy 71:108–110.

Ford, J. K. B. 1983. Group-specific dialects of killer whales (*Orcinus orca*) in British Columbia. Pp. 129–161. *In* R. Payne (ed.), Communication and Behavior of Whales. American Association for the Advancement of Science Symposium 76. Westview Press, Boulder, Colorado.

Forman, R. T. T., and M. Godron. 1986. Landscape Ecology. Wiley, New York.

Foster, J. B., and A. I. Dagg. 1972. Notes on the biology of the giraffe. East African Wildlife Journal 10:1–16.

Foster, J. B., and D. Kearney. 1967. Nairobi National Park game census, 1966. East African Wildlife Journal 5:112–120.

Foster, M. S. 1982. The research natural history museum: pertinent or passé? Biologist 64:1–12.

Frame, L. H., J. R. Malcolm, G. W. Frame, and H. van Lawick. 1979. Social organization of African wild dogs (*Lycaon pictus*) on the Serengeti Plains, Tanzania 1967–1978. Zeitschrift für Tierpsychologie 50:225–249.

Francis, C. M. 1989. A comparison of mist nets and two designs of harp traps for capturing bats. Journal of Mammalogy 70:865–870.

———. 1994. Vertical stratification of fruit bats (Pteropodidae) in lowland dipterocarp rainforest in Malaysia. Journal of Tropical Ecology 10:523–530.

Francis, C. M., E. L .P. Anthony, J. A. Brunton, and T. H. Kunz. 1994. Lactation in male fruit bats. Nature 367:691-692.

Franklin, W. L. 1983. Contrasting socioecologies of South America's wild camelids: the vicuña and guanaco. Pp. 573-629. *In* J. F. Eisenberg and D.G. Kleiman (eds.), Advances in the Study of Mammalian Behavior. American Society of Mammalogists Special Publication 7.

Frantz, S. C. 1989. Remote videography of small mammals under conditions of dim illumination or darkness. Pp. 41-51. *In* K. A. Fagerstone and R. D. Curnow (eds.), Vertebrate Pest Control and Management Materials. Vol. 6 American Society for Testing Materials Special Technical Publication 1055. Philadelphia, Pennsylvania.

Frey, J. K., T. L. Yates, D. W. Duszynski, W. L. Gannon, and S. L. Gardner. 1992. Designation and curatorial management of type host specimens (symbiotypes) for new parasite species. Journal of Parasitology 78:930-932.

Friend, M. 1967. Some observations regarding eye-lens weight, as a criterion of age in animals. New York Fish and Game Journal 14:91-121.

Frith, H. J. 1973. Wildlife Conservation. Angus Robertson, Sydney, New South Wales, Australia.

Fritschen, L. J., and L. W. Gay. 1979. Environmental Instrumentation. Springer-Verlag, New York.

Frost, D. R. 1985. Amphibian Species of the World. Allen Press and Association of Systematics Collections, Lawrence, Kansas.

Frylestam, B., and T. von Schantz. 1977. Age determination of European hares based on periosteal growth lines. Mammal Review 7:151-154.

Fuller, T. K. 1991. Do pellet counts index white-tailed deer numbers and population change? Journal of Wildlife Management 55:393-396.

Fuller, T. K., and B. A. Sampson. 1988. Evaluation of a simulated howling survey for wolves. Journal of Wildlife Management 52:60-63.

Gaisler, J., and J. Kolibac. 1992. Summer occurrence of bats in agrocoenoses. Folia Zoologica 41:19-27.

Gaisler, J., and M. Nevrly. 1961. The use of coloured bands in investigating bats. Acta Societatis Zoologicae Bohemoslovacae 25:135-141.

Gaisler, J., V. Hanak, and J. Dungel. 1979. A contribution to the population ecology of *Nyctalus noctula* (Mammalia: Chiroptera). Acta Scientiarum Naturalium Academiae Scientiarum Bohemoslovacae Brno 13:1-38.

Gannon, W. L., and W. T. Stanley. 1991. Chip vocalization of Palmer's chipmunk (*Tamias palmeri*). Southwestern Naturalist 36:315-317.

Gardner, A. L. 1993a. Order Didelphimorphia. Pp. 15-23. *In* D. E. Wilson and D. M. Reeder (eds.), Mammal Species of the World: A Taxonomic and Geographic Reference. 2d ed. Smithsonian Institution Press, Washington, D.C.

———. 1993b. Order Paucituberculata. P. 25. *In* D. E. Wilson and D. M. Reeder (eds.), Mammal Species of the World: A Taxonomic and Geographic Reference. 2d ed. Smithsonian Institution Press, Washington, D.C.

———. 1993c. Order Microbiotheria. P. 27. *In* D. E. Wilson and D. M. Reeder (eds.), Mammal Species of the World: A Taxonomic and Geographic Reference. 2d ed. Smithsonian Institution Press, Washington, D.C.

———. 1993d. Order Xenarthra. Pp. 63-68. *In* D. E. Wilson and D. M. Reeder (eds.), Mammal Species of the World: A Taxonomic and Geographic Reference. 2d ed. Smithsonian Institution Press, Washington, D.C.

Gardner, J. E., J. D. Garner, and J. E. Hoffmann. 1989. A portable mist netting system for capturing bats with emphasis on *Myotis sodalis* (Indiana bat). Bat Research News 30:1-8.

Gardner, S. L. 1991. Phyletic coevolution between subterranean rodents of the genus *Ctenomys* (Rodentia: Hystricognathi) and nematodes of the genus *Paraspidodera* (Heterakoidea: Aspidoderidae) in the neotropics: temporal and evolutionary implications. Zoological Journal of the Linnean Society 102:169-201.

Gardner, S. L., and M. L. Campbell. 1992a. Parasites as probes for biodiversity. Journal of Parasitology 78:596-600.

———. 1992b. A new species of *Linstowia* (Cestoda: Anoplocephalidae) from marsupials in Bolivia. Journal of Parasitology 78:795-799.

Gardner, S. L., and D. W. Duszynski. 1990. Polymorphism of eimerian oocysts can be a problem in naturally infected hosts: an example from subterranean rodents in Bolivia. Journal of Parasitology 76:805-811.

Gardner, S. L., S. J. Upton, C. R. Lambert, and O. C. Jordan. 1991. Redescription of *Eimeria escomeli* (Rastegaieff, 1930) from *Myrmecophaga tridactyla,* and a first report from Bolivia. Journal of the Helminthological Society of Washington 58:16-18.

Garrott, R. A., and R. W. Hayes. 1984. A radio-controlled device for triggering traps. Wildlife Society Bulletin 12:320–322.

Garrott, R. A., and L. Taylor. 1990. Dynamics of a feral horse population in Montana. Journal of Wildlife Management 54:603–612.

Garrott, R. A., and G. C. White. 1982. Age and sex selectivity in trapping mule deer. Journal of Wildlife Management 46:1083–1086.

Gaskell, B. H. 1984. Flying fruit-bat faunas of the upper canopy in two paleotropical rain-forests. P. 303. *In* A. C. Chadwick and S. L. Sutton (eds.), Tropical Rain-Forest: The Leeds Symposium. Leeds Philosophical and Literary Society, Leeds, United Kingdom.

Gaskin, D. E. 1982. The Ecology of Whales and Dolphins. Heinemann Educational Books, London, England.

Gates, C. E. 1980. LINETRAN, a general computer program for analyzing line-transect data. Journal of Wildlife Management 44:658–661.

Geibel, J. J., and D. J. Miller. 1984. Estimation of sea otter, *Enhydra lutris,* population, with confidence bounds, from air and ground counts. California Fish and Game 70:225–233.

Geist, V. 1966. The evolution of horn-like organs. Behaviour 27:175–214.

George, S. B. 1989. Sorex trowbridgii. Mammalian Species 337:1–5.

Georgiadis, N. 1985. Growth patterns, sexual dimorphism and reproduction in African ruminants. African Journal of Ecology 23:75–87.

Gerrard, D. J., and H. C. Chiang. 1970. Density estimation of corn rootworm egg populations based upon frequency of occurrence. Ecology 51:237–245.

Getz, L. L., and M. L. Prather. 1975. A method to prevent removal of trap bait by insects. Journal of Mammalogy 56:955.

Gifford, C. E., and D. R. Griffin. 1960. Notes on homing and migratory behavior of bats. Ecology 41:378–381.

Gillman, J. 1935. The cyclical changes in the external genital organs of the baboon (*P. porcarius*). South African Journal of Science 32:342–355.

Gittleman, J. L. (ed.). 1989. Carnivore Behavior, Ecology, and Evolution. Cornell University Press, Ithaca, New York.

Goddard, J. 1969. Aerial census of black rhinoceros using stratified random sampling. East African Wildlife Journal 7:105–114.

Godfrey, G. K. 1954. Tracing field voles (*Microtus agrestis*) with a Geiger-Müller counter. Ecology 35:5–10.

–––. 1955. A field study of the activity of the mole (*Talpa europaea*). Ecology 36:678–685.

Goetz, R. C. 1981. A photographic system for multiple automatic exposures under field conditions. Journal of Wildlife Management 45:273–276.

Gogan, P. J., S. C. Thompson, W. Pierce, and R. H. Barrett. 1986. Line-transect censuses of fallow and black-tailed deer on the Point Reyes Peninsula. California Fish and Game 72:47–61.

González, G. C., and C. G. Leal. 1984. Mamíferos silvestres de la Cuenca de México. Programme on Man and the Biosphere and Instituto de Ecología y Museo de Historia Natural de la Ciudad de México, México.

Goodchild, M. F., B. O. Parks, and L. T. Steyaert. 1993. Environmental Modeling with GIS. Oxford University Press, New York.

Goodwin, G. G. 1942. Mammals of Honduras. Bulletin of the American Museum of Natural History 79:107–195.

Goos, M. 1990. The global positioning system and its GIS applications. Pp. 254–259. *In* D. H. Parker (ed.), GIS Sourcebook. 2d ed. GIS World, Fort Collins, Colorado.

Gorman, M. L., and B. J. Trowbridge. 1989. The role of odor in the social lives of carnivores. Pp. 57–88. *In* J. L. Gittleman (ed.), Carnivore Behavior, Ecology, and Evolution. Cornell University Press, Ithaca, New York.

Goundie, T. R., and S. H. Vessey. 1986. Survival and dispersal of young white-footed mice born in nest boxes. Journal of Mammalogy 67:53–60.

Graham, A., and R. Bell. 1969. Factors influencing the countability of animals. East African Agricultural and Forestry Journal 34:38–43.

–––. 1989. Investigating observer bias in aerial survey by simultaneous double-counts. Journal of Wildlife Management 53:1009–1016.

Grassle, J. F., G. P. Patil, W. Smith, and C. Taillie (eds.). 1979. Ecological Diversity in Theory and Practice. International Co-operative Publishing House, Fairland, Maryland.

Grau, G. A., G. C. Sanderson, and J. P. Rogers. 1970. Age determination in raccoons. Journal of Wildlife Management 34:364–372.

Greenhall, A. M., and J. L. Paradiso. 1968. Bats and Bat Banding. Bureau of Sport Fisheries and Wildlife Resource Publication 72. Washington, D.C.

Greenlaw, J. S., and J. Swinebroad. 1967. A method for constructing and erecting aerial-nets in a forest. Bird-Banding 38:114–119.

Gregory, T. 1939. Eyes in the Night. Thomas Y. Crowell, New York.

Griffin, D. R. 1934. Marking bats. Journal of Mammalogy 15:202-207.

———. 1940. Migration of New England bats. Bulletin of the Museum of Comparative Zoology 86:217-246.

———. 1971. The importance of atmospheric attenuation for the echolocation of bats (Chiroptera). Animal Behaviour 19:55-61.

Griffiths, M. 1978. The Biology of the Monotremes. Academic Press, New York.

Griffiths, M., and C. P. van Schaik. 1993. The impact of human traffic on the abundance and activity periods of Sumatran rain forest wildlife. Conservation Biology 7:623-626.

Gromov, I. M., and G. I. Baranova (eds.). 1981. Katalog Mlekopitaiushchikh SSSR [Catalog of Mammals of the USSR]. Nauka, Leningrad, Russia (in Russian).

Groves, C. P. 1971. Pongo pygmaeus. Mammalian Species 4:1-6.

———. 1993a. Order Monotremata. P. 13. *In* D. E. Wilson and D. M. Reeder (eds.), Mammal Species of the World: A Taxonomic and Geographic Reference. 2d ed. Smithsonian Institution Press, Washington, D.C.

———. 1993b. Order Dasyuromorphia. Pp. 29-37. *In* D. E. Wilson and D. M. Reeder (eds.), Mammal Species of the World: A Taxonomic and Geographic Reference. 2d ed. Smithsonian Institution Press, Washington, D.C.

———. 1993c. Order Peramelemorphia. Pp. 39-42. *In* D. E. Wilson and D. M. Reeder (eds.), Mammal Species of the World: A Taxonomic and Geographic Reference. 2d ed. Smithsonian Institution Press, Washington, D.C.

———. 1993d. Order Notoryctemorphia. P. 43. *In* D. E. Wilson and D. M. Reeder (eds.), Mammal Species of the World: A Taxonomic and Geographic Reference. 2d ed. Smithsonian Institution Press, Washington, D.C.

———. 1993e. Order Diprotodontia. Pp. 45-62. *In* D. E. Wilson and D. M. Reeder (eds.), Mammal Species of the World: A Taxonomic and Geographic Reference. 2d ed. Smithsonian Institution Press, Washington, D.C.

———. 1993f. Order Primates. Pp. 243-277. *In* D. E. Wilson and D. M. Reeder (eds.), Mammal Species of the World: A Taxonomic and Geographic Reference. 2d ed. Smithsonian Institution Press, Washington, D.C.

Grubb, P. 1993a. Order Perissodactyla. Pp. 369-372. *In* D. E. Wilson and D. M. Reeder (eds.), Mammal Species of the World: A Taxonomic and Geographic Reference. 2d ed. Smithsonian Institution Press, Washington, D.C.

———. 1993b. Order Artiodactyla. Pp. 377-414. *In* D. E. Wilson and D. M. Reeder (eds.), Mammal Species of the World: A Taxonomic and Geographic Reference. 2d ed. Smithsonian Institution Press, Washington, D.C.

Grue, H., and B. Jensen. 1979. Review of the formation of incremental lines in tooth cementum of terrestrial mammals. Danish Review of Game Biology 11:1-48.

Grzimek, B. (ed). 1975. Grzimek's Animal Life Encyclopedia. Mammals 3. Van Nostrand Reinholt, New York.

Guiler, E. R. 1970. Observations on the Tasmanian devil, *Sarcophilus harrisii* (Marsupialia: Dasyuridae). I. Numbers, home range, movements, and food in two populations. Australian Journal of Zoology 18:49-62.

Haas, C. C. 1990. Alternative maternal-care patterns in two herds of bighorn sheep. Journal of Mammalogy 71:24-35.

Hadow, H. H. 1972. Freeze-branding: a permanent marking technique for pigmented mammals. Journal of Wildlife Management 36:645-649.

Hafner, D. J., J. C. Hafner, and M. S. Hafner. 1984. Skin-plus-skeleton preparation as the standard mammalian museum specimen. Curator 27:141-145.

Hafner, M. S., and D. J. Hafner. 1979. Vocalizations of grasshopper mice (genus *Onychomys*). Journal of Mammalogy 60:85-94.

Hale, J. B. 1949. Aging cottontail rabbits by bone growth. Journal of Wildlife Management 13:216-225.

Hall, A. 1988. Malaria. Pp. 101-111. *In* R. Dawood (ed.), How to Stay Healthy Abroad. Penguin Books, New York.

Hall, E. R. 1962. Collecting and preparing study skins of vertebrates. University of Kansas, Museum of Natural History, Miscellaneous Publications 30. Lawrence, Kansas.

———. 1981. The Mammals of North America. 2d ed. Wiley, New York.

Halliday, W. R. 1982. American Caves and Caving. Harper Row, New York.

Haltenorth, T., and H. Diller. 1980. A Field Guide to the Mammals of Africa including Madagascar. Collins, London, England.

Hamilton, W. J., Jr. 1929. Breeding habits of the short-tailed shrew, *Blarina brevicauda*. Journal of Mammalogy 10:125-134.

Hammond, P. S. 1984. On the application of line transect sampling to the estimation of the number of bowhead whales passing the Point Barrow ice-camp. Reports of the International Whaling Commission 34:465–467.

Handley, C. O., Jr. 1966. Checklist of the mammals of Panama. Pp. 753–795. *In* R. L. Wenzel and V. J. Tipton (eds.), Ectoparasites of Panama. Field Museum of Natural History, Chicago, Illinois.

———. 1967. Bats of the canopy of an Amazonian forest. Atas do Simposio sobre Biota Amazonia (Zoologia) 5:211–215.

———. 1968. Capturing bats with mist nets. Pp. 15–19. *In* A. M. Greenhall and J. L. Paradiso (eds.), Bats and Bat Banding. Bureau of Sport Fisheries and Wildlife Resource Publication 72. Washington, D.C.

Handley, C. O., Jr., and E. K. V. Kalko. 1993. A short history of pitfall trapping in America, with a review of methods currently used for small mammals. Virginia Journal of Science 44:19–26.

Handley, C. O., Jr., D. E. Wilson, and A. L. Gardner (eds.). 1991. Demography and Natural History of the Common Fruit Bat, *Artibeus jamaicensis* on Barro Colorado Island, Panamá. Smithsonian Institution Press, Washington, D.C.

Hansen, R. M. 1978. Use of dung pH to differentiate herbivore species. Journal of Wildlife Management 42:441–444.

Harcourt, A. H., and A. F. H. Groom. 1972. Gorilla census. Oryx 11:355–363.

Hardy, A. R., and K. D. Taylor. 1980. Radio tracking of *Rattus norvegicus* on farms. Pp. 657–665. *In* C.J. Amlaner and D.W. Macdonald (eds.), A Handbook on Biotelemetry and Radio Tracking. Pergamon Press, Oxford, England.

Hardy, D. L. 1992. A review of first aid measures for pitviper bite in North America with an appraisal of Extractor™ suction and stun gun electroshock. *In* J. A. Campbell and E. D. Brodie, Jr. (eds.), Biology of the Pitvipers. Selva, Tyler, Texas.

———. 1994a. *Bothrops asper* (Viperidae) snakebite and field researchers in Middle America. Biotropica 26:198–207.

———. 1994b. Snakebite and field biologists in México and Central America: report on ten cases with recommendations for field management. Herpetological Natural History 2:67–82.

Harper, J. A., and W. C. Lightfoot. 1966. Tagging devices for Roosevelt elk and mule deer. Journal of Wildlife Management 30:461–466.

Harrington, F. H., and L. D. Mech. 1982. An analysis of howling response parameters useful for wolf pack censusing. Journal of Wildlife Management 46:686–693.

Harrington, R., and P. Wilson. 1974. Immobilon-Rompun in deer. Veterinary Record 94:362–363.

Harris, S. 1978. Age determination in the red fox (*Vulpes vulpes*) — an evaluation of technique efficiency as applied to a sample of suburban foxes. Journal of Zoology 184:91–117.

Harrison Matthews, L. 1935. The estrous cycle and intersexuality in the female mole (*Talpa europea* Linn.). Proceedings of the Zoological Society of London 1935:347–383.

Hartman, G. D., and T. L. Yates. 1985. Scapanus orarius. Mammalian Species 253:1–5.

Harvey, M. J. 1965. Detecting animals tagged with Co^{60} through air, soil, water, wood, and stone. Transactions of the Kentucky Academy of Sciences 26:63–66.

Hausfater, G. 1975. Dominance and Reproduction in Baboons (*Papio cynocephalus*): A Quantitative Analysis. S. Karger, Basel, Switzerland.

Hawkins, R. E., D. C. Autry, and W. D. Klimstra. 1967a. Comparison of methods used to capture white-tailed deer. Journal of Wildlife Management 31:460–464.

Hawkins, R. E., W. D. Klimstra, G. Fooks, and J. Davis. 1967b. Improved collar for white-tailed deer. Journal of Wildlife Management 31:356–359.

Hawkins, R. E., L. D. Martoglio, and G. G. Montgomery. 1968. Cannon-netting deer. Journal of Wildlife Management 32:191–195.

Hay, K. 1982. Aerial line-transect estimates of abundance of humpback, fin, and long-finned pilot whales in the Newfoundland-Labrador area. Reports of the International Whaling Commission 32:475–486.

Hay, K. G. 1958. Beaver census methods in the Rocky Mountain region. Journal of Wildlife Management 22:395–402.

Hayes, R. J, and S. T. Buckland. 1983. Radial-distance models for the line-transect method. Biometrics 39:29–42.

Hayssen, V., A. van Tienhoven, and A. van Tienhoven. 1993. Asdell's Patterns of Mammalian Reproduction: A Compendium of Species-Specific Data. Cornell University Press, Ithaca, New York.

Headstrom, R. 1983. Identifying Animal Tracks: Mammals, Birds, and Other Animals of the Eastern United States. Dover Publications, New York.

Healy, W. M., and C. J. E. Welsh. 1992. Evaluating line transects to monitor gray squirrel populations. Wildlife Society Bulletin 20:83–90.

Heaney, L. R., P. C. Gonzales, and A. C. Acala. 1987. An annotated checklist of the taxonomic and conservation status of land mammals in the Philippines. Silliman Journal 34:32-66.

Heideman, P. D., and K. R. Erickson. 1992. A new task for an old tool: using crochet hooks to remove bats from mist nets. Bat Research News 33:7.

Heideman, P. D., and L. R. Heaney. 1989. Population biology and estimates of fruit bats (Pteropodidae) in Philippine submontane rainforest. Journal of Zoology (London) 218:565-586.

Helman, P., and S. Churchill. 1986. Bat capture techniques and their use in surveys. Macroderma 2:32-53.

Heltshe, J. F., and N. E. Forrester. 1983. Estimating species richness using the jackknife procedure. Biometrics 39:1-11.

Hench, J. E., G. L. Kirkland, H. W. Setzer, and L. W. Douglas. 1984. Age classification for the gray squirrel based on eruption, replacement, and wear of molariform teeth. Journal of Wildlife Management 48:1409-1414. ·

Henderson, J. R., and T. C. Johanos. 1988. Effects of tagging on weaned Hawaiian monk seal pups. Wildlife Society Bulletin 16:312-317.

Herreid, C. F., R. B. Davis, and H. L. Short. 1960. Injuries due to bat banding. Journal of Mammalogy 41:398-400.

Hershkovitz, P. 1972. The recent mammals of the neotropical region: a zoogeographic and ecological review. Pp. 311-431. *In* A. Keast, F. C. Erk, and B. Glass (eds.), Evolution, Mammals, and Southern Continents. State University of New York, Albany, New York.

Heyer, W. R., M. A. Donnelly, R. W. McDiarmid, L. C. Hayek, and M. S. Foster. 1994. Measuring and Monitoring Biological Diversity: Standard Methods for Amphibians. Smithsonian Institution Press, Washington, D.C.

Heyning, J. E., and M. E. Dahlheim. 1988. Orcinus orca. Mammalian Species 304:1-9.

Hill, G. J. E. 1981. A study of grey kangaroo density using pellet counts. Australian Wildlife Research 8:237-243.

———. 1982. Seasonal movement patterns of the eastern grey kangaroo in southern Queensland. Australian Wildlife Research 9:373-387.

Hill, S. B., and D. H. Clayton. 1985. Wildlife after Dark: A Review of Nocturnal Observation Techniques. Bell Museum of Natural History Occasional Paper 17. Minneapolis, Minnesota.

Hitchcock, H. B. 1957. The use of bird bands on bats. Journal of Mammalogy 38:402-405.

Hoffmann, R. S. 1993. Order Lagomorpha. Pp. 807-827. *In* D. E. Wilson and D. M. Reeder (eds.), Mammal Species of the World: A Taxonomic and Geographic Reference. 2d ed. Smithsonian Institution Press, Washington, D.C.

Hoffmann, R. S., C. G. Anderson, R. W. Thorington, Jr., and L. R. Heaney. 1993. Family Sciuridae. Pp. 419-465. *In* D. E. Wilson and D. M. Reeder (eds.), Mammal Species of the World: A Taxonomic and Geographic Reference. 2d ed. Smithsonian Institution Press, Washington, D.C.

Holden, M. E. 1993a. Family Dipodidae. Pp. 487-499. *In* D. E. Wilson and D. M. Reeder (eds.), Mammal Species of the World: A Taxonomic and Geographic Reference. 2d ed. Smithsonian Institution Press, Washington, D.C.

———. 1993b. Family Myoxidae. Pp. 763-770. *In* D. E. Wilson and D. M. Reeder (eds.), Mammal Species of the World: A Taxonomic and Geographic Reference. 2d ed. Smithsonian Institution Press, Washington, D.C.

Hölzenbein, S., and G. Schwede. 1989. Activity and movements of female white-tailed deer during the rut. Journal of Wildlife Management 53:219-223.

Hone, J. 1988. A test of the accuracy of line and strip transect estimators in aerial survey. Australian Wildlife Research 15:493-497.

Houston, D. B. 1969. Immobilization of the Shiras moose. Journal of Wildlife Management 33:534-537.

Howard, R. A., Jr., and J. W. Kelley. 1977. Trapping Furbearers: Managing and Using a Renewable Resource. New York College of Agriculture and Life Sciences, Cornell University, Ithaca, New York.

Howard, V. W., Jr. 1967. Identifying fecal groups by pH analysis. Journal of Wildlife Management 31:190-191.

Howard, W. E. 1952. A live trap for pocket gophers. Journal of Mammalogy 33:61-65.

Huber, H. R., A. C. Rovetta, L. A. Fry, and S. Johnston. 1991. Age-specific natality of northern elephant seals at the South Farallon Islands, California. Journal of Mammalogy 72:525-534.

Hudson, R. J. 1982. Bighorn sheep. P. 271. *In* D. E. Davis (ed.), CRC Handbook of Census Methods for Terrestrial Vertebrates. CRC Press, Boca Raton, Florida.

Hudson, W. S., and D. E. Wilson. 1986. Macroderma gigas. Mammalian Species 260:1-4.

Humphrey, P. S., D. Bridge, and T. E. Lovejoy. 1968. A technique for mist-netting in the forest canopy. Bird-Banding 39:43-50.

Humphrey, S. R. 1971. Photographic estimation of population size of the Mexican free-tailed bat,

Tadarida brasiliensis. American Midland Naturalist 86:220–223.

Humphrey, S. R., and T. L. Zinn. 1982. Seasonal habitat use by river otters and Everglades mink in Florida. Journal of Wildlife Management 46:375–381.

Humphrey, S. R., A. R. Richter, and J. B. Cope. 1977. Summer habitat and ecology of the endangered Indiana bat, *Myotis sodalis.* Journal of Mammalogy 58:334–346.

Hunt, P. 1980. Experimental choice. Pp. 63–75. *In* Animal Experimentation Research Department, RSPCA (ed.), The Reduction and Prevention of Suffering in Animal Experiments. Royal Society for the Prevention of Cruelty to Animals, Horsham, West Sussex, England.

Huntingford, F. A. 1984. Some ethical issues raised by studies of predation and aggression. Animal Behaviour 32:210–215.

Hurn, J. 1989. GPS: A Guide to the Next Utility. Trimble Navigation, Sunnyvale, California.

Hutterer, R. 1993. Order Insectivora. Pp. 69–130. *In* D. E. Wilson and D. M. Reeder (eds.), Mammal Species of the World: A Taxonomic and Geographic Reference. 2d ed. Smithsonian Institution Press, Washington, D.C.

Iason, G. R. 1988. Age determination of mountain hares (*Lepus timidus*): a rapid method and when to use it. Journal of Applied Ecology 25:389–395.

Ingle, N. R. 1990. Natural History of Bats in a Philippine Rainforest. Unpubl. M.S. thesis. Cornell University, Ithaca, New York.

Ingles, L. G. 1954. Mammals of California and Its Coastal Waters. Stanford University Press, Stanford, California.

International Union for the Conservation of Nature and Natural Resources. 1990. 1990 IUCN Red List of Threatened Animals. IUCN, Gland, Switzerland.

Irvine, A. B., and H. W. Campbell. 1978. Aerial census of the West Indian manatee, *Trichechus manatus,* in the southeastern United States. Journal of Mammalogy 59:613–617.

Jachmann, H. 1991. Evaluation of four survey methods for estimating elephant densities. African Journal of Ecology 29:188–195.

Jameson, E. W., Jr. 1947. Natural history of the prairie vole (mammalian genus *Microtus*). University of Kansas Publications, Museum of Natural History 1:125–151.

———. 1950. Determining fecundity in male small mammals. Journal of Mammalogy 31:433–436.

Janis, C. 1982. Evolution of horns in ungulates: ecology and paleoecology. Biological Reviews 57:261–318.

Jarman, P. J. 1991. Social behavior and organization in the Macropodoidea. Pp. 1–50. *In* P. J. B. Slater, J. S. Rosenblatt, C. Beer, and M. Milinski (eds.), Advances in the Study of Behavior. Vol. 20. Academic Press, San Diego, California.

Jarman, P. J., M. E. Jones, C. N. Johnson, C. J. Southwell, R. I. Stuart-Dick, K. B. Higginbottam, and J. L. Clarke. 1989. Macropod studies at Wallaby Creek. 8. Individual recognition of kangaroos and wallabies. Australian Wildlife Research 16:179–185.

Jenkins, R. E., Jr. 1988. Information management for the conservation of biodiversity. Pp. 231–239. *In* E. O. Wilson and F. M. Peter (eds.), Biodiversity. National Academy Press, Washington, D.C.

Jenkins, S. H., and P. E. Busher. 1979. Castor canadensis. Mammalian Species 120:1–8.

Jensen, I. M. 1982. A new live trap for moles. Journal of Wildlife Management 46:249–252.

Jett, D. A., and J. D. Nichols. 1987. A field companion of nested grid and trapping web density estimators. Journal of Mammalogy 68:888–892.

Jike, L., G. O. Batzli, and L. L. Getz. 1988. Home ranges of prairie voles as determined by radiotracking and by powdertracking. Journal of Mammalogy 69:183–186.

Johnson, B. K., F. G. Lindzey, and R. J. Guenzel. 1991. Use of aerial line transect surveys to estimate pronghorn populations in Wyoming. Wildlife Society Bulletin 19:315–321.

Johnson, C. N., and P. J. Jarman. 1987. Macropod studies at Wallaby Creek. 6. A validation of the use of dung-pellet counts for measuring absolute densities of populations of macropodids. Australian Wildlife Research 14:139–145.

Johnson, K. A. 1977. Methods for the Census of Wallaby and Possum in Tasmania. Tasmanian National Parks and Wildlife Service Wildlife Division Technical Report 77/2.

Johnson, K. G., and M. R. Pelton. 1981. A survey of procedures to determine relative abundance of furbearers in the southeastern United States. Proceedings of the Southeastern Association of Fish and Game Wildlife Agencies 35:261–272.

Johnson, L. B. 1990. Analyzing spatial and temporal phenomena using geographical information systems: a review of ecological applications. Landscape Ecology 4:31–43.

———. 1993. Ecological analyses using geographic information systems. Pp. 27–38. *In* S. B. McLaren and J. K. Braun (eds.), GIS Applications in Mammalogy. Special Publication of the Oklahoma Museum of Natural History, Norman, Oklahoma.

Johnston, C. A., and R. J. Naiman. 1990. The use of a geographic information system to analyze long-term landscape alteration by beaver. Landscape Ecology 4:5–19.

Jolly, G. M. 1965. Explicit estimates from capture-recapture data with both death and immigration-stochastic model. Biometrika 52:225–247.

Jolly, G. M., and J. M. Dickson. 1983. The problem of unequal catchability in mark-recapture estimation of small mammal populations. Canadian Journal of Zoology 61:922–927.

Jones, E. M., and R. D. Owen. 1987. Fluid preservation of specimens. Pp. 51–63. *In* H. H. Genoways, C. Jones, and O. L. Rossolima (eds.), Mammal Collection Management. Texas Tech University Press, Lubbock, Texas.

Jong, E. C. 1993. Immunizations. Pp. 7–20. *In* M. S. Wolfe (ed.), Health Hints for the Tropics. American Committee on Tropical Medicine and Travelers' Health of the American Society of Tropical Medicine and Hygiene, Northbrook, Illinois.

Jongejan, G., P. Arcese, and A. R. E. Sinclair. 1991. Growth, size and the timing of births in an individually identified population of oribi. African Journal of Ecology 29:340–352.

Jonsgard, A. 1969. Age determination in marine mammals. Pp. 1–30. *In* H. T. Anderson (ed.), The Biology of Marine Mammals. Academic Press, New York.

Joslin, P. 1977. Night stalking. Photo Life 2:34–35.

———. 1986. A phototrapline for cold temperatures. Pp. 121–128. *In* H. Freeman (ed.), Proceedings of the Fifth International Snow Leopard Symposium. International Snow Leopard Trust and Wildlife Institute of India, Sprinagar, India.

Joyce, G. G., S. Nakanishi, T. Hata, and L. Pastene. 1985. Preliminary evaluation of an apparatus for improved sighting angle estimation. Reports of the International Whaling Commission 35:437–439.

Justice, K. E. 1961. A new method for measuring home ranges of small animals. Journal of Mammalogy 42:462–470.

Kalko, E. K. V., and C. O. Handley, Jr. 1992. Comparative studies of small mammal populations with transects of snap traps and pitfall arrays in southwest Virginia. Virginia Journal of Science 44:3–18.

Kammerer, W. S. 1993. Malaria prevention. Pp. 21–30. *In* M. S. Wolfe (ed.), Health Hints for the Tropics. American Committee on Tropical Medicine and Travelers' Health of the American Society of Tropical Medicine and Hygiene, Northbrook, Illinois.

Karanth, K. U. 1995. Estimating tiger *Panthera tigris* populations from camera-trap data using capture-recapture models. Biological Conservation 71:333–338.

Karr, J. R., S. K. Robinson, J. G. Blake, and R. O. Bierregaard, Jr. 1990. Birds of four neotropical forests. Pp. 237–269. *In* A. H. Gentry (ed.), Four Neotropical Forests. Yale University Press, New Haven, Connecticut.

Kattel, B., and A. W. Alldredge. 1991. Capturing and handling of the Himalayan musk deer. Wildlife Society Bulletin 19:397–399.

Kauffman, R. G., W. H. Norton, B. G. Harmon, and B. C. Breidenstein. 1967. Growth of the porcine eye lens as an index to chronological age. Journal of Animal Science 26:31–35.

Kaufman, G. A. 1989. Use of fluorescent pigments to study social interactions in a small nocturnal rodent, *Peromyscus maniculatus*. Journal of Mammalogy 70:171–174.

Kaufman, G. A., and D. W. Kaufman. 1989. An artificial burrow for the study of natural populations of small mammals. Journal of Mammalogy 70:656–659.

Kaufmann, J. 1965. A three-year study of mating behavior in a freeranging band of rhesus monkeys. Ecology 46:500–512.

Keith, L. B., E. C. Meslow, and O. J. Rongstad. 1968. Techniques for snowshoe hare population studies. Journal of Wildlife Management 32:801–812.

Kenagy, G. J. 1979. Rapid surgical technique for measurement of testis size in small mammals. Journal of Mammalogy 60:636–638.

Kendall, W. L., K. H. Pollock, and C. Brownie. 1995. A likelihood-based approach to capture-recapture estimation of demographic parameters under the robust design. Biometrics 51:293–308.

Kennedy, M. L., G. Baumgardner, M. E. Cope, F. R. Tabatabai, and O. S. Fuller. 1986. Raccoon (*Procyon lotor*) density as estimated by the census-assessment line technique. Journal of Mammalogy 67:166–168.

Kent, D. M., M. P. Harris, G. L. Abegg, and T. H. Kunz. 1985. A new moulage technique for casting animal tracks. American Biology Teacher 47:432–433.

Keystone, J. S. 1993. Malaria prevention. Pp. 21–30. *In* M. S. Wolfe (ed.), Health Hints for the Tropics. American Committee on Tropical Medicine and Travelers' Health of the American Society of Tropical Medicine and Hygiene, Northbrook, Illinois.

Kingdon, J. 1971–1982. East African Mammals: An Atlas of Evolution in Africa. Vol. 1–7. Academic Press, London, England.

———. 1979. East African Mammals: An Atlas of Evolution in Africa. Vol. 3, Part B (Large Mammals). Academic Press, London, England.

Kingsley, K. J., Y. Petryszyn, and F. Reichenbacher. 1991. Protocol for Conducting Surveys for Lesser Long-nosed Bats and Other Bats in Inactive Mines. Unpubl. report. ASARCO, Silver Bell Unit, Marana, Arizona.

Kirkland, G. L., Jr., and P. K. Sheppard. 1994. Proposed standard protocol for sampling of small mammal communities. Pp. 277–283. *In* J. F. Merritt, G. L. Kirkland, Jr., and R. K. Rose (eds.), Advances in the Biology of Shrews. Special Publication of the Carnegie Museum of Natural History 18. Pittsburgh, Pennsylvania.

Kirsch, J. A. W., and P. F. Waller. 1979. Notes on the trapping and behavior of the Caenolestidae (Marsupialia). Journal of Mammalogy 60:390–395.

Kishino, H. 1986. On parallel ship experiments and the line transect method. Reports of the International Whaling Commission 36:491–495.

Kiwia, H. D. 1989. Black rhinoceros (*Diceros bicornis* (L.)): population size and structure in Ngorogoro Crater, Tanzania. African Journal of Ecology 27:1–6.

Kleiman, D. G. (ed.). 1977. The Biology and Conservation of the Callitrichidae. Smithsonian Institution Press, Washington, D.C.

Kleiman, D. G., and T. M. Davis. 1974. Punch-mark renewal in bats of the genus *Carollia*. Bat Research News 15:29–30.

Klevezal, G. A., and S. E. Kleinenberg. 1967. Age Determination of Mammals from Annual Layers in Teeth and Bones. Academy of Sciences of the U.S.S.R. Severtsov Institute of Animal Morphology, Moscow, U.S.S.R. (English Translation, Israel Program for Scientific Translations, Jerusalem, Israel, 1969.)

Klingle, H. 1965. Notes on the biology of the plains zebra *Equus guagga boehmi* Matschie. East African Wildlife Journal 3:86–88.

Koeppl, J. W., R. S. Hoffmann, and C. F. Nadler. 1978. Pattern analysis of acoustical behavior in four species of ground squirrels. Journal of Mammalogy 59:677–696.

Koopman, K. F. 1993. Order Chiroptera. Pp. 137–241. *In* D. E. Wilson and D. M. Reeder (eds.), Mammal Species of the World: A Taxonomic and Geographic Reference. 2d ed. Smithsonian Institution Press, Washington, D.C.

Korytko, A. I., and S. H. Vessey. 1991. Agonistic and spacing behaviour in white-footed mice, *Peromyscus leucopus*. Animal Behaviour 42:913–919.

Koster, S. H., and J. A. Hart. 1988. Methods of estimating ungulate populations in tropical forests. African Journal Ecology 26:117–126.

Krebs, C. J., B. S. Gilbert, S. Boutin, and R. Boonstra. 1987. Estimation of snowshoe hare population density from turd transects. Canadian Journal of Zoology 65:565–567.

Kroodsma, D. E. 1989. Suggested experimental designs for song playbacks. Animal Behaviour 37:600–609.

Kruuk, H. 1972. The Spotted Hyena: A Study of Predation and Social Behavior. University of Chicago Press, Chicago, Illinois.

Kufeld, R. C., J. H. Olterman, and D. C. Bowden. 1980. A helicopter quadrat census for mule deer on Uncompahgre Plateau, Colorado. Journal of Wildlife Management 44:632–639.

Kunz, T. H. 1974. Feeding ecology of a temperate insectivorous bat (*Myotis velifer*). Ecology 55:693–711.

———. 1982. Roosting ecology of bats. Pp. 1–55. *In* T. H. Kunz (ed.), Ecology of Bats. Plenum, New York.

———. 1987. Post-natal growth and energetics of suckling bats. Pp. 395–420. *In* M. B. Fenton, P. Racey, and J. M. V. Rayner (eds.), Recent Advances in the Study of Bats. Cambridge University Press, Cambridge, England.

——— (ed.). 1988. Ecological and Behavioral Methods for the Study of Bats. Smithsonian Institution Press, Washington, D.C.

Kunz, T. H., and E. L. P. Anthony. In press. Variation in nightly emergence behavior in the little brown bat, *Myotis lucifugus* (Chiroptera: Verspertilionidae). *In* H. H. Genoways and R. J. Baker (eds.), J. Knox Jones, Jr. Memorial. Texas Tech University Press, Lubbock, Texas.

Kunz, T. H., and C. E. Brock. 1975. A comparison of mist nets and ultrasonic detectors for monitoring flight activity of bats. Journal of Mammalogy 56:907–911.

Kunz, T. H., and A. Kurta. 1988. Capture methods and holding devices. Pp. 1–30. *In* T. H. Kunz (ed.), Ecological and Behavioral Methods for the Study of Bats. Smithsonian Institution Press, Washington, D.C.

Kunz, T. H., P. V. August, and C. D. Burnett. 1983. Harem social organization in cave roosting *Artibeus jamaicensis* (Chiroptera: Phyllostomidae). Biotropica 15:133–138.

Kunz, T. H., M. S. Fujita, A. P. Brooke, and G. F. McCracken. 1994. Convergence in tent architecture and tent-making behavior among neotropical and paleotropical bats. Journal of Mammalian Evolution 2:57–78.

Kurta, A., D. King, J. A. Teramino, J. A. Stribley, and K. J. Williams. 1993. Summer roosts of the endangered bat (Myotis sodalis) on the northern edge of its range. American Midland Naturalist 129:132–138.

Kusinitz, M. 1990. Tropical Medicine. Chelsea House Publishers, New York.

Laake, J. L. 1992. Catch-per-unit-effort models: an application to an elk population in Colorado. Pp. 44–45. *In* D. R. McCullough and R. H. Barrett (eds.), Wildlife 2001: Populations. Elsevier, New York.

Laake, J. L., S. T. Buckland, D. R. Anderson, and K. P. Burnham. 1991. DISTANCE User's Guide. Colorado State University, Fort Collins, Colorado.

Laake, J. L., K. P. Burnham, and D. R. Anderson. 1979. User's Manual for Program TRANSECT. Utah State University Press, Logan, Utah.

Lancia, R. A., J. D. Nichols, and K. H. Pollock. 1994. Estimating the number of animals in wildlife populations. Pp. 215–253. *In* T. A. Bookhout (ed.), Research and Management Techniques for Wildlife and Habitats. 5th ed. Wildlife Society, Bethesda, Maryland.

Lancia, R. A., K. H. Pollock, J. W. Bishir, and M. C. Conner. 1988. A white-tailed deer harvesting strategy. Journal of Wildlife Management 52:589–595.

Laundré, J. W. 1981. Temporal variation in coyote vocalization rates. Journal of Wildlife Management 45:767–769.

Laundré, J. W., and B. L. Keller. 1984. Home-range size of coyotes: a critical review. Journal of Wildlife Management 48:127–139.

Laurie, W. A., E. M. Lang, and C. P. Groves. 1983. Rhinoceros unicornis. Mammalian Species 211:1–6.

LaVal, R. K., and H. S. Fitch. 1977. Structure, movements and reproduction in three Costa Rican bat communities. University of Kansas, Museum of Natural History, Occasional Papers 69. Lawrence, Kansas.

LaVal, R. K., and M. L. LaVal. 1980. Ecological Studies and Management of Missouri Bats, with Emphasis on Cave-dwelling Species. Terrestrial Series 8. Missouri Department of Conservation, Jefferson City, Missouri.

Lawrence, B. D., and J. A. Simmons. 1982. Measurements of atmospheric attenuation at ultrasonic frequencies and the significance for echolocation by bats. Journal of the Acoustical Society of America 71:585–590.

Lawrence, M. J., and R. W. Brown. 1967. Mammals of Britain—Their Tracks, Trails, and Signs. Blanford Press, London, England.

Layne, J. N. 1987. An enclosure for protecting small mammal traps from disturbance. Journal of Mammalogy 68:666–668.

Leatherwood, S., J. R. Gilbert, and D. G. Chapman. 1978. An evaluation of some techniques for aerial censuses of bottlenosed dolphins. Journal of Wildlife Management 42:239–250.

Leatherwood, S., I. T. Show, Jr., T. R. Reeves, and M. B. Wright. 1982. Proposed modification of transect models to estimate population size from aircraft with obstructed downward visibility. Reports of the International Whaling Commission 32:577–580.

Lechleitner, R. R. 1959. Sex ratio, age classes and reproduction of the black-tailed jack rabbit. Journal of Mammalogy 40:63–81.

Lee, W. L., B. M. Bell, and J. F. Sutton (eds.). 1982. Guidelines for Acquisition and Management of Biological Specimens. Association of Systematic Collections, Lawrence, Kansas.

Lefebvre, L. W., and H. I. Kochman. 1991. An evaluation of aerial survey replicate methodology to determine trends in manatee abundance. Wildlife Society Bulletin 19:298–309.

Leimgruber, P., W. J. McShea, and J. H. Rappole. 1994. Predation on artificial nests in large forest blocks. Journal of Wildlife Management 58:254–260.

Lekagul, B., and J. A. McNeely. 1977. Mammals of Thailand. Association for the Conservation of Wildlife, Bangkok, Thailand.

Lemen, C. A., and P. W. Freeman. 1985. Tracking mammals with fluorescent pigments: a new technique. Journal of Mammalogy 66:134–136.

Lentfer, J. W. 1968. A technique for immobilizing and marking polar bears. Journal of Wildlife Management 32:317–321.

Leslie, P. H., and D. H. S. Davis. 1939. An attempt to determine the absolute number of rats in a given area. Journal of Animal Ecology 8:94–113.

Leutscher, A. 1960. Tracks and Signs of British Animals. Cleaver-Hume Press, London, England.

Lieberman, D. E., and R. H. Meadow. 1992. The biology of cementum increments (with an archaeological application). Mammal Review 22:57–77.

Limpens, H. J. G. A., W. Helmer, A. van Winden, and K. Mostert. 1989. Vleermuizen (Chiroptera) en Lintvormige Landschapselementen. Lutra 32:1–20.

Lincoln, F. C. 1930. Calculating waterfowl abundance on the basis of banding returns. U.S. Department of Agriculture Circular 118. Washington, D.C.

Linduska, J. P. 1942. A new technique for marking fox squirrels. Journal of Wildlife Management 6:93–94.

Lindzey, F. G., and E. C. Meslow. 1977. Home range and habitat use of black bears in southwestern Washington. Journal of Wildlife Management 41:413–425.

Lindzey, F. G., S. K. Thompson, and J. I. Hodges. 1977. Scent station index of black bear abundance. Journal of Wildlife Management 40:408–415.

Ling, J. K., D. G. Nicholls, and C. D. B. Thomas. 1967. Immobilization of southern elephant seals with succinylcholine chloride. Journal of Wildlife Management 31:468–479.

Linhart, S. B., and G. J. Dasch. 1992. Improved performance of padded jaw traps for capturing coyotes. Wildlife Society Bulletin 20:63–66.

Linhart, S. B., and F. F. Knowlton. 1975. Determining the relative abundance of coyotes by scent station lines. Wildlife Society Bulletin 3:119–124.

Linhart, S. B., G. J. Dasch, C. B. Male, and R. M. Engeman. 1986. Efficiency of unpadded and padded steel foothold traps for capturing coyotes. Wildlife Society Bulletin 14:212–218.

Link, W. A., and R. J. Barker. 1994. Density estimation using the trapping web design: a geometric analysis. Biometrics 50:733–745.

Lishak, R. S. 1977. Censusing 13-lined ground squirrels with adult and young alarm calls. Journal of Wildlife Management 41:755–759.

———. 1982. Thirteen-lined ground squirrel. Pp. 156–158. *In* D. E. Davis (ed.), CRC Handbook of Census Methods for Terrestrial Vertebrates. CRC Press, Boca Raton, Florida.

Lombard, A. T., P. V. August, and W. R. Siegfried. 1992. A proposed geographic information system for assessing the optimal dispersion of protected areas in South Africa. South African Journal of Science 88:136–140.

Lomolino, M. V. 1994. Species richness of mammals inhabiting nearshore archipelagoes: area, isolation, and immigration filters. Journal of Mammalogy 75:39–49.

Lord, R. D., A. M. Vilches, J. I. Maiztegui, and C. A. Soldini. 1970. The tracking board: a relative census technique for studying rodents. Journal of Mammalogy 51:828–829.

Loughlin, T. R. 1980. Home range and territoriality of sea otters near Monterey, California. Journal of Wildlife Management 44:576–582.

Lowe, K. W. 1989. The Australian Bird Bander's Manual. Australian Bird and Bat Banding Schemes. Australian National Parks and Wildlife, Canberra, New South Wales, Australia.

Lunetta, R. S., R. G. Congalton, L. K. Fenstermaker, J. R. Jensen, K. C. McGwire, and L. R. Tinney.
1991. Remote sensing and geographic information system data integration: error sources and research issues. Photogrammetric Engineering and Remote Sensing 57:677–687.

Lunney, D., J. Barker, and D. Priddel. 1985. Movements and day roosts of the chocolate wattled bat *Chalinolobus morio* (Gray) (Megachiroptera: Vespertilionidae) in a logged forest. Australian Mammalogy 8:313–317.

Macdonald, D. (ed.) 1984. The Encyclopedia of Mammals. Facts on File Publications, New York.

MacDonald, K., E. Matsui, R. Stevens, and M. B. Fenton. 1994. Echolocation calls and field identification of the eastern pipistrelle (*Pipistrellus subflavus:* Chiroptera: Vespertilionidae), using ultrasonic bat detectors. Journal of Mammalogy 75:462–465.

MacKinnon, J., and K. MacKinnon. 1980. The behavior of wild spectral tarsiers. International Journal of Primatology 1:361–379.

Mackison, F. W., R. S. Stricoff, and L. J. Partridge, Jr. (eds.). 1981. NIOSH/OSHA Occupational Health Guidelines for Chemical Hazards. U.S. Department of Health and Human Services (NIOSH) Publication 81–123 (3 vol.). U.S. Government Printing Office, Washington, D.C.

Madison, D. M. 1985. Activity rhythms and spacing. Pp. 373–419. *In* R. H. Tamarin (ed.), Biology of New World *Microtus*. American Society of Mammalogists Special Publication 8.

Madsen, R. M. 1967. Age Determination of Wildlife. A Bibliography. U.S. Department of Interior, Washington, D.C.

Maffini, G. 1987. Raster versus vector data encoding and handling: a commentary. Photogrammetric Engineering and Remote Sensing 53:1397–1398.

Maguire, D. J., M. F. Goodchild, and D. W. Rhind. 1991. Geographical Information Systems: Principles and Applications. Wiley, New York.

Malcolm, J. R. 1992. Use of tooth impression to identify and age live *Proechimys guyannensis* and *P. cuvieri* (Rodentia: Echimyidae). Journal of Zoology (London) 227:537–546.

Malcolm, J. R., and R. J. Brooks. 1985. Influence of photoperiod and photoperiod reversal on growth, mortality, and indicators of age of *Dicrostonyx groenlandicus*. Canadian Journal of Zoology 63:1497–1509.

Maly, M. S., and J. A. Cranford. 1985. Relative capture efficiency of large and small Sherman live traps. Acta Theriologica 30:165–167.

Mann, T. 1964. Biochemistry of Semen and the Male Reproductive Tract. Methuen, London, England.

Manson-Bahr, C. 1988. Leishmaniasis. Pp. 123–129. *In* R. Dawood (ed.), How to Stay Healthy Abroad. Penguin Books, New York.

Manville, C. J., S. A. Barnum, and J. R. Tetser. 1992. Influence of bait on arboreal behavior of *Peromyscus leucopus*. Journal of Mammalogy 73:335–336.

Mark, D. M., and E. Zubrow. 1993. Join the GIS-L electronic community. GIS World 6:56–57.

Marsh, H., and D. F. Sinclair. 1989a. Correcting for visibility bias in strip transect aerial surveys of aquatic fauna. Journal of Wildlife Management 53:1017–1024.

———. 1989b. An experimental evaluation of dugong and sea turtle aerial survey techniques. Australian Wildlife Research 16:639–650.

Marshall, A. G., and A. N. McWilliam. 1982. Ecological observations on epomophorine fruit-bats (Megachiroptera) in West Africa savanna woodland. Journal of Zoology (London) 198:53–67.

Marshall, L. G. 1978a. Dromiciops australis. Mammalian Species 99:1–5.

———. 1978b. Chironectes minimus. Mammalian Species 109:1–6.

Marten, K., and P. Marler. 1977. Sound transmission and its significance for animal vocalization. I. Temperate habitats. Behavioral Ecology and Sociobiology 2:271–290.

Marten, K., D. Quine, and P. Marler. 1977. Sound transmission and its significance for animal vocalization. 2. Tropical forest habitats. Behavioral Ecology and Sociobiology 2:291–302.

Martin, L. 1987. Flying-fox (Chiroptera: Pteropodidae) research: future needs and priorities. Australian Mammalogy 10:153.

Martinka, C. J. 1968. Habitat relationships of white-tailed deer and mule deer in northern Montana. Journal of Wildlife Management 32:558–565.

———. 1969. Population ecology of summer resident elk in Jackson Hole, Wyoming. Journal of Wildlife Management 33:465–481.

Mason, D. R. 1985. Dentition and age determination of the warthog *Phacochoerus aethiopicus* in Zululand, South Africa. Koedoe 27:79–119.

Mattfeld, G. F., J. E. Wiley, and D. F. Behrend. 1972. Salt versus browse — seasonal baits for deer trapping. Journal of Wildlife Management 36:996–998.

McBee, K., and R. J. Baker. 1982. Dasypus novemcinctus. Mammalian Species 162:1–9.

McCarley, H. 1966. Annual cycle, population dynamics and adaptive behavior of *Citellus tridecemilineatus*. Journal of Mammalogy 47:294–316.

McComb, K. 1991. Female choice for high roaring rates in red deer, *Cervus elaphus*. Animal Behaviour 41:79–88.

McComb, W. C., R. G. Anthony, and K. McGarigal. 1991. Differential vulnerability of small mammals and amphibians to two trap types and two trap baits in Pacific Northwest forests. Northwest Science 65:109–115.

McCracken, G. F. 1984. Communal nursing in Mexican free-tailed bat maternity colonies. Science 223:1090–1091.

McCracken, G. F., and J. W. Bradbury. 1981. Social organization and kinship in the polygynous bat *Phyllostomus hastatus*. Behavioral Ecology and Sociobiology 8:11–34.

McCracken, G. F., and M. K. Gustin. 1991. Nursing behavior in Mexican free-tailed bat maternity colonies. Ethology 89:305–321.

McCravy, K. W., and R. K. Rose. 1992. An analysis of external features as predictors of reproductive status in small mammals. Journal of Mammalogy 73:151–159.

McCullough, D. R. 1982. Evaluation of night spotlighting as a deer study technique. Journal of Wildlife Management 46:963–973.

McManus, J. J. 1974. Didelphis virginiana. Mammalian Species 40:1–6.

McShea, W. J., and J. Rappole. 1992. White-tailed deer as keystone species within forest habitats of Virginia. Virginia Journal of Science 42:173.

Mead, J. G., and R. L. Brownell, Jr. 1993. Order Cetacea. Pp. 349–364. *In* D. E. Wilson and D. M. Reeder (eds.), Mammal Species of the World: A Taxonomic and Geographic Reference. 2d ed. Smithsonian Institution Press, Washington, D.C.

Mech, L. D. 1983. Handbook of Animal Radio-Tracking. University of Minnesota Press, Minneapolis, Minnesota.

Medway, L. 1983. The Wild Mammals of Malaya. 2d ed. Oxford University Press, Kuala Lumpur, Malaysia.

Melchior, H. R., and F. A. Iwen. 1965. Trapping, restraining, and marking arctic ground squirrels for behavioral observations. Journal of Wildlife Management 29: 671–678.

Mengak, M. T., and D. G. Guynn, Jr. 1987. Pitfalls and snap traps for sampling small mammals and herpetofauna. American Midland Naturalist 118:284–288.

Meunier, M., A. Solari, and L. Martinet. 1982. Common vole (*Microtus arvalis*) (France). P. 194. *In* D. E. Davis (ed.), CRC Handbook of Census Methods for Terrestrial Vertebrates. CRC Press, Boca Raton, Florida.

Michener, G. R. 1980. Estrous and gestation periods in Richardson's ground squirrels. Journal of Mammalogy 61:531–534.

———. 1984. Age, sex, and species differences in the annual cycles of ground-dwelling sciurids: implications for sociality. Pp. 81–107. *In* J. O. Murie and G. R. Michener (eds.), The Biology of Ground-Dwelling Squirrels. University of Nebraska Press, Lincoln, Nebraska.

Mickleburgh, S. P., A. M. Hutson, and P. A. Racey. 1992. Old World Fruit Bats: An Action Plan for Their Conservation. IUCN, Gland, Switzerland.

Midgley, M. 1981. Why knowledge matters. Pp. 319–336. *In* D. Sperlinger (ed.), Animals in Research: New Perspectives in Animal Experimentation. Wiley, New York.

Miller, F. L. 1968. Immobilization of free-ranging black-tailed deer with succinylcholine chloride. Journal of Wildlife Management 32:195–197.

Miller, G. S. 1932. Directions for preparing specimens of mammals. Bulletin of the U.S. National Museum 39, Part N. Washington, D.C.

Miller, G. T., Jr. 1992. Living in the Environment: An Introduction to Environmental Science. 7th ed. Wadsworth, Belmont, California.

Miller, L. A., and B. B. Andersen. 1983. Studying bat echolocation signals using ultrasonic detectors. Zeitschrift für Säugetierkunde 49:6–13.

Miller, M. N., and J. A. Byers. 1991. Energetic costs of locomotor play in pronghorns. Animal Behaviour 41:1007–1013.

Mills, M. G. L. 1982a. Hyaena brunnea. Mammalian Species 194:1–5.

———. 1982b. Notes on age determination, growth and measurements of brown hyaenas *Hyaena brunnea* from the Kalahari Gemsbok National Park. Koedoe 25:55–61.

Mims, F. M. 1982. Experimenting with Kodak's disk camera. Part 1. Modifying the camera for electronic triggering. Computers and Electronics 20:111–115.

Mingoti, S. A., and G. Meeden. 1992. Estimating the total number of distinct species using presence and absence data. Biometrics 48:863–875.

Minta, S., and M. Mangel. 1989. A simple population estimate based on simulation for capture-recapture and capture-resight data. Ecology 70:1738–1751.

Mitchell, J. 1978. Incremental growth layers in the dentine of dugong incisors (*Dugong dugon* (Müller)) and their application to age determination. Zoological Journal of the Linnean Society 62:317–348.

Mitchell-Jones, A. J. 1987. Survey and monitoring. Pp. 12–16. *In* A. J. Mitchell-Jones (ed.), The Bat Worker's Manual. Nature Conservancy Council, Peterborough, United Kingdom.

Mohr, C. E. 1934. Marking bats for later recognition. Proceedings of the Pennsylvania Academy of Science 8:26–30.

Molina, D. M., and J. A. Oporto. 1993. Comparative study of dentine staining techniques to estimate age in the Chilean dolphin *Cephalorhynchus eutropia* Gray 1846. Aquatic Mammals 19:45–48.

Mones, A., and J. Ojasti. 1986. Hydrochoerus hydrochaeris. Mammalian Species 264:1–7.

Monfort, S. L., C. C. Schwartz, and S. K. Wasser. 1993. Monitoring reproduction in captive moose using urinary and fecal steroid metabolites. Journal of Wildlife Management 57:400–407.

Monmonier, M., and G. A. Schnell. 1988. Map Appreciation. Prentice-Hall, Englewood Cliffs, New Jersey.

Montalbano, F., III, P. W. Glanz, M. W. Olinde, and L. S. Perrin. 1985. A solar-powered time-lapse camera to record wildlife activity. Wildlife Society Bulletin 13:178–182.

Montgomery, G. G. (ed.) 1985. The Evolution and Ecology of Armadillos, Sloths, and Vermilinguas. Smithsonian Institution Press, Washington, D.C.

Moore, D. W., and T. L. Yates. 1983. Rate of protein inactivation in selected mammals following death. Journal of Wildlife Management 47:1166–1169.

Morgan, D. G. 1986. Estimating Vertebrate Population Densities by Line Transect Methods. Melbourne College Advanced Education Occasional Paper 11. Carlton, Victoria, Australia.

Morgan, G. S. 1989. Geocapromys thoracatus. Mammalian Species 341:1–5.

Morris, P. 1972. A review of mammalian age determination methods. Mammal Review 2:69–104.

Morris, W., M. Fainstat, T. Robinson, and R. Hoo. 1994. Hemorrhagic fever with renal syndrome in California. Western Journal of Medicine 161:418–421.

Morrison, D. H. 1980. Foraging and day-roosting dynamics of canopy fruit bats in Panama. Journal of Mammalogy 61:20–29.

Morse, M. A., and D. S. Balser. 1961. Fox calling as a hunting technique. Journal of Wildlife Management 25:148–154.

Mosby, H. S. 1955. Live Trapping Objectionable Animals. Virginia Polytechnic Institute Agriculture Extension Service Circular 667. Blacksburg, Virginia.

Mudge, G. P., S. J. Aspinall, and C. H. Crooke. 1987. A photographic study of seabird attendance at

Moray Firth colonies outside the breeding season. Bird Study 34:28–36.

Mullican, T. R. 1988. Radio telemetry and fluorescent pigments: a comparison of techniques. Journal of Wildlife Management 52:627–631.

Munn, C. A. 1991. Tropical canopy netting and shooting lines over tall trees. Journal of Field Ornithology 62:454–463.

Murie, O. J. 1974. A Field Guide to Animal Tracks. Peterson Field Guide Series. 2d ed. Houghton Mifflin, Boston, Massachusetts.

Musser, G. G. 1987. The mammals of Sulawesi. Pp. 73–93. *In* T. C. Whitmore (ed.), Biogeographical Evolution of the Malay Archipelago. Oxford University Press, Oxford, England.

Musser, G. G., and M. D. Carleton. 1993. Family Muridae. Pp. 501–755. *In* D. E. Wilson and D. M. Reeder (eds.), Mammal Species of the World: A Taxonomic and Geographic Reference. 2d ed. Smithsonian Institution Press, Washington, D.C.

Mutere, F. A. 1980. *Eidolon helvum* revisited. Pp. 145–150. *In* D. E. Wilson and A. L. Gardner (eds.), Proceedings of the 5th International Bat Research Conference. Texas Tech University Press, Lubbock, Texas.

Myers, P. 1978a. Sexual dimorphism in size of vespertilionid bats. American Naturalist 112:701–711.

———. 1978b. A method for determining the age of living small mammals. Journal of Zoology (London) 186:551–556.

Nachtigall, P. E., and P. W. B. Moore (eds.). 1988. Animal Sonar: Processes and Performance. Plenum, New York.

Nadeau, J. H. 1985. Ontogeny. Pp. 254–285. *In* R. H. Tamarin (ed.), Biology of New World *Microtus*. American Society of Mammalogists Special Publication 8.

Nadkarni, H. N. 1988. Use of portable platform for observations of tropical forest canopy animals. Biotropica 20:350–351.

Nagorsen, D. W., and R. L. Peterson. 1980. Mammal Collector's Manual. Life Sciences Miscellaneous Publications. Royal Ontario Museum, Toronto, Ontario, Canada.

National Research Council. 1981. Techniques for the Study of Primate Population Ecology. National Academy Press, Washington, D.C.

———. 1985. Guide for the Care and Use of Laboratory Animals. A Report of the Institute of Laboratory Animal Resources Committee on the Care and Use of Laboratory Animals. NIH Publication 85-23. U.S. Department of Health and Human Services, Washington, D.C.

———. 1993. A Biological Survey for the Nation. National Academy Press, Washington, D.C.

Neal, A. K., G. C. White, R. B. Gill, D. F. Reed, and J. H. Olterman. 1993. Evaluation of mark-resight assumptions for estimating mountain sheep numbers. Journal of Wildlife Management 57:436–450.

Neff, D. J. 1968. The pellet-group count technique for big game trend, census, and distribution: a review. Journal of Wildlife Management 32:597–614.

Nellis, C. H. 1968. Some methods for capturing coyotes alive. Journal of Wildlife Management 32:402–405.

Nellis, C. H., C. J. Terry, and R. D. Taber. 1974. A conical pitfall trap for small mammals. Northwest Science 48:102–104.

Nelson, J. 1965a. Techniques and equipment. Hanging of mist nets. Australian Bat Research News 4:1–2.

Nelson, J. E. 1965b. Movements of Australian flying foxes (Pteropodidae: Megachiroptera). Australian Journal of Zoology 13:53–73.

Nesbit, W. 1926. How to Hunt with the Camera; A Complete Guide to All Forms of Outdoor Photography. E. P. Dutton, New York.

Newsom, J. D., and J. S. Sullivan, Jr. 1968. Cryobranding, a Marking Technique for White-tailed Deer. Louisiana Cooperative Wildlife Research Unit, Louisiana State University, Baton Rouge, Louisiana.

Newsome, A. E., M. L. Dudzinski, and W. A. Low. 1981. Measuring Bias in Aerial Surveys Due to Dispersion of Animals. ILCA Monograph 4. Nairobi, Kenya.

Newsome, A. E., J. McIlroy, and P. Catling. 1975. The effect of an extensive wildfire on populations of twenty ground vertebrates in southeast Australia. Proceedings of the Ecological Society of Australia 9:107–123.

Nichols, J. D. 1986. On the use of enumeration estimators for interspecific comparisons, with comments on a "trappability" estimator. Journal of Mammalogy 67:590–593.

Nichols, J. D., and K. H. Pollock. 1983. Estimation methodology in contemporary small mammal capture-recapture studies. Journal of Mammalogy 64:253–260.

Nicoll, M. E., and P. A. Racey. 1981. The Seychelles fruit bat, *Pteropus seychellensis seychellensis*. African Journal of Ecology 19:361–364.

Niethammer, J., and F. Krapp (eds.). 1990. Handbuch der Säugetiere Europas, 3/I. Aula-Verlag, Wiesbaden, Germany.

Nixon, C. M., M. W. McClain, and R. W. Donohoe. 1980. Effects of clear-cutting on gray squirrels. Journal of Wildlife Management 44:403–412.

Norton-Griffiths, M. 1974. Reducing counting bias in aerial censuses by photography. East African Wildlife Journal 12:245–248.

———. 1975. The numbers and distribution of large mammals in Ruaha National Park, Tanzania. East African Wildlife Journal 13:121–140.

———. 1978. Counting Animals. Rev. 2d ed. African Wildlife Leadership Foundation, Nairobi, Kenya.

Noss, R. F. 1990. Indicators for monitoring biodiversity: a hierarchical approach. Conservation Biology 4:355–364.

Novak, J. M., K. T. Scribner, W. D. Dupont, and M. H. Smith. 1991. Catch-effort estimation of white-tailed deer population size. Journal of Wildlife Management 55:31–38.

Nowak, R. M. 1991. Walker's Mammals of the World. 5th ed. Johns Hopkins University Press, Baltimore, Maryland.

Numan, M. 1988. Maternal behavior. Pp. 1569–1645. In E. Knobil and J. D. Neill (eds.), The Physiology of Reproduction. Raven Press, New York.

Nyholm, E. S. 1965. Zur Ökologie von Myotis mystacinus (Leisl.) und M. daubentoni (Leisl.) (Chiroptera). Annales Zoologica Fennica 2:77–123.

Odell, D. K. 1982. California sea lion. Pp. 239–240. In D. E. Davis (ed.), CRC Handbook for Census Methods for Terrestrial Vertebrates. CRC Press, Boca Raton, Florida.

Odum, E. P. 1971. Fundamentals of Ecology. 3d ed. W. B. Saunders, Philadelphia, Pennsylvania.

O'Farrell, M. J., D. W. Kaufman, and D. W. Lundahl. 1977. Use of live-trapping with the assessment line method for density estimation. Journal of Mammalogy 58:575–582.

O'Farrell, M. J., W. A. Clark, F. H. Emmerson, S. M. Juarez, F. R. Kay, T. M. O'Farrell, and T. Y. Goodlett. 1994. Use of a mesh live trap for small mammals: are results from Sherman live traps deceptive? Journal of Mammalogy 75:692–699.

Ohtaishi, N., N. Hachiya, and Y. Shibata. 1976. Age determination of the hare from annual layers in the mandibular bone. Acta Theriologia 21:168–171.

Ohtaishi, N., K. Kaji, S. Miura, and J. Wu. 1990. Age determination of the white-lipped deer Cervus albirostris by dental cementum and molar wear. Journal of the Mammalogical Society of Japan 15:15–24.

Olsen, G. H., S. B. Linhart, R. A. Holmes, G. J. Dasch, and C. B. Male. 1986. Injuries to coyotes caught in padded and unpadded steel foothold traps. Wildlife Society Bulletin 14:219–223.

Onderka, D. K., D. L. Skinner, and A. W. Todd. 1990. Injuries to coyotes and other species caused by four models of footholding devices. Wildlife Society Bulletin 18:175–182.

Oppenheimer, J. R. 1969. Changes in forehead patterns and group composition of the white-faced monkey (Cebus capucinus). Pp. 36–42. In C. R. Carpenter (ed.), Proceedings of the Second International Congress of Primatology. Karger, Basel, Switzerland.

Otis, D. L., K. P. Burnham, G. C. White, and D. R. Anderson. 1978. Statistical Inference from Capture Data on Closed Animal Populations. Wildlife Monographs 62.

Ottichilo, W. K., J. W. Kufwafwa, and J. G. Stelfox. 1987. Elephant population trends in Kenya: 1977–1981. African Journal of Ecology 25:9–18.

Otto, M. C., and K. H. Pollock. 1990. Size bias in line transect sampling: a field test. Biometrics 46:239–245.

Overton, W. S., and D. E. Davis. 1969. Estimating the numbers of animals in wildlife populations. Pp. 403–455. In R. H. Giles, Jr. (ed.), Wildlife Management Techniques. Rev. 3d ed. Wildlife Society, Washington, D.C.

Packard, J. M., R. C. Summers, and L. B. Barnes. 1985. Variation of visibility bias during aerial surveys of manatees. Journal of Wildlife Management 49:347–351.

Packer, C. 1983. Sexual dimorphism: the horns of African antelopes. Science 221:1191–1193.

Page, R. J. C. 1993. X-ray method for determination of the age of live badgers (Meles meles) in the field. Mammalia 57:123–126.

Palmeirim, J., and K. Etheridge. 1985. The influence of man-made trails on foraging by tropical frugivorous bats. Biotropica 17:82–83.

Palmeirim, J. M., and L. Rodrigues. 1989. Using stereo photographs to estimate the size of bat colonies. Macroderma 5:26.

Palmisano, A. W., and H. H. Dupuie. 1975. An evaluation of steel traps for taking fur animals in coastal Louisiana. Proceedings of the Southeastern Association of Game and Fish Commissioners 29:342–347.

Parker, D. H. (ed.). 1991. GIS Sourcebook. 3d ed. GIS World, Fort Collins, Colorado.

Parry-Jones, K., and M. L. Augee. 1992. Movements of grey-headed foxes (Pteropus poliocephalus) to and from a colony on the central coast of New South Wales. Wildlife Research 19:331–340.

Parry-Jones, K., and L. Martin. 1987. Open forum on movements and feeding patterns of flying foxes (Chiroptera: Pteropodidae). Australian Mammalogy 10:129-132.

Patterson, B. D., and M. H. Gallardo. 1987. Rhyncholestes raphanurus. Mammalian Species 286:1-5.

Patton, J. L. 1993a. Family Geomyidae. Pp. 469-476. *In* D. E. Wilson and D. M. Reeder (eds.), Mammal Species of the World: A Taxonomic and Geographic Reference. 2d ed. Smithsonian Institution Press, Washington, D.C.

———. 1993b. Family Heteromyidae. Pp. 477-486. *In* D. E. Wilson and D. M. Reeder (eds.), Mammal Species of the World: A Taxonomic and Geographic Reference. 2d ed. Smithsonian Institution Press, Washington, D.C.

Paulik, G. J., and D. S. Robson. 1969. Statistical calculations for change-in-ratio estimators of population parameters. Journal of Wildlife Management 33:1-27.

Payne, J., C. M. Francis, and K. Phillips. 1985. A Field Guide to the Mammals of Borneo. Sabah Society, Sabah, Malaysia.

Payne, R. (ed.) 1983. Communication and Behavior of Whales. American Association for the Advancement of Science Symposium 76. Westview Press, Boulder, Colorado.

Pearcy, R. W. 1989. Field data acquisition. Pp. 15-27. *In* R. W. Pearcy, J. R. Ehleringer, H. A. Mooney, and P. W. Rundel (eds.), Plant Physiological Ecology Field Methods and Instrumentation. Chapman Hall, London, England.

Pearson, O. P. 1959. A traffic survey of *Microtus-Reithrodontomys* runways. Journal of Mammalogy 40:169-180.

Pelton, M. R., and L. C. Marcum. 1977. Potential use of radioisotopes for determining densities of black bears and other carnivores. *In* R. L. Phillips and C. Jonkel (eds.), Proceedings of the 1975 Predator Symposium. University of Montana, Missoula, Montana.

Pennycuick, C. J., and J. Rudnai. 1970. A method of identifying individual lions *Panthera leo* with an analysis of the reliability of identification. Journal of Zoology 160:497-508.

Pereira, J. M. C., and R. M. Itami. 1991. GIS-based habitat modelling using logistic multiple regression: a study of the Mt. Graham red squirrel. Photogrammetric Engineering and Remote Sensing 57:1475-1486.

Perkins, H. M. 1954. Animal Tracks: The Standard Guide for Identification and Characteristics. Stackpole, Harrisburg, Pennsylvania.

Perry, A. E., and G. Beckett. 1966. Skeletal damage as a result of band injury in bats. Journal of Mammalogy 47:131-132.

Perry, D. R. 1978. A method of access into the crowns of emergent and canopy trees. Biotropica 10:155-157.

Perry, D. R., and J. Williams. 1981. The tropical forest canopy: a method providing total access. Biotropica 13:283-285.

Perry, R. J., and M. L. Braysher. 1986. A technique for estimating the numbers of eastern grey kangaroos, *Macropus giganteus,* grazing a given area of pasture. Australian Wildlife Research 13:335-338.

Person, D. K., and D. H. Hirth. 1991. Home range and habitat use of coyotes in a farm region of Vermont. Journal of Wildlife Management 32:628-629.

Phillips, C. J. , B. Steinberg, and T. H. Kunz. 1982. Dentin, cementum, and age determination in bats: a critical evaluation. Journal of Mammalogy 63:197-207.

Phillips, W. W. A. 1980-1984. Manual of the Mammals of Sri Lanka. Rev. 2d ed. Vol. 1-3. Wildlife and Nature Protection Society of Sri Lanka. Colombo, Sri Lanka.

Picman, J. 1987. An inexpensive camera setup for the study of egg predation at artificial nests. Journal of Field Ornithology 58:372-382.

Pielou, E. C. 1977. Mathematical Ecology. 2d ed. Wiley, New York.

Pienaar, U. De V. 1975. The drug immobilization of antelope species. Pp. 35-50. *In* E. Young (ed.), The Capture and Care of Wild Animals. Ralph Curtis Books, Hollywood, Florida.

Pierson, E. D., W. E. Rainey, and D. M. Koontz. 1991. Bats and mines: experimental mitigation for Townsend's big-eared bat at the McLaughlin Mine in California. Pp. 31-42. *In* Proceedings V: Issues and Technology in Management of Impacted Wildlife. Thorne Ecological Institute, Boulder, Colorado.

Plage, D., and M. Plage. 1985. Return of Java's wildlife. National Geographic 167:750-771.

Poduschka, W., and C. Wemmer. 1986. Observations on chemical communication and its glandular sources in selected insectivora. Pp. 609-616. *In* D. Duvall, D. Müller-Schwarze, and R. M. Silverstein (eds.), Chemical Signals in Vertebrates 4. Plenum, New York.

Pollock, K. H. 1975. A K-sample tag-recapture model allowing for unequal survival and catchability. Biometrika 62:577-583.

———. 1981. Capture-recapture models allowing for age-dependent survival and capture rates. Biometrics 37:521-529.

————. 1982. A capture-recapture design robust to unequal probability of capture. Journal of Wildlife Management 46:752-757.

Pollock, K. H., and W. L. Kendall. 1987. Visibility bias in aerial surveys: a review of estimation procedures. Journal of Wildlife Management 51:502-510.

Pollock, K. H., and M. C. Otto. 1983. Robust estimation of population size in closed animal populations from capture-recapture experiments. Biometrics 39:1035-1049.

Pollock, K. H., J. E. Hines, and J. D. Nichols. 1984. The use of auxiliary variables in capture-recapture and removal experiments. Biometrics 40:329-340.

Pollock, K. H., R. A. Lancia, M. C. Conner, and D. L. Wood. 1985. A new change-in-ratio procedure robust to unequal catchability of types of animal. Biometrics 41:653-662.

Pollock, K. H., J. D. Nichols, C. Brownie, and J. E. Hines. 1990. Statistical Inference for Capture-Recapture Experiments. Wildlife Monographs 107.

Poole, J. H. 1989. Mate guarding, reproductive success and female choice in African elephants. Animal Behaviour 37:842-849.

Poole, W. E. 1982. Macropus giganteus. Mammalian Species 187:1-8.

Prater, S. H. 1980. The Book of Indian Animals. Bombay Natural History Society, Bombay, India.

Preno, W. L., and R. F. Labisky. 1971. Abundance and Harvest of Doves, Pheasants, Bobwhites, Squirrels, and Cottontails in Illinois, 1956-1969. Department of Conservation, Springfield, Illinois.

Pritchard, M. H, and G. O. W. Kruse. 1982. The Collection and Preservation of Animal Parasites. Harold W. Manter Laboratory Technical Bulletin 1. University of Nebraska Press, Lincoln, Nebraska.

Progulske, D. R. 1957. A collar for identification of big game. Journal of Wildlife Management 21:251-252.

Proulx, G., and F. F. Gilbert. 1984. Estimating muskrat trends by house counts. Journal of Wildlife Management 48:917-922.

Proulx, G., I. M. Pawlina, D. K. Onderka, M. J. Badry, and K. Seidel. 1994. Field evaluation of the number 1 1/2 steel-jawed leghold and the Sauvageau 2001-8 traps to humanely capture arctic fox. Wildlife Society Bulletin 22:179-183.

Pucek, Z., and V. P. W. Lowe. 1975. Age criteria in small mammals. Pp. 55-72. In F. B. Golley, K. Petrusewicz, and L. Ryszkowski (eds.), Small Mammals: Their Productivity and Population Dynamics. Cambridge University Press, Cambridge, England.

Punt, A., and P. J. van Nieuwenhoven. 1957. The use of radioactive bands in tracing hibernating bats. Experientia 13:51-54.

Puterski, R., J. Carter, M. Hewitt, H. F. Stone, L. T. Fisher, and E. T. Slonecker. 1990. Global positioning systems technology and its application to environmental programs. U.S. Environmental Protection Agency, Office of Research and Development, Environmental Monitoring Systems Laboratory, GIS Technical Memorandum 3. Washington, D.C.

Putman, R. J. 1984. Facts from faeces. Mammal Review 14:79-97.

Pye, J. D. 1983. Techniques for studying ultrasound. Pp. 39-65. In B. Lewis (ed.), Bioacoustics: A Comparative Approach. Academic Press, London, England.

Quang, P. X. 1991. A nonparametric approach to size-biased line transect sampling. Biometrics 47:269-279.

Queal, L. M., and B. D. Hlavachick. 1968. A modified marking technique for young ungulates. Journal of Wildlife Management 32:628-629.

Racey, P. A. 1969. Diagnosis of pregnancy and experimental extension of gestation in the pipistrelle bat, *Pipistrellus pipistrellus*. Journal of Reproduction and Fertility 19:465-474.

————. 1972. Viability of bat spermatozoa after prolonged storage in the epididymis. Journal of Reproduction and Fertility 28:309-311.

————. 1974. Ageing and assessment of reproductive status of pipistrelle bats, *Pipistrellus pipistrellus*. Journal of Zoology (London) 173:264-271.

————. 1979. Two bats in the Seychelles. Oryx 15:148-152.

————. 1987. Keeping, handling, and releasing. Pp. 36-41. In A. J. Mitchell-Jones (ed.), The Bat Worker's Manual. Nature Conservancy Council, Peterborough, United Kingdom.

————. 1988. Reproductive assessment in bats. Pp. 31-45. In T. H. Kunz (ed.), Ecological and Behavioral Methods for the Study of Bats. Smithsonian Institution Press, Washington, D.C.

Ralls, K. 1976. Mammals in which females are larger than males. Quarterly Review of Biology 51:245-276.

Ramirez, J. R. 1991. Understanding universal exchange formats. Photogrammetric Engineering and Remote Sensing 57:89-92.

Ramírez-Pulido, J., R. López-Wilchis, C. Müdespacher, and I. Lira. 1982. Catálogo de los Mamíferos Terrestres Nativos de México. Universidad Autónoma Metropolitana-Iztapalapa, Iztapalapa, D.F., Mexico.

Ramsay, M. A., and I. Stirling. 1986. Long-term effects of drugging and handling free-ranging polar bears. Journal of Wildlife Management 50:619-626.

Ramsey, C. W. 1968. A drop-net deer trap. Journal of Wildlife Management 32:187-190.

Randall, J. A. 1989. Individual footdrumming signatures in banner-tailed kangaroo rats *Dipodomys spectabilis*. Animal Behaviour 38:620-630.

Randall, R. M. 1979. Perineal gland marking by free-ranging african civets, *Civettictis civetta*. Journal of Mammalogy 60:622-627.

Raphael, M. G., C. A. Taylor, and R. H. Barrett. 1986. Smoked aluminum track stations record flying squirrel occurrence. U.S. Department of Agriculture, Forest Service Research Note PSW-384:1-3. Washington, D.C.

Rappole, J. H., D. N. Lopez, M. Tewes, and D. Everett. 1985. Remote trip cameras as a means for surveying for nocturnal felids. Pp. 45-49. *In* R. P. Brooks (ed.), Nocturnal Mammals: Techniques for Study. School of Forest Resources, Pennsylvania State University, University Park, Pennsylvania.

Ratcliffe, F. N. 1931. The flying fox (*Pteropus*) in Australia. Council for Scientific and Industrial Research Bulletin 53. Melbourne, New South Wales, Australia.

Rautenbach, I. L. 1985. A new technique for the efficient use of macro-mistnets. Koedoe 28:81-86.

Redford, K. H., and J. F. Eisenberg. 1992. Mammals of the Neotropics. Vol. 2. University of Chicago Press, Chicago, Illinois.

Renfree, M. B., E. M. Russell, and R. D. Wooller. 1984. Reproduction and life history of the honey possum, *Tarsipes rostratus*. Pp. 427-437. *In* A. Smith and I. Hume (eds.), Possums and Gliders. Surrey Beatty Sons, Chipping Norton, New South Wales, Australia.

Rexstad, E., and K. Burnham. 1991. Users' Guide for Interactive Program CAPTURE. Colorado Cooperative Fish and Wildlife Research Unit, Fort Collins, Colorado.

Rice, W. R., and J. D. Harder. 1977. Application of multiple aerial sampling to a mark-recapture census of white-tailed deer. Journal of Wildlife Management 41:197-206.

Richards, G. C. 1987. Aspects of the ecology of spectacled flying foxes, *Pteropus conspicillatus* (Chiroptera: Pteropodidae) in tropical Queensland. Australian Mammalogy 10:87-88.

———. 1989. Nocturnal activity of insectivorous bats relative to temperature and prey availability in tropical Queensland. Australian Wildlife Research 16:151-158.

———. 1990. Rainforest bat conservation: unique problems in a unique environment. Australian Zoologist 26:44-46.

Richards, G. C., R. E. Smyth, and P. A. Walker. 1992. Patterns in the Distribution of Cape York Bats: Results of GIS Modelling and Conservation Implications. Proceedings of Conservation Biology in Australia and Oceania, Brisbane, Queensland, Australia.

Richter, W. C. 1955. A technique for night identification of animals. Journal of Wildlife Management 19:159-160.

Ricker, W. E. 1958. Handbook of computations for biological statistics of fish populations. Bulletin of the Fisheries Research Board of Canada 119. Ottawa, Ontario, Canada.

Ride, W. D. L. 1970. A Guide to the Native Mammals of Australia. Oxford University Press, Melbourne, New South Wales, Australia.

Riedman, M. 1990. The Pinnipeds: Seals, Sealions, and Walruses. University of California Press, Berkeley, California.

Rieger, J. F., and E. M. Jakob. 1988. The use of olfaction in food location by frugivorous bats. Biotropica 20:161-164.

Risler, L., S. K. Wasser, and G. P. Sackett. 1987. Measurement of excreted steroids in *Macaca nemestrina*. American Journal of Primatology 12:91-100.

Roberts, J. D. 1978. Variation in coyote age determination from annuli in different teeth. Journal of Wildlife Management 42:454-456.

Roberts, T. J. 1977. The Mammals of Pakistan. Ernest Benn, London, England.

Robinson, J. G. 1988. Seasonal variation in the use of time and space by the wedge-capped capuchin monkey *Cebus olicaceous*: implications to foraging theory. Smithsonian Contributions to Zoology 431:1-60.

Rodriguez-Duran, A., and A. R. Lewis. 1987. Patterns of population size, diet, and activity time for a multispecies assemblage of bats at a cave in Puerto Rico. Caribbean Journal of Science 23:352-360.

Roeder, K., B. Dennis, and E. O. Garton. 1987. Estimating density from variable circular plot censuses. Journal of Wildlife Management 51:224-230

Rood, J. P. 1975. Population dynamics and food habits of banded mongoose. East African Wildlife Journal 13:89-113.

———. 1983. The social system of the dwarf mongoose. Pp. 454-488. *In* J. F. Eisenberg and D. G.

Kleiman (eds.), Advances in the Study of Mammalian Behavior. American Society of Mammalogists Special Publication 7.

Rood, J. P., and D. W. Nellis. 1980. Freeze marking mongooses. Journal of Wildlife Management 44:500–502.

Rose, R. K., N. A. Slade, and J. H. Honacki. 1977. Live trap preference among grassland mammals. Acta Theriologia 22:297–307.

Rose, S. R. 1992. International Travel Health Guide. Travel Medicine, Northampton, Massachusetts.

Rothstein, A., and J. G. Griswold. 1991. Age and sex preferences for social partners by juvenile bison bulls, *Bison bison.* Animal Behaviour 41:227–237.

Roughgarden, J., S. W. Running, and P. A. Matson. 1991. What does remote sensing do for ecology? Ecology 72:1918–1922.

Roughton, R. D., and M. W. Sweeny. 1982. Refinements in scent-station methodology for assessing trends in carnivore populations. Journal of Wildlife Management 46:217–229.

Roussel, Y. E., and R. Patenaude. 1975. Some physiological effects of M99 etorphine on immobilized free-ranging moose. Journal of Wildlife Management 39:634–636.

Rowlands, I. W. (ed.). 1966. Comparative biology of reproduction in mammals. Symposia of the Zoological Society of London 15. Zoological Society of London, London, England.

Rowsemitt, C., T. H. Kunz, and R. H. Tamarin. 1975. The timing and pattern of molt in *Microtus breweri.* University of Kansas, Museum of Natural History, Occasional Papers 34. Lawrence, Kansas.

Rudran, R. 1973. Adult male replacement in one-male troops of purple-faced langurs (*Presbytis senex senex*) and its effect on population structure. Folia Primatology 19:166–192.

———. 1978. Socioecology of blue monkeys (*Cercopithecus mitis stuhlmanni*) of the Kibale Forest, Uganda. Smithsonian Contributions to Zoology 249:1–88.

———. 1979. The demography and social mobility of a red howler (*Alouatta seniculus*) population in Venezuela. Pp. 107–126. *In* J. F. Eisenberg (ed.), Vertebrate Ecology in the Northern Neotropics. Smithsonian Institution Press, Washington, D.C.

Rumiz, D. I. 1990. *Alouatta caraya:* population density and demography in northern Argentina. American Journal of Primatology 21:279–294.

Rutnagur, R. S., M. D. Burns, and M. H. Hanseli. 1990. A simple electronic timing device for triggering time-lapse cameras and its use in the instantaneous time sampling of rookery attendance. Animal Behaviour 90:899–900.

Rydell, J. 1990. Behavioural variation in echolocation pulses of the northern bat, *Eptesicus nilssoni.* Ethology 85:103–113.

Ryel, L. A. 1959. Deer Pellet Group Surveys on an Area of Known Herd Size. Michigan Department of Conservation, Game Division Report 2252. Lansing, Michigan.

———. 1971. Evaluation of Pellet Group Surveys for Estimating Deer Populations in Michigan. Unpubl. Ph.D. thesis, Michigan State University, East Lansing, Michigan.

Samuel, M. D., E. O. Garton, M. W. Schlegel, and R. G. Carson. 1987. Visibility bias during aerial surveys of elk in northcentral Idaho. Journal of Wildlife Management 51:622–630.

Sanderson, G. C. 1950. Methods of measuring productivity in raccoons. Journal of Wildlife Management 14:389–402.

Sarrazin, J-P. R., and J. R. Bider. 1973. Activity, a neglected parameter in population estimates — the development of a new technique. Journal of Mammalogy 54:369–382.

Savidge, J. A., and T. F. Seibert. 1988. An infrared trigger and camera to identify predators at artificial nests. Journal of Wildlife Management 52:291–294.

Schaller, G. B. 1967. The Deer and the Tiger: A Study of Wildlife in India. University of Chicago Press, Chicago, Illinois.

———. 1972. The Serengeti Lion. University of Chicago Press, Chicago, Illinois.

Scheffer, V. B. 1950. Growth layers on the teeth of Pinnipedia as indication of age. Science 112:309–311.

Schemnitz, S. D. (ed.). 1980. Wildlife Management Techniques Manual. 4th ed. Wildlife Society, Washington, D.C.

Schlitter, D. A. 1993a. Order Hyracoidea. Pp. 373–374. *In* D. E. Wilson and D. M. Reeder (eds.), Mammal Species of the World: A Taxonomic and Geographic Reference. 2d ed. Smithsonian Institution Press, Washington, D.C.

———. 1993b. Order Tubulidentata. P. 375. *In* D. E. Wilson and D. M. Reeder (eds.), Mammal Species of the World: A Taxonomic and Geographic Reference. 2d ed. Smithsonian Institution Press, Washington, D.C.

———. 1993c. Order Pholidota. P. 415. *In* D. E. Wilson and D. M. Reeder (eds.), Mammal Species of the World: A Taxonomic and Geographic Reference. 2d ed. Smithsonian Institution Press, Washington, D.C.

———. 1993d. Order Macroscelidea. Pp. 829–830. *In* D. E. Wilson and D. M. Reeder (eds.), Mammal Species of the World: A Taxonomic and Geographic Reference. 2d ed. Smithsonian Institution Press, Washington, D.C.

Schowalter, D. B., L. D. Harder, and B. H. Treichel. 1978. Age composition of some vespertilionid bats as determined by dental annuli. Canadian Journal of Zoology 56:355–358.

Schreiber, A., R. Wirth, M. Riffel, and H. Van Rompaey. 1989. Weasels, Civets, Mongooses and Their Relatives — An Action Plan for the Conservation of Mustelids and Viverrids. IUCN, Gland, Switzerland.

Schwarz, C. R. 1989. North American Datum of 1983. NOAA Professional Paper 2. National Geodetic Survey, National Oceanic and Atmospheric Administration, Rockville, Maryland.

Scott, J. M., B. Csuti, and S. Caicco. 1991. Gap analysis: assessing protection needs. Pp. 15–26. *In* W. E. Hudson (ed.), Landscape Linkages and Biodiversity. Island Press, Washington, D.C.

Seabloom, R. W. 1985. Endocrinology. Pp. 685–724. *In* R. H. Tamarin (ed.), Biology of New World *Microtus*. American Society of Mammalogists Special Publication 8.

Sebeok, T. A. (ed.). 1977. How Animals Communicate. Indiana University Press, Bloomington, Indiana.

Seber, G. A. F. 1965. A note on the multiple-recapture census. Biometrika 52:249–259.

———. 1970. The effects of trap response on tag-recapture estimates. Biometrics 26:13–22.

———. 1982. The Estimation of Animal Abundance and Related Parameters. 2d ed. MacMillan, New York.

———. 1986. A review of estimating animal abundance. Biometrics 42:267–292.

———. 1992. A review of estimating animal abundance II. International Statistical Review 60:129–166.

Seidensticker, J., K. M. Tamang, and C. W. Gray. 1974. The use of CI-744 to immobilize free-ranging tigers and leopards. Journal of Zoo Animal Medicine 5:22–25.

Seton, E. T. 1958. Animal Tracks and Hunter Signs. Doubleday, New York.

Seydack, A. H. W. 1984. Application of a photo-recording device in the census of larger rain-forest mammals. South African Journal of Wildlife Research 14:10–14.

Shepherd, R. C. H., I. F. Nolan, and J. W. Edmonds. 1978. A review of methods to capture live wild rabbits in Victoria. Journal of Wildlife Management 42:179–184.

Sherman, P. W., M. L. Morton, L. M. Hoopes, J. Bochantin, and J. M. Watt. 1985. The use of tail collagen strength to estimate age in Belding's ground squirrels *Spermophilus beldingi*. Journal of Wildlife Management 49:874–883.

Shiras, G. 1936. Hunting Wild Life with Camera and Flashlight: A Record of Sixty-five Years' Visits to the Woods and Waters of North America. National Geographic Society, Washington, D.C.

Short, J., and P. Bayliss. 1985. Bias in aerial survey estimates of kangaroo density. Journal of Applied Ecology 22:415–422.

Short, J., and J. Hone. 1988. Calibrating aerial surveys of kangaroos by comparison with drive counts. Australian Wildlife Research 15:277–284.

Shriner, W. M., and P. B. Stacey. 1991. Spatial relationships and dispersal patterns in the rock squirrel, *Spermophilus variegatus*. Journal of Mammalogy 72:601–606.

Siglin, R. J. 1966. Marking mule deer with an automatic tagging device. Journal of Wildlife Management 30:631–633.

Simmons, N. M., and J. L. Phillips. 1966. Modifications of a dye-spraying device for marking desert bighorn sheep. Journal of Wildlife Management 30:208–209.

Sinclair, A. R. E. 1972. Long term monitoring of mammal populations in the Serengeti: census of non-migratory ungulates. East African Wildlife Journal 10:287–297.

———. 1973. Population increases of buffalo and wildebeest in the Serengeti. East African Wildlife Journal 11:93–107.

Singer, F. J. 1978. Behavior of mountain goats in relation to U.S. Highway 2, Glacier National Park, Montana. Journal of Wildlife Management 42:591–597.

Siniff, D. B., and R. O. Skoog. 1964. Aerial censusing of caribou using stratified random sampling. Journal of Wildlife Management 28:391–401.

Sinnary, A. S. M., and J. J. Hebrard. 1991. A new approach to detecting visibility bias for fixed-width transect method. African Journal of Ecology 29:222–228.

Skalski, J. R. 1991. Using sign counts to quantify animal abundance. Journal of Wildlife Management 55:705–715.

Skalski, J. R., and D. S. Robson. 1992. Techniques for Wildlife Investigations. Design and Analysis of Capture Data. Academic Press, San Diego, California.

Skalski, J. R., D. S. Robson, and M. A. Simmons. 1983. Comparative census procedures using single mark-recapture methods. Ecology 64:752–760.

Skinner, J. D., and R. H. N. Smithers. 1990. The Mammals of the Southern African Subregion. University of Pretoria, Pretoria, South Africa.

Slade, N. A., M. A. Eifler, N. M. Gruenhagen, and A. L. Davelos. 1993. Differential effectiveness of standard and long Sherman livetraps in capturing small mammals. Journal of Mammalogy 74:156–161.

Slobodchikoff, C. N., and R. Coast. 1980. Dialects in the alarm calls of prairie dogs. Behavioral Ecology and Sociobiology 7:49–53.

Slobodchikoff, C. N., J. Kiriazis, C. Fischer, and E. Creef. 1991. Semantic information distinguishing individual predators in the alarm calls of Gunnison's prairie dogs. Animal Behaviour 42:713–719.

Slonecker, E. T., and J. A. Carter. 1990. GIS applications of global positioning system technology. GPS World 1:50–55.

Smith, A. P., D. Lindenmayer, R. J. Begg, M. A. Macfarlane, J. H. Seebeck, and G. C. Suckling. 1989. Evaluation of stagwatching technique for census of possums and gliders in tall open forest. Australian Wildlife Research 16:575–580.

Smith, A. T., and W. X. Gao. 1991. Social relationships of black-lipped pikas (*Ochotona curzoniae*). Journal of Mammalogy 72:231–247.

Smith, A. T., and M. L. Weston. 1990. Ochotona princeps. Mammalian Species 352:1–8.

Smith, E. P., and G. van Belle. 1984. Nonparametric estimation of species richness. Biometrics 40:119–129.

Smith, G. C., D. W. Kaufman, R. M. Jones, J. B. Gentry, and M. H. Smith. 1971. The relative effectiveness of two types of snap traps. Acta Theriologia 16:284–288.

Smith, G. W., and N. C. Nydegger. 1985. A spotlight, line-transect method for surveying jack rabbits. Journal of Wildlife Management 49:699–702.

Smith, J. L. D., C. McDougal, and D. Miquelle. 1989. Scent marking in free-ranging tigers, *Panthera tigris*. Animal Behaviour 37:1–10.

Smith, L. M., I. L. Brisbin, and G. C. White. 1984. An evaluation of total trapline captures as estimates of furbearer abundance. Journal of Wildlife Management 48:1452–1455.

Smith, M. J. 1973. Petaurus breviceps. Mammalian Species 30:1–5.

Smith, T. D. 1981. Line-transect techniques for estimating density of porpoise schools. Journal of Wildlife Management 45:650–657.

Smithers, R. H. N. 1983. The Mammals of the Southern African Subregion. University of Pretoria, Pretoria, South Africa.

Smythe, N. 1978. The natural history of the Central American Agouti (*Dasyprocta punctata*). Smithsonian Contributions to Zoology 257:1–51.

Snyder, J. P. 1983. Map Projections Used by the U.S. Geological Survey. Geological Survey Bulletin 1532. U.S. Geological Survey, Washington, D.C.

Southwell, C. 1989. Techniques for monitoring the abundance of kangaroo and wallaby populations. Pp. 659–693. *In* G. Grigg, P. Jarman, and I. Hume (eds.), Kangaroos, Wallabies and Rat-kangaroos. Surrey Beatty Sons, Chipping Norton, New South Wales, Australia.

———. 1994. Evaluation of walked line transect counts for estimating macropod density. Journal of Wildlife Management 58:348–356.

Southwell, C., and M. Fletcher. 1990. The use of roads and tracks as transect routes for surveying the abundance of whiptail wallabies, *Macropus parryi* (Marsupialia: Macropodidae). Australian Mammalogy 13:223–226.

Southwick, C. H., M. A. Beg, and M. R. Siddiqi. 1961. A population survey of rhesus monkeys in villages, towns and temples of northern India. Ecology 42:538–547.

Sowler, S. 1984. Growth and reproduction in the fruit bat, *Epomophorus walbergi*. Unpubl. Ph.D. thesis, University of Natal, Pietermaritzburg, Republic of South Africa.

Sowls, L. K. 1984. The Peccaries. University of Arizona Press, Tucson, Arizona.

Sparling, D. W., and J. D. Williams. 1978. Multivariate analysis of avian vocalizations. Journal of Theoretical Biology 74:83–107.

Speakman, J. R., D. J. Bullock, L. A. Eales, and P. A. Racey. 1992. A problem defining temporal pattern in animal behaviour: clustering in the emergence behaviour of bats from maternity roosts. Animal Behaviour 43:491–500.

Spencer, H. J., and T. H. Fleming. 1989. Roosting and foraging behaviour of the Queensland tube-nosed bat, *Nyctimene robinsoni* (Pteropodidae). Preliminary radio-tracking observations. Wildlife Research 16:413–420.

Spencer, H. J., C. Palmer, and K. Parry-Jones. 1991. Movements of fruit-bats in eastern Australia, determined by using radio-tracking. Wildlife Research 18:463–468.

Spillett, J. J., and R. S. Zobell. 1967. Innovations in trapping and handling pronghorn antelope. Journal of Wildlife Management 31:347–351.

Spinage, C. A. 1973. A review of age determination of mammals by means of teeth, with special reference to Africa. East African Wildlife Journal 11:165–187.

Star, J., and J. Estes. 1990. Geographic Information Systems: An Introduction. Prentice-Hall, Englewood Cliffs, New Jersey.

Statham, H. 1983. Browsing Damage in Tasmanian Forest Areas and Effects of 1080 Poisoning. Forestry Commission Tasmania Bulletin 7.

Stebbings, R. E. 1987. Ringing and marking. Pp. 33–35. *In* A. J. Mitchell-Jones (ed.), The Bat Worker's Manual. Nature Conservancy Council, Peterborough, United Kingdom.

Steigers, W. D., and J. T. Flinders. 1980. Mortality and movements of mule deer fawns in Washington. Journal of Wildlife Management 44:381–388.

Stein, F. J. 1978. Sex determination in the common marmoset (*Callithrix jacchus*). Laboratory Animal Science 28:75–80.

Stelfox, J. G., and J. R. Robertson. 1976. Immobilizing bighorn sheep with succinylcholine chloride and phencyclidine hydrochloride. Journal of Wildlife Management 40:174–176.

Stephenson, A. B. 1977. Age determination and morphological variation of Ontario otters. Canadian Journal of Zoology 55:1577–1583.

Stewart, A. 1982. Evaluation of kangaroo pellet counts. Bulletin of the Australian Mammal Society 7:25.

Still, A. W. 1982. On the number of subjects used in animal behaviour experiments. Animal Behaviour 30:873–880.

Stokes, D., and L. Stokes. 1986. A Guide to Animal Tracking and Behavior. Little Brown, Boston, Massachusetts.

Stoms, D. M. 1992. Effects of habitat map generalization in biodiversity assessment. Photogrammetric Engineering and Remote Sensing 58:1587–1591.

Storm, G. L., R. D. Andrews, R. L. Phillips, R. A. Bishop, D. B. Siniff, and J. R. Tester. 1976. Morphology, Reproduction, Dispersal, and Mortality of Midwestern Red Fox Populations. Wildlife Monographs 49.

Strahan, R. (ed.) 1983. Complete Book of Australian Mammals. Angus Robertson, London, England.

Strandgaard, H. 1967. Reliability of the Petersen method tested on a roe-deer population. Journal of Wildlife Management 31:643–651.

Strathearn, S. M., J. S. Lotimer, G. B. Kolenosky, and W. M. Lintack. 1984. An expanding break-away radio collar for black bear. Journal of Wildlife Management 48:939–942.

Struhsaker, T. T. 1967. Ecology of vervet monkeys (*Cercopithecus aethiops*) in the Masai-Amboseli Game Reserve, Kenya. Ecology 48:891–904.

———. 1975. The Red Colobus Monkey. University of Chicago Press, Chicago, Illinois.

Sugiyama, Y. 1967. Social organization of Hanuman langurs. Pp. 221–236. *In* S. A. Altmann (ed.), Social Communication among Primates. University of Chicago Press, Chicago, Illinois.

Sullivan, J. B., C. A. DeYoung, S. L. Beasom, J. R. Heffelfinger, S. P. Coughlin, and M. W. Hellickson. 1991. Drive-netting deer: incidence of mortality. Wildlife Society Bulletin 19:393–396.

Szaro, R. C., L. H. Simons, and S. C. Belfit. 1988. Comparative effectiveness of pitfalls and live-traps in measuring small mammal community structure. Pp. 282–288. *In* R. C. Szaro, K. E. Severson, and D. R. Patton (tech. coordinators), Management of Amphibians, Reptiles, and Small Mammals in North America. U.S. Department of Agriculture Forest Service General Technical Report RM-166. Washington, D.C.

Taitt, M. J., and C. J. Krebs. 1985. Population dynamics and cycles. Pp. 587–620. *In* R. H. Tamarin (ed.), Biology of New World *Microtus*. American Society of Mammalogists Special Publication 8.

Takos, M. J. 1943. Trapping and banding muskrats. Journal of Wildlife Management 7:400–407.

Tamura, N., and H. S. Yong. 1993. Vocalizations in response to predators in three species of Malaysian *Callosciurus* (Sciuridae). Journal of Mammalogy 74:703–714.

Taylor, P. J., J. U. M. Jarvis, T. M. Crowe, and K. C. Davies. 1985. Age determination in the Cape molerat *Georychus capensis*. South Africa Journal of Zoology 20:261–267.

Teer, J. G. 1982. White-tailed deer (Texas). Pp. 251–253. *In* D. E. Davis (ed.), CRC Handbook for Census Methods for Terrestrial Vertebrates. CRC Press, Boca Raton, Florida.

Tembo, A. 1987. Population status of the hippopotamus on the Luangwa River, Zambia. African Journal of Ecology 25:71–77.

Thomas, D. C. 1977. Metachromatic staining of dental cementum for mammalian age determination. Journal of Wildlife Management 41:207–210.

Thomas, D. W. 1988. The distribution of bats in different ages of Douglas-fir forests. Journal of Wildlife Management 52:619–626.

Thomas, D. W., and M. B. Fenton. 1978. Notes on the dry season roosting and foraging behaviour of *Epomophorus gambianus* and *Rousettus aegyptiacus* (Chiroptera: Pteropodidae). Journal of Zoology (London) 186:403–406.

Thomas, D. W., and R. K. LaVal. 1988. Survey and census methods. Pp. 77–89. *In* T. H. Kunz (ed.), Ecological and Behavioral Methods for the Study of Bats. Smithsonian Institution Press, Washington, D.C.

Thomas, D. W., and S. D. West. 1989. Wildlife habitat relationships: sampling procedures for Pacific Northwest vertebrates. Sampling Methods for Bats. U.S. Forest Service, General Technical Report, Pacific Northwest No. 243. Portland, Oregon.

Thomas, D. W., G. P. Bell, and M. B. Fenton. 1987. Variation in echolocation call frequencies recorded from North American vespertilionid bats: a cautionary note. Journal of Mammalogy 68:842–847.

Thompson, D., and A. R. Hiby. 1985. The use of scale binoculars for distance estimation and a time lapse camera for angle estimation during the 1983/84 IDCR minke whale assessment cruise. Report of the International Whaling Commission 35:309–314.

Thompson, H. V., and C. J. Armour. 1954. Methods of marking wild rabbits. Journal of Wildlife Management 18:411–414.

Thomsen, H. P., and O. A. Mortensen. 1946. Bone growth as an age criterion in the cottontail rabbit. Journal of Wildlife Management 10:171–174.

Thorington, R. W., Jr., R. Rudran, and D. Mack. 1979. Sexual dimorphism of *Alouatta seniculus* and observations on capture techniques. Pp. 97–106. *In* J. F. Eisenberg (ed.), Vertebrate Ecology in the Northern Neotropics. Smithsonian Institution Press, Washington, D.C.

Tidemann, C. R. 1985. A Study of the Status, Habitat Requirements and Management of the Two Species of Bats on Christmas Island (Indian Ocean). Report to Australian National Parks and Wildlife Service, Canberra, New South Wales, Australia.

———. 1987. Notes on the flying-fox, *Pteropus melanotus* (Chiroptera: Pteropodidae), on Christmas Island, Indian Ocean. Australian Mammalogy 10:89–91.

Tidemann, C. R., and S. C. Flavel. 1987. Factors affecting choice of diurnal roost site by tree-hole bats (Microchiroptera) in south-eastern Australia. Australian Wildlife Research 14:459–473.

Tidemann, C. R., and R. A. Loughland. 1993. A harp trap for large megachiropterans. Wildlife Research 20:607–611.

Tidemann, C. R., and D. P. Woodside. 1978. A collapsible bat-trap and a comparison of results obtained with the trap and with mist nets. Australian Wildlife Research 5:355–362.

Tilton, B. 1994. Backcountry First Aid and Extended Care. ICS Books, Merrillville, Indiana.

Timm, R. M. 1987. Tent construction by bats of the genera *Artibeus* and *Uroderma*. Pp. 187–212. *In* B. D. Patterson and R. M. Timm (eds.), Studies in Neotropical Mammalogy: Essays in Honor of Philip Hershkovitz. Fieldiana: Zoology (n.s.) 39:187–212.

Tinbergen, L. 1960. The natural control of insects in pinewoods. 1. Factors influencing the intensity of predation by songbirds. Archives Néerlandaises de Zoologie 13:265–335.

Tinbergen, von N. 1965. Von den vorraatskammern des ruotfuchses (*Vulpes vulpes* L.) Zeitschrift für Tierpsychologie 22:119–149.

Tomlin, C. D. 1990. Geographic Information Systems and Cartographic Modeling. Prentice-Hall, Englewood Cliffs, New Jersey.

Tomlinson, R. F. 1987. Current and potential uses of geographic information systems: the North American experience. International Journal of Geographical Information Systems 1:203–218.

Troyer, W. A., R. J. Hensel, and K. E. Durley. 1962. Live-trapping and handling of brown bears. Journal of Wildlife Management 26:330–331.

Turkowski, F. J., M. L. Popelka, and R. W. Bullard. 1983. Efficacy of odor lures and baits for coyotes. Wildlife Society Bulletin 11:136–145.

Turner, D. C. 1975. The Vampire Bat. Johns Hopkins University Press, Baltimore, Maryland.

Tuttle, M. D. 1974a. An improved trap for bats. Journal of Mammalogy 55:475–477.

———. 1974b. Bat trapping: results and suggestions. Bat Research News 15:4–7.

———. 1975. Population ecology of the gray bat (*Myotis grisescens*): factors influencing early growth and development. University of Kansas, Museum of Natural History, Occasional Papers 36. Lawrence, Kansas.

———. 1976a. Population ecology of the gray bat (*Myotis grisescens*): philopatry, timing and patterns of movement, weight loss during migration, and seasonal adaptive strategies. University of Kansas, Museum of Natural History, Occasional Papers 54. Lawrence, Kansas.

———. 1976b. Collecting techniques. Pp. 71–88. *In* R. J. Baker, J. K. Jones, Jr., and D. C. Carter (eds.), Biology of Bats of the New World Family Phyllostomatidae. Part 1. Special Publications, The Museum, Texas Tech University. Texas Tech Press, Lubbock, Texas.

———. 1979. Status, causes of decline, and management of endangered gray bats. Journal of Wildlife Management 43:1–17.

Twente, J. W., Jr. 1955. Aspects of a population study of cavern-dwelling bats. Journal of Mammalogy 36:379–390.

Twigg, G. I. 1975. Marking mammals. Mammal Review 5:101–116.

Tyson, E. L. 1959. A deer drive vs. track census. Transactions of the North American Wildlife Conference 24:457–464.

Udevitz, M. S. 1989. Change-in-Ratio Methods for Estimating the Size of Closed Populations. Unpubl. Ph.D. thesis. North Carolina State University, Raleigh, North Carolina.

Udevitz, M. S., and K. H. Pollock. 1991. Change-in-ratio methods for estimating population size. Pp. 90–101. *In* D. R. McCullough and R. H. Barrett (eds.), Wildlife 2001: Populations. Elsevier Applied Science, London, England.

Uhlig, H. G. 1956. The Gray Squirrel in West Virginia. The Conservation Commission of West Virginia, Division Game Management Bulletin 3. Charleston, West Virginia.

Uexkull, J., and G. Kriszat. 1934. Streifzuege durch die Umwelten von Tieren und Menschen. Springer-Verlag, Hamburg, Germany.

van Aarde, R. J. 1985. Age determination of Cape porcupines *Hystrix africaeaustralis*. South African Journal of Zoology 20:232–236.

Van Dyke, F. G., R. H. Brocke, and H. G. Shaw. 1986. Use of road track counts as indices of mountain lion presence. Journal of Wildlife Management 50:102–109.

van Heerdt, P. F., and J. W. Sluiter. 1958. The Polish ear tags for the Dutch bats. De Levende Natuur 61:216.

van Lavieren, L. P. 1982. Wildlife Management in the Tropics with Special Emphasis on South East Asia, A Guidebook for the Warden. Part 1. School of Environmental Conservation Management, Ciawi, Bogor, Indonesia.

van Lawick-Goodall, H., and J. van Lawick-Goodall. 1970. Innocent Killers. Houghton Mifflin, Boston, Massachusetts.

van Lawick-Goodall, J. 1967. My Friends the Wild Chimpanzees. National Geographic Society, Washington, D.C.

Van Sickle, W. D., and F. G. Lindzey. 1991. Evaluation of a cougar population estimator based on

probability sampling. Journal of Wildlife Management 55:738–743.

Varland, K. L. 1976. Techniques for elk immobilization with succinylcholine chloride. Proceedings of the Iowa Academy of Science 82:194–197.

Vaughan, T. A. 1978. Mammalogy. W. B. Saunders, Philadelphia, Pennsylvania.

Verme, L. J. 1962. An automatic tagging device for deer. Journal of Wildlife Management 26:387–392.

Vincent, F. 1963. Acoustic signals for auto-information or echolocation. Pp. 183–227. *In* R.-G. Busnel (ed.), Acoustic Behaviour of Animals. Elsevier Scientific Publishing, Amsterdam, Netherlands.

von Frisch, K. 1974. Animal Architecture. Harcourt Brace Jovanovich, New York.

Walhovd, H. 1966. Reliability of age criteria for Danish hares (*Lepus europaeus* Pallas). Danish Review of Game Biology 4:105–127.

Walter, H. 1973. Vegetation of the Earth in Relation to Climate and the Eco-Physiological Conditions. Springer-Verlag, New York.

Walther, F. R., E. C. Mungall, and G. A. Grau. 1983. Gazelles and Their Relatives, a Study in Territorial Behavior. Noyes Publications, Park Ridge, New Jersey.

Walton, R., and B. J. Trowbridge. 1983. The use of radio-telemetry in studying the foraging behaviour of the Indian flying fox (*Pteropus giganteus*). Journal of Zoology (London) 201:575–595.

Warrell, D. A. 1988. Animal bites; rabies; venomous bites and stings. Pp. 167–191. *In* R. Dawood (ed.), How to Stay Healthy Abroad. Penguin Books, New York.

Waser, P. M., and C. H. Brown. 1984. Is there a "sound window" for primate communication? Behavioral Ecology and Sociobiology 15:73–76.

Wasser, S., K. L. Risler, and R. A. Steiner. 1988. Excreted steroids in primate feces over the menstrual cycle and pregnancy. Biology of Reproduction 39:862–872.

Watson, J. S., and C. H. Tyndale-Biscoe. 1953. The apophyseal line as an age indicator for the wild rabbit, *Oryctolagus cuniculus* (L.). New Zealand Journal of Science and Technology 34:427–435.

Watts, C. H. S., and H. J. Aslin. 1981. The Rodents of Australia. Angus Robertson, London, England.

Weir, B. J. 1974. Reproductive characteristics of hystricomorph rodents. Symposia of the Zoological Society of London 34:265–301.

Welsby, P. 1988. Yellow fever, dengue, and other arboviruses. Pp. 112–118. *In* R. Dawood (ed.), How to Stay Healthy Abroad. Penguin Books, New York.

Wemmer, C., and D. Watling. 1986. Ecology and status of the Sulawesi palm civet *Macrogalidia musschenbroeki* Schlegel. Biological Conservation 35:1-17.

Wemmer, C. M. (ed). 1987. Biology and Management of the Cervidae. Smithsonian Institution Press, Washington, D.C.

Werner, D. 1994. Where There Is No Doctor: A Village Health Care Handbook. Rev. ed. Hesperian Foundation, Palo Alto, California.

Wheelock, W., and R. Robbins. 1988. Ropes, Knots and Slings for Climbers. La Siesta Press, Glendale, California.

Whitacre, D. F. 1981. Additional techniques and safety hints for climbing tall trees, and some equipment and information sources. Biotropica 13:286-291.

White, G. C. 1992. Do pellet counts index white-tailed deer numbers and population change?: a comment. Journal of Wildlife Management 56:611-612.

White, G. C., and L. E. Eberhardt. 1980. Statistical analysis of deer and elk pellet-group data. Journal of Wildlife Management 44:121-131.

White, G. C., and R. A. Garrott. 1990. Analysis of Wildlife Radio-tracking Data. Academic Press, New York.

White, G. C., D. R. Anderson, K. P. Burnham, and D. L. Otis. 1982. Capture-Recapture and Removal Methods for Sampling Closed Populations. Los Alamos National Laboratory Publication LA-8787-NERP. Los Alamos, New Mexico.

White, G. C., R. M. Bartmann, L. H. Carpenter, and R. A. Garrott. 1989. Evaluation of aerial line transects for estimating mule deer densities. Journal of Wildlife Management 53:625-635.

White, T. G., and M. S. Alberico. 1992. Dinomys branickii. Mammalian Species 410:1-5.

Whitehouse, S. 1980. Radiotracking in Australia. Pp. 733-739. *In* C. J. Amlaner and D.W. Macdonald (eds.), A Handbook on Biotelemetry and Radiotelemetry. Pergamon Press, Oxford, England.

Whittow, G. C., and E. Gould. 1976. Body temperature and oxygen consumption of the pentail tree shrew (*Ptilocercus lowii*). Journal of Mammalogy 57:754-756.

Wiener, J. G., and M. H. Smith. 1972. Relative efficiencies of four small mammal traps. Journal of Mammalogy 53:868-873.

Wilcox, B. A. 1984. In situ conservation of genetic resources: determinants of minimum area requirements. Pp. 639-647. *In* J. A. McNeely and K. R. Miller (eds.), National Parks, Conservation and Development: The Role of Protected Areas in Sustaining Society. Smithsonian Institution Press, Washington, D.C.

Wiles, G. J. 1980. Faeces deterioration rates of four wild ungulates in Thailand. Natural History Bulletin of the Siam Society 28:121-134.

———. 1987. The status of fruit bats on Guam. Pacific Science 41:148-157.

Wiles, G. J., J. Engbring, and M. V. C. Falanruw. 1991. Population status and natural history of *Pteropus mariannus* on Ulithi Atoll, Caroline Islands. Pacific Science 45:76-84.

Wiles, G. J., T. O. Lemke, and N. H. Payne. 1989. Population estimates of fruit bats (*Pteropus mariannus*) in the Mariana Islands. Conservation Biology 3:66-76.

Wiley, R. H., and D. G. Richards. 1982. Adaptations for acoustic communication in birds: sound transmission and signal detection. Pp. 131-181. *In* D. E. Kroodsma, E. H. Miller, and H. Ouellet (eds.), Acoustic Communication in Birds: Production, Perception, and Design Features of Sounds. Academic Press, New York.

Wilkie, D. S. 1989. Performance of a backpack GPS in a tropical rain forest. Photogrammetric Engineering and Remote Sensing 55:1747-1749.

Williams, D. F., and J. S. Findley. 1979. Sexual size dimorphism in vespertilionid bats. American Midland Naturalist 102:113-126.

Williams, J. M. 1976. Determination of age of Polynesian rats (*Rattus exulans*). Proceedings of the New Zealand Ecological Society 23:79-82.

Williams, S. L., and C. A. Hanks. 1987. History of preparation materials used for recent mammal specimens. Pp. 21-49. *In* H. H. Genoways, C. Jones, and O. L. Rossolimo (eds.), Mammal Collection Management. Texas Tech University Press, Lubbock, Texas.

Williamson, D. T. 1994. Social behaviour and organization of red lechwe in the Linyanti Swamp. African Journal of Ecology 32:130-141.

Willner, G. R., K. R. Dixon, and J. A. Chapman. 1983. Age determination and mortality of the nutria (*Myocastor coypus*) in Maryland, U.S.A. Zeitschrift für Säugetierkunde 48:19-34.

Wilson, C. C., and W. L. Wilson. 1975. Influence of selective logging on primates and some other animals in East Kalimantan. Folia Primatologica. 23:245-274.

Wilson, D. E. 1973. Bat faunas: a trophic comparison. Systematic Zoology 22:14-29.

———. 1983. Checklist of mammals. Pp. 443–447. *In* D. H. Janzen (ed.), Costa Rican Natural History. University of Chicago Press, Chicago, Illinois.

———. 1989. Bats. Pp. 365–382. *In* H. Lieth and M. J. A. Werger (eds.), Tropical Rain Forest Ecosystems. Elsevier Scientific Publishing, Amsterdam, Netherlands.

———. 1993a. Order Scandentia. Pp. 131–133. *In* D. E. Wilson and D. M. Reeder (eds.), Mammal Species of the World: A Taxonomic and Geographic Reference. 2d ed. Smithsonian Institution Press, Washington, D.C.

———. 1993b. Order Dermoptera. P. 135. *In* D. E. Wilson and D. M. Reeder (eds.), Mammal Species of the World: A Taxonomic and Geographic Reference. 2d ed. Smithsonian Institution Press, Washington, D.C.

———. 1993c. Order Sirenia. Pp. 365–366. *In* D. E. Wilson and D. M. Reeder (eds.), Mammal Species of the World: A Taxonomic and Geographic Reference. 2d ed. Smithsonian Institution Press, Washington, D.C.

———. 1993d. Order Proboscidea. P. 367. *In* D. E. Wilson and D. M. Reeder (eds.), Mammal Species of the World: A Taxonomic and Geographic Reference. 2d ed. Smithsonian Institution Press, Washington, D.C.

———. 1993e. Family Aplodontidae. P. 417. *In* D. E. Wilson and D. M. Reeder (eds.), Mammal Species of the World: A Taxonomic and Geographic Reference. 2d ed. Smithsonian Institution Press, Washington, D.C.

———. 1993f. Family Castoridae. P. 467. *In* D. E. Wilson and D. M. Reeder (eds.), Mammal Species of the World: A Taxonomic and Geographic Reference. 2d ed. Smithsonian Institution Press, Washington, D.C.

Wilson, D. E., and G. L. Graham (eds.). 1992. Pacific Island Flying Foxes: Proceedings of an International Conservation Conference. U.S. Fish and Wildlife Service Biological Report 90. Washington, D.C.

Wilson, D. E., and D. M. Reeder. 1993. Mammal Species of the World: A Taxonomic and Geographic Reference. 2d ed. Smithsonian Institution Press, Washington, D.C.

Wilson, E. O. 1992. The Diversity of Life. Belknap, Cambridge, Massachusetts.

Wilson, E. O., and F. M. Peter (eds.). 1988. Biodiversity. National Academy Press, Washington, D.C.

Wilson, K. R., and D. R. Anderson. 1985. Evaluation of a density estimator based on a trapping web and distance sampling theory. Ecology 66:1185–1194.

Wilson, L. O., J. Day, J. Helvie, G. Gates, T. L. Hailey, and G. K. Tsukamoto. 1973. Guidelines for capturing and re-establishing desert bighorns. Transactions of the Desert Bighorn Council 17:137–154.

Wolfe, M. L., and J. F. Kimball. 1989. Comparison of bison population estimates with a total count. Journal of Wildlife Management 53:593–596.

Wolfe, M. S. 1993. En route and after returning. Pp. 42–47. *In* M. S. Wolfe (ed.), Health Hints for the Tropics. American Committee on Tropical Medicine and Travelers' Health of the American Society of Tropical Medicine and Hygiene, Northbrook, Illinois.

Wood, J. E. 1959. Relative estimates of fox population levels. Journal of Wildlife Management 23:53–63.

Wood, J. E., and E. P. Odum. 1964. A nine-year history of furbearer populations on the AEC Savannah River Plant area. Journal of Mammalogy 45:540–551.

Woodcock, C. E., C. H. Sham, and B. Shaw. 1990. Comments on selecting a geographic information system for environmental management. Environmental Management 14:307–315.

Woods, C. A. 1973. Erethizon dorsatum. Mammalian Species 29:1–6.

———. 1993. Suborder Hystricognathi. Pp. 771–806. *In* D. E. Wilson and D. M. Reeder (eds.), Mammal Species of the World: A Taxonomic and Geographic Reference. 2d ed. Smithsonian Institution Press, Washington, D.C.

Woods, C. A., and D. K. Boraker. 1975. Octodon degus. Mammalian Species 67:1–5.

Woods, C. A., L. Contreras, G. Willner-Chapman, and H. P. Whidden. 1992. Myocastor coypus. Mammalian Species 398:1–8.

Woodside, D. P., and K. J. Taylor. 1985. Echolocation calls of fourteen bats from eastern New South Wales. Australian Mammalogy 8:279–297.

Woolley, P. 1966. Reproduction in *Antechinus* spp. and other dasyurid marsupials. Symposia of the Zoological Society of London 15:281–294.

———. 1974. The pouch of Planigale subtilissima and other dasyurid marsupials. Journal of the Royal Society of Western Australia 57:11–15.

Wozencraft, W. C. 1993. Order Carnivora. Pp. 279–348. *In* D. E. Wilson and D. M. Reeder (eds.), Mammal Species of the World: A Taxonomic and Geographic Reference. 2d ed. Smithsonian Institution Press, Washington, D.C.

Yates, T. L. 1985. The role of voucher specimens in mammal collections: characterization and funding responsibilities. Acta Zoologica Fennica 170:81–82.

Yates, T. L., W. R. Barber, and D. M. Armstrong. 1987. Survey of North American Collections of Recent Mammals. Journal of Mammalogy 68 (suppl.).

Youngson, W. K., and N. L. McKenzie. 1977. An improved bat-collecting technique. Australian Mammal Society Bulletin 3:20–21.

Zhenhuang, S. (ed.). 1962. Chung-kuo ching chi tung wu chih [Chinese economic zoology: mammal section]. Ko Hsueh chu pan she [Scientific Publications Office, Beijing, China]. (in Chinese)

Ziegler, A. C. 1982. An ecological check-list of New Guinea recent mammals. Pp. 863–894. *In* J. L. Gressitt (ed.), Biogeography and Ecology of New Guinea. Vol. 2. Monographiae Biologicae 42. W. Junk, The Hague, Netherlands.

Index

Addresses of Authors and Contributors

Peter August
Department of Natural Resources Science
University of Rhode Island
Kingston, Rhode Island 02881
USA

Carol Baker
Department of Natural Resources Science
University of Rhode Island
Kingston, Rhode Island 02881
USA

Kyle R. Barbehenn
8208 Thoreau Drive
Bethesda, Maryland 20817
USA

John W. Bishir
Department of Mathematics
North Carolina State University
Raleigh, North Carolina 27695
USA

Stuart C. Cairns
Department of Zoology
University of New England
Armidale, New South Wales 2351
Australia

Timothy F. Clancy
Department of Environment and Heritage
P.O. Box 155
North Quay, Queensland 4470
Australia

F. Russell Cole
Department of Biology
Colby College
Waterville, Maine 04901
USA

Michael J. Conroy
National Biological Service
Georgia Cooperative Fish and Wildlife
    Research Unit
School of Forest Resources
University of Georgia
Athens, Georgia 30602
USA

Joseph A. Cook
University of Alaska Museum
907 Yukon Drive
Fairbanks, Alaska 99775
USA

Ronald I. Crombie
Division of Amphibians and Reptiles
NHB Mail Stop 162
Smithsonian Institution
Washington, DC 20560-0162
USA

Martha L. Crump
Department of Biological Sciences
Box 5640
Northern Arizona University
Flagstaff, Arizona 86011
USA

Martin Denny
Oorong
Bathurst Road
Oberon, New South Wales 2787
Australia

Chris R. Dickman
Department of Zoology
University of Sidney

Sidney, New South Wales 2006
Australia

Michael E. Dorcas
Department of Biological Sciences
Idaho State University
Pocatello, Idaho 83209
USA

Louise H. Emmons
Division of Mammals
NHB Mail Stop 108
Smithsonian Institution
Washington, DC 20560-0108
USA

Mercedes S. Foster
National Biological Service
NHB Mail Stop 111
National Museum of Natural History
Washington, DC 20560-0111
USA

Thomas H. Fritts
National Biological Service
NHB Mail Stop 111
National Museum of Natural History
Washington, DC 20560-0111
USA

William L. Gannon
Department of Biology
University of New Mexico
Albuquerque, New Mexico 87131-1091
USA

Scott Lyell Gardner
Museum of Natural History
University of Nebraska
Lincoln, Nebraska 68588-0514
USA

Sarah B. George
Museum of Natural History
University of Utah
Salt Lake City, Utah 84112
USA

Thomas Grant
10 Allison Road
Cronulla
Sydney, New South Wales 2230
Australia

Gregory Gurri-Glass
Department of Immunology and Infectious
  Disease
Johns Hopkins University
615 North Wolfe Street
Baltimore, Maryland 21205
USA

Lee-Ann C. Hayek
Statistics and Mathematics
NHB Mail Stop 136
Smithsonian Institution
Washington, DC 20560-0136
USA

Virginia Hayssen
Animal Physiology Section
University of Nottingham
Sutton Bonington
North Loughborough LE12 5RD
United Kingdom

Robert F. Inger
Amphibians and Reptiles
Field Museum of Natural History
Roosevelt Road at Lake Shore Drive
Chicago, Illinois 60605
USA

Peter Jarman
Department of Ecosystem Management
University of New England
Armidale, New South Wales 2351
Australia

Clyde Jones
The Museum
Texas Tech University

Lubbock, Texas 79409
USA

J. Edward Kautz
New York State Department of Environmental
  Conservation
700 Troy-Schenectady Road
Latham, New York 12110-2400
USA

Gordon W. Kirkland, Jr.
Vertebrate Museum
Shippensburg University
Shippensburg, Pennsylvania 17257
USA

Thomas H. Kunz
Department of Biology
5 Cummington Street
Boston University
Boston, Massachusetts 02215
USA

Charles LaBash
Department of Natural Resources Science
University of Rhode Island
Kingston, Rhode Island 02881
USA

Richard A. Lancia
Department of Forestry
College of Forest Resources
North Carolina State University
Raleigh, North Carolina 27695
USA

Geoffrey Lundie-Jenkins
Conservation Commission of the Northern
  Territory
P.O. Box 1046
Alice Springs, Northern Territories 0871
Australia

Roy W. McDiarmid
National Biological Service
NHB Mail Stop 111
National Museum of Natural History
Washington, DC 20560-0111
USA

William J. McShea
National Zoological Park
Conservation and Research Center
Smithsonian Institution
Front Royal, Virginia 22630
USA

Helene Marsh
Zoology Department
James Cook University of North Queensland
Townville, Queensland 4811
Australia

Dale G. Miquelle
National Zoological Park
Conservation and Research Center
Smithsonian Institution
Front Royal, Virginia 22630
USA

James D. Nichols
National Biological Service
Patuxent Wildlife Research Center
11510 American Holly Drive
Laurel, Maryland 20708-4017
USA

Thomas J. O'Shea
National Biological Service
Mid-Continent Ecological Science Center
Ft. Collins, Colorado 80524
USA

Elizabeth D. Pierson
Wildlife Resources Center
University of California

Berkeley, California 94720
USA

Paul A. Racey
Department of Zoology
University of Aberdeen
Aberdeen, Scotland ABS 5B6
United Kingdom

William E. Rainey
Wildlife Resources Center
University of California
Berkeley, California 94720
USA

Galen B. Rathbun
National Biological Service
Piedras Blancas Research Station
P.O. Box 70
San Simeon, California 93452
USA

Robert P. Reynolds
National Biological Service
NHB Mail Stop 111
National Museum of Natural History
Washington, DC 20560-0111
USA

Gregory C. Richards
Division of Wildlife and Ecology, CSIRO
Lyneham, Australian Capital Territory 2602
Australia

Rasanayagam Rudran
National Zoological Park
Smithsonian Institution
Washington, DC 20008-2598
USA

Andrew P. Smith
Department of Ecosystem Management
University of New England
Armidale, New South Wales 2351
Australia

Christopher Smith
Department of Natural Resources Science
University of Rhode Island
Kingston, Rhode Island 02881
USA

Colin Southwell
Wildlife Monitoring Unit
Australian National Parks and Wildlife Service
P.O. Box 636
Canberra, Australian Capital Territory 2601
Australia

Donald W. Thomas
Département de Biologie
Université de Sherbrooke
Montreal, Quebec PQ J1K 2R1
Canada

Christopher R. Tidemann
Department of Forestry
Australian National University

Canberra, Australian Capital Territory 2601
Australia

Christen Wemmer
National Zoological Park
Conservation and Research Center
Smithsonian Institution
Front Royal, Virginia 22630
USA

Don E. Wilson
Biodiversity Programs
NHB Mail Stop 180
National Museum of Natural History
Smithsonian Institution
Washington, DC 20560-0180
USA

Terry L. Yates
Department of Biology
University of New Mexico
Albuquerque, New Mexico 87131
USA